Cabling Part 2
Fiber-Optic Cabling and Components

5th Edition

Cabling Part 2
Fiber-Optic Cabling and Components

5th Edition

Bill Woodward

SYBEX®
A Wiley Brand

Acquisitions Editor: Mariann Barsolo

Development Editor: David Clark

Technical Editor: Charlie Husson

Production Editor: Rebecca Anderson

Copy Editor: Elizabeth Welch

Editorial Manager: Pete Gaughan

Vice President and Executive Group Publisher: Richard Swadley

Associate Publisher: Chris Webb

Book Designers: Maureen Forys, Happenstance Type-O-Rama; Judy Fung

Compositor: Maureen Forys, Happenstance Type-O-Rama

Proofreader: Kim Wimpsett

Indexer: Ted Laux

Project Coordinator, Cover: Todd Klemme

Cover Designer: Ryan Sneed/Wiley

Cover Image: Courtesy of MicroCare Corporation, used with permission

Dear Reader,

Thank you for choosing *Cabling Part 2: Fiber-Optic Cabling and Components*. This book is part of a family of premium-quality Sybex books, all of which are written by outstanding authors who combine practical experience with a gift for teaching.

Sybex was founded in 1976. More than 30 years later, we're still committed to producing consistently exceptional books. With each of our titles, we're working hard to set a new standard for the industry. From the paper we print on to the authors we work with, our goal is to bring you the best books available.

I hope you see all that reflected in these pages. I'd be very interested to hear your comments and get your feedback on how we're doing. Feel free to let me know what you think about this or any other Sybex book by sending me an email at contactus@wiley.com. If you think you've found a technical error in this book, please visit http://sybex.custhelp.com. Customer feedback is critical to our efforts at Sybex.

Best regards,

Chris Webb
Associate Publisher, Sybex

In memory of Frank J. Grabo, teacher, coach, and mentor.

—BW

Acknowledgments

Writing a book is a team effort that takes a dedicated group of professionals. I am very fortunate to have been able to work with this team of talented and dedicated individuals.

First, I would like to thank Sybex for giving me the opportunity to write this book. Special thanks to Acquisitions Editor Mariann Barsolo, Production Editor Becca Anderson, Developmental Editor David Clark, Editorial Manager Pete Gaughan, and editorial staff Connor O'Brien, Rebekah Worthman, Rayna Erlick, and Jenni Housh for the outstanding job you did guiding me through this project from start to finish.

Thanks to Chuck Schue, Randy Hall, and Pat McGillvray at UrsaNav, Inc., for all your support with this project.

Thanks, Charlie Husson, for the outstanding job with the technical edits. You are an exceptional engineer, great mentor, and friend. I have learned so much from you over the years and look forward to working with you on future projects.

Many companies also provided technical information, equipment, and photographs. Special thanks to Donald Stone from KITCO Fiber Optics, Jay S. Tourigny from MicroCare, Mark Messer from Carlisle Interconnect Technologies, Dede Starnes and Ryan Spillane from Corning Cable Systems, Bob Scharf from Moog Protokraft, Bill Reid from Amphenol Fiber Systems International, Earle Olson from TE Connectivity, Peter Koudelka from PROMET International Inc., Chuck Casbeer from Infotec IT and Leadership Training, Bruno Huttner from Luciol Instruments, Laurence N. Wesson from Aurora Optics Inc., Art Schweiss from Electronic Manufacturers' Agents Inc., Kevin Lefebvre from EigenLight Corporation, Matt Krutsch from COTSWORKS, Ed Forrest from ITW Chemtronics, Mike Gleason from Panduit, Scott Kale from Norfolk Wire, Christine Pons from OptiConcepts, and Dave Edwards from W.R. Systems.

Dick Glass has been a friend, mentor, and co-worker for many years; he has spent many hours guiding me through various writing projects. I feel very blessed to have met Dick and greatly appreciate his guidance over the years and his assistance with this project.

Thanks to the host of people behind the scenes who I did not mention for all your efforts to make this book the best that it can be.

Last but not least, thank you to my family—to the love of my life, my beautiful wife Susan, for making this possible; to my children, Mike, Brandon, Eric, Nathan, and Kathryn; and to my grandchildren for your patience, inspiration, encouragement, and prayers. I am the luckiest man alive to have all of you in my life.

—*Bill Woodward*

About the Authors

Bill Woodward is the director of C5ISR Engineering Products with UrsaNav, Inc., an engineering services company. Bill has been teaching fiber optics and other technical courses since 1992. He has more than 25 years of experience in the design, operation, maintenance, troubleshooting, and repair of electronic and electrical systems.

Bill is licensed in the Commonwealth of Virginia as a professional electrical engineer. He is chairman of SAE International's Aerospace Fiber Optics and Applied Photonics Committee, AS-3, as well as chairman of the AS-3B2 Education and Design Working Group. He is also a member of the Electronics Technicians Association (ETA) International; he has served four terms as chairman of the ETA and has been chair of the Fiber Optic Committee for over a decade.

Contents at a Glance

Contents

Introduction

The term "broadband" commonly refers to high-speed Internet access that is always on and faster than the traditional dial-up access. Without fiber optics, broadband as we know it today would not exist. Fiber optics is the backbone of the global telecommunications system. No other transmission medium can move the high rates of data over the long distances required to support the global telecommunications system. This technology works so well that the typical user may not be aware that it even exists.

This book focuses on building a solid foundation in fiber-optic theory and application. It describes in great detail fiber-optic cable technology, connectorization, splicing, and passive devices. It examines the electronic technology built into fiber-optic receivers, transmitters, and test equipment that makes incredible broadband download and upload speeds possible. In addition, many current industry standards pertaining to optical fiber, connector, splice, and network performance are discussed in detail.

This book is an excellent reference for anyone currently working in fiber optics as well as those who are just starting to learn about fiber optics. The book covers in detail all of the competencies of the Electronics Technicians Association International (ETA) fiber optic installer (FOI) and fiber optic technician (FOT) certification.

ETA's FOI and FOT Programs

The ETA's FOI and FOT programs are the most comprehensive in the industry. Each program requires students to attend an ETA-approved training school. Each student must achieve a score of 75% or greater on the written exam and satisfactorily complete all the hands-on requirements. Those who are interested in obtaining ETA FOI or FOT certification can visit the ETA's website at www.eta-i.org and get the most up-to-date information on the program and a list of approved training schools.

The ETA FOI certification requires no prerequisite and is designed for anyone who is interested in learning how to become a fiber-optic installer. The FOI certification is recommended as a prerequisite for the FOT certification, for those who want to learn how to test a fiber-optic link to the current industry standards and how to troubleshoot. Fiber-optic certification demonstrates to your employer that you have the knowledge and hands-on skills required to install, test, and troubleshoot fiber-optic links and systems. With the push to bring fiber optics to every home, these skills are highly sought after.

About This Book

This book's topics run the gamut of fiber-optics technology and application; they include the following:

- The history of fiber optics and broadband access

- The principles of fiber-optic transmission

- The basic principles of light

- Optical fiber construction and theory

- Optical fiber characteristics

- Safety

- Fiber-optic cables

- Fusion and mechanical splicing

- Connectors

- Fiber-optic light sources and transmitters

- Fiber-optic detectors and receivers

- Passive components and multiplexers

- Passive optical networks

- Cable installation and hardware

- Fiber-optic system design considerations

- Test equipment and link/cable testing

- Troubleshooting and restoration

A cabling glossary is included at the end of the book so you can look up unfamiliar terms. The Solutions to the Master It questions in The Bottom Line sections at the end of each chapter are gathered in Appendix A, and Appendix B lists the knowledge competencies for the information about ETA's line of cabling certifications.

Who Is This Book For?

If you are standing in your neighborhood bookstore browsing through this book, you may be asking yourself whether you should buy it. The procedures in this book are illustrated and written in English rather than "technospeak." That's because this book was designed specifically to help unlock the mysteries of fiber optics. Fiber optics can be a confusing topic; it has its own language, acronyms, and standards. This book was developed with the following types of people in mind:

♦ Information technology (IT) professionals who can use this book to gain a better understanding and appreciation of a structured fiber-optic cabling system

♦ IT managers who are preparing to install a new computer system

♦ Do-it-yourselfers who need to install a few new cabling runs in their facility and want to get it right the first time

♦ New cable installers who want to learn more than just what it takes to pull a cable through the ceiling and terminate it to the patch panel

♦ Students taking introductory courses in LANs and cabling

♦ Students preparing for the ETA fiber optic installer (FOI) or fiber optic technician (FOT) certifications

In addition, this book is an excellent reference for anyone currently working in the field of fiber optics.

How to Use This Book

To understand the way this book is put together, you must learn about a few of the special conventions that were used. Here are some of the items you will commonly see.

Italicized words indicate new terms. After each italicized term, you will find a definition.

TIP   *Tips* will be formatted like this. A tip is a special bit of information that can make your work easier or make an installation go more smoothly.

NOTE   *Notes* are formatted like this. When you see a note, it usually indicates some special circumstance to make note of. Notes often include out-of-the-ordinary information about working with a telecommunications infrastructure.

WARNING   *Warnings* are found within the text whenever a technical situation arises that may cause damage to a component or cause a system failure of some kind. Additionally, warnings are placed in the text to call particular attention to a potentially dangerous situation.

KEYTERM   *Key terms* are used to introduce a new word or term that you should be aware of. Just as in the worlds of networking, software, and programming, the world of cabling and telecommunications has its own language.

SIDEBARS

This special formatting indicates a sidebar. *Sidebars* are entire paragraphs of information that, although related to the topic being discussed, fit better into a standalone discussion. They are just what their name suggests: a sidebar discussion.

CABLING @ WORK SIDEBARS

These special sidebars are used to give real-life examples of situations that actually occurred in the cabling world.

Enjoy!

Have fun reading this book—it has been fun writing it. I hope that it will be a valuable resource to you and will answer at least some of your questions on fiber optics. As always, I love to hear from readers, you can reach Bill Woodward at wrwoodward2013@gmail.com.

Part II

Fiber-Optic Cabling and Components

Chapter 1

History of Fiber Optics and Broadband Access

Like many technological achievements, fiber-optic communications grew out of a succession of quests, some of them apparently unrelated. It is important to study the history of fiber optics to understand that the technology as it exists today is relatively new and still evolving.

This chapter discusses the major accomplishments that led to the creation of high-quality optical fibers and their use in high-speed communications and data transfer, as well as their integration into existing communications networks.

In this chapter, you will learn to:

♦ Recognize the refraction of light

♦ Identify total internal reflection

♦ Detect crosstalk between multiple optical fibers

♦ Recognize attenuation in an optical fiber

Evolution of Light in Communication

Hundreds of millions of years ago, the first bioluminescent creatures began attracting mates and luring food by starting and stopping chemical reactions in specialized cells. Over time, these animals began to develop distinctive binary, or on-off, patterns to distinguish one another and communicate intentions quickly and accurately. Some of them have evolved complex systems of flashing lights and colors to carry as much information as possible in a single glance. These creatures were the first to communicate with light, a feat instinctive to them but tantalizing and elusive to modern civilization until recently.

Early Forms of Light Communication

Some of the first human efforts to communicate with light consisted of signal fires lit on hill-tops or towers to warn of advancing armies, and lighthouses that marked dangerous coasts for ancient ships and gave them reference points in their journeys. To the creators of these signals, light's tremendous speed (approximately 300,000 kilometers per second) made its travel over great distances seem instantaneous.

An early advance in these primitive signals was the introduction of relay systems to extend their range. In some cases, towers were spread out over hundreds of kilometers, each one in the line of sight of the next. With this system, a beacon could be relayed in the time it took each tower guard to light a fire—a matter of minutes—while the fastest transportation might have taken days. Because each tower only needed in its line of sight the sending and receiving towers, the light, which normally travels in a straight line, could be guided around obstacles such as mountains as well as over the horizon. As early as the fourth century A.D., Empress Helena, the mother of Constantine, was believed to have sent a signal from Jerusalem to Constantinople in a single day using a relay system.

NOTE The principle behind signal relay towers is still used today in the form of repeaters, which amplify signals attenuated by travel over long distances through optical fibers.

Early signal towers and lighthouses, for all their usefulness, were still able to convey only very simple messages. Generally, no light meant one state, whereas a light signaled a change in that state. The next advance needed was the ability to send more detailed information with the light. A simple but notable example is the signal that prompted Paul Revere's ride at the start of the American Revolution. By prearranged code, one light hung in the tower of Boston's Old North Church signaled a British attack by land; two lights meant an invasion by sea. The two lamps that shone in the tower not only conveyed a change in state, but also provided a critical detail about that change.

The Quest for Data Transmission

Until the 1800s, light had proven to be a speedy way to transmit simple information across great distances, but until new technologies were available, its uses were limited. It took a series of seemingly unrelated discoveries and inventions to harness the properties of light through optical fibers.

The first of these discoveries was made by Willebrord van Roijen Snell, a Dutch mathematician who in 1621 wrote the formula for the principle of *refraction*, or the bending of light as it passes from one material into another. The phenomenon is easily observed by placing a stick into a glass of water. When viewed from above, the stick appears to bend because light travels more slowly through the water than through the air. Snell's formula, which was published 70 years after his death, stated that every transparent substance had a particular *index of refraction*, and that the amount that the light would bend was based on the relative refractive indices of the two materials through which the light was passing. Air has an approximate refractive index of 1 and water has a refractive index of 1.33.

The next breakthrough came from Jean-Daniel Colladon, a Swiss physicist, and Jacques Babinet, a French physicist. In 1840, Colladon and Babinet demonstrated that bright light could be guided through jets of water through the principle of *total internal reflection*. In their demonstration, light from an arc lamp was used to illuminate a container of water. Near the bottom of the container was a hole through which the water could escape. As the water poured out of the hole, the light shining into the container followed the stream of water. Their use of this discovery, however, was limited to illuminating decorative fountains and special effects in operas. It took John Tyndall, a natural philosopher and physicist from Ireland, to bring the phenomenon to greater attention. In 1854, Tyndall performed the demonstration before the British Royal Society and made it part of his published works in 1871, casting a shadow over the contribution of Colladon and Babinet. Tyndall is now widely credited with discovering total internal reflection, although Colladon and Babinet had demonstrated it 14 years previously.

Total internal reflection takes place when light passing through a material with a higher index of refraction (the water in the experiment) hits a boundary layer with a material that has a lower index of refraction (the air). When this takes place, the boundary layer becomes reflective, and the light bounces off the boundary layer, remaining contained within the material with the higher index of refraction.

Shortly after Tyndall, Colladon, and Babinet laid the groundwork for routing light through a curved material, another experiment took place that showed how light could be used to carry higher volumes of data.

In 1880, Alexander Graham Bell demonstrated his photophone, one of the first true attempts to carry complex signals with light. It was also the first device to transmit signals wirelessly. The photophone gathered sunlight onto a mirror attached to a mouthpiece that vibrated when a user spoke into it. The vibrating mirror reflected the light onto a receiver coated with selenium, which produced a *modulated* electrical signal that varied with the light coming from the sending device. The electrical signal went to headphones where the original voice input was reproduced.

Bell's invention suffered from the fact that outside influences such as dust or stray light confused the signals, and clouds or other obstructions to light rendered the device inoperable. Although Bell had succeeded in transmitting a modulated light signal nearly 200 meters, the photophone's limitations had already fated it to be eclipsed by Bell's earlier invention, the telephone. Until the light could be modulated and guided as well as electricity could, inventions such as the photophone would continue to enjoy only novelty status.

Evolution of Optical Fiber Manufacturing Technology

John Tyndall's experiment in total internal reflection had led to attempts to guide light with more control than could be achieved in a stream of water. One such effort by William Wheeler in 1880, the same year that Bell's photophone made its debut, used pipes with a reflective coating inside that guided light from a central arc lamp throughout a house. As with other efforts of the time, there was no attempt to send meaningful information through these conduits—merely to guide light for novelty or decorative purposes. The first determined efforts to use guided light to carry information came out of the medical industry.

Controlling the Course of Light

Doctors and researchers had long tried to create a device that would allow them to see inside the body with minimal intrusion. They had begun experimenting with bent glass and quartz rods, bringing them tantalizingly close to their goal. These tools could transmit light into the body, but they were extremely uncomfortable and sometimes dangerous for the patient, and there was no way yet to carry an image from the inside of the body out to doctors. What they needed was a flexible substance or *medium* that could carry whole images for about half a meter.

One such material was in fact pioneered for quite a different purpose. Charles Vernon Boys was a British physics teacher who needed extremely sensitive instruments for his continuing research in heat and gravity. In 1887, to provide the materials he needed, he began drawing fine fibers out of molten silica. Using an improvised miniature crossbow, he shot a needle that dragged the molten material out of a heat source at high speed. The resulting fiber—more like quartz in its crystalline structure than glass—was finer than any that had been made to date, and was also remarkably even in its thickness. Even though glass fibers had already been available for decades before this, Boys' ultra-fine fibers were the first to be designed for scientific purposes and were also the strongest and smallest that had been made to date. He did not, however, pursue research into the optical qualities of his fibers.

Over the next four decades, attempts to use total internal reflection in the medical industry yielded some novel products, including glass rods designed by Viennese researchers Roth and Reuss to illuminate internal organs in 1888, and an illuminated dental probe patented in 1898 by David Smith. A truly flexible system for illuminating or conveying images of the inside of the body remained elusive, however.

The next step forward in the optical use of fibers occurred in 1926. In that year, Clarence Weston Hansell, an electrical engineer doing research related to the development of television at RCA, filed a patent for a device that would use parallel quartz fibers to transmit a lighted image over a short distance. The device remained in the conceptual stage, however, until a German medical student, Heinrich Lamm, developed the idea independently in an attempt to form a flexible gastroscope. In 1930, Lamm bundled commercially produced fibers and managed to transmit a rough image through a short stretch of the first fiber-optic cable. The process had several problems, however, including the fact that the fiber ends were not arranged exactly, and they were not properly cut and polished. Another issue was to prove more daunting. The image quality suffered from the fact that the quartz fibers were bundled against each other. This meant that the individual fibers were no longer surrounded by a medium with a lower index of refraction. Much of the light from the image was lost to *crosstalk*. Crosstalk or optical coupling is the result of light leaking out of one fiber into another fiber.

The poor focus and resolution of Lamm's experimental image meant that a great deal more work would be needed, but Lamm was confident enough to write a paper on the experiment. The rise of the Nazis, however, forced Lamm, a Jew, to leave Germany and abandon his research. The dream of Hansell and Lamm languished until a way could be found to solve the problems that came with the materials available at the time.

Also in 1930, the chemical company DuPont invented a clear plastic material that it branded Lucite. This new material quickly replaced glass as the medium of choice for lighted medical probes. The ease of shaping Lucite pushed aside experiments with bundles of glass fiber, along with the efforts to solve the problems inherent in Lamm's probe.

The problems surfaced again 20 years later, when the Dutch government began looking for better periscopes for its submarines. They turned to Abraham van Heel, who was at the time the president of the International Commission of Optics and a professor of physics at the Technical University of Delft, the Netherlands. Van Heel and his assistant, William Brouwer, revived the idea of using fiber bundles as an image-transmission medium. Fiber bundles, Brouwer pointed out, had the added advantage of being able to scramble and then unscramble an image—an attractive feature to Dutch security officials.

When van Heel attempted to build his image carrier, however, he rediscovered the problem that Lamm had faced. The refractive index of adjacent fibers reduced a fiber's ability to achieve total internal reflection, and the system lost a great deal of light over a short distance. At one point, van Heel even tried coating the fibers with silver to improve their reflectivity, but the effort provided little benefit.

At his government's suggestion, van Heel approached Brian O'Brien, president of the Optical Society of America, in 1951. O'Brien suggested a procedure that is still the basis for fiber optics today: surrounding, or "cladding," the fiber with a layer of material with a lower refractive index.

Following O'Brien's suggestion, van Heel ran the fibers through a liquid plastic that coated them, and in April 1952, he succeeded in transmitting an image through a 400-fiber bundle over a distance of half a meter.

Van Heel's innovation—along with research performed by Narinder Singh Kapany, who also coined the term *fiber optics*, and Harold Hopkins—helped make the 1950s the pivotal decade in the development of modern fiber optics.

Working in England, Kapany and Hopkins developed a method for ensuring that the fibers at each end of a cable were in precise alignment. They wound a single fine strand several thousand times in a figure-eight pattern and sealed a section in clear epoxy to bind the fibers together throughout the bundle. They then sawed the sealed portion in half, leaving the fiber ends bonded in exact alignment. The image transmitted with this arrangement was clearly an improvement, but the brightness degraded quickly since the fibers were unclad.

Extending Fiber's Reach

In January 1954, the British journal *Nature* chanced to publish papers on the findings of van Heel as well as Kapany and Hopkins in the same issue. Although their placement in the journal was apparently coincidental, the two advancements were precisely the right combination of ideas for Professor Basil Hirschowitz, a gastrosurgeon from South Africa who was working on a fellowship at the University of Michigan. Hirschowitz assembled a team to study the uses of these new findings as a way to finally build a flexible endoscope for peering inside the body. Assisting Hirschowitz were physicist C. Wilbur Peters and a young graduate student named Lawrence Curtiss.

Curtiss studied the work of Kapany and Hopkins and used their winding method to create a workable fiber bundle, but his first attempt at cladding used van Heel's suggestion of cladding glass fibers with plastic. The results were disappointing.

In 1956, Curtiss began working with a new type of glass from Corning, one with a lower refractive index than the glass he was using in his fibers. He placed a tube made of the new glass around a core made from the higher refractive index glass and melted the two together. The cladded glass fiber that he drew from this combination was a success. On December 8, 1956, Curtiss made a fiber with light-carrying ability far superior to that of any fiber before it. Even when he was 12 meters away from the glass furnace, he could see the glow of the fire inside the fiber that was being drawn from it. By early 1957, Hirschowitz and Curtiss had created a working endoscope, complete with lighting and optics. This event marked the first practical use of optical fibers to transmit complex information.

Curtiss' fibers were well suited for medical applications, but their ability to carry light was limited. Suffering a signal loss of one *decibel* per meter, the fibers were still not useful for long-distance communications. Many thought that glass was inherently unusable for communications, and research in this area remained at a minimal level for nearly a decade.

In the meantime, the electronic communications industry had been experimenting with methods of improving bandwidth for the higher volumes of traffic they expected to carry. The obvious choice for increasing the amount of information a signal could carry was to increase the frequency, and throughout the 1950s, researchers had pushed frequencies into the tens of *gigahertz*, which produced *wavelengths* of only a few millimeters. Frequencies in this range—just below the lowest infrared frequencies—required hollow pipes to be used as *waveguides*, because the signals were easily disturbed by atmospheric conditions such as fog or dust.

With the invention of the laser in 1960, the potential for increasing communication bandwidths literally increased exponentially. Wavelengths had been slashed from the millimeter range to the micrometer range, and true optical communications seemed within reach. The problems of atmospheric transmission remained, however, and waveguides used for lower frequencies were proving inadequate for optical wavelengths unless they were perfectly straight. Optical fibers, too, were all but ruled out as a transmission medium because of the loss of light or *attenuation*. The loss of 1000 decibels per kilometer was still too great.

One researcher did not give up on fiber, however. Charles K. Kao, working at Standard Telecommunications Laboratories, began studying the problems encountered in optical fibers. His conclusions revived interest in the medium after he announced in 1966 that signal losses in glass fibers were not caused by inherent deficiencies of the material, but by flaws in the manufacturing process. Kao proposed that improved manufacturing processes could lower attenuation to levels of 20 decibels per kilometer or better, while providing the ability to carry up to 200,000 telephone channels in a single fiber.

Kao's pronouncement sparked a race to find the lower limit of signal loss in optical fibers. In 1970, Corning used pure silica to create a fiber with a loss that achieved Kao's target of only 20 decibels per kilometer. That was just the beginning. Six years later, the threshold had dropped to just half a decibel per kilometer, and in 1979 the new low was 0.2 decibel per kilometer. Optical fiber had passed well into the realm of practicality for communications and could begin showing its promise as a superior medium to copper.

Evolution of Optical Fiber Integration and Application

Once signal losses in fiber dropped below Kao's projected figure of 20 decibels per kilometer, communications companies began looking seriously at fiber optics as a new transmission medium. The technology required for this fledgling medium was still expensive, however, and fiber-optic communications systems remained in closed-circuit, experimental stages until 1973. In that year, the U.S. Navy installed a fiber-optic telephone link aboard the *USS Little Rock*. Fiber optics had left the lab and started working in the field. Further military tests showed fiber's advantages over copper in weight and information-carrying capacity.

The first full-scale commercial application of fiber-optic communication systems occurred in 1977, when both AT&T and GTE began using fiber-optic telephone systems for commercial customers. During this period, the U.S. government breakup of the Bell Telephone system monopoly ushered in a boom time for smaller companies seeking to market long-distance service. A number of companies had positioned themselves to build microwave towers throughout the country to create high-speed long-distance networks. With the rise of fiber-optic technology, however, the towers were obsolete before they had even been built. Plans for the towers were scrapped in the early 1980s in favor of fiber-optic links between major cities. These links were then connected to local telephone companies that leased their capacity from the operators. The result was a bandwidth feeding frenzy. The fiber-optic links had such high capacities that extra bandwidth was leased to other local and long-distance carriers, which often undercut the owners of the lines, driving some out of business.

Following the success of fiber optics in the telecommunications industry, other sectors began taking advantage of this medium. During the 1990s, fiber-optic networks began to dominate in the fields of industrial controls, computers, and information systems. Improvements in lasers and fiber manufacturing continued to drive data rates higher and bring down operating costs.

Today, fiber optics have become commonplace in many areas as the technology continues to improve. Until recently, the transition to fiber optics was cost effective only for business and industry; equipment upgrades made it too expensive for telephone and cable companies to run fiber to every home. Manufacturing improvements have reduced costs, however, so that running fiber to the home is now an affordable alternative for telephone and cable companies.

Broadband since the Turn of the Century

A search on the Internet for the definition of broadband will yield many different results. Which result is correct? For this chapter, the definition published by the Federal Communications Commission (FCC) is correct. The FCC states that the term broadband commonly refers to high-speed Internet access that is always on and faster than the traditional dial-up access.

Broadband can be accessed using different high-speed transmission technologies over different mediums. Typical broadband connections include:

- Fiber optics
- Wireless
- Cable modem
- Satellite
- Digital Subscriber Line (DSL)
- Broadband over Power Lines (BPL)

In June of 2013, the United States Office of Science and Technology Policy and The National Economic Council published a report entitled *Four Years of Broadband Growth*. Many of the facts and definitions presented in this section of the chapter were obtained from that report.

The Role of Optical Fiber in Broadband

Today broadband can be accessed using different transmission technologies over different mediums. On the road, you may access the Internet using your cell phone and *fourth-generation* (4G) technology. At your local coffee shop, that same phone may access the Internet using the coffee shop's *Wi-Fi* connection. When you arrive home, your phone connects to your wireless router that provides Internet access over a *cable modem*. Later in the day, you place the phone on the charger and turn on a high-definition television with Internet capabilities that connects to your router with a cable. While your favorite show is playing in the background, you turn on your laptop and check email over your wireless connection.

The state of broadband technology today makes all this connectivity relatively easy, and to most users it is completely transparent. In other words, you do not need to understand anything about the infrastructure that supports the global telecommunications system to communicate and share information with nearly anyone in the world.

Without fiber optics, broadband as we know it today would not exist. Fiber optics is the backbone of the global telecommunications system. No other transmission medium can move the high rates of data over the long distances required to support the global telecommunications system. This technology works so well that the typical user may not be aware that it even exists.

Cell phone towers like the one shown in Figure 1.1 are everywhere. From a distance, you can see the antennas at the top of the tower. As you get closer to the tower, you can see the copper cables running up the tower to the antennas. However, what you do not see are the optical fibers typically buried underground moving the data to and from the cell tower.

FIGURE 1.1
Cell phone tower

Broadband Speed and Access at the Turn of the Century and Today

As stated in the Four Years of Broadband Growth Report, at the turn of the century broadband speed was considered 200,000 bits per second or 200 kilobits per second (kbps), while dial-up Internet connections were typically 28.8kbps or 56kbps. Only 4.4 percent of the households in America had a broadband connection to their home. However, 41.5 percent had a dial-up Internet connection.

In 2013, the basic broadband speed was defined as 3,000,000 bits per second or 3 megabits per second (Mbps) *downstream* and 768kbps *upstream*. Downstream describes the number of bits that travel from the Internet service provider (ISP) to the person accessing broadband. This is often referred to as *download speed*. Upstream describes the number of bits being sent to the ISP. This is often referred to as *upload speed*.

While the basic broadband speed was defined with a 3Mbps download speed, more than 94 percent of the homes in America exceed 10Mbps. More than 75 percent have download speeds greater than 50Mbps, 47 percent have download speeds greater than 100Mbps, and more than 3 percent enjoy download speeds greater than 1 billion bits, or a gigabit, per second (Gbps).

The Bottom Line

Recognize the refraction of light.   Refraction is the bending of light as it passes from one material into another.

> **Master It**   You are cleaning your pool with a small net at the end of a pole when you notice a large bug that appears to be 2′ below the surface. You place the net where you believe the bug to be and move it through the water. When you lift the net from the pool, the bug is not in the net. Why did the net miss the bug?

Identify total internal reflection.   In 1840, Colladon and Babinet demonstrated that bright light could be guided through jets of water through the principle of total internal reflection.

> **Master It**   You just removed your fish from a dirty 10-gallon aquarium you are preparing to clean when a friend shows up with a laser pointer. Your friend energizes the laser pointer and directs the light into the side of the tank. The laser light illuminates the small dirty particles in the tank and you and your friend observe the light entering one end of the tank and exiting the other. As you friend aims the laser pointer at an angle toward the surface of the water, the light does not exit; instead it bounces off the surface of the water at an angle. Why did this happen?

Detect crosstalk between multiple optical fibers.   Crosstalk or optical coupling is the result of light leaking out of one fiber into another fiber.

> **Master It**   You have bundled six flexible clear plastic strands together in an effort to make a fiber-optic scope that will allow you to look into the defroster vent in your car in hope of locating your missing Bluetooth headset. You insert your fiber-optic scope into the defroster vent and are disappointed with the image you see. What is one possible cause for the poor performance of your fiber-optic scope?

Recognize attenuation in an optical fiber.   In 1960 after the invention of the laser, optical fibers were all but ruled out as a transmission medium because of the loss of light or attenuation.

> **Master It**   You are troubleshooting a clog in a drainpipe and because of the size of your flashlight and the location of the drain; you cannot illuminate the drain adequately to see the clog. You decide to modify a flashlight to illuminate the inside of drainpipe in hope of identifying the source of the clog. You purchase a small diameter flexible clear plastic rod approximately 12″ in length and secure one end over the bright LEDs in the center of the flashlight. After powering up the flashlight, you are disappointed that the light exiting the optical fiber is much dimmer than you expected; however, you still attempt to identify the clog. Why did the fiber-optic flashlight fail to illuminate the inside of the drainpipe?

Chapter 2

Principles of Fiber-Optic Transmission

Like Bell's photophone, the purpose of fiber optics is to convert a signal to light, move the light over a distance, and then reconstruct the original signal from the light. The equipment used to do this job has to overcome all of the same problems that Bell encountered, while carrying more data over a much greater distance.

In this chapter, you will learn about the basic components that transmit, receive, and carry the optical signal. You will also learn some of the methods used to convert signals to light and light back to the original signals, as well as how the light is carried over the distances required.

In this chapter, you will learn to:

♦ Calculate the decibel value of a gain or loss in power

♦ Calculate the gain or loss in power from a known decibel value

♦ Calculate the gain or loss in power using the dB rules of thumb

♦ Convert dBm to a power measurement

♦ Convert a power measurement to dBm

The Fiber-Optic Link

A *link* is a transmission pathway between two points using some kind of generic cable. The pathway includes a means to couple the signal to the cable and a way to receive it at the other end in a useful way.

Any time we send a signal from one point to another over a wire, we are using a link. A simple intercom, for example, consists of the sending station (which converts voice into electrical signals), the wire over which the signals are transmitted, and the receiving station (which converts the electrical signal back into voice).

Links are often described in terms of their ability to send and receive signals as part of a communication system. When described in these terms, they are broken down into *simplex* and *duplex*. Simplex means that the link can only send at one end and receive at the other end. In other words, the signal goes only one way. An example is the signal from a radio station. Duplex means that the link has a transmitter and a receiver at each end. A *half-duplex* system allows signals to go only one way at a time—an example is an intercom system. A *full-duplex* system allows users to send and receive at the same time. A telephone is a common example of a full-duplex system.

A fiber-optic link, shown in Figure 2.1, is like any other link, except that it uses optical fiber instead of wire. A fiber-optic link consists of four basic components:

◆ Transmitter that converts a signal into light and sends the light

◆ Receiver that captures the light and converts it back to a signal

◆ The optical fiber that carries the light

◆ The connectors that couple the optical fiber to the transmitter and receiver

FIGURE 2.1
The fiber-optic link

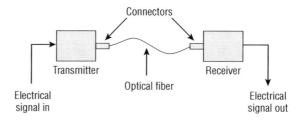

Now let's look at each component in a little more detail.

Transmitter

The transmitter, shown in Figure 2.2, converts an electrical signal into light energy to be carried through the optical fiber. The signal can be generated by many sources, such as a computer, a voice over a telephone, or data from an industrial sensor.

FIGURE 2.2
The fiber-optic
transmitter

Receiver

The receiver is an electronic device that collects light energy from the optical fiber and converts it into electrical energy, which can then be converted into its original form, as shown in Figure 2.3. The receiver typically consists of a *photodiode* to convert the received light into electricity, and circuitry to amplify and process the signal.

FIGURE 2.3
The fiber-optic
receiver

Optical Fibers

Optical fibers carry light energy from the transmitter to the receiver. An optical fiber may be made of glass or plastic, depending on the requirements of the job that it will perform. The advantage of light transmission through optical fiber as compared to the transmission of light through air is that the fiber can carry light around corners and over great distances.

Many fibers used in a fiber-optic link have a *core* between 8 and 62.5 microns (millionths of a meter) in diameter. For comparison, a typical human hair is about 100 microns in diameter. The *cladding* that surrounds the core is typically 125 microns in diameter.

The optical fiber's coating protects the cladding from abrasion. The thickness of the coating is typically half the diameter of the cladding, which increases the overall size of the optical fiber to 250 microns. Even with the additional thickness of the coating, optical fiber cabling is much smaller and lighter than copper cabling, as shown in Figure 2.4, and can carry many times the information.

FIGURE 2.4

Comparison of copper cable (top) and fiber cable (bottom)

Connectors

The connector is attached to the optical fiber and the fiber-optic cable. The connector allows the optical fiber to be mated to the transmitter or receiver. Transmitters or receivers typically have a *receptacle* that securely holds the connector and provides solid contact between the optical fiber and the optical subassembly of the device. The connector must align the fiber end precisely with the light source or photodiode to minimize signal loss.

The connectors, shown in Figure 2.5, could be considered the elements that make it possible for us to use fiber optics because they allow large hands to handle the small, fragile fibers. They are also the only place in the link where the optical fiber is exposed.

Now that you've seen the components required for a fiber-optic link, let's look at some of the methods that make it possible to transmit data with light.

FIGURE 2.5
Fiber-optic
connectors

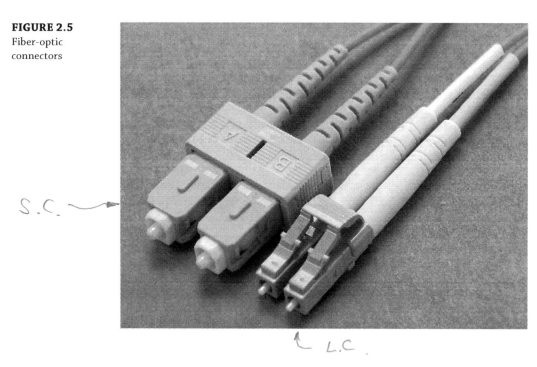

Amplitude Modulation

One method used for converting electrical signals into light signals for transmission is *amplitude modulation (AM)*. Amplitude refers to the strength of a signal, represented by a waveform, as shown in Figure 2.6. In amplitude modulation, electrical energy with continuously varying voltage is converted into light with continuously varying brightness.

FIGURE 2.6
Amplitude on a
waveform

Amplitude modulation requires two components: a carrier and a signal that is imposed on the carrier—also known as the *intelligence*—to change it in some way. When we speak, we impose the intelligence created by the vibration of our vocal cords on air, which is the carrier. Similarly, Bell's photophone used sound to vibrate a mirror, which *modulated* the light reflected from it. At the receiving end, a similar arrangement worked in reverse to *demodulate* the light, retrieving the intelligence from it and creating the sound again.

To modulate the amplitude of the light in a fiber-optic transmitter, the intelligence is sent through a circuit that changes it to a continuously varying voltage. As the intelligence changes, the voltage controlling the light rises and falls, varying the light's intensity to match the intelligence. Figure 2.7 shows the basic process of amplitude modulation.

FIGURE 2.7
Amplitude
modulation

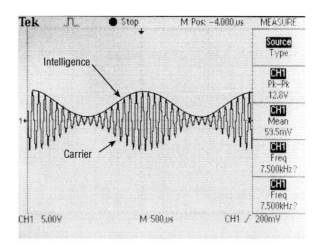

The intelligence imposed on the light changes the amplitude of the light, but not its wavelength. Amplitude modulation suffers from two problems that can affect the quality of the signal: attenuation and noise.

Attenuation is the loss of optical power as the signal passes through the optical fiber or interconnections. Attenuation in the optical fiber occurs as light is absorbed or scattered by impurities in the fiber. Attenuation from interconnections can be caused by several factors, which are covered in depth in later chapters.

As an amplitude-modulated signal is attenuated, its power decreases and small differences in amplitude become even smaller, or disappear entirely, as shown in Figure 2.8. When the light energy is converted back to electrical energy, these small differences are lost and cannot be reconstructed.

FIGURE 2.8
Attenuation of an
AM signal

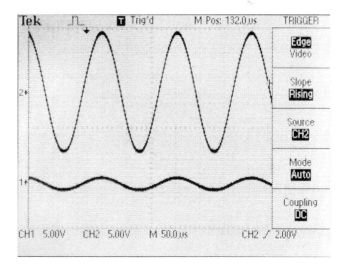

Noise is the introduction of unwanted energy into a signal. An example is static on an AM radio, especially when passing near high-voltage power lines. The unwanted energy changes the amplitude of the signal, sometimes rendering it unusable if the noise is great enough in comparison to the original signal.

Analog Transmission

Amplitude modulation is a form of *analog* transmission. An analog signal is one that varies continuously through time in response to an input. In addition, the response is infinitely variable within the specified range. In other words, a smooth change in the input will produce a smooth change in the signal.

A common example of an analog system is an electrical temperature sensor such as a thermocouple, which generates a small voltage that changes with the temperature. As the temperature rises, the thermocouple senses the temperature change, and the voltage increases. Because the relationship between temperature and voltage from the device is predictable, the thermocouple's output can be translated into a temperature reading. A reading of 3 millivolts (mV), for example, could indicate a temperature of 140° F.

When amplitude modulation is used with fiber optics, the amplitude of the optical transmission changes in relation to the strength of the incoming signal. Because of their infinitely variable response within a given range, analog signals are commonly used in RF-over-fiber applications.

Digital Data Transmission

In spite of the problems caused by noise, analog signals are still used in fiber-optic communications. If information is to be stored, carried, or manipulated by computers, however, it must be in a *digital* form—that is, represented by a series of on-off or high-low voltage readings. Figure 2.9 shows a digital waveform. The voltage readings are often represented as ones and zeros, with the high or on state being a one, and the low or off state being a zero. Because only two states—or digits—are used, the numbering system is referred to as *binary*.

FIGURE 2.9
Digital waveform

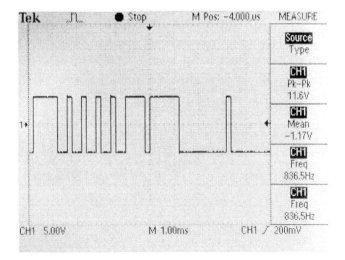

Recall that early signal fires, a form of digital communication, could announce a change in state by being lit but could not communicate complex information. To make digital information more detailed, *binary digits,* or *bits*, are combined into eight-place sequences called *bytes*. A byte can be used to represent a single number in the same way that a voltage reading would be used in an analog transmission. For example, the temperature reading of 141° F might be transmitted digitally as 10001101, the binary equivalent of 141.

Analog Data Transmission vs. Digital Data Transmission

One of the reasons that digital transmission is chosen over analog transmission is the fact that a digital signal is not affected by noise or attenuation the way an analog signal is.

Digital information can be stored and transmitted accurately because noise that would interfere with the analog data does not affect digital data. Each voltage in the sequence is either high or low, and voltages that do not match either the high or the low level do not change the meaning of the digital sequence.

The difference between the two is like the difference between a tape recording of a musical performance and a CD of the same performance. The analog recording may carry the same detail, but it would also contain a certain amount of hiss caused by electrical noise. The CD would be free of hiss, because the stray voltages do not register as either high or low signals.

More and more, digital transmissions are replacing analog transmissions, even in radio and television. Many radio stations now broadcast digital signals to receivers. In addition to carrying the regular programming as digital data, the broadcast can carry digital data for display on the receiver, such as program details, announcers' names, and song titles. All full-power television stations in the United States transitioned from analog to digital in June 2009. Canadian television stations began transitioning from analog to digital in August 2011. Mexico has also begun the process of transitioning from analog to digital and will complete it by the year 2021. In fiber-optic transmission, digital signals make it possible to carry many thousands of conversations over a single fiber through the use of multiplexing, which will be explained later in this chapter.

Analog to Digital (A/D) Conversion

To transmit an analog signal such as a voice through a digital system, it is necessary to *digitize*, or *encode*, it. This is also known as analog to digital, or A/D, conversion.

In A/D conversion, the smooth, continuously variable analog signal is translated into a digital signal that carries the same information. To do this, the analog signal's voltage is "sampled" at regular intervals and converted into binary numbers that represent the voltage at each interval. In Figure 2.10, for example, each vertical line represents a sampling of the analog signal at a given time.

FIGURE 2.10
Sampling an analog
signal

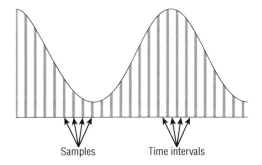

Samples   Time intervals

As with frequency measurements, the sample rate or sampling frequency is measured in terms of cycles per second, or hertz, so a rate of one sample per second would be designated 1Hz. A rate of 1,000 samples per second would be 1 kilohertz, or 1kHz.

Two factors affect the quality of the digital sample: sample rate and quantizing error.

Sample Rate

When an analog signal is digitized, any information between the samples is lost, so instead of a smooth transition over time, the digital information jumps from one voltage to the next in the signal. To smooth out the transitions and retain more of the information from the original analog signal, more samples must be taken over time. The higher the sampling rate, the more accurately the original analog signal can be digitized, as shown in Figure 2.11. Typically, audio signals for CDs and other digital music are sampled at 44.1kHz or 48kHz.

FIGURE 2.11
Low sample rate vs.
high sample rate

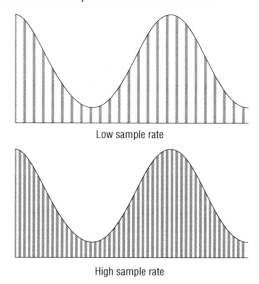

Low sample rate

High sample rate

Quantizing Error

The second factor affecting digital signal quality is called *quantizing error*. Quantizing error is caused by the inability of a binary number to capture the exact voltage of a digital sample.

Because an analog signal is infinitely variable, the sample's voltage could be any number within a specified range. If the binary number used to represent the voltage does not have enough bits, it cannot represent the voltage accurately.

In Figure 2.12, for example, a 4-bit number can represent 16 voltage levels—from 0 to 15 with 15 discrete steps or increments. Therefore, on a scale from 0V to +15V, each binary number represents a change of 1 whole volt. If a sample returned a reading of 1.5V, the binary number would still read 0001, or 1V. You can calculate the maximum error by dividing the number of discrete steps or the voltage range by the number of increments. In this case, 15 ÷ 15 = 1, so you have a maximum error of 1V, and the average error is one-half of that, or 0.5V.

FIGURE 2.12

Sampling with a
4-bit number

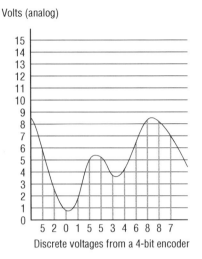

Volts (analog)

Discrete voltages from a 4-bit encoder

Increase the number of bits to eight, however, and you have 255 increments plus 0. The voltage between increments, and the maximum error, is now $15 \div 255 = 58.82 \times 10^{-3}$, or 58.82mV. Now, a reading of 1.5V is 0001 1001, or 25 steps from 0 instead of just one. Multiplying the number of increments by the voltage between them gives us $25 \times 58.82\text{mV} = 1.4705\text{V}$. This result is much closer to the analog reading of 1.5V.

As with the sample rate, the more bits used in encoding, the more accurate each sample can be. CD-quality audio signals are usually encoded at 16 bits, which means that there are 65,535 increments available, plus 0.

Digital-to-Analog (D/A) Conversion

When digital information is used to control analog devices such as temperature controls, or when analog information has been converted to digital data for transmission and must be converted back to analog data, digital-to-analog (D/A) conversion is used.

When digital data is converted to analog, two processes take place. First, a digital-to-analog converter converts each sequential binary sample to a proportional voltage. From our previous example of a 0V to +15V range represented by an 8-bit number, the binary sample 0001 1001 would be converted to 1.4705V. If the same binary sample were applied to a different voltage range, the result would be proportional to that range. The D/A converter outputs a stepped version of the analog signal, as shown in Figure 2.13. When reconstructing an encoded analog signal, the higher the sampling rate and the greater the number of bits in each sample, the more accurate the analog reconstruction can be.

Next, the steps between each digital sample must be smoothed out to provide a transition from one voltage to another. No matter how many samples are used, the digital output will always produce a signal that jumps from one voltage to another, and then holds each voltage for the amount of time between samples. When a smooth analog signal is required, D/A converters have circuits that filter the stairstep voltage into a smooth waveform, as shown in Figure 2.14.

FIGURE 2.13
Converting analog
to digital

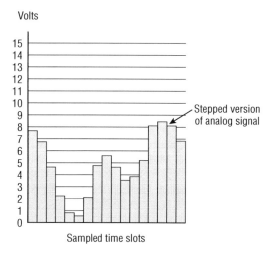

FIGURE 2.14
Filters convert
stairstep volt-
age to a smooth
waveform.

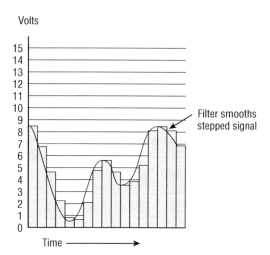

Pulse Code Modulation (PCM)

When an analog signal has been digitally encoded for transmission over a fiber-optic link, it has undergone a process known as *pulse code modulation (PCM)*. Pulse code modulation is a common method of digitizing analog data such as telephone conversations for transmission over a fiber-optic link. The analog voice data is sampled at regular intervals by the A/D converter and converted into a series of binary bits.

Data transmission using PCM in fiber optics is typically *serial*, which means that the binary bits are sent one after another in the order they were generated. The circuitry that converts the data also sends a timing, or *clock*, signal so the receiver can synchronize itself with the data that is being transmitted and reconstruct it accurately. Figure 2.15 shows a typical PCM sequence with a clock pulse burst.

FIGURE 2.15
In this PCM transmission, the binary numbers are sent along with a clock signal.

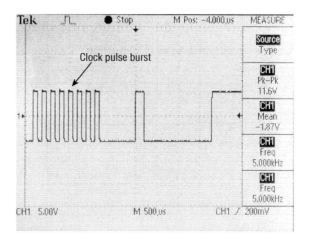

In order for pulse code modulation to be effective, an analog signal must be sampled at a rate that is at least twice the highest expected frequency. This number is referred to as the *Nyquist Minimum*. In practice, though, the sampling rate is usually closer to three times or more the highest expected frequency. This formula ensures that sampling will capture some portion of even the highest frequencies. For example, in a telephone conversation, the highest frequency encountered is about 4kHz. That means sampling must take place at the Nyquist Minimum of 8kHz to maintain a basic signal quality.

Multiplexing

Most fiber-optic data transmission systems can send data at rates that far exceed the requirements of a single stream of information. To take advantage of this fact, multiplexing can be used to carry several information channels, such as telephone conversations, nearly simultaneously. *Multiplexing* is the process of transmitting many channels of information over one link or circuit.

There are many multiplexing schemes or processes, which are discussed in detail in Chapter 12, "Passive Components and Multiplexers." Figure 2.16 shows a simple multiplexing process that may be used for the transmission of multiple telephone conversations. A *multiplexer* first divides each channel into several parts, each of which could represent a byte of voice data. In a process known as *interleaving*, the multiplexer sends the first part or byte of each channel, then the second part of each channel, continuing the process until all of the transmissions are completed. At the receiving end, a *demultiplexer* separates the transmissions into their individual channels and reassembles them in their proper order.

FIGURE 2.16
Multiplexing allows thousands of conversations to be carried in a single fiber.

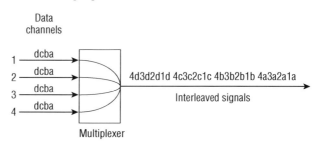

Decibels (dB)

As light travels away from its source through the link, it loses energy. Energy loss can be caused by several factors, such as the absorption or scattering of light by impurities in the fiber, or by light passing through the core and cladding and being absorbed in the coating.

It is important to be able to measure the amount of light energy lost in a fiber-optic link. Knowing the loss allows us to predict the strength of the light energy at the receiving equipment. The receiving equipment needs a minimum amount of light energy to accurately convert the light energy to the original signal. In addition, an understanding of how and where light is lost in a link can be helpful when troubleshooting the link.

One of the more common terms used when discussing the quality of a signal in fiber optics is the *decibel (dB)*. The decibel was originally used to measure the strength of sounds as perceived by the human ear. Its name means "one-tenth of a bel."

NOTE   A *bel* is a sound measurement named for telephone inventor Alexander Graham Bell.

Calculating dB Power Loss and Power Gain

The decibel can be used to express power gain or power loss relative to a known value. In fiber optics, the decibel is most commonly used to describe optical power, which is also known as signal power. Optical power is typically measured with an *optical power meter* (OPM). Measuring optical power with an OPM is cover in detail in Chapter 16, "Test Equipment and Link/Cable Testing."

Recall that in the 1960s, a signal loss of 20 decibels per kilometer was considered the goal for making fiber optics practical for communications. A 20-decibel loss means that of the original power put into the signal, only 1 percent remains.

Modern optical fiber has very low signal power loss and many fiber-optic links do not require the signal to be amplified. The decibel is used to express the signal power loss in all fiber-optic links. In links that extend many kilometers, it is used to express the signal power gain provided by the amplifier.

To calculate the decibel value of a gain or loss in optical power, use the following equation:

$$dB = 10Log_{10} (P_{out} \div P_{in})$$

Let's apply this equation to the loss associated with a length of optical fiber. The transmitter couples 10 microwatts (µW) into the optical fiber and the power at the receiver is 3µW. Since both values are in microwatts, we do not have to use microwatts in the equation. The equation would be written as follows:

$$dB = 10Log_{10} (3 \div 10)$$

$$dB = 10Log_{10} 0.3$$

$$dB = -5.23$$

$$Loss = 5.23dB$$

Because the calculated value is negative, it is a loss. When a loss is stated, the negative sign is dropped. The loss is 5.23dB.

As we learned earlier, most fiber-optic applications do not require amplification. However, amplification may be required for long transmission distances. If the input power and output power of the amplifier are known, the gain in dB can be calculated. We can solve for the gain of an amplifier where the input power is 1μW and the output power is 23μW by solving the equation as shown here:

$dB = 10Log_{10} (23 ÷ 1)$

$dB = 10Log_{10} 23$

$dB = 13.6$

$Gain = 13.6dB$

If you know the decibel value and want to calculate the gain or loss, you will have to rearrange the equation as shown here:

$(P_{out} ÷ P_{in}) = antilog (dB ÷ 10)$

Let's apply this equation to the loss associated with a length of optical fiber. To solve for the gain or loss we do not need to know the input power or output power; we just need the gain or loss in dB. Remember the loss in dB is a ratio of power out divided by power in. This length of optical fiber has a loss of 3.5dB. The equation would be written as follows:

$(P_{out} ÷ P_{in}) = antilog (dB ÷ 10)$

$(P_{out} ÷ P_{in}) = antilog (-3.5 ÷ 10)$

$(P_{out} ÷ P_{in}) = antilog -0.35$

$(P_{out} ÷ P_{in}) = 0.447$

If one power is known, the other power can be calculated. In this example, the input power is 13mW. To calculate the output power the equation would be written as follows:

$(P_{out} ÷ 13mW) = 0.447$

$P_{out} = 0.447 × 13mW$

$P_{out} = 5.8mW$

Remember that signals may be decreased, or attenuated, just about anywhere in the link. In addition to attenuation in the optical fiber itself, connectors, splices, and bends in the fiber-optic cable can also cause loss in the signal—sometimes considerable loss.

Expressing dB in Percentages

When measuring signal power gain or loss, decibels are calculated relative to the original power, rather than as an absolute number. For example, a loss of 0.1 decibel means that the signal has 97.7 percent of its power remaining, and a loss of 3 decibels means that only 50 percent of the original power remains. This relationship is logarithmic, rather than linear, meaning that with each 10 decibels of loss, the power is 10 percent of what it was (as shown in Table 2.1).

With each 10 decibels of gain, the power is 10 times what it was (as shown in Table 2.2).

TABLE 2.1:   Decibel losses expressed in percentages

LOSS IN dB	% POWER LOST	% POWER REMAINING
0.1	2.3%	97.7%
0.2	4.5%	95.5%
0.3	6.7%	93.3%
0.4	8.8%	91.2%
0.5	10.9%	89.1%
0.6	12.9%	87.1%
0.75	15.9%	84.1%
0.8	16.8%	83.2%
0.9	18.7%	81.3%
1	21%	79%
3	50%	50%
6	75%	25%
7	80%	20%
9	87%	13%
10	90%	10%
13	95.0%	5%
16	97%	3%
17	98.0%	2.0%
19	98.7%	1.3%
20	99.00%	1.00%
23	99.50%	0.50%
30	99.9%	0.1%
33	99.95%	0.05%
40	99.99%	0.01%
50	99.999%	0.001%
60	99.9999%	0.0001%
70	99.99999%	0.00001%

TABLE 2.2:   Decibel gains expressed in percentages

GAIN IN dB	% POWER INCREASE	% TOTAL POWER
0.1	2.3%	102.3%
0.2	4.7%	104.7%
0.3	7.2%	107.2%
0.4	9.6%	109.6%
0.5	12.2%	112.2%
0.6	14.8%	114.8%
0.75	18.9%	118.9%
0.8	20.2%	120.2%
0.9	23.0%	123.0%
1	26%	126%
3	100%	200%
6	298%	398%
7	401%	501%
9	694%	794%
10	900%	1,000%
13	1,895%	1,995%
16	3,881%	3,981%
17	4,912%	5,012%
19	7,843%	7,943%
20	9,900%	10,000%
23	19,853%	19,953%
30	99,900%	100,000%
33	199,426%	199,526%
40	999,900%	1,000,000%
50	9,999,900%	10,000,000%
60	99,999,900%	100,000,000%
70	999,999,900%	1,000,000,000%

One of the advantages of using decibels when calculating gain and loss is their relative ease of use compared to percentages. When measuring loss through different components, you can algebraically add the decibel loss from each component and arrive at a total signal loss for the system. If you were to use percentages, you would have to calculate the remaining power after the signal passed through each component, then take another percentage off for the next component, and so on, until the loss through the entire system had been calculated.

The Rules of Thumb

When calculating gain or loss in a system, it is useful to remember the three rules of thumb shown in Table 2.3, which make it easier to perform some common decibel calculations. These rules help you calculate how much power you have after the indicated decibel gain or loss.

TABLE 2.3:   Rules of thumb

DECIBEL	LOSS	GAIN
3 dB	½	×2
7 dB	$^4/_5$	×5
10 dB	$^9/_{10}$	×10

For example, a loss of 3 decibels means that you have about ½ of the original power. A gain of 3 decibels means that you have about twice the original power. These rules are intended for rough calculations only, because a 3 decibel loss actually leaves 50.1187 percent of the original power, and a 7 decibel loss leaves 19.953 percent of the original power.

Because decibels can be algebraically added and subtracted, you can use combinations of the decibel values to determine total gains or losses. For example, the rules of thumb can be applied to find the power output from an amplifier with an input of 10μW and a gain of 17dB. First, apply the 10dB rule and multiply 10μW by 10. Then apply the 7dB rule and multiple the result of the first calculation by 5. That value is the power remaining as shown in the following equations:

$P_{out} = 10μW \times 10$

$P_{out} = 100μW$

$P_{out} = 100μW \times 5$

$P_{out} = 500μW$

The optical power at the output to the amplifier is 500μW.

Keep in mind that the dB rules of thumb cannot calculate the answer for every scenario. If the gain for the amplifier had been 18dB, we would not have been able to accurately calculate the optical power output of the amplifier. However, we would have been able to estimate that value by calculating for the closest value possible 17dB. The estimated output would be greater than 500μW but less than 1mW. The 1mW value is the result of solving the above equation a gain of 20dB, the closest value above 18dB that can be calculated using the dB rules of thumb.

Absolute Power

Taken by itself, the decibel is only a relative number, and it has meaning only when it is referenced to a known input or output power. In many cases, however, it is important to have an absolute value to use for comparison or for equipment specifications. When such a value is required in fiber optics, the decibel is referenced to 1 milliwatt (mW). When the decibel is referenced to 1mW, a lowercase "m" or sometimes "mW" is inserted after dB. The decibel referenced to 1mW is written as 0dBm.

When using dBm, power levels below 1mW are negative and power levels 1mW or greater are positive. The positive sign is typically implied and only the negative sign is used. As an example, a value of –10dBm means that 1mW of power has been attenuated by 10dB, so only 10 percent of its power, or 100μW, remains. A value of 10dBm means that 1mW of power has been increased by 10dB, which has value of 10mW. Table 2.4 shows how dBm values convert to power measurements.

TABLE 2.4:   Converting power to dBm

OPTICAL POWER IN WATTS	OPTICAL POWER IN dBm
100mW	20dBm
20mW	13dBm
10mW	10dBm
8mW	9dBm
5mW	7dBm
4mW	6dBm
2mW	3dBm
1mW	0dBm
500μW	–3dBm
250μW	–6dBm
200μW	–7dBm
125μW	–9dBm
100μW	–10dBm
50μW	–13dBm
25μW	–16dBm
20μW	–17dBm

TABLE 2.4:   Converting power to dBm   *(continued)*

OPTICAL POWER IN WATTS	OPTICAL POWER IN dBm
12.5µW	–19dBm
10µW	–20dBm
5µW	–23dBm
1µW	–30dBm
500nW	–33dBm

The formula used to convert dBm to a power measurement is:

$dBm = 10Log_{10} (P \div 1mW)$

This equation is similar to the equation for finding a decibel value from an input power and an output power, except that in this case the input power is fixed at 1mW. Let's apply this equation to calculate optical power from a dBm value. In this example the power meter measures –11 dBm. To calculate the power in watts, the equation would be written as follows:

$-11 = 10Log_{10} (P \div 1mW)$

$-1.1 = Log_{10} (P \div 1mW)$

$0.0794 = P \div 1mW$

$P = 0.0794 \times 1mW$

$P = 0.0794mW$

We can also convert a power measurement into dBm. Remember that when using dBm, power levels below 1mW are negative and power levels 1mW or greater are positive. The formula used to convert a power measurement to dBm:

$dBm = 10Log_{10} (P \div 1mW)$

Let's apply this equation to calculate dBm from an optical power value of 10µW. To calculate dBm from the power in watts, the equation would be written as follows:

$dBm = 10Log_{10} (10µW \div 1mW)$

$dBm = 10Log_{10} (0.01)$

$dBm = 10 \times -2$

$dBm = -20$

The power value is –20dBm.

This answer can be checked by locating 10µW in the first column of Table 2.4.

The Bottom Line

Calculate the decibel value of a gain or loss in power.   The decibel is used to express gain or loss relative to a known value. In fiber optics, the decibel is most commonly used to describe signal loss through the link after the light has left the transmitter.

> **Master It**   The output power of a transmitter is 100µW and the power measured at the input to the receiver is 12.5µW. Calculate the loss in dB.

Calculate the gain or loss in power from a known decibel value.   If the gain or loss in power is described in dB, the gain or loss can be calculated from the dB value. If the input power is known, the output power can be calculated, and vice versa.

> **Master It**   The loss for a length of optical fiber is 4dB. Calculate the loss in power from the optical fiber and the output power of the optical fiber for an input power of 250µW.

Calculate the gain or loss in power using the dB rules of thumb.   When calculating gain or loss in a system, it is useful to remember the three rules of thumb shown in Table 2.3.

> **Master It**   A fiber-optic link has 7dB of attenuation. The output power of the transmitter is 100µW. Using the dB rules of thumb calculate the power at the input to the receiver.

Convert dBm to a power measurement.   In fiber optics, dBm is referenced to 1mW of power.

> **Master It**   A fiber-optic transmitter has an output power of –8dBm. Calculate the output power in watts.

Convert a power measurement to dBm.   In fiber optics, dBm is referenced to 1mW of power.

> **Master It**   A fiber-optic transmitter has an output power of 4mW. Calculate the output power in dBm.

Chapter 3

Basic Principles of Light

Even the simplest fiber-optic link is a triumph of innovative design, manufacturing precision, and technical creativity. One of the factors contributing to the high data rates and long transmission distances of fiber optics is knowledge of the principles of light. While the optical fiber is the transmission medium, light is the carrier on which the signals are imposed. Its small wavelengths and high transmission velocity make possible bandwidths that would be unimaginable using other forms of transmission. To get the best performance from every part of the fiber-optic link, you must understand the characteristics of light and the factors within the fiber-optic link that affect it.

This chapter describes the basic characteristics of light, especially the type of light used in fiber-optic communications. It also discusses the principles and materials that make fiber-optic communications possible, along with some of the problems that must be overcome.

In this chapter, you will learn to:

- ◆ Convert various wavelengths to corresponding frequencies
- ◆ Convert various frequencies to corresponding wavelengths
- ◆ Calculate the amount of energy in a photon using Planck's constant
- ◆ Calculate the speed of light through a transparent medium using its refractive index
- ◆ Use Snell's law to calculate the critical angle of incidence
- ◆ Calculate the loss in decibels from a Fresnel reflection

Light as Electromagnetic Energy

Whether it comes from the sun, an electric bulb, or a laser, all light is a form of electromagnetic energy. Electromagnetic energy is emitted by any object that has a temperature above absolute zero ($-273.15°$ C), which means that the atoms in the object are in motion. The electrons orbiting the atoms pick up energy from the motion, and the energy causes them to move to higher orbits, or energy levels. As they drop back to their original energy levels, they release the energy again. The energy takes two forms: an electrical field and a magnetic field, formed at right angles to each other and at right angles to their path of travel, as shown in Figure 3.1.

FIGURE 3.1
The three-dimensional nature of electromagnetic energy

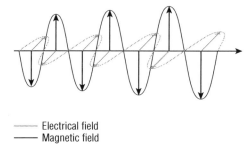

---------- Electrical field
———— Magnetic field

The combination of these electrical and magnetic fields is an electromagnetic wave, which travels through open space or air at approximately 300,000km/s (kilometers per second). It is important to understand the three-dimensional nature of electromagnetic waves because different types of fiber-optic transmission take advantage of these characteristics. How this is done will be discussed in later chapters.

Wavelength and Frequency

In fiber optics, wavelength (λ) and frequency (f) are used to describe electromagnetic energy. Wavelength is typically used to describe the light source and frequency is used to describe the channel spacing in dense wavelength division multiplexing (DWDM) systems. The wavelengths of commonly used light sources are described in detail in Chapter 10, "Fiber-Optic Light Sources and Transmitters," and DWDM systems are covered in Chapter 12, "Passive Components and Multiplexers."

Although electromagnetic energy exists three-dimensionally, it is often represented as a two-dimensional sine wave, where the wavelength is the distance between corresponding points on two consecutive waves as shown in Figure 3.2. Depending on its wavelength, electromagnetic energy can occur as radio waves, microwaves, light waves, X-rays, and more.

FIGURE 3.2
Electromagnetic energy is often shown as a sine wave.

Wavelength

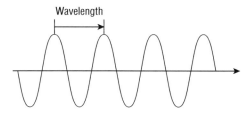

Wavelength is important because it allows us to calculate the electromagnetic energy's frequency, which is the number of waves that pass a given point in one second. Frequency is measured in cycles per second, or hertz (Hz).

A way to express the relationship between wavelength and frequency is that wavelength equals the velocity of the wave (v) divided by its frequency, or

$$\lambda = v \div f$$

Note that v is usually the speed of light in a vacuum or in open air, approximately 300,000km/s.

Likewise, you can calculate the frequency using the wavelength with the equation:

$$f = v \div \lambda$$

For example, one of the infrared light sources used in fiber optics has a wavelength of 1310 nanometers (billionths of a meter), which translates to a frequency of 229 terahertz (THz), or 229 trillion Hz.

SOME USEFUL TERMS

Many of the terms used to express electromagnetic wavelengths and frequencies describe large multiples of cycles or small fractions of meters. To understand these terms, it is helpful to know the prefixes used with them.

Each prefix expresses a multiple of 10 or 1 divided by a multiple of 10 (as in 1/1000) applied to the measurement unit, such as meters or cycles. Some terms may already be familiar, such as *kilo*meter, for a thousand meters, or *centi*meter, for a hundredth of a meter.

Here is a list of prefixes, along with their translations and mathematical equivalents, in descending order of magnitude:

Prefix	Meaning	Magnitude	Symbol
peta-	Quadrillion	10^{15}	P
tera-	Trillion	10^{12}	T
giga-	Billion	10^{9}	G
mega-	Million	10^{6}	M
kilo-	Thousand	10^{3}	k
centi-	Hundredth	10^{-2}	c
mili-	Thousandth	10^{-3}	m
micro-	Millionth	10^{-6}	μ
nano-	Billionth	10^{-9}	n
pico-	Trillionth	10^{-12}	p
femto-	Quadrillionth	10^{-15}	f

Note in Figure 3.3 that as the wavelength becomes smaller, more waves will occur in one second, which means that the frequency will increase as wavelength decreases. However, the reverse happens if the wavelength were to become longer. The frequency will decrease as the wavelength increases.

FIGURE 3.3
As wavelength
decreases, fre-
quency increases.

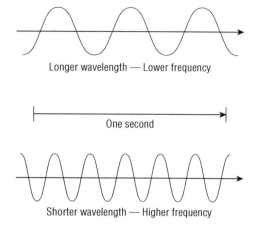

Longer wavelength — Lower frequency

One second

Shorter wavelength — Higher frequency

To illustrate this, the signal from an AM radio station transmitting at 890 kilohertz (kHz), or 890,000Hz, has a wavelength of 337.1 meters. If we choose a higher frequency, such as 960kHz, the wavelength decreases to 312.5 meters and if we choose a lower frequency such as 560kHz the wavelength increases to 535.7 meters.

Characteristics of Electromagnetic Radiation

Electromagnetic radiation, like most radiated energy, has characteristics of both waves and particles. As a wave, it propagates through a medium and transfers energy without permanently displacing the medium. However, as a massless particle, called a *photon*, it travels in a wavelike pattern moving at the speed of light.

A photon, which is emitted from an electron as it changes energy levels, is the basic unit, or *quantum*, of electromagnetic energy. The amount of energy in each photon, however, depends on the electromagnetic energy's frequency: the higher the frequency, the more energy in the particle.

To express the amount of energy in a photon, we use the equation

$E = hf$

where E is the energy expressed in watts, h is Planck's constant, or 6.626×10^{-34} joule-seconds, and f is the frequency of the electromagnetic energy.

So, to find the energy of a photon of infrared light at 229 THz:

$(2.29 \times 10^{14})(6.626 \times 10^{-34}) = 1.517 \times 10^{-19}$ W

NOTE Much of what we take for granted in the field of fiber optics comes from work done by pioneers in the field of physics in the late 1800s and early 1900s. The *joule* is a unit of energy named for James Prescott Joule, who studied the relationship between heat and mechanical work. One joule is equal to the amount of work required to produce 1 watt in 1 second. Planck's constant was defined in 1899 by Nobel Laureate Max Planck, who is regarded as the founder of quantum theory.

The Electromagnetic Spectrum

In 1964, scientists at a Bell Laboratories facility in New Jersey accidentally discovered electromagnetic radiation associated with the very beginnings of our universe. The radiation, predicted by certain cosmological theories, was emitted by hydrogen atoms at a temperature of just about 3° on the *Kelvin* (K) scale, or –270.15° C, and is practical evidence that anything with a temperature above absolute zero puts out electromagnetic energy.

The wavelengths emitted by the cosmic hydrogen atoms, about 0.5mm to 5mm, are in the *microwave* range of the electromagnetic spectrum, just above radio waves in frequency, and just below infrared light.

Some characteristics of electromagnetic energy in free space are dependent on wavelength. Free space typically implies the medium is a vacuum; however, in free-space optical communication it could also be referring to air or some similar medium. As we have learned, optical communication can also take place in a controlled space such as an optical fiber. In a free space transmission, you may not be able to control the medium as you can in a controlled space transmission. For example, the maximum transmitting distance for the optical signal on a foggy day will be less than a clear day. However, the transmission through an optical fiber is not affected by weather.

Longer wavelengths require less energy for propagation than shorter wavelengths of the same amplitude, making them useful for long-distance communication. By the same rule, particles of higher frequency electromagnetic radiation have more energy than particles from lower frequency emissions of the same amplitude. This fact is important when determining which wavelengths to use in fiber-optic transmissions.

USING PREFIXES

Very wide ranges of numbers are used in fiber optics. You may be describing the length of a cable in kilometers and the wavelength of a light source in nanometers. The following is a list of the most common prefixes used in fiber optics and an example of how they may be used to describe something.

Prefix	Used to describe
tera-	The frequency of light in terahertz (THz)
giga-	The bandwidth of a link in gigabits per second (Gbps)
mega-	The bandwidth of an optical fiber in megahertz (MHz)
kilo-	The length of a fiber-optic cable in kilometers (km)
centi-	The bend radius of a fiber-optic cable in centimeters (cm)
mili-	The diameter of a fiber-optic cable in millimeters (mm)
micro-	The diameter of the optical fiber in micrometers (μm)
nano-	The wavelength of light in nanometers (nm)
pico-	The dispersion of light in picoseconds (ps)

These transmissions consist of turning the carrier, or the light, on and off at high switching rates. At the higher rates desired for communications, higher frequencies of light are needed if an extremely short "on" cycle is to have enough energy to be detected.

Remember also that higher frequencies can carry more data, because more waves per second allow more bits per second to be carried. This fact extends the principle that had been applied first to radio and then to television, which relied on ever higher frequencies to carry greater amounts of information.

What we commonly call light is just one small part of the electromagnetic spectrum, shown in Figure 3.4. Visible light is an even smaller component, bordered by infrared, with longer wavelengths, and ultraviolet, with shorter wavelengths.

FIGURE 3.4

The electromagnetic spectrum

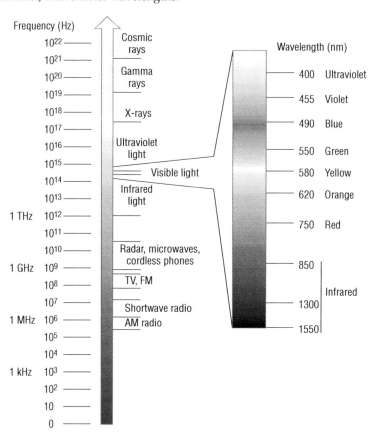

The wavelengths most commonly used for fiber optics are in the infrared range, at *windows* of 850 nanometers (nm), 1300nm, and 1550nm. The spectrum range of these wavelengths provides an important combination of characteristics: it is high enough to make high data rates possible, but low enough to require relatively low power for transmission over long distances.

NOTE   Transmission windows were created as a way of standardizing useful wavelengths in fiber optics, making it easier to build interoperable equipment.

The specific wavelength windows have been selected because they provide the best possible characteristics for transmission. Even within the range between 850nm and 1550nm, certain regions have high losses due to materials in the fiber, such as stray water molecules, absorbing light at a wavelength of 1380nm. Other wavelengths, such as 1550nm, are favored because they have a low loss, allowing longer transmission distances. On the other hand, wavelengths near 1300nm suffer less from *dispersion*, which will be discussed at length in Chapter 5, "Optical Fiber Characteristics."

Refraction

Refraction, or the change in the direction of light as it changes speeds passing from one material into another, is a key component in fiber-optic transmissions. The principles that cause an object in water to look like it is bent, as seen in Figure 3.5, are the same principles that keep light contained within the core of an optical fiber even through curves.

FIGURE 3.5
Refraction of light
through water

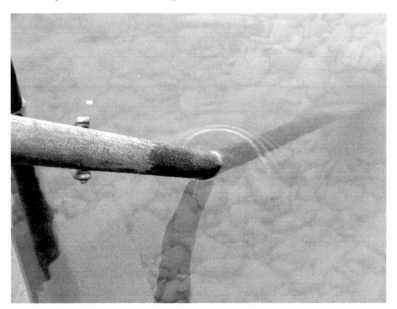

What Causes Refraction?

Refraction occurs when light waves change speed as they travel between two materials, each with a different refractive index or *index of refraction* (*n*). We commonly think of the speed of light as constant at about 300,000km/s (299,792.458km/s, to be precise), but that figure is actually the theoretical top speed of light, which only applies to its speed in a vacuum. In reality, light travels at lower velocities in various materials or media such as the earth's atmosphere, glass, plastic, and water. A *medium* is any material or space through which electromagnetic radiation can travel.

How quickly light travels through a medium is determined by its refractive index. Light travels more slowly in a material with a high refractive index and faster in a material with a low refractive index. It is important to note that even though light slows down when passing from a medium with a lower refractive index into a medium with a higher refractive index, it speeds up again when it passes into a medium with a lower refractive index (see Figure 3.6).

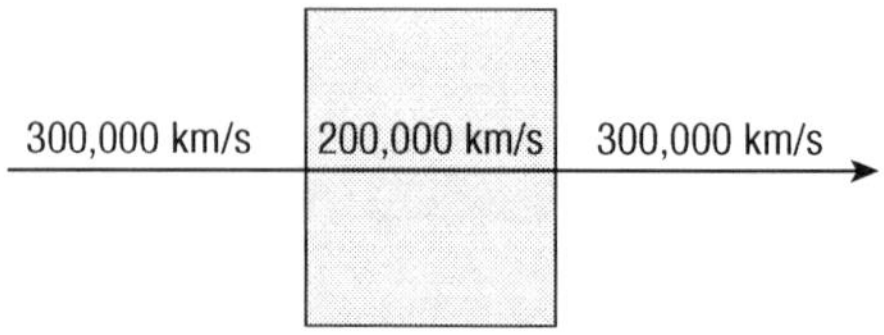

FIGURE 3.6
Light changes speeds as it passes through mediums with different refractive indexes.

300,000 km/s | 200,000 km/s | 300,000 km/s

When light changes velocity as it travels from one medium into another, refraction occurs. The amount of refraction is determined by the relative difference in the light's velocity in each of the media. The greater the difference, the greater the refraction.

In Figure 3.5, shown earlier, the portion of the oar in the water looks bent because the light rays reflected from it are bent as they pass through the boundary between the water and the air, as seen in Figure 3.7. Notice also that the portion of the oar that is underwater is distorted because the light rays reflected from it have been bent.

FIGURE 3.7
Light rays from the oar are refracted as they pass from water into the air.

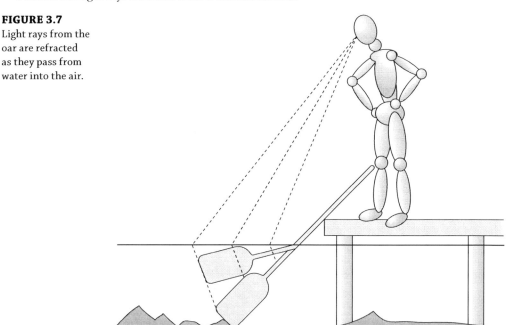

To explain refraction, we have to look at the wave nature of light. Recall that light consists of two perpendicular waves, and that the light travels in a path that forms right angles with both waves.

As the light wave meets the boundary between media with different refractive indexes, that portion of the light wave changes velocity or it experiences a *phase velocity* change, while the rest of the light wave maintains its original velocity. A change in the phase velocity of a light wave typically causes the light wave to change direction as described by Snell's law, which is discussed in detail later in this chapter. Figure 3.8 shows light waves changing direction because of a change in phase velocity as multiple light waves cross from a lower refractive index into a higher refractive index.

FIGURE 3.8
Light waves chang-
ing direction
because of a change
in phase velocity

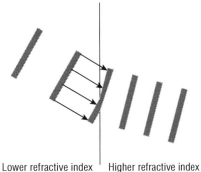

Lower refractive index | Higher refractive index

The velocity of light through different media such as glass depends on the light's wavelength. One of the most common ways to observe this is by seeing how white light is refracted through a prism and broken up into its component wavelengths, as shown in Figure 3.9. Notice that violet, which has the shortest visible wavelength, is refracted more than red, which has the longest visible wavelength. In this example, the red light changed direction the least and is traveling the fastest through the prism, while the violet light changed direction the most and is traveling the most slowly through the prism.

FIGURE 3.9
Refraction of white
light into compo-
nent colors

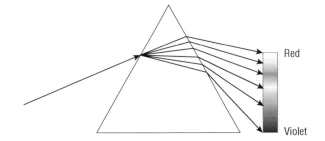

Red

Violet

NOTE It is important to remember that different wavelengths of light travel at different speeds in an optical fiber.

Calculating the Index of Refraction

We know the velocity of light can change and its velocity depends on its wavelength and the index of refraction of the medium it is passing through. We also know that the higher the refractive index, the slower the light travels. In other words, the speed of light (c) in a medium is inversely proportional to its index of refraction.

It is important to be able to assign an index of refraction to different materials. Standard refractive index measurements are taken at a wavelength of 589nm. The refractive index of light wavelengths above or below 589nm will vary slightly. Table 3.1 lists the speed of light and index of refraction for some common materials.

TABLE 3.1:   Refractive indexes of different materials

MATERIAL	SPEED OF LIGHT TRAVELING THROUGH MATERIAL (km/s)	INDEX OF REFRACTION
Vacuum	300,000	1.0000
Air	300,000	1.0003
Water	225,056	1.333
Ethyl alcohol	220,588	1.36
Optical fiber cladding	205,479	1.46
Optical fiber core	204,082	1.47
Glass	200,000	1.5

The index of refraction is a relative value and is based on the speed of light in a vacuum, which has an index of refraction of 1. We can calculate the index of refraction with the equation

$n = c/v$

where n is the index of refraction of the material, c is the velocity of light through vacuum, and v is the light's velocity through the material. So if light passes through a theoretical material at 230,000km/s, the material's index of refraction is

$n = 300000 \div 230000 = 1.3$

To calculate the amount of refraction that will take place when light passes from one material to another, we'll need a model and some basic terms. The model, shown in Figure 3.10, illustrates light passing from a medium with a lower index of refraction (n_1) to a medium with a higher index of refraction (n_2). The *interface* is represented by a horizontal line. A path perpendicular to the interface is known as the *normal*. Light traveling along the normal will change speed but will not change direction.

The *angle of incidence* represents the angle between the incident ray and normal. The *angle of refraction* represents the angle between the refracted ray and normal. Note that a small amount of light is also reflected from the interface at an angle equal to the angle of incidence.

FIGURE 3.10
Model used to calculate refraction

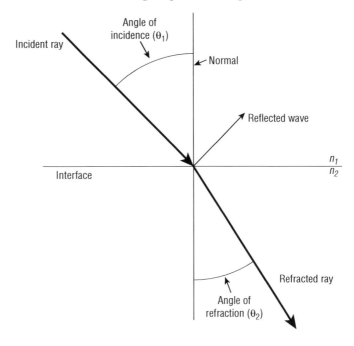

Total Internal Reflection

As shown earlier in Figure 3.10, light passing from a lower index of refraction to a higher index of refraction refracts toward normal. This is illustrated by the angle of refraction (θ_2), which is smaller, then the angle of incidence (θ_1). When light passes from a higher index of refraction to a lower index of refraction, as shown in Figure 3.11, the light refracts away from normal.

We can calculate the amount of refraction using Snell's law, which shows the relationship between incident light and refracted light:

$$n_1\sin\theta_1 = n_2\sin\theta_2$$

where *n1* and n_2 are the index of refraction values of each material, θ_1 is the angle of incidence, and θ_2 is the angle of refraction.

FIGURE 3.11

Light passing from higher index of refraction to lower index of refraction refracts away from normal.

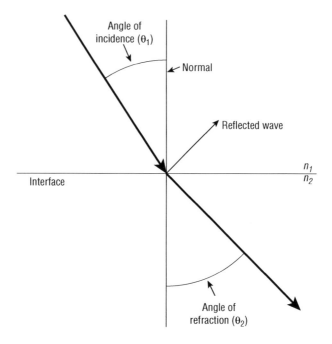

Recall that the phenomenon that makes fiber-optic transmission possible, *total internal reflection* (TIR), is caused by the same principles that cause refraction. In the case of TIR, the light is passing from a higher index of refraction to a lower index of refraction at an angle that causes all of the light to be reflected. What has happened is that the angle of refraction has exceeded 90° from normal.

The incident angle required to produce a refracted angle of 90° is called the *critical angle*. As the incident ray moves from normal toward the critical angle, less and less of the incident ray's energy is carried into the refracted ray. At the critical angle all of the incident ray's energy is refracted along the interface. As the incident angle exceeds 90°, the light is reflected, as shown in Figure 3.12.

To find the critical angle of two materials, we can use a modified version of Snell's equation:

$$\theta_c = \arcsin(n_2 \div n_1)$$

where θ_c produces a refractive angle of 90° from normal.

So if we want to know the critical angle of an optical fiber having a core index of refraction of $n_1 = 1.5$ and a cladding index of refraction of $n_2 = 1.46$, we solve this equation:

$$\theta_c = \arcsin(1.46 \div 1.5) = 76.7°$$

FIGURE 3.12
Reflection of light
exceeding critical
angle

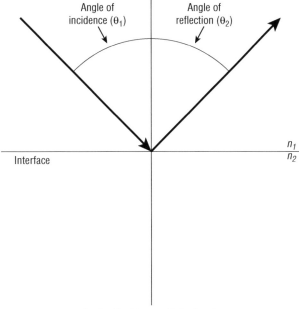

FIGURE 3.13
Total internal
reflection

In this example, if light is passing through the core at an angle greater than 76.7°, it will be reflected from the interface with the cladding. As long as the interface remains parallel, the light will continue to reflect at the same angle, as shown in Figure 3.13.

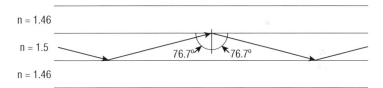

Fresnel Reflections

Recall that when light passes from one medium into another, refraction occurs as described by Snell's law. However, a reflection also occurs. This reflection is known as a *Fresnel reflection*. A Fresnel reflection occurs when light changes speed as it moves from a material with a given index of refraction into a material with a different index of refraction. The greater the difference in the index of refraction between the two materials, the greater the amount of light reflected. You experience Fresnel reflection whenever you look through a window and see a faint reflection of yourself in the glass.

Augustin-Jean Fresnel determined how to calculate the amount of light lost through a Fresnel reflection when light passes from one medium into another with the following equation:

$$\rho = ((n_1 - n_2) \div (n_1 + n_2))^2$$

where ρ is the amount of light reflected and n is the index of refraction of the medium.

If we calculate the reflection value for the interface between air, with an index of refraction of 1, and glass, with an index of refraction of 1.5, we get the following result:

$$\rho = ((1.5 - 1) \div (1.5 + 1))^2 = 0.2^2 = 0.04$$

To calculate the loss in decibels, we use the equation

$$dB = 10Log_{10}(1 - \rho)$$

plugging in the Fresnel reflection value from earlier,

$$dB = 10Log_{10}\, 0.96 = 10 \times -0.018 = -0.18$$

which gives us a loss of 0.18dB.

It is important to understand that a Fresnel reflection occurs whenever there is a velocity change. It does not matter whether the light is increasing speed or decreasing speed. Light traveling in the core of an optical fiber travels slower than it does in air. If there were an air gap at an interconnection, a Fresnel reflection would occur as light left the core of the optical fiber and entered the air. A Fresnel reflection of equal magnitude would also occur as light traveling in the air entered the core of the other optical fiber. The reflected light is lost and never makes it to its intended destination. Some of the reflected light may travel in the opposite direction in the core of the optical fiber back toward the source. Depending on the light source, this could cause a problem such as a shift in wavelength or a reduction in output power.

REDUCING FRESNEL REFLECTIONS AND MAKING THE INTERCONNECTION WORK

The connector endface is the part of the fiber-optic connector where the optical fiber is exposed. From time to time, the optical fiber at the connector endface is damaged, and typically the person causing the damage is unaware of it. The damage may occur because there was debris on one connector endface when it was mated with another connector. It may occur when a damaged connector is mated with an undamaged connector. Regardless of how it occurs, the end result is typically a fiber-optic link that does not function properly.

Critical fiber-optic links need to be restored as quickly as possible. This means that replacing the damaged connector(s) at this time may not be an option. Although it is not a common practice to use index matching gel in an interconnection, using it to reduce the Fresnel reflections can sometimes restore a damaged or poorly performing interconnection. This could mean the difference between a successful and a failed restoration. (*Index matching gel* is a substance that fills the air space within a connection with a material matching the refractive index of the fiber core.)

The Bottom Line

Convert various wavelengths to corresponding frequencies. The relationship between wavelength and frequency can be expressed with the formula $\lambda = v \div f$.

Master It If an electromagnetic wave has a frequency of 94.7MHz, what is its wavelength?

Convert various frequencies to corresponding wavelengths. The relationship between frequency and wavelength can be expressed with the formula $f = v \div \lambda$.

Master It If an electromagnetic wave has a wavelength of 0.19 meters, what is its frequency?

Calculate the amount of energy in a photon using Planck's constant. The amount of energy in each photon depends on the electromagnetic energy's frequency: the higher the frequency, the more energy in the particle. To express the amount of energy in a photon, we use the equation $E = hf$.

Master It Calculate the energy of a photon at a frequency of 193.55THz.

Calculate the speed of light through a transparent medium using its refractive index. The index of refraction is a relative value, and it is based on the speed of light in a vacuum, which has an index of refraction of 1. The index of refraction can be calculated using the equation $n = c/v$.

Master It If light is passing through a medium with a refractive index of 1.33, what will the velocity of the light through that medium be?

Use Snell's law to calculate the critical angle of incidence. The incident angle required to produce a refracted angle of 90° is called the critical angle. To find the critical angle of two materials, we can use a modified version of Snell's equation:

$$\theta_c = \arcsin(n_2 \div n_1)$$

Master It Calculate the critical angle for two materials where n_2 is 1.45 and n_1 is 1.47.

Calculate the loss in decibels from a Fresnel reflection. Augustin-Jean Fresnel determined how to calculate the amount of light lost from a Fresnel reflection using the equation $\rho = ((n_1 - n_2) \div (n_1 + n_2))^2$.

Master It Calculate the loss in decibels from a Fresnel reflection that is the result of light passing from a refractive index of 1.51 into a refractive index of 1.45.

Chapter 4

Optical Fiber Construction and Theory

Optical fibers are called on to operate in a wide variety of conditions. Some of them must carry high volumes of data over many miles, whereas others carry smaller amounts of data inside an office building or aboard an aircraft. The type of job an optical fiber will do determines the type of fiber you'll choose to install. It is important to understand the types of fibers that are available and the ways in which they are built so that you can select and use them properly.

This chapter describes the construction of optical fibers and the components that make them up. We will discuss some of the important factors that must be considered in the manufacture and use of optical fibers, as well as the designs used to optimize them for different types of data transmission. Finally, the chapter will introduce you to some of the commonly used commercial optical fibers and describe their features.

In this chapter, you will learn to:

- ♦ Select the proper optical fiber coating for a harsh environment
- ♦ Identify an industry standard that defines the specific performance characteristics on an optical fiber
- ♦ Identify an optical fiber from its refractive index profile
- ♦ Calculate the numerical aperture of an optical fiber
- ♦ Calculate the number of modes in an optical fiber

Optical Fiber Components

Today's standard optical fiber is the product of precision manufacturing techniques and exacting standards. Make no mistake: even though it is found in almost any data or communications link, optical fiber is a finely tuned instrument requiring care in its production, handling, and installation.

As shown in Figure 4.1, a typical optical fiber comprises three main components: the core, which carries the light; the cladding, which surrounds the core with a lower refractive index and contains the light; and the coating, which protects the fragile fiber within.

Let's look at these components individually.

FIGURE 4.1
Optical fiber components include the core, cladding, and coating.

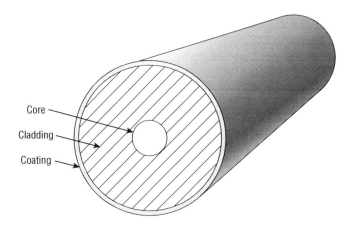

Core

Cladding

Coating

Core

The *core*, which carries the light, is the smallest part of the optical fiber. The optical fiber core is usually made of glass, although some are made of plastic. The glass used in the core is extremely pure silicon dioxide (SiO_2), a material so clear that you could look through 5 miles of it as though you were looking through a household window.

In the manufacturing process, dopants such as germania, phosphorus pentoxide, and alumina are used to raise the refractive index under controlled conditions.

Optical fiber cores are manufactured in different diameters for different applications. Typical glass cores range from as small as 3.7µm up to 200µm. Core sizes commonly used in telecommunications are 9µm, 50µm, and 62.5µm. Plastic optical fiber cores can be much larger than glass. A popular plastic core size is 980µm.

Cladding

Surrounding the core, and providing the lower refractive index to make the optical fiber work, is the *cladding*. When glass cladding is used, the cladding and the core are manufactured together from the same silicon dioxide–based material in a permanently fused state. The manufacturing process adds different amounts of dopants to the core and the cladding to maintain a difference in refractive indexes between them of about 1 percent. A typical core may have a refractive index of 1.49 at 1300nm while the cladding may have a refractive index of 1.47. These numbers, however, are wavelength dependent. The core of the same fiber will have a different refractive index at a different wavelength. At 850nm the core refractive index will increase slightly.

Like the core, cladding is manufactured in standard diameters. The two most commonly used diameters are 125µm and 140µm. The 125µm cladding typically supports core sizes of 9µm, 50µm, 62.5µm, and 85µm. The 140µm cladding typically has a 100µm core.

Coating

The *coating* is the true protective layer of the optical fiber. The coating absorbs the shocks, nicks, scrapes, and even moisture that could damage the cladding. Without the coating, the optical fiber is very fragile. A single microscopic nick in the cladding could cause the optical fiber to break when it's bent. Coating is essential for all-glass fibers, and they are not sold without it.

The coating is solely protective. It does not contribute to the light-carrying ability of the optical fiber in any way. The outside diameter of the coating is typically either 250µm or 500µm. The coating is typically colorless. In some applications, however, the coating is colored, as shown in Figure 4.2, so that individual optical fibers in a group of optical fibers can be identified.

FIGURE 4.2
The fiber coating can be color-coded to help identify it.

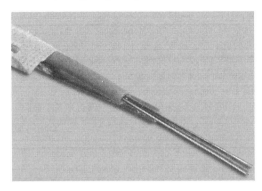

The coating found on an optical fiber is selected for a specific type of performance or environment. One of the most common types of coatings is *acrylate*. This coating is typically applied in two layers. The *primary coating* is applied directly on the cladding. This coating is soft and provides a cushion for the optical fiber when it is bent. The *secondary coating* is harder than the primary coating and provides a hard outer surface. Acrylate, however, is limited in temperature performance. A typical acrylate coating has a maximum operating temperature of 100° C while some high-temperature acrylates may perform at temperatures up to 125° C.

Silicone, carbon, and *polyimide* are coatings that may be found on optical fibers that are used in harsh environments such as those associated with avionics, aerospace, and space. They may also be used on optical fibers designed for mining, or oil and gas drilling.

Silicone is a soft material with a higher temperature capability then acrylate, typically 200° C. Like acrylate, silicone cushions the optical fiber during bending, but it is too soft to be the sole coating. It must be used in combination with a buffer or harder thermoplastic and is generally thicker than acrylate, with diameters up to 500µm. Silicone has other advantages besides a high-temperature capability; it has a high resistance to water absorption and low flammability.

Silicone or acrylate coatings must be removed prior to terminating the optical fiber. Both of these coatings can leave a residue, so extra care must be used when removing silicone. Terminating optical fibers is discussed in depth in Chapter 9, "Connectors."

Polyimide is a popular coating for aerospace applications because it can operate in temperatures up to 350° C. This coating is much thinner than acrylate or silicone, typically only 15µm thick. Beside a very high operational temperature, polyimide is resistant to abrasion and chemicals. However, polyimide is difficult to remove. If it must be removed before termination, hot sulfuric acid or a portable polyimide-stripping machine is required.

Carbon is the thinnest coating of all, typically only a fraction of a micron thick. Carbon hermetically seals the glass surface, protecting it from moisture and increasing the optical fiber's fatigue resistance. Carbon cannot be used by itself; it must be combined with another coating such as silicone or polyimide. Unlike the other coatings discussed, carbon coating does not have to be removed before termination.

Standards

While many combinations of core and cladding sizes are possible, standards are necessary to ensure that connectors and equipment can be matched properly. This is especially important when dealing with components as small as those used in fiber optics, where even slight misalignments can render the entire system useless.

There are many standards related to optical fiber. This chapter focuses on the standards that define the performance of optical fibers used in the telecommunications industry published by the Telecommunications Industry Association (TIA), the International Telecommunications Union (ITU), the International Organization for Standardization (ISO), and the International Electrotechnical Commission (IEC). While TIA, ITU, ISO, and IEC publish many standards, the key standards that you should be familiar with are ANSI/TIA-568-C, ITU-T G.651.1, ITU-T G.652, ITU-T G.655, ITU-T G.657, ISO/IEC 11801, IEC 60793-2-10, and IEC 60793-2-50.

The ANSI/TIA-568-C standards and the ISO/IEC 11801 standard are applicable to *premises* cabling components. Premises are a telecommunications term for the space occupied by a customer or an authorized/joint user in a building on continuous or contiguous property that is not separated by a public road or highway. Premises cabling standards are typically optimized for cabling distances up to 2000 meters. These standards can define the performance of both multimode and single-mode optical fibers.

The ITU standards are applicable to multimode and single-mode optical fiber and cable. Here are their descriptions:

♦ ITU-T G.651.1: Characteristics of a 50/125 μm multimode graded index optical fiber cable for the optical access network

♦ ITU-T G.652: Characteristics of a single-mode optical fiber and cable

♦ ITU-T G.655: Characteristics of a non-zero dispersion-shifted single-mode optical fiber and cable

♦ ITU-T G.657: Characteristics of a bending loss-insensitive single-mode optical fiber and cable for the access network

The IEC 60793-2 series standards address optical fiber product specifications. Here are their descriptions:

♦ IEC 60793-2-10: Category A1 multimode optical fiber product specifications

♦ IEC 60793-2-50: Class B single-mode optical fiber product specifications

These standards contain important information that defines the performance of the optical fiber, fiber-optic cable, and components such as connectors and splices. Subsequent chapters will discuss these standards in detail.

Materials

Optical fibers are commonly made with a glass core and glass cladding, but other materials may be used if the fiber's performance must be balanced with the cost of installing the fiber, fitting it with connectors, and ensuring that it is properly protected from damage. In many cases, fibers must run only a short distance, and the benefits of high-quality all-glass fibers become less important than simply saving money. There are also circumstances in which the fibers are exposed to harsh conditions, such as vibration, extreme temperature, repeated handling, or constant movement. Different fiber classifications have evolved to suit different conditions, cost factors, and performance requirements. Here are the major fiber classifications by material:

Glass fibers   These have a glass core and glass cladding. They are used when high data rates, long transmission distances, or a combination of both are required. As you will see in later chapters, connectors for glass fibers must be built with a high degree of precision and can take time to apply correctly. Glass fibers are the most fragile of the various types available, and as a result they must be installed in environments where they will not be subjected to a great deal of abuse, or they must be protected by special cables or enclosures to ensure that they are not damaged.

Glass fibers are commonly found in long-distance data and interbuilding and interoffice networking applications.

Glass, glass, polymer (GGP)   These fibers have a glass core and a hybrid cladding. The inner cladding is glass and is identical to a glass optical fiber. However, the outer cladding is a polymer that is bonded to the inner cladding. They can be manufactured to offer the same optical performance as glass fibers. The outer cladding polymer coating is far more resistant to mechanical damage than an all-glass fiber. This unique property prevents breakage during the termination process.

GGP fibers can be used for long-distance data and premises applications.

Plastic-clad silica (PCS)   These fibers have a glass core and plastic cladding. The core is larger than all-glass fiber—typically, 200µm with a cladding thickness of 50µm. Like a silicone-coated glass optical fiber, the plastic coating of a PCS optical fiber is commonly used with a thermoplastic buffer that surrounds the plastic cladding. A typical PCS fiber specification would be 200/300µm. The plastic cladding also serves as a protective layer for the glass core, so the coating normally found on all-glass fiber is not included on PCS fibers. PCS fibers are generally used for industrial sensing applications and medical/dental applications.

Hard-clad silica (HCS)   These fibers are similar to PCS fiber but they have a glass core with cladding made of a hard polymer or other material, typically stronger than other cladding materials. Hard-clad silica fiber is commonly used in locations where ruggedness is a prime consideration, such as manufacturing, factory automation, and other areas where shock and vibration would render standard glass fibers unreliable.

HCS optical fibers are typically much larger than glass optical fibers. A very popular size is 200/230µm.

Plastic fiber   These fibers have a plastic core and plastic cladding. They are selected for their low cost, ruggedness, and ease of use, and are installed where high bandwidth and long transmission distances are not required. While plastic fibers are unsuited for long-distance, high-performance transmissions, they can still carry signals with useful data rates over distances of less than 100m. Plastic fibers are much larger in diameter than glass fibers, as shown in the comparison in Figure 4.3. A very popular size is 980/1000µm. Plastic fiber is typically designed for visible wavelengths in the 650nm range. Some typical locations for plastic fiber include home entertainment systems, automotive, and manufacturing control systems. They may also be used in links between computers and peripherals and in medical equipment.

FIGURE 4.3
Comparison of fiber
core and cladding
diameters

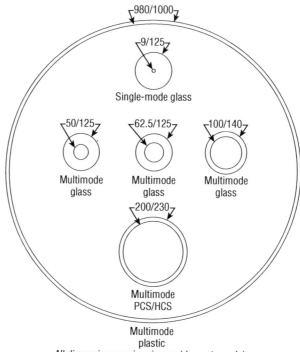

All dimensions are in microns (drawn to scale).

THE ADVANTAGES OF LARGE CORE PLASTIC OPTICAL FIBER

It is easy to get excited about the high bandwidth and long-distance transmission capabilities of glass optical fiber. It clearly outperforms any other medium. However, many applications do not require a high bandwidth over great distances. There are many applications for optical fiber in your home. You may already have a home entertainment system that uses plastic optical fiber, or you may own a car that uses plastic optical fiber to connect audio devices or a DVD changer. None of these applications requires high bandwidth over great distances. These applications are ideal for large core plastic optical fiber.

As you learned in this chapter, plastic optical fiber is typically designed to operate at a visible wavelength around the 650nm range. Being able to see the light as it exits the optical fiber has a significant advantage; no expensive test equipment is required. A power meter is needed to measure the light exiting a glass optical fiber operating in the infrared range. Power meters can cost more than your home entertainment system.

The large core of the plastic optical fiber has another advantage over small glass fibers: it is easy to align with another fiber or a light source or detector. Imagine aligning two human hairs so that the ends touch and are perfectly centered. Now imagine doing the same thing with two uncooked spaghetti noodles. Which do you think would be easier?

Plastic optical fiber is a great choice for audio and video electronics being integrated into homes and vehicles.

Tensile Strength

In addition to the coating, optical fibers may have several layers of protection such as buffers, jackets, armor, and other materials designed to protect the fiber and keep it together in cables. Like copper, though, optical fiber is still subject to hazards caused by handling, installation, careless digging, and bad weather.

One characteristic of optical fiber that deserves special attention is its tensile strength, or the amount of stress from pulling that the optical fiber can handle before it breaks. Optical fiber has a high tensile strength and resists stretching.

Tensile strength is important for several reasons. It affects the way fiber must be handled during installation, the amount of curvature it can take, and the way it will perform throughout its lifespan.

To understand optical fiber's tensile strength, let's look at a standard piece of plate glass. To cut the glass, all you have to do is scribe a sharp line through the surface layer. Once the strength of that layer is compromised, the glass snaps easily along the scribe, even if it follows a curve.

Optical fiber follows the same rule. The outer layer of the cladding provides much of the fiber's tensile strength, which is often measured in thousands of pounds per square inch (kpsi). A typical breaking strength for an optical fiber is 100kpsi. That means that a typical fiber with a cladding thickness of 125µm can withstand a pull of about 1.9 pounds. However, the maximum pull should never be applied when removing the buffer or coating. It is a good practice to limit the amount of pull to around a half of pound while removing the buffer or coating.

Once the outer layer is scratched or cracked, however, the tensile strength is gone at that location. Like a scribed line, a scratch or crack compromises the integrity of the glass and allows the fiber to break more easily under stress. Scratching or cracking can occur due to mishandling, sharp blows, or bending beyond the fiber's *minimum bend radius*, especially if the bending takes place while the fiber is under tension. The minimum bend radius is the minimum radius the optical fiber can be bent without damaging it.

Manufacturing Optical Fiber

Optical fiber is manufactured to very high standards, because the core diameter and refractive indexes of the core and cladding must remain consistent over stretches of up to 80km.

Four methods are commonly used to make optical fiber:

♦ Modified chemical vapor deposition (MCVD)

♦ Outside vapor deposition (OVD)

♦ Vapor axial deposition (VAD)

♦ Plasma chemical vapor deposition (PCVD)

Each method uses variations on a process to create a *preform*, a short, thick glass rod that has a similar cross section to the final fiber. As shown in the block diagram in Figure 4.4, the preform is heated to 1900° C in a drawing tower until it begins to melt and a blob falls from the end, drawing a small thread of glass after it. The thread contains the core and cladding of the optical fiber, their relative thicknesses preset by the preform's construction.

FIGURE 4.4
Drawing the pre-
form into fiber

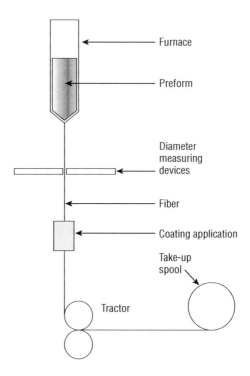

This thread is taken up by a pulling machine, or *tractor*, at a constant rate to maintain a consistent thickness. A thinner, longer fiber can be created by speeding up the draw rate. The entire process is closely monitored by laser measuring devices to ensure that the thickness remains consistent over the entire length of the fiber, which can be anywhere from 10 to 80km, depending on the size of the preform.

The fiber is then drawn through another process, which deposits and cures the coating. The coating is cured using ultraviolet lamps or thermal ovens. After the coating is applied, measurements are taken and the optical fiber is taken up on a spool.

Each of the preceding methods is best suited for different types of fiber, depending on the type of signals it will carry, the distance it will cover, and other factors, which will be discussed later in this chapter.

Let's look at the differences in the manufacturing methods.

Modified Chemical Vapor Deposition (MCVD)

Fibers manufactured using MCVD begin with a hollow glass tube about 2.5cm in diameter, as shown in Figure 4.5.

The cladding is created first by placing the hollow tube, or *bait*, on a lathe and spinning it rapidly. As the bait is spinning, it is heated by an oxygen/hydrogen torch passing lengthwise underneath as a gaseous mixture of vaporized silicon dioxide is introduced into the tube. The ultra-pure gas, mixed with carefully controlled impurities or *dopants*, forms a soot, which fuses to the inside of the bait in successive layers as the heat passes beneath it. The dopants are used to increase the refractive index of the fused material.

FIGURE 4.5
Modified chemical
vapor deposition

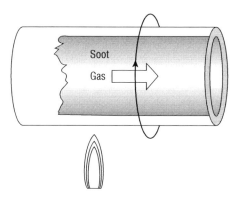

Every time the flame passes beneath the tube to heat it, another thin layer of soot adheres to the inside, building a thicker and thicker layer of glass.

The core is formed next by changing the gas/dopant mixture to create a vapor with a higher refractive index. This, too, is introduced into the spinning, heated tube, inside the first layers that were laid down. When the deposited material has reached the desired thickness, the tube is heated to an even higher temperature to consolidate the soot into glass without melting it, a process known as *sintering*. A drying gas is passed through the core to remove water contamination, and the tube is compressed into a preform and taken to the drawing tower.

Outside Vapor Deposition (OVD)

The OVD method is similar to MCVD, except that the fiber preform is built from the inside out.

In the OVD method, a glass target rod serves as the bait, as shown in Figure 4.6. The rod is placed on the lathe and spun as heat is applied to it. The core is laid down first by introducing a gas mixture between the heat and the rod. When the core layer is thick enough, the mixture is changed and the cladding is laid down.

FIGURE 4.6
Outside vapor
deposition

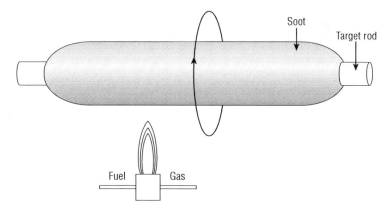

After the layers have been deposited, the rod is removed and the layers are dried and sintered and then collapsed into the preform.

Vapor Axial Deposition (VAD)

As with MCVD and OVD, the VAD process uses a heated glass bait to collect a soot of silicon dioxide and dopant. With VAD, however, the glass rod is suspended vertically and the heat source is at the lower end, as shown in Figure 4.7.

FIGURE 4.7
Vapor axial
deposition

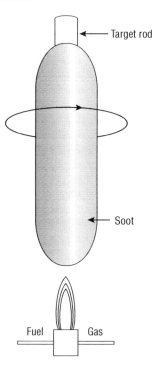

Target rod

Soot

Fuel   Gas

The gas is introduced between the end of the rod and the heat source, and the soot builds up in a radial pattern. This method gives the manufacturer a great deal of control over not only the refractive index of the various layers, but also the pattern in which they are laid down.

As with OVD, the rod is then removed, and the layers are dried, sintered, and collapsed into the preform before being taken to the drawing tower.

Plasma Chemical Vapor Deposition (PCVD)

PCVD is similar to MCVD, but allows a finer layer of material to be deposited.

In PCVD, the gas particles are heated with microwaves, causing plasma to form inside the tube. The plasma contains electrons with extremely high energy levels, approximating those found in a gas heated to 60,000° C, even though the temperature inside the tube is no higher than it is with MCVD—about 1200° C. As the high-energy electrons meet the soot forming inside the tube, they transfer their energy to it in the form of heat, causing it to fuse into the final glass form on the tube walls, making the step of consolidation unnecessary.

As with the other methods, the tube can then be compressed and taken to the drawing tower.

Mode

One of the most important characteristics used to distinguish types of fiber is the number of potential paths light can take through it. It may seem that light would go straight through the fiber core, following all of its curves, until it comes out the other end. However, the light itself is a complex combination of electrical and magnetic waves, and their wavelengths can be many times smaller than the core of the fiber. Like a rubber ball shot through a sewer pipe, the light actually has more than one path or *mode*. There are many potential paths or modes it can follow, depending on the size of the core, wavelength of the light source, and *numerical aperture*.

To understand modes in optical fibers, let's take another look at the basis for fiber-optic transmission: total internal reflection.

Recall that total internal reflection depends on the principle that light passing through a medium of a given refractive index will be reflected off the interface with a medium of a lower refractive index if it hits the interface at or above the critical angle, which is determined by the difference between the refractive indexes of the two media.

To satisfy this requirement, optical fiber is manufactured with a core having a refractive index slightly higher than that of the cladding surrounding it. As a result, light entering the end of the fiber will be reflected off the interface with the cladding and guided through the fiber. By definition, each reflection will be at the same angle as the angle of incidence.

The number of modes possible in a length of fiber depends on the diameter of the core, the wavelength of the light, and the core's numerical aperture.

Calculating the Numerical Aperture and Modes

You can find the numerical aperture using the following formula:

$$ NA = \sqrt{n^2_{core} - n^2_{cladding}} $$

For example, if the core has a refractive index of 1.485 and the cladding has a refractive index of 1.47, we can calculate:

$$ NA = \sqrt{1.485^2 - 1.47^2} = 0.211 $$

Note that the numerical aperture has no dimension. It is intended as a relative indication of the light-gathering capacity of the fiber core.

To find the number of modes, we can use the equation

$$ M = (D \times \pi \times NA/\lambda)^2/2 $$

where D is the diameter of the core and λ is the wavelength of the light. For example, let's use our previously derived value for the numerical aperture with a core diameter of 50μm and a light wavelength of 850nm:

$$ M = (50 \times 10^{-6} \times \pi \times 0.211/850 \times 10^{-9})^2/2 = 760.2 $$

Because light cannot have part of a mode, we must round down to the nearest whole number if we come up with a decimal. So the answer is 760 modes, or potential paths for the light to follow.

Refractive Index Profiles

What happens when the light rays entering a fiber take slightly different paths, some entering at sharper angles, some at shallower angles? As shown in Figure 4.8, the light can follow modes ranging from a straight line through the fiber (*zero-order* mode) to a low number of reflections (*low-order* mode) to a high number of reflections (*high-order* mode). Depending on a fiber's construction and the wavelength of the light source, the fiber may support one mode or more than a million modes (in fibers with large cores such as those found in plastic optical fibers). The path or paths that light takes through an optical fiber can be understood by examining the fiber's *refractive index profile*.

FIGURE 4.8
The three types of light modes

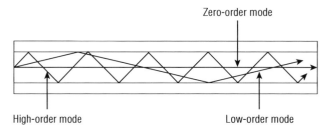

The refractive index profile graphically represents the relationship between the refractive index of the core and that of the cladding. Several common profiles are found on optical fibers used in telecommunications, avionics, aerospace, and space applications:

♦ Step index

♦ Graded index

♦ Depressed clad

These refractive index profiles are shown in Figure 4.9.

FIGURE 4.9
Refractive index profiles

So far in this text, we have described a simple relationship in which the core has one refractive index and the cladding another, and there is a single interface between the two. This is known as *step-index* fiber. If a step index fiber has a small core and only permits one path for light, it is referred to as a *single-mode* optical fiber. However, if the fiber has a large enough core diameter with room for the light to take a number of different possible paths, or modes, by way of reflection, it is referred to as *multimode* fiber.

MULTIMODE STEP-INDEX FIBER

Also known simply as step-index fiber, *multimode step-index fiber*, as already discussed, has a core with a single refractive index and a cladding with another, slightly lower refractive index. These are separated by a single interface, which reflects light that hits it at any angle greater than its critical angle. The large core of a step index multimode optical fiber allows a ray of light to take many possible paths, as shown in Figure 4.10.

FIGURE 4.10
Light propagation
profiles

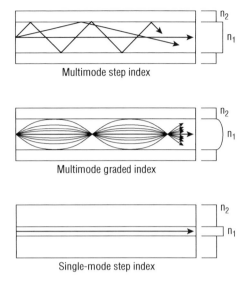

Multimode step index

Multimode graded index

Single-mode step index

The result is that even though all of the rays pass through the same length of fiber, the reflections create a longer path for the light to follow. The more reflections, the longer the path. You would get the same result if you and a friend walked down the same road at the same rate of speed, with you walking down the middle of the road and your friend zigzagging from one side of the road to the other. Even if your friend started out before you did, you would eventually pass him, because he is taking a longer path. As a result, you would arrive at the end of the path before your friend, even though you started out together.

The same effect occurs as light is reflected through the fiber. Even though light rays enter the fiber in the order 1, 2, 3, they may arrive in the order 3, 1, 2, or even overlap each other as they arrive, causing the information they carry to become muddled and useless. This effect is called *modal dispersion*, and is an important factor limiting bandwidth in optical fiber.

Typically, modal dispersion in step-index fiber offsets light rays by about 15 to 30ns/km, depending on the diameter of the fiber core and the wavelength of the light. In other words, if one ray takes the most direct path through the fiber, and another ray takes the longest possible path by reflecting off the fiber walls, the ray taking the longest path will follow the ray taking the shortest path by anywhere from 15 to 30 nanoseconds (15 to 30 billionths of a second) for every kilometer of fiber length. That means that for every kilometer of fiber length, each bit would have to be separated from the one before it by at least 30ns, or roughly 6m, to ensure that the bits would not overlap. This results in a data rate limit of 33.33MHz for a 1km fiber, 16.67MHz for a 2km fiber, 11.11MHz for a 3km fiber, and so on—far below the capabilities of transmitters, receivers, and processors, and completely unsuited for long-distance fiber-optics transmission. For this reason multimode step index optical fibers are not used in telecommunications and their performance is not addressed in any of the standards discussed in this text. PCS, HCS, and plastic optical fibers are typically step-index.

Two methods used to reduce modal dispersion are multimode graded-index fiber and single-mode step-index fiber.

MULTIMODE GRADED-INDEX FIBER

Multimode graded-index fiber tackles the problem of modal dispersion by increasing the speed of the high-order mode light rays, allowing them to keep up with the low-order mode rays.

The fiber accomplishes this by using the very principles upon which fiber optics is based: the laws of refraction.

Remember that when light passes from a higher refractive index medium to a lower refractive index medium, it gains velocity. A *graded-index* fiber core actually consists of many concentric glass layers with refractive indexes that decrease with the distance from the center. Viewed on end, these layers resemble the rings of a tree.

Any light that passes straight through the fiber travels at a constant speed. If a light ray enters at an angle, however, it passes through the graded layers. As it does, two things happen.

First, the light is refracted away from normal, because the refractive index has decreased. Second, the light propagation velocity increases. This happens in every new layer the light traverses, until it has reached the cladding or has been bent to the critical angle for one of the layers. At this point, the light is reflected and begins its new direction, this time refracting toward normal and slowing down until it reaches the highest refractive index, at the center of the optical fiber core. It then passes through the center and begins the cycle again on the opposite side of the fiber core. The path resembles a segmented sine wave, as shown in Figure 4.10.

Note that as the light follows a longer path, its average velocity increases, offsetting the extra distance it must travel. The time it takes to move from one end of the fiber to the other is now much closer to that of light following a straight path through the fiber. In fact, modal dispersion in graded-index fiber can be reduced to as little as 1ns/km.

SINGLE-MODE STEP-INDEX FIBER

Single-mode step-index fiber uses another approach to reduce modal dispersion. It doesn't give the light enough space to follow anything but a single path through the fiber.

Single-mode fiber uses a core so small that light can only travel in one mode. The diameter of the single-mode fiber is typically only about 8 to 10μm, but its performance also depends on the wavelength of the light.

For example, if a fiber is designed to carry a wavelength of 1310nm in only one mode, there is room for light at 850nm to travel in several modes, so the fiber has become a multimode fiber at the shorter wavelength. The shortest wavelength at which the fiber carries only one mode is called the *cutoff wavelength*. The cutoff wavelength for a fiber designed to carry 1310nm in single mode is about 1260nm.

Even in single-mode fiber, light does not follow a straight path. Because of the way light propagates, it follows a corkscrew-like path through the fiber core and propagates in a portion of the cladding, as shown in Figure 4.11. As a result, manufacturers must make single-mode fiber with cladding that will carry the light with less attenuation.

The unique light propagation characteristics of single-mode fiber also make it necessary to take the light propagated in the cladding into account when matching fiber sizes for connections, so the light in the cladding is not lost. Single-mode fibers are typically specified by their *mode field diameter* rather than their core and cladding size. As you can see in Figure 4.11, the mode field diameter is the real estate used by the light within the core and the cladding as it propagates.

FIGURE 4.11

Light propagation
in single-mode fiber

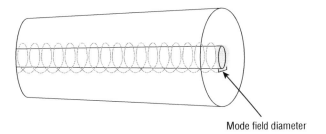

Mode field diameter

Depressed-Clad Fiber

Multimode and single-mode depressed clad optical fibers are commonly referred to as *bend insensitive*. Bringing fiber to the home or an apartment complex can require the cable to be bent sharply. Bending an optical fiber can cause loss, and single-mode optical fibers can experience considerable loss at longer wavelengths such as 1550nm.

To reduce attenuation from bending, fiber manufacturers created multimode and single-mode optical fibers that have very little loss when bent sharply. These optical fibers feature proprietary refractive index profiles similar to the depressed clad refractive index profiles shown in Figure 4.9.

These fibers are manufactured with an optical trench as shown in Figure 4.12. The *optical trench* surrounds the core of the optical fiber. It has a lower refractive index than the core or the cladding. The low refractive index acts as a barrier and prevents light from escaping the core of the optical fiber when it is bent reducing the amount of attenuation. The exact location, size, and refractive index value of the trench is typically proprietary and not published.

FIGURE 4.12

The optical trench
surrounding the
core of multimode
and single-mode
optical fibers

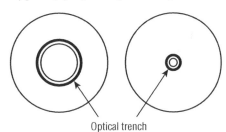

Optical trench

Single-mode depressed-clad optical fiber has three levels of refractive indexes. Like all optical fibers, the core has the highest refractive index. However, the cladding has two refractive indexes, the refractive index of the cladding and the refractive index of the optical trench. This refractive index profile resists attenuation from bends with a small radius better than step index.

Multimode depressed-clad optical fiber is typically referred to as *bend insensitive multimode optical fiber* (BIMMF). Like single-mode depressed-clad optical fiber, an optical trench surrounds the core. These optical fibers typically offer a tenfold reduction in bending loss without any impact on bandwidth performance.

GETTING TO KNOW THE STANDARDS

If you are a sports fan, you know the statistics of your favorite team and players. If you plan to be the person everyone turns to when there is a question about optical fiber, you need to know the standards just as well as the sports fan who recites the statistics of their favorite team and players. The nice thing about standards, however, is they do not change nearly as fast as your favorite team's record does. Standards can take years to develop and typically have a five-year review cycle.

The Bottom Line

Select the proper optical fiber coating for a harsh environment. The coating is the true protective layer of the optical fiber; it absorbs the shocks, nicks, scrapes, and even moisture that could damage the cladding. Without the coating, the optical fiber is very fragile. The coating found on an optical fiber is selected for a specific type of performance or environment.

Master It Choose a coating for an optical fiber that must operate in temperatures as high as 175° C.

Identify an industry standard that defines the specific performance characteristics on an optical fiber. There are many standards related to optical fiber. This chapter focused on the standards that define the performance of optical fibers used in the telecommunications industry published by TIA, ITU, ISO, and the IEC.

Master It You have been placed in charge of selecting a contractor to install fiber optic cabling in a new building. Which standard or standards should you review prior to selecting the best contractor?

Identify an optical fiber from its refractive index profile. The refractive index profile graphically represents the relationship between the refractive index of the core and that of the cladding. Several common profiles are found on optical fibers used in telecommunications, avionics, aerospace, and space applications.

Master It The refractive index profile of an optical fiber is shown in the following graphic. Determine the type of optical fiber.

Calculate the numerical aperture of an optical fiber. To find the numerical aperture of an optical fiber, you need to know the refractive index of the core and the cladding.

Master It The core of an optical fiber has a refractive index of 1.51 and the cladding has a refractive index of 1.46. Calculate the numerical aperture.

Calculate the number of modes in an optical fiber.   To find the number of modes in an optical fiber, you need to know the diameter of the core, the wavelength of the light source, and the numerical aperture.

Master It   Calculate the number of possible modes in an optical fiber with the following specifications:

Core diameter = 62.5μm

Core refractive index = 1.48

Cladding refractive index = 1.46

Wavelength of the light source = 1300nm

Chapter 5

Optical Fiber Characteristics

In addition to factors such as construction, materials, and size as discussed in the previous chapter, optical fibers have certain performance or operational characteristics that define them. These characteristics may describe limitations or features of the fiber with regard to its light-carrying ability under various conditions, and are generally affected by its physical properties.

In this chapter, we describe the characteristics of optical fiber that affect the way it is selected, handled, installed, and used. We'll cover in detail how these characteristics change a fiber's ability to carry light, as well as the methods used to take advantage of some characteristics while minimizing the effects of others.

In this chapter, you will learn to:

♦   Calculate the attenuation in dB for a length of optical fiber

♦   Calculate the usable bandwidth for a length of optical fiber

♦   Calculate the total macrobending loss in a single-mode fiber-optic cable

♦   Calculate the total macrobending loss in a multimode fiber-optic cable

♦   Calculate the acceptance angle for an optical fiber

♦   Determine the latest published revision of a standard using the Internet

It All Adds Up

If you've ever looked at your bank balance and wondered where it all went, you have some idea of why it's important to understand optical fiber characteristics in great detail.

Chances are, when you took another look at your account, you remembered that some of it went to necessities, some to entertainment, some to service charges, and so on. It didn't all go away at once: it was spent little by little, in many different places. Any single expenditure may have gone almost unnoticed, but over time all of those expenses can add up to a significant amount.

This is typically the fate of light as it passes through an optical fiber. By the time it reaches the other end, it is diminished in several ways—most notably in its power and its ability to carry a signal. In spite of the purity of the materials that go into fibers, they are still not perfect—in part by design, and in part because there are factors that simply cannot be overcome with current technology.

Going back to your bank account, if you know where the money goes, you can take steps to make it last longer or make sure that when it is spent, you get more use out of it.

The same goes for our light beam. If you understand the characteristics in an optical fiber that take away from its ability to carry a signal at high speed and over long distances, it is possible to overcome them or use them to your advantage. Remember, though, that propagation of light within a fiber is a complex mix of influences, and each of the characteristics that we'll be discussing affects different portions of this mix.

Some aspects of these characteristics have been covered in previous chapters to describe why fiber is constructed in certain ways. Now we are going to look at them more closely.

Dispersion

In general, *dispersion* is the spreading of light as it travels away from its source. The light spreads because different components of it travel at slightly different velocities or take slightly different paths, depending on the conditions in the medium through which it is traveling and the wavelengths that make up the light. There are different kinds of dispersion, however, and the kind that is taking place depends on several factors in the optical fiber and in the light itself.

The greatest effect of dispersion is that as the light spreads, it can degrade or destroy the distinct pulses of the digital signals in the light by making them overlap each other, as shown in Figure 5.1, blurring and blending them to the point that they are unusable. The effect grows more pronounced as the distance the light travels increases.

FIGURE 5.1

The effects of dispersion on a signal

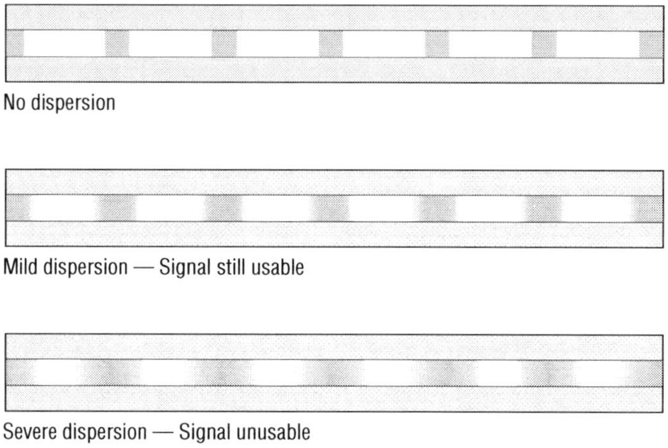

No dispersion

Mild dispersion — Signal still usable

Severe dispersion — Signal unusable

The effect is similar to looking into a hallway through a frosted glass window. If people are moving through the hallway close together, the glass spreads their images so much that they merge with one another and look like a single mass rather than individuals. If they spread out far enough from each other, however, you can see each person moving past the window. The images are still spread out, but the space between each person is great enough to see.

To prevent signal degradation due to dispersion, it is necessary to keep the pulses far enough apart to ensure that they do not overlap. This limits the signals to a bit rate that is low enough to be only minimally affected.

Dispersion limits the optical fiber's *bandwidth*, or the amount of information it can carry. Fortunately, optical fiber manufacturers and standards organizations provide information about the bandwidth or the information-carrying capacity of optical fibers at specific wavelengths.

The types of dispersion that affect optical fiber performance are:

- Modal dispersion
- Material dispersion
- Waveguide dispersion
- Chromatic dispersion
- Polarization-mode dispersion

Let's look at these types of dispersion more closely.

Modal Dispersion

We mentioned modal dispersion in the previous chapter to explain why fibers are classified as multimode or single-mode. It will help to review some of the important points.

Modal dispersion results from light taking different paths, or modes, as it passes through the optical fiber. The number of potential modes the light can take is determined by the diameter of the optical fiber core, the refractive indices of the fiber core and cladding, and the wavelength of the light.

A mode can be a straight line through the fiber, or the light can follow an angular path, resulting in reflections every time the light meets the interface between the core and the cladding. The more reflections, the longer the path through the fiber, and the longer light takes to pass through it.

Depending on the mode, some parts of the light will pass through the fiber in less time than others. The difference in travel time can cause parts of the light pulses to overlap each other, or in extreme cases to arrive in a different order from the order they were transmitted. If this happens, the signal is no longer usable.

Methods for overcoming modal dispersion include:

Lower bit rate   Lowering the bit rate increases the gap between bits in the signal. Although dispersion will still affect the bits, they will not overlap one another, and the signal will still be usable.

Graded index fiber   *Graded index fiber* gradually reduces the refractive index of the fiber core from the center toward the cladding, allowing the light that follows a more angled path to speed up as it leaves the center and causing it to slow down again as it reaches the center. This effect reduces the difference in travel time between modes and allows greater bandwidths and faster data transmission. Graded index fiber is a moderately priced solution that allows wider bandwidths than multimode step index fiber. In addition, the gradual change of indexes as the light heads for the cladding causes the light to curve back into the core of the fiber before it has a chance to approach the cladding at a penetrating angle.

Reducing the core size or increasing the wavelength   In the previous chapter, we calculated the number of potential modes in a multimode optical fiber. For a given wavelength, a smaller core optical fiber will have fewer modes than a larger core optical fiber. For a given core size, a multimode optical fiber will have fewer modes at a longer wavelength than a shorter wavelength.

Single-mode fiber   Single-mode fiber has a core that is small enough for only one mode to propagate, eliminating the problems caused by multiple modes. This type of fiber requires more expensive connectors and equipment because of the small core size and is typically used when very wide bandwidth requirements over long distances justify the cost.

Material Dispersion

Material dispersion occurs when different wavelengths of light travel at different velocities in the optical fiber. Even the light from a laser is made up of different wavelengths within a narrow range called the *spectral width*, which varies depending on the light source. Depending on the wavelength, a *light-emitting diode (LED)* may have a spectral width as narrow as 20nm or as wide as 170nm. A laser's spectral width is much narrower than that of the LED. Depending on the laser type, the spectral width can range from 0.1nm to 3nm.

Recall the formula for determining the refractive index of a material:

$$n = c/v$$

where n is the refractive index, c is the speed of light in a vacuum, and v is the speed of the wavelength of light through the material. In this equation, n changes with the wavelength of the light passing through the material. Remember that this is why white light breaks into its component colors in a prism.

When the different wavelengths travel at different velocities, the slower wavelengths begin to lag behind as the light travels down the optical fiber core, causing the light to spread as shown in Figure 5.2. If the light pulses must travel a great distance, the lag in the slower wavelengths can cause them to overlap the faster wavelengths of the bits following them. As with modal dispersion, these overlaps can degrade and ultimately destroy the signal.

FIGURE 5.2
Material dispersion in fiber causes some wavelengths to travel more slowly than others.

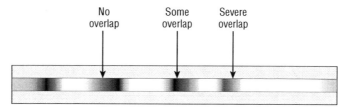

Because the wavelengths used in fiber-optic transmissions have a narrow spectral width, material dispersion takes place on a much smaller scale than modal dispersion. Its effects in a fiber are measured in picoseconds per nanometer of spectral width per kilometer (ps/nm/km) and are insignificant in a multimode fiber when compared to the effects of modal dispersion.

Material dispersion becomes a problem only when modal dispersion is overcome with single-mode fiber and higher data rates are used over long distances.

Waveguide Dispersion

Waveguide dispersion occurs in single-mode fiber as the light passes through not only the core, but also part of the cladding, as shown in Figure 5.3. By design, the core has a higher refractive index than the cladding, so the light will travel more slowly through the core than through the cladding.

FIGURE 5.3
Waveguide dispersion in optical fiber

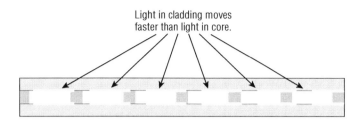

Light in cladding moves faster than light in core.

While the difference in refractive indexes of single-mode fiber core and cladding are minuscule, they can still become a factor over great distances. In addition, waveguide dispersion can combine with material dispersion to create another problem for single-mode fiber: chromatic dispersion (discussed in the next section).

Various tweaks in the design of single-mode fiber can be used to overcome waveguide dispersion, and manufacturers are constantly refining their processes to reduce its effects.

Chromatic Dispersion

Chromatic dispersion occurs in single-mode optical fiber and results from the combination of effects from material dispersion and waveguide dispersion.

When chromatic dispersion occurs, the effects of material dispersion, as shown in Figure 5.4, compound the effects of waveguide dispersion. At lower data rates and in multimode fiber, the effects of chromatic dispersion are so small as to be unnoticed, especially when buried under modal dispersion. It is mostly a problem in single-mode fiber carrying high bit rates over long distances, where the detrimental effects build up.

FIGURE 5.4
Waveguide dispersion and material dispersion combine to create chromatic dispersion.

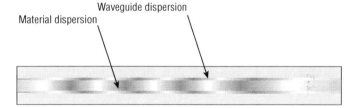

Material dispersion

Waveguide dispersion

One way to reduce chromatic dispersion is to take advantage of the fact that the relationship between wavelength, refractive index, and velocity is not linear. In the infrared range of most fiber-optic transmissions, the light's velocity through the medium drops as the wavelength increases until it reaches the range between 1300nm and 1550nm. At wavelengths greater than 1550nm, the longer wavelengths have a higher velocity. Somewhere in the 1300nm to 1550nm range there is a crossover where, depending on the specific composition of the fiber, the refractive index is the same for the wavelengths within the narrow spectral width of the light being transmitted. In other words, as shown in Figure 5.5, as the wavelength approaches this range, dispersion drops to zero. This *zero-dispersion point* normally occurs at 1300nm in a standard single-mode fiber. Unfortunately, other characteristics of the optical fiber attenuate the signal at this wavelength, making it unusable for long-distance runs.

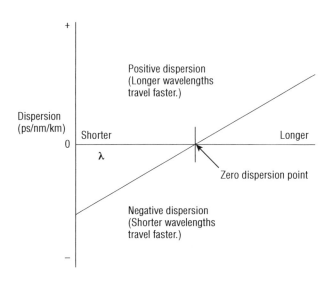

FIGURE 5.5
Dispersion profile of a typical optical fiber

There are two approaches to reduce chromatic dispersion. One approach is to use *dispersion-shifted fiber* and the other is to reduce the spectral width of the light source. The spectral width of the light source is discussed in Chapter 10, "Fiber-Optic Light Sources and Transmitters."

Dispersion-shifted fiber, sometimes called *zero-dispersion-shifted fiber*, is made so that the zero dispersion point is shifted from 1300nm to 1550nm, where the attenuation is lowest, as shown in Figure 5.6. Shifting the zero dispersion point to 1550nm allows longer transmission distances because of less attenuation. Dispersion-shifted fiber is used when high data rates over long distances are required.

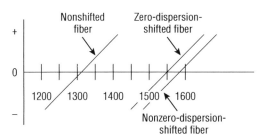

FIGURE 5.6
Zero dispersion points of nonshifted, dispersion-shifted, and non-zero-dispersion-shifted fibers

A variation of the dispersion-shifted fiber, *non-zero-dispersion-shifted fiber (NZ-DSF)*, is used when multiple frequencies are being used to send more than one channel through the fiber, a process known as *wavelength division multiplexing (WDM)*. When WDM takes place at the zero dispersion point, an effect known as *four-wave mixing* occurs, in which multiple wavelengths combine to form new wavelengths. These new wavelengths can interfere with the signal carrying wavelengths, reducing their power and introducing noise into the system.

To reduce four-wave mixing, NZ-DSF is used to move the dispersion point just off the 1550nm mark. The slight amount of dispersion introduced minimizes four-wave mixing while preserving most of the benefits of dispersion-shifted fiber.

The refractive index profile of dispersion-shifted fiber is shown in Figure 5.7. The peak in the center of the profile reveals an inner core that has its highest refractive index at the center. The refractive index gradually decreases toward a thin inner layer of cladding. The smaller peaks represent a ring of glass with a higher refractive index surrounding the inner cladding, slowing the light that would normally increase its velocity in the cladding. This effect reduces wave-guide dispersion, which in turn reduces chromatic dispersion.

FIGURE 5.7
Refractive
index profile of
dispersion-shifted
fiber

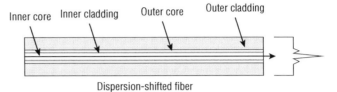

Inner core   Inner cladding   Outer core   Outer cladding

Dispersion-shifted fiber

MIXING IT UP—THE PROBLEM OF FOUR-WAVE MIXING

It seems that nobody can leave a good thing alone. Once engineers overcame the problem of chromatic dispersion in single-mode optical fiber using dispersion shifting, they decided to squeeze all the use they could out of it by piling on different wavelengths to create multiple transmission channels. The idea behind this is that different wavelengths can actually occupy the same space but remain distinct from one another until they are sorted out at the other end of the fiber link.

It seemed to make very good sense, but then another problem cropped up. The wavelengths used in the multiple channels must stay near the zero-dispersion range of 1550nm, so you end up with individual channels only 2nm apart, typically at 1546, 1548, 1550, and 1552nm, for example. It's difficult for any two things to be only 2nm apart and not interact, so interact they do. In fact, they create new wavelengths that can interfere with the wavelengths that are part of the transmission.

The problem gets exponentially worse as the number of wavelengths being transmitted increases. The formula for predicting the number of new waves created is:

$$FWM = (n^2 (n-1))/2$$

where *FWM* is the number of waves created and *n* is the number of wavelengths being transmitted through the fiber. So if 2 wavelengths are being used, an extra 2 wavelengths will appear. That's not too bad in itself. But if you are transmitting 4 original wavelengths, 24 extra wavelengths are created:

$$FWM = (16 \times 3)/2 = 24$$

And 8 original wavelengths will produce 224 extra wavelengths:

$$FWM = (64 \times 7)/2 = 224$$

The solution to four-wave mixing actually involves creating just enough dispersion in the fiber to render the newly created wavelengths harmless to the signals while leaving the original signal clear enough to use. The fiber created for this purpose is *non-zero-dispersion-shifted fiber (NZ-DSF)*. Non-zero-dispersion shifting moves the zero-dispersion point slightly away from the wavelength used for the transmission, usually about 10nm above or below the transmission wavelength, so there is sufficient dispersion to keep the effects of four-wave mixing to minimum.

REDUCED SPECTRAL WIDTH

Because material dispersion is caused by an overabundance of wavelengths in the optical signal, the simplest solution is to reduce the number of wavelengths by reducing the spectral width. Recall that dispersion is expressed as picoseconds per nanometer of spectral width per kilometer of fiber, so any reduction in spectral width will have a significant effect on the amount of material dispersion.

Polarization-Mode Dispersion

Polarization-mode dispersion (PMD) is masked by other forms of dispersion until the bit rate exceeds 2.5Gbps. In order to understand PMD, we must look at an information pulse more closely.

Recall that light is an electromagnetic wave, consisting of an electrical and a magnetic wave traveling at right angles to one another. The orientation of the two waves along the path of propagation determines the light's *polarization mode*, or *polarity*.

As shown in Figure 5.8, it is possible to have different polarities of light traveling through the fiber in a signal, occupying different parts of the fiber as they pass through it. Because no fiber is perfect, there will be obstacles in one part of the fiber that are not present in another. As a result, the light having one polarity may pass an area without interference, while light having another polarity may pass through a defective region, slowing it down.

FIGURE 5.8
Polarized light shown in a cross section of optical fiber

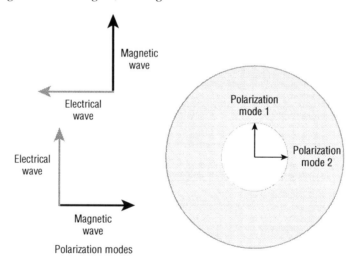

Polarization modes

Polarization-mode dispersion is not so much a function of the fiber's overall characteristics as it is a result of irregularities, damage, or environmental conditions such as temperature. Small areas of damage called *microbends* can cause PMD, as can fiber that is not perfectly round or concentric.

Because PMD is caused by specific conditions within the fiber, and not the fiber's overall characteristics, it is difficult to assign a consistent PMD value to a length of fiber. The exact amount of PMD changes with external conditions, the physical condition of the fiber, and the polarization state of the light passing through it at any given moment. For this reason, PMD is measured in terms of the total difference in the travel time between the two polarization states, referred to as the *differential group delay (DGD)* and measured in picoseconds (ps).

How Dispersion Affects Bandwidth

While different types of dispersion have different causes and are measured different ways (ns/km, ps/nm/km), all of them have one effect in common: they place a limit on the bandwidth of optical fibers. The bandwidth of a specific brand of optical fiber can be found in the manufacturer's data sheet. The bandwidth of a type of optical fiber can also be found in an industry standard. Remember, an industry standard defines the minimum performance levels; optical fibers are available that exceed industry standards.

The product of bandwidth and length (MHz · km) expresses the information carrying capacity of a multimode optical fiber. Bandwidth is measured in megahertz (MHz) and the length is measured in kilometers. The MHz/km figure expresses how much bandwidth the fiber can carry per kilometer of its length. The fiber's designation must always be greater than or equal to the product of the bandwidth and the length of the optical fiber.

The minimum bandwidth-length product for a 50/125µm laser optimized OM4 multimode optical fiber as defined in ANSI/TIA-568-C.3 is 3500MHz · km with an overfilled launch at a wavelength of 850nm. To find the usable bandwidth for this optical fiber at a length of 1250 meters or 1.25km you must change the equation around and solve as shown here:

Bandwidth-length product = 3500MHz · km

Optical fiber length = 1.25km

Bandwidth in MHz = 3500 ÷ 1.25

Bandwidth = 2800MHz

As you can see, the bandwidth decreases as the optical fiber length increases. However, the opposite is true when the length of the optical fiber decreases as shown here:

Bandwidth-length product = 3500MHz · km

Optical fiber length = 0.125km

Bandwidth in MHz = 3500 ÷ 0.125

Bandwidth = 28000MHz

Bandwidth-length product is defined in ANSI/TIA-568-C.3 for multimode optical fiber, but not for single-mode optical fiber. The bandwidth for a length of a single-mode optical fiber is specified by dispersion for a specific wavelength. The spectral width of the light source will affect the bandwidth. For a given length of optical fiber, a laser with a wide spectral width will have less bandwidth than a laser with a narrower spectral width. Spectral width can vary tremendously among lasers. Laser sources are covered in detail in Chapter 10.

Attenuation

The *attenuation* of a fiber-optic signal is the loss of optical power as the signal travels through the fiber. There are two types of attenuation: intrinsic and extrinsic. *Intrinsic attenuation* occurs because no manufacturing process can produce a perfectly pure fiber. Either by accident or by design, the fiber will always have some characteristics that attenuate the signal passing through it. *Extrinsic attenuation* is caused by external mechanisms that bend the optical fiber. These bends are broken into two categories: *microbends* and *macrobends*, which are discussed in detail later in this chapter.

The wavelength of the light passing through the fiber also affects attenuation. In general, attenuation decreases as wavelength increases, but there are certain wavelengths that are more easily absorbed in plastic and glass fibers than others. One of the reasons for establishing standard operating wavelengths of 850nm, 1300nm, and 1550nm in glass fiber and in the visible range of 650nm for plastic is because the wavelengths in between are considered high-loss regions. Specifically, these wavelengths for glass optical fiber are in the range of 730, 950, 1250, and 1380nm.

NOTE Attenuation provides a good example of the superiority of fiber over copper for carrying signals. When an electrical signal is carried through copper wire, attenuation increases with the data rate of the signal, requiring an increase in transmission power or, more often, the use of repeaters. Attenuation per unit length in an optical signal for an optical fiber of a given type is constant no matter what the data rate.

Remember from Chapter 2, "Principles of Fiber-Optic Transmission," that attenuation is measured in *decibels* (dB). Decibels help us account for the loss of power when we are measuring attenuation in an optical fiber. In glass optical fiber, attenuation is lowest at long wavelengths and highest at the short wavelengths. This is not true for plastic optical fiber. Plastic optical fiber is typically designed for visible light sources with a center wavelength of 650nm. Attenuation increases dramatically above or below this wavelength.

Attenuation values for a length of optical fiber can be calculated using the attenuation coefficient for a specific type of optical fiber. This information can be found in the manufacturer's data sheet or in a standard. The "Fiber Specifications and Standards" section, later in this chapter, provides attenuation information for multimode and single-mode optical fibers.

You can calculate the maximum attenuation for a length of optical fiber using this formula:

Attenuation for the length of optical fiber = Attenuation coefficient × Optical fiber length

Here we calculate the maximum attenuation for a 1.75km length of optical fiber with and attenuation coefficient of 1.5dB/km.

Attenuation coefficient = 1.5dB/km

Optical fiber length = 1.75km

Attenuation for the length of optical fiber = 1.75 × 1.5 = 2.625

Maximum attenuation for 1.75km of optical fiber is 2.625dB

Remember that the attenuation is wavelength dependent. The shorter wavelengths will attenuate more and the longer wavelengths less. Also remember that the attenuation coefficient is a "not to exceed" number. In other words, the attenuation should be no greater than that. The measured attenuation will typically be less.

While decibels are useful in measuring *total attenuation*, we can also divide attenuation into two types: *absorption* and *scattering*.

Absorption

The absorption of light is a form of intrinsic attenuation. Absorption accounts for a very small percentage of attenuation in an optical fiber—typically between 3 and 5 percent.

All materials, even the clearest glass, absorb some light. The amount of absorption depends on the type of material and the wavelength of the light passing through it. You can see absorption easily in sunglasses. Even on the brightest days, only a fraction of the light energy passes through the tinted lenses. The wavelengths that do not pass through are mostly absorbed by impurities that have been placed in, or coated on, the lens material.

In an optical fiber, absorption occurs when impurities such as water or ions of materials such as copper or chromium absorb certain wavelengths, as shown in Figure 5.9. By keeping these impurities as low as possible, manufacturers can produce fibers with a minimum of attenuation.

FIGURE 5.9
Absorption in optical fiber

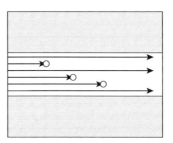

Scattering

The scattering of light is a form of intrinsic attenuation. Scattering accounts for the greatest percentage of attenuation in an optical fiber—typically between 95 to 97 percent.

Scattering is caused by atomic structures and particles in the fiber redirecting light that hits them, as shown in Figure 5.10. The process is called *Rayleigh scattering*, for Lord Rayleigh, a British physicist who first described the phenomenon in the late nineteenth century.

FIGURE 5.10
Scattering in optical fiber

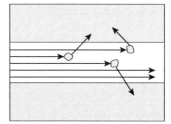

Rayleigh scattering is also the answer to the age-old question "Why is the sky blue?" The blue that we see is actually the more prevalent blue wavelengths of light from the sun being scattered by particles in the atmosphere. As the sun moves toward the horizon and the light must pass through more of the atmosphere, the scattering increases to the point where the blue light is almost completely attenuated, leaving the red wavelengths, which are less affected by the scattering for reasons that we'll see shortly.

Rayleigh scattering depends on the relationship between wavelength and the size of the structures in the fiber. Scattering decreases as the wavelength of the light approaches the size of the structures, which means that in fiber-optic communications, shorter wavelengths are more likely to be scattered than the longer wavelengths. This is one of the main reasons that infrared wavelengths are used in fiber optics. The relatively long wavelengths of infrared are less subject to scattering than visible wavelengths. It also explains why the sun turns red on the horizon. The shorter blue wavelengths are more likely to be scattered by the similarly sized particles in the atmosphere than are the red wavelengths.

Total Attenuation

Total attenuation is the combination of the intrinsic effects of absorption and scattering in a fiber. Figure 5.11 shows a typical attenuation curve for an optical fiber with the effects of absorption and scattering combined. Note that the general curve is caused by the effects of scattering, while the irregularities in the plot are caused by specific impurities, such as hydroxyl molecules, absorbing light in those wavelengths.

FIGURE 5.11

An optical fiber's attenuation curve

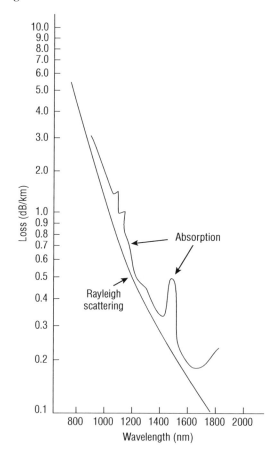

Note also the windows at the 850, 1300, and 1550nm ranges. Remember that while the 1300nm range is better in terms of dispersion, it still has a higher attenuation than the 1550nm range, which is the reason for dispersion-shifted fiber.

Bending Losses

In addition to intrinsic losses within the optical fiber, the actual condition of the fiber can lead to losses. These losses are referred to as *extrinsic losses* and are caused by bends in the optical fiber. These bends, depending on their radius, are referred to as *microbends* or *macrobends*.

Microbends

Microbends are small distortions of the boundary layer between the core and cladding caused by crushing or pressure. Microbends are very small and may not be visible when looking at the fiber-optic cable.

Microbends change the angle of incidence within the fiber, as shown in Figure 5.12. Changing the angle of incidence forces high-order light rays to reflect at angles that prevent further reflection, causing them to be lost in the cladding and absorbed.

FIGURE 5.12
Losses caused by
microbending and
macrobending

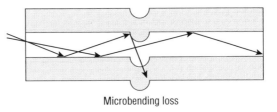

Microbending loss

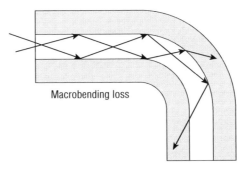

Macrobending loss

Microbends can be caused by the coating, cabling, packaging, and installation of the cable. Remember from Chapter 4, "Optical Fiber Construction and Theory," that one of the most common types of coatings is *acrylate*. Acrylate is limited in temperature performance. Exceeding the operating temperature range of the acrylate coating can cause microbending, which can result in significant attenuation.

Macrobends

Compared to a microbend, a macrobend has a much larger radius. Macrobends occur when the fiber is bent around a radius that can be measured in millimeters. As shown in Figure 5.12, these tight radii change the angle of incidence within the fiber, causing some of the light rays to reflect outside of the fiber and, as with microbending, be lost in the cladding and absorbed.

At long wavelengths such as 1550nm to 1625nm, single-mode optical fiber can experience severe attenuation from macrobends. This is not the case with multimode optical fiber. A multimode optical fiber is less sensitive to bending loss than single-mode. However, bend-insensitive single-mode and multimode optical fibers are available for installations where tight bends may be required.

As we learned in Chapter 4, multimode and single-mode depressed clad optical fibers are commonly referred to as bend insensitive. These optical fibers are manufactured with an optical trench that surrounds the core and feature proprietary refractive index profiles.

Single-mode Macrobending Performance

As you will learn in the "Fiber Specifications and Standards" section later in this chapter, single-mode optical fiber can be broken into two groups: bend sensitive and bend insensitive. There is a considerable difference in the bending attenuation of the two fiber types. ITU Standards G.652, G.655 define macrobending loss for bend-sensitive optical fiber whereas G.657 defines macrobending loss for bend-insensitive fiber.

The macrobending loss in each standard is for an optical fiber that has not been placed in a cable. The macrobending loss will vary when the optical fiber is placed in a cable and that information must be obtained from the cable manufacturer. However, to demonstrate how macrobending loss is calculated it is assumed that optical fiber values listed in the standards are the same as the cable loss.

ITU-T G.652 allows for a maximum attenuation of 0.1dB at 1550nm with 100 turns around a 30mm radius mandrel. The 30mm radius of the mandrel is slightly less than the radius of a tennis ball. Reducing the radius of the bend will increase the attenuation. Optical fiber with a macrobending specification like this is not a good choice for installations that may require several sharp bends especially if they are much less than 30mm.

ITU-T G.657 defines the macrobending loss for bend-insensitive single-mode optical fiber. This type of optical fiber is typically employed in *passive optical networks* (PONs) that run to your home or residence. Installing a fiber-optic cable in the wall of a home or apartment may require one or more sharp bends. A single sharp 90-degree bend with an optical fiber that is not bend insensitive could cause enough attenuation to prevent the link from functioning.

Macrobending loss is defined several ways in ITU-T G.657. Tables 5.7 through 5.10 in the section "Fiber Specifications and Standards" list the parameters. In Table 5.7, the maximum macrobending loss for one turn around a 10mm radius mandrel at a wavelength of 1550nm is 0.75dB. With this information, it is possible to estimate the macrobending loss for a typical installation. If the number of bends and the radius of each bend are known, the macrobending losses can be calculated.

Let's look at a single-mode ITU-T G.657 Category A2 bend-insensitive fiber-optic cable installed in a home with three small radius 90 bends. Two bends have a radius of 10mm and one bend has a radius of 7.5mm. The wavelength of the light source is 1550nm. Referring to Table 5.8, we see that the loss for each single turn 10mm bend is 0.1dB and the loss for the single turn 7.5mm bend is 0.5dB.

The total macrobending loss (MBL) will be the sum of the three losses; however, we cannot simply add up the dB values from Table 5.8 because each bend is not a full 360 turn. Both 10mm bends are 90 and the 7.5mm bend is 180. First have to calculate the loss for the partial bends using the following formula:

(Bend in degrees ÷ 360) × dB loss per turn = MBL

(90 ÷ 360) × 0.1dB = 0.025dB

(180 ÷ 360) × 0.5dB = 0.25dB

Then we calculate the total loss using this formula:

MBL 1 + MBL 2 + MBL 3 = Total MBL

0.025dB + 0.025dB + 0.25dB = 0.3dB

Total macrobending loss = 0.3dB

When the bending loss is not listed for a single turn but a number of turns, you must first establish the single turn loss using the following formula:

MBL ÷ number of turns = dB loss per turn

MULTIMODE MACROBENDING LOSS

Unlike single-mode optical fiber, there are currently no standards for bend insensitive multi-mode optical fiber (BIMMF). However, as of this writing they are in development and offered by multiple manufacturers with extremely low bending loss at 850nm and 1300nm. Table 5.1 describes the bending loss performance of a generic 50/125µm BIMMF. The macrobending loss is defined for a number of turns of optical fiber around a mandrel at a specified radius and wavelength.

TABLE 5.1: Generic 50/125µm BIMMF bending loss

RADIUS (MM)	NUMBER OF TURNS	ATTENUATION (DB) AT 850NM	ATTENUATION (DB) AT 1300NM
37.5	100	≤0.05	≤0.15
15	2	≤0.1	≤0.3
7.5	2	≤0.2	≤0.5

Non-bend insensitive multimode optical fiber macrobending loss is defined in ITU-T G.651.1 and IEC 60793-2-10. It will be covered in greater detail in the "Fiber Specifications and Standards" section later in this chapter.

Numerical Aperture

As we saw in Chapter 4, many of a fiber's characteristics are determined by the relationship between the core and the cladding. The *numerical aperture (NA)* expresses the light-gathering ability of the fiber. The NA is a dimensionless number, meaning that it is to be used as a variable in determining other characteristics of the fiber, or as a means of comparing two fibers.

As discussed earlier, the numerical aperture is determined by the refractive indexes of the core and the cladding:

$$NA = \sqrt{[(n_1)^2 - (n_2)^2]}$$

where n_1 is the refractive index of the core and n_2 is the refractive index of the cladding.

Recall that in order for light to be contained within a multimode fiber it must stay above the *critical angle*, or the angle at which it reflects off the boundary between the core and the cladding, rather than penetrating the boundary and refracting through the cladding.

To maintain the critical angle, light must enter within a specified range defined by the *acceptance cone* or *cone of acceptance*. As shown in Figure 5.13, this region is defined by a cone extending outside the optical fiber's core. Light entering the core from outside of the cone will either miss the core or enter at an angle that will allow it to pass through the boundary with the cladding and be lost. The *acceptance angle* defines the acceptance cone. Light entering the core of the optical fiber at an angle greater than the acceptance angle may not propagate the length of the optical fiber. For light to propagate the length of the optical fiber, it must enter the core at an angle that does not exceed the acceptance angle.

FIGURE 5.13

The cone of acceptance

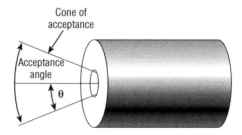

The cone of acceptance is determined using the numerical aperture:

$NA = \sin\theta$

where θ is the angle defining the cone of acceptance.

To determine the acceptance angle of a fiber with a core refractive index of 1.48 and a cladding refractive index of 1.47, we must determine the NA of the fiber:

$$NA = \sqrt{n^2_{core} - n^2_{cladding}} = \sqrt{(2.19 - 2.16)} = \sqrt{0.03} = 0.17$$

Next, we can use the NA to determine the acceptance angle. Remember that the acceptance angle is the value of θ.

$NA = \sin\theta$

So

θ = arcsin NA = arcsin 0.17 = 9.79

The acceptance angle is 9.79°.

The acceptance angle is also used to determine how light emerges from a fiber. The light that comes out of a fiber end is the light that has not been scattered, absorbed or lost in the cladding, so with the exception of a small percentage of light that has propagated in the boundary between the core and the cladding, what emerges is coming out at an angle equal to or less than the acceptance angle, as shown in Figure 5.14.

FIGURE 5.14
Light emerges from the fiber in the cone of acceptance.

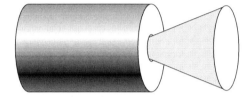

The numerical aperture varies depending on the optical fiber type. Single-mode optical fibers have the smallest NA and narrowest acceptance cone whereas large core plastic optical fibers have the largest NA and widest acceptance cone. NA is not defined in ANSI/TIA-568-C.3, ITU-T G.652, ITU-T G.655, or ITU-T G.657. However, it is defined in ITU-T G.651.1 and IEC 60793-2-10 for some multimode optical fibers. Table 5.2 lists some of the typical NA values for various single-mode and multimode optical fibers.

TABLE 5.2:   Optical fiber numerical aperture (NA)

FIBER TYPE	NA
Single-mode	0.14
Multimode 50/125μm	0.20
Multimode 62.5/125μm	0.275
Multimode 100/140μm	0.29
Multimode 200/230μm	0.37
Multimode 980/1000μm	0.47

Equilibrium Mode Distribution

Equilibrium mode distribution (*EMD*) is a condition in which the light traveling through the fiber populates the available modes in an orderly way.

When light energy first enters a fiber, it does not automatically fill every available mode with an equal amount of energy. Some modes will carry a high amount of light energy, whereas others may carry nothing at all. Over distance, this condition balances out as light transfers between modes because of imperfections or bends in the fiber. Until EMD is achieved, however, light may be traveling in inefficient modes that will eventually lose energy. These modes may include paths that carry the light through the cladding, or high-order modes that produce a high number of reflections through the fiber core.

Let's back up to where the light enters the fiber and see how EMD is reached.

A light source rarely, if ever, perfectly matches the numerical aperture of the fiber. Typically, the beam will either be larger or smaller than the NA. If it is larger, as is the case with most LEDs, the fiber will be overfilled. This means that the light energy will occupy most of the modes, including those that are less efficient and are doomed to loss somewhere down the length of the fiber. The *overfilled launch* is discussed in detail in Chapter 10.

If the source has a smaller beam, as with most lasers, the fiber will be underfilled. The light energy will occupy only a few of the available modes until twists, turns, and imperfections in the fiber distribute some of the light energy into higher-order modes.

The effect is similar to traffic entering a highway. If eight lanes of cars are trying to get onto a six-lane highway, you'll see cars jockeying for position in the available lanes, and possibly even a few driving on the shoulder until they are attenuated by the highway patrol. This condition will persist until the drivers settle down into their new pattern. On the other hand, if a single lane of cars enters the same six-lane highway, they will first occupy only a single lane but eventually spread out to a more evenly distributed pattern.

Even when EMD is achieved, the effect is short-lived due to minute changes in the fiber characteristics and the effects of connectors. It does not take much to throw the modes out of equilibrium.

The distance required to achieve EMD depends on the fiber material. Light passing through plastic fibers reaches EMD in a few meters, but light passing through glass fibers may reach equilibrium only after several kilometers.

It is important to understand EMD because it affects the measurement of light energy in a fiber. In an overfilled fiber, light that has not yet reached EMD still has much of the energy traveling in the inefficient or high-order modes, and this energy can be measured over a short distance. Over a longer distance, however, once EMD has been reached, the energy in the less efficient modes has been lost, and the energy reading may drop significantly.

Once EMD has been reached, however, energy loss drops off, because the light is traveling in more efficient modes and is less likely to be lost. For this reason, loss before EMD is proportional to the length of the fiber, and loss after EMD has been reached is proportional to the square root of the fiber length. In other words, if the energy loss before EMD is 0.2dB/km, the loss over 2km would be 0.4dB. After reaching EMD, however, the loss would be $0.2 \times \sqrt{2}$.

Fiber Specifications and Standards

As you learned in Chapter 1, "History of Fiber Optics and Broadband Access," the first full-scale commercial application of fiber-optic communication systems occurred in 1977, when both AT&T and GTE began using fiber-optic telephone systems for commercial customers. Twenty-three years later as published in The National Economic Council report entitled Four Years of Broadband Growth, 4.4 percent of the households in America had a broadband connection to their home with a download speed of just 200kbps. By 2013, basic broadband download speed was 3Mbps; however, more than 94 percent of the homes in America exceed 10Mbps and more than 3 percent enjoy download speeds greater than 1Gbps.

In just 13 years, download speeds increased 5,000 times. A download in the year 2000 that would take one hour can be accomplished today in less than one second. Creating a global fiber-optic infrastructure that supports this type of rapid change requires coordinated global efforts that standardize physical and optical properties.

As download rates have evolved since the turn of the century, so have optical fibers, transmitters, and receivers. In fact, there has been so much change it is difficult even for those working in the industry to keep pace with the changes and new developments, let alone a novice or someone just learning about fiber optics.

In this section, the major industry standards that define optical fiber performance are presented. Keep in mind that standards define a minimum level of performance and many manufacturers offer products that exceed the minimum performance levels. Often the working groups within standards organizations are playing catch-up, creating standards for products that currently exist and may have been in widespread use for several years or more.

In Chapter 4, some of the many standards related to optical fiber were introduced along with four major standards organizations that publish standards that define the performance of optical fibers used in the telecommunications industry. The four major standards organizations are the Telecommunications Industry Association (TIA), the International Telecommunications Union (ITU), the International Organization for Standardization (ISO), and the International Electrotechnical Commission (IEC). These organizations publish many different standards on optical fiber, fiber-optic cable, and testing.

The TIA standards covered in this text were prepared in accordance with the policies and procedures of the American National Standards Institute (ANSI). The mission of ANSI is to "enhance both the global competitiveness of U.S. business and the U.S. quality of life by promoting and facilitating voluntary consensus standards and conformity assessment systems, and safeguarding their integrity." Because these TIA standards were prepared in accordance with ANSI policies and procedures, each standard contains the ANSI prefix.

Revisions and Addendums

Standards can take many years to develop, and the organizations that publish the standards typically do not give them away. The cost for a standard can be significant. For example, as of this writing the cost for the ANSI/TIA-568-C.3 base document is $219. It is for this reason that companies that require access to many standards typically pay a significant annual fee for an online service that allows access to standards published by many different organizations.

The ANSI/TIA-568-C series of standards contains documents C.0, C.1, C.2, C.3, and C4. The letter "C" represents the major revision. When the "C" revisions were published, they replaced the "B" revisions. As of this writing, the "D" revisions are in development. In addition to the revision letter, the date the standard was approved can be found in the upper-right corner of the cover page. The cover page will also list the month and year the standard was published.

When a TIA standard is updated without a major revision, an addendum is included. The addition of the addendum also includes the addition of a suffix. For example, ANSI/TIA-568-C.3 was published in June 2008; Addendum 1 was published in December 2011. The addendum is known as ANSI/TIA-568-C.3-1.

It is important to be aware of standards that have been superseded. Companies typically do not rip out their old cabling just because a standard changed. Fiber-optic cabling installed at the turn of the century would have complied with ANSI/TIA/EIA-568-B.3. More than likely that cabling is still in use today. You will need access to the ANSI/TIA/EIA-568-B.3 optical fiber cable performance parameters to determine compatibility with newer fiber-optic cables, transmitters, and receivers.

The ISO, IEC, and ITU standards covered in this section do not use a letter to represent the revision; each has a date on the cover page. The IEC standards also state the edition number on the cover page just as the publisher of this book does.

Unless you work with industry standards on a regular basis and keep track of the progress of the many standards working groups, it is difficult to know which standard is the most current. The best way to determine this is to visit each organization's website. The following links will assist you in staying up to date:

www.tiaonline.org

www.iec.ch

www.itu.int

www.iso.org

www.ansi.org

Mirroring of Performance Specifications

It is not uncommon to find the performance specifications for an optical fiber mirrored in another standard; after all, the goal is inter-compatibility. While it may be the intent of a standards organization to produce specifications that can be utilized globally, they may not be the recognized organization for specific geographic locations. The TIA headquarters is located in Arlington, Virginia and the IEC central office is located in Geneva, Switzerland.

When ANSI/TIA/EIA-568-B.3 was first published in March 2000, there were no references to ISO/IEC optical fiber cable performance parameters in Section 4, Table 1. However, when ANSI/TIA-568-C.3 was published in June 2008 the ISO/IEC 11801 fiber designations and optical fiber cable performance parameters were included in Section 4, Table 1. This is an example of the performance specifications for optical fibers being mirrored in multiple standards.

Fiber Designations

While optical fiber performance parameters may be mirrored in different standards, the fiber designations typically are not. ISO/IEC 11810 uses the acronym "OM" to describe multimode optical fiber and "OS" to describe single-mode. IEC 60793-2-10 uses the letter "A" to describe multimode optical fiber and IEC 60793-2-50 uses the letter "B" to describe single-mode. In the ANSI/TIA-568-C standard optical fiber is designated using the ANSI/TIA-492 series of documents where the letter "A" after 492 designates multimode optical fiber and "C" designates single-mode. ITU does not use letters only numbers; G.651.1 is a multimode standard while G.652, G.655, and G.657 are single-mode. These designations are broken out in Table 5.3.

TABLE 5.3: Fiber designation by industry standard

INDUSTRY STANDARD	MULTIMODE	SINGLE-MODE
ISO/IEC 11810	OM	OS
IEC 60793-2-10/50	A	B
ANSI/TIA	TIA-492A	TIA-492C
ITU	G.651.1	G.652, G.655, G.675

Premises Standards

The ANSI/TIA-568-C standards and the ISO/IEC 11801 standard are applicable to premises cabling components; this includes copper and fiber-optic cabling. However, in this section only optical fiber cable transmission performance parameters will be addressed. Optical fiber cable transmission performance parameters can be found in Section 4 of ANSI/TIA-568-C.3, Section 9 of ISO/IEC 11801, and the annex's of IEC-60793-2-10. To simplify presentation, the performance parameters for all three standards are combined into Table 5.4.

Table 5.4 describes the attenuation and bandwidth characteristics for multimode and single-mode optical fibers. Unlike ANSI/TIA-568-C.3 and ISO/IEC 11801, IEC-60793-2-10 does not list a minimum value for attenuation and bandwidth performance, it lists a value range. In Table 5.4, IEC-60793-2-10 value ranges have been correlated with the ANSI/TIA-568-C.3 and ISO/IEC 11801 designations.

Two laser-optimized multimode optical fibers that were not previously described in ANSI/TIA/EIA-568-B.3 are introduced; they will be covered in detail later in this chapter. New bandwidth requirements for multimode optical fiber are also defined. Bandwidth is defined with an overfilled launch (OFL) and a restricted mode launch (RML), also known as effective laser launch in ISO/IEC 11801.

Up until 1996 when TIA formed the TIA FO-2.2.1 Task Group on Modal Dependence of Bandwidth to study the interaction between multimode optical fiber and a laser source, multimode optical fiber bandwidth was defined with an overfilled launch. However, as laser technology improved and IEEE began developing a Gigabit Ethernet Standard (IEEE 802.3z) for high-speed local area networks (LANs) using lasers, a better indicator of bandwidth performance in a laser-based system was required.

Since 1996 the FO-2.2.1 Task Group has developed two Fiber Optic Test Procedures (FOTPs). FOTP 203 defines a standard procedure for measuring the launch power distribution of a laser-based multimode fiber-optic transmitter. FOTP 204 describes the methods used to measure the information-carrying capacity of a multimode optical fiber using an OFL or RML. As you learned earlier, bandwidth-length product (MHz · km) is used to define the information-carrying capacity of a multimode optical fiber.

TABLE 5.4: Characteristics of ANSI/TIA-568-C.3 and ISO/IEC 11801-recognized optical fibers

OPTICAL FIBER CABLE TYPE	WAVELENGTH (NM)	MAXIMUM ATTENUATION (DB/KM)	MINIMUM OVERFILLED MODAL BANDWIDTH-LENGTH PRODUCT (MHZ · KM)	MINIMUM EFFECTIVE MODAL BANDWIDTH-LENGTH PRODUCT (MHZ · KM)
62.5/125µm Multimode OM1 Type A1b TIA 492AAAA	850 1300	3.5 1.5	200 500	Not Required Not Required
50/125µm Multimode OM2 TIA 492AAAB Type A1a.1	850 1300	3.5 1.5	500 500	Not Required Not Required
850nm Laser-Optimized 50/125µm Multimode OM3 Type A1a.2 TIA 492AAAC	850 1300	3.5 1.5	1500 500	2000 Not Required
850nm Laser-Optimized 50/125µm Multimode OM4 Type A1a.3 TIA 492AAAD	850 1300	3.5 1.5	3500 500	4700 Not Required

TABLE 5.4:   Characteristics of ANSI/TIA-568-C.3 and ISO/IEC 11801-recognized optical fibers   *(continued)*

OPTICAL FIBER CABLE TYPE	WAVELENGTH (NM)	MAXIMUM ATTENUATION (DB/KM)	MINIMUM OVERFILLED MODAL BANDWIDTH-LENGTH PRODUCT (MHZ · KM)	MINIMUM EFFECTIVE MODAL BANDWIDTH-LENGTH PRODUCT (MHZ · KM)
Single-mode	1310		N/A	N/A
Indoor-Outdoor	1550		N/A	N/A
OS1				
Type B1.1				
TIA 492CAAA				
OS2				
Type B1.3				
TIA 492CAAB				
Single-mode	1310	1.0	N/A	N/A
Inside Plant	1550	1.0	N/A	N/A
OS1				
Type B1.1				
TIA 492CAAA				
OS2				
Type B1.3				
TIA 492CAAB				
Single-mode	1310	0.5	N/A	N/A
Outside Plant	1550	0.5	N/A	N/A
OS1				
Type B1.1				
TIA 492CAAA				
OS2				
Type B1.3				
TIA 492CAAB				

As noted in Table 5.4, laser optimized multimode optical fiber offers the greatest bandwidth performance at 850nm. This is because it is designed to be used with the vertical-cavity surface-emitting lasers (VCSELs). VCSEL technology has improved tremendously since the turn of the century and is the light source of choice for gigabit plus LANs. The VCSEL and other light sources are covered in detail in Chapter 10.

As you can see in Table 5.4, bandwidth-length product is not defined for single-mode optical fiber. As you learned earlier, the bandwidth of a single-mode optical fiber depends on the dispersion for a given length at a given wavelength. It is also impacted by the spectral width of the light source. A light source with a narrow spectral width will provide a greater bandwidth over a given distance than a light source with a wide spectral width.

Single-mode ITU Standards

ITU-T G.652, ITU-T G.655, and ITU-T G.657 are three standards typically used to describe the performance characteristics of single-mode optical fiber:

- ITU-T G.652 defines the characteristics of a single-mode optical fiber and cable.

- ITU-T G.655 defines the characteristics of a non-zero-dispersion-shifted single-mode optical fiber and cable.

- ITU-T G.657 defines the characteristics of a bending loss-insensitive single-mode optical fiber and cable for the access network.

ITU-T G.652 contains several tables that list the performance characteristics of single-mode optical fiber and cable. The values in each table are identical for many of the fiber and cable attributes. Table 5.5 lists commonly used optical and geometric characteristics for single-mode optical fiber and cable.

TABLE 5.5: ITU-T G.652, single-mode optical fiber and cable

FIBER ATTRIBUTES, G.652.A, G.652.B, G.652.C, AND G.652.D		
Mode field diameter	Wavelength	1310nm
	Range of nominal values	8.6–9.5µm
	Tolerance	±0.6µm
Cladding diameter	Nominal	125.0µm
	Tolerance	±1µm
Core concentricity error	Maximum	0.6µm
Cladding noncircularity	Maximum	1%
Cable cut-off wavelength	Maximum	1260nm
Macrobend loss	Radius (mm)	30
	Number of turns	100
	Maximum at 1550nm or 1625nm	0.1dB

TABLE 5.5:   ITU-T G.652, single-mode optical fiber and cable *(continued)*

FIBER ATTRIBUTES, G.652.A, G.652.B, G.652.C, AND G.652.D		
Cable attributes, G.652.A		
Attenuation coefficient	Maximum at 1310nm	0.5dB/km
	Maximum at 1550nm	0.4dB/km
Cable attributes, G.652.B		
Attenuation coefficient	Maximum at 1310nm	0.4dB/km
	Maximum at 1550nm	0.35dB/km
	Maximum at 1625nm	0.4dB/km
Cable attributes, G.652.C		
Attenuation coefficient	Maximum from 1310nm to 1625nm	0.4dB/km
	Maximum at 1550nm	0.3dB/km
Cable attributes, G.652.D		
Attenuation coefficient	Maximum from 1310nm to 1625nm	0.4dB/km
	Maximum at 1550nm	0.3dB/km

ITU-T G.655 contains several tables that list the characteristics of a non-zero-dispersion-shifted single-mode optical fiber and cable. Unlike ITU-T G.652, several of the fiber attributes are not similar among the cable types. Table 5.6 lists commonly used optical and geometric characteristics for non-zero-dispersion-shifted single-mode optical fiber and cable.

TABLE 5.6:   ITU-T G.655.C, G.655.D, and G.655.E, non-zero-dispersion-shifted single-mode optical fiber and cable

FIBER ATTRIBUTES		G.655.C	G.655.D and G.655.E
Mode field diameter	Wavelength	1550nm	1550nm
	Range of nominal values	8–11µm	8–11µm
	Tolerance	±0.7µm	±0.6µm

TABLE 5.6:   ITU-T G.655.C, G.655.D, and G.655.E, non-zero-dispersion-shifted single-mode optical fiber and cable   *(continued)*

FIBER ATTRIBUTES		G.655.C	G.655.D and G.655.E
Cladding diameter	Nominal	125.0μm	125.0μm
	Tolerance	±1μm	±1μm
Core concentricity error	Maximum	0.8μm	0.6μm
Cladding noncircularity	Maximum	2%	1%
Cable cut-off wavelength	Maximum	1450nm	1450nm
Macrobend loss	Radius (mm)	30	30
	Number of turns	100	100
	Maximum at 1625nm	0.5dB	0.1dB
Cable attributes, G.655.C, G.655.D, and G.655.E			
Attenuation coefficient	Maximum at 1550nm	0.35dB/km	
	Maximum at 1625nm	0.4dB/km	

ITU-T G.657 defines the characteristics of two category "A" and "B" bending loss-insensitive single-mode optical fibers and cables for the access network. Like ITU-T G.655, several of the fiber attributes are not similar among the cable types, primarily in the area of bend radius. Table 5.7 through Table 5.10 list commonly used optical and geometric characteristics for bending loss-insensitive single-mode optical fiber and cable for the access network.

TABLE 5.7:   ITU-T G.657, specifications for G.657 category A1 bending loss-insensitive single-mode optical fiber and cable

FIBER ATTRIBUTES, G.657 CATEGORY A1			
Mode field diameter	Wavelength	1310nm	
	Range of nominal values	8.6–9.5μm	
	Tolerance	±0.4μm	
Cladding diameter	Nominal	125.0μm	
	Tolerance	±0.7μm	
Core concentricity error	Maximum	0.5μm	
Cladding noncircularity	Maximum	1%	
Cable cut-off wavelength	Maximum	1260nm	
Macrobend loss	Radius (mm)	15	10
	Number of turns	10	1
	Maximum at 1550nm	0.25dB	0.75dB
	Maximum at 1625nm	1.0dB	1.5dB
Cable attributes			
Attenuation coefficient	Maximum from 1310 to 1625nm	0.4dB/km	
	Maximum at 1550nm	0.3dB/km	

TABLE 5.8: ITU-T G.657, specifications for G.657 category A2 bending loss-insensitive single-mode optical fiber and cable

FIBER ATTRIBUTES, G.657 CATEGORY A2				
Mode field diameter	Wavelength	1310nm		
	Range of nominal values	8.6–9.5µm		
	Tolerance	±0.4µm		
Cladding diameter	Nominal	125.0µm		
	Tolerance	±0.7µm		
Core concentricity error	Maximum	0.5µm		
Cladding noncircularity	Maximum	1%		
Cable cut-off wavelength	Maximum	1260nm		
Macrobend loss	Radius (mm)	15	10	7.5
	Number of turns	10	1	1
	Maximum at 1550nm	0.03dB	0.1dB	0.5dB
	Maximum at 1625nm	0.1dB	0.2dB	1.0dB
Cable attributes				
Attenuation coefficient	Maximum from 1310 to 1625nm	0.4dB/km		
	Maximum at 1550nm	0.3dB/km		

TABLE 5.9: ITU-T G.657, specifications for G.657 category B2 bending loss-insensitive single-mode optical fiber and cable

FIBER ATTRIBUTES, G.657 CATEGORY B2				
Mode field diameter	Wavelength	1310nm		
	Range of nominal values	8.6–9.5μm		
	Tolerance	±0.4μm		
Cladding diameter	Nominal	125.0μm		
	Tolerance	±0.7μm		
Core concentricity error	Maximum	0.5μm		
Cladding noncircularity	Maximum	1%		
Cable cut-off wavelength	Maximum	1260nm		
Macrobend loss	Radius (mm)	15	10	7.5
	Number of turns	10	1	1
	Maximum at 1550nm	0.03dB	0.1dB	0.5dB
	Maximum at 1625nm	0.1dB	0.2dB	1.0dB
Cable attributes				
Attenuation coefficient	Maximum at 1310nm to 1625nm	0.4dB/km		
	Maximum at 1550nm	0.3dB/km		

TABLE 5.10: ITU-T G.657, specifications for G.657 category B3 bending loss-insensitive single-mode optical fiber and cable

FIBER ATTRIBUTES, G.657 CATEGORY B3				
Mode field diameter	Wavelength	1310nm		
	Range of nominal values	8.6–9.5μm		
	Tolerance	±0.4μm		
Cladding diameter	Nominal	125.0μm		
	Tolerance	±0.7μm		
Core concentricity error	Maximum	0.5μm		
Cladding noncircularity	Maximum	1%		
Cable cut-off wavelength	Maximum	1260nm		
Macrobend loss	Radius (mm)	10	7.5	5
	Number of turns	1	1	1
	Maximum at 1550nm	0.03dB	0.08dB	0.15dB
	Maximum at 1625nm	0.1dB	0.25dB	0.45dB
Cable attributes				
Attenuation coefficient	Maximum at 1310nm	0.4dB/km		
	Maximum at 1550nm	0.3dB/km		

Multimode ITU and IEC Standards

As you have learned in this chapter, ITU-T G.651.1 and IEC 60793-2-10 describe the performance characteristics of multimode optical fiber and many of the optical performance specifications are covered in Table 5.4. This section of the chapter describes only the macrobending performance of these optical fibers in Tables 5.11 and 5.12.

TABLE 5.11: ITU-T G.651.1 multimode optical fiber macrobending loss

RADIUS (MM)	NUMBER OF TURNS	MAXIMUM ATTENUATION AT 850NM	MAXIMUM ATTENUATION AT 1300NM
15	2	1dB	1dB

TABLE 5.12: IEC 60793-2-10 types A1a and A1b multimode optical fiber macrobending loss

RADIUS (MM)	NUMBER OF TURNS	MAXIMUM ATTENUATION AT 850NM	MAXIMUM ATTENUATION AT 1300NM
37.5	100	0.5dB	0.5dB

Specialty Optical Fibers

Table 5.13 describes the typical attenuation and bandwidth characteristics for plastic fibers and HCS (hard-clad silica) and PCS (plastic-clad silica) fibers. These optical fibers are sometimes referred to as specialty optical fibers. The performance characteristics of these optical fibers are not defined in the TIA, ISO, IEC, or ITU standards covered in this text. The data contained in this table was obtained from manufacturers' data sheets and can vary from manufacturer to manufacturer.

TABLE 5.13: Characteristics of HCS/PCS and plastic fibers

OPTICAL FIBER CABLE TYPE	WAVELENGTH (NM)	MAXIMUM ATTENUATION	MAXIMUM INFORMATION TRANSMISSION CAPACITY FOR OVERFILLED LAUNCH (MHZ · KM)
200/230μm HCS/PCS	650	10dB/km	17
	850	8dB/km	20
1mm plastic fiber	650	0.19dB/m	1.5

The Bottom Line

Calculate the attenuation in dB for a length of optical fiber. The attenuation values for a length of optical fiber cable can be calculated using the attenuation coefficient for a specific type of optical fiber cable. This information can be found in the manufacturer's data sheet or in a standard.

Master It Calculate the maximum attenuation for a 22.5km ITU-T G.657.B.3 optical fiber cable at 1550nm.

Calculate the usable bandwidth for a length of optical fiber. The product of bandwidth and length (MHz · km) expresses the information carrying capacity of a multimode optical fiber. Bandwidth is measured in megahertz (MHz) and the length is measured in kilometers. The MHz · km figure expresses how much bandwidth the fiber can carry per kilometer of its length. The fiber's designation must always be greater than or equal to the product of the bandwidth and the length of the optical fiber.

Master It Refer to Table 5.4 and calculate the usable bandwidth for 425 meters of 50/125μm OM3 multimode optical fiber using an OFL at a wavelength of 850nm.

Calculate the total macrobending loss in a single-mode fiber-optic cable. The macrobending loss is defined for a number of turns of optical fiber around a mandrel at a specified radius and wavelength.

Master It A single-mode ITU-T G.657.A2 bend-insensitive fiber-optic cable is installed in a home with two small 15mm 90° radius bends. The wavelength of the light source is 1625nm.

Calculate the total macrobending loss in a multimode fiber-optic cable. The macrobending loss is defined for a number of turns of optical fiber around a mandrel at a specified radius and wavelength.

Master It A generic multimode bend-insensitive fiber-optic cable is installed in a home with three small 7.5mm radius, 120° bends. The wavelength of the light source is 850nm.

Calculate the acceptance angle for an optical fiber. The acceptance angle defines the acceptance cone. Light entering the core of the optical fiber at an angle greater than the acceptance angle will not propagate the length of the optical fiber. For light to propagate the length of the optical fiber, it must be at an angle that does not exceed the acceptance angle.

Master It Determine the acceptance angle of an optical fiber with a core refractive index of 1.48 and a cladding refractive index of 1.45.

Determine the latest published revision of a standard using the Internet. It is important to be aware of standards that have been superseded.

Master It Using the Internet determine the latest revision and publication date for ANSI/TIA-568-C.3.

Chapter 6

Safety

Whether you work as a technician or an installer, your work with fiber optics can expose you to several workplace hazards that are defined and regulated by the Occupational Safety and Health Administration (OSHA). OSHA has published numerous regulations on workplace hazards ranging from laser light sources to ladders, and employers are required to be familiar with these regulations and follow them to keep the workplace safe.

You are responsible for your own safety as well as for the safety of your co-workers. It is up to you to know and incorporate safe work practices in everything you do.

This chapter describes the types of hazards that you will encounter as you work with fiber optics. Some of the hazards are unique to fiber optics work, but others are more common. This chapter discusses the dangers that these hazards create, and explains different methods of working safely around them.

This chapter provides a general overview on safety; it is not the catch-all for safety. Each company and individual is responsible for maintaining a safe environment. The availability of safety information is also the responsibility of the company and individual.

In this chapter, you will learn to:

♦ Classify a light source based on optical output power and wavelength

♦ Identify the symptoms of exposure to solvents

♦ Calculate the proper distance the base of a ladder should be from a wall

Basic Safety

Whenever you work in a hazardous environment, such as a construction site, lab, or production facility, you must always be aware of the potential dangers you face. Your workplace is required by law to provide you with equipment and facilities that meet standards set by OSHA, but you will have to be an active participant in your own safety.

You can use three lines of defense to help you get through the day safely: engineering controls, personal protective equipment (PPE), and good work habits.

Engineering Controls

Engineering controls are the mechanisms that your facility has established to make a hazardous situation safer. They may include ventilation in the form of exhaust fans or hoods, special cabinets for storing flammables, or workstations that minimize the hazards of specialized work, such as cutting optical fibers.

Do not ignore or try to get around the engineering controls set up in your workplace. By doing so, you only endanger yourself and others. Make sure that fans and ventilation systems are working properly. If they are not, report any problems to your facility supervisor immediately.

Do not try to alter or modify the engineering controls unless the modifications have been approved by your safety officer. Improper modifications could reduce the effectiveness of the controls and create a greater hazard.

Personal Protective Equipment (PPE)

PPE consists of anything that you would wear to protect yourself from materials or situations. It can include protective gloves and eyewear for cutting and grinding operations, respirators for working with chemicals that put out harmful vapors, and specialized goggles for working with lasers.

Your PPE protects you not only from short-term accidents, such as cuts or flying shards of glass, but also from damage that can build up over time. Such damage may include dust from construction operations such as drywall sanding that can build up over time in your lungs and cause diseases such as silicosis, or exposure to chemicals such as solvents that can have harmful long-term or chronic effects as well as harmful short-term effects.

Whenever you use PPE, inspect it carefully to ensure that it is in good condition. Look for cuts, tears, or other signs of damage in protective outerwear such as gloves or aprons. Inspect eyewear for cracks or pitting. If you use goggles designed to protect you from certain light wavelengths, make sure they are clean and free from scratches that could reduce their effectiveness.

If you wear contact lenses, be sure your facility allows them in your work area. If you work with adhesives or solvents, you should avoid wearing them anyway, because splashed chemicals could be trapped by the lens and be more difficult to wash out. You may be able to obtain safety goggles with prescription lenses if you have to use them on a regular basis.

If you work with a respirator, test it every time you put it on. Cover the canisters with your hands and try to inhale, then cover the exhaust port and try to exhale. The respirator should form a good seal with your face, and no air should leak through the canisters or exhaust.

Some construction areas may require hardhats. Do not take these warnings lightly. Even a small hand tool dropped from a few feet can injure or kill you if you are not protected. Hardhats are designed to absorb the shock from falling objects so your head doesn't have to. To make sure the hardhat fits properly, adjust the inner band so that it fits snugly against your forehead and does not allow the hat to move around on your head. Make sure there is enough room between the suspension and the hardhat shell to absorb any blows.

Good Work Habits

Good work habits are in some ways the simplest and most effective means to working safely. Good work habits can help you prevent accidents and spot potential problems in time to correct them.

Here are some general rules for working safely:

♦ Keep a clean workspace. Clean up at the end of your work day and store tools properly. A "rat's nest" can hide problems and add to confusion.

♦ Observe your surroundings. Look up from what you are doing once in a while to make sure everything around you is the way it should be.

♦ Use tools for the job they were designed to perform. Misuse of tools is one of the most common causes of accidents in the workplace.

♦ Do not eat or drink in the work area. In addition to accidentally drinking from the wrong bottle, you could accidentally ingest glass fiber or other dangerous materials that might get mixed in with your food.

♦ Report problems or injuries immediately. Let your facility supervisors know about hazards so they can correct them as soon as possible.

♦ Know how to reach emergency personnel. Have emergency numbers posted by the nearest telephone so you don't waste time fumbling through a directory in an emergency.

♦ Put your emergency contact information in your cell phone.

Let's look at some of the hazards directly related to your work with fiber optics.

Light Sources

Even though most lasers and LEDs used in fiber optics operate in the near-infrared and infrared (IR) wavelengths and are invisible to the eye, they can still cause damage if they are delivered at high intensity or if the exposure is long enough. The possibility of damage is even greater because you cannot see the beam, and in many cases, the damage is done before you know it.

A laser can be especially dangerous because it can concentrate a great amount of power into a small beam of coherent light. Many lasers used in fiber optics operate below dangerous levels, but some, such as those used for transmission over long distances, put out enough power to cause damage in a very short time.

Injuries from the infrared wavelengths output by lasers used in fiber-optic systems and test equipment include but are not limited to cataracts and corneal and retinal burns.

Federal Regulations and International Standards

Federal regulations and international standards have been created to prevent injuries from laser radiation.

In Chapter 4, "Optical Fiber Construction and Theory," TIA and ITU were introduced as organizations that publish standards on the performance of optical fibers used in the telecommunications industry. There is a difference between a standard and a federal regulation. A federal regulation is a law. If there are federal regulations for a specific type of product sold in the United States, such as a laser, that product must meet federal regulations. However, it does not need to meet any standards. A standard only provides guidance; it is not a law.

FEDERAL REGULATIONS

The U.S. Food and Drug Administration (FDA) Center for Devices and Radiological Health (CDRH) is responsible for eliminating unnecessary human exposure to man-made radiation from medical, occupational, and consumer products. The FDA has had performance standards for light-emitting products since 1976. These performance standards are described in the Code of Federal Regulations 21 CFR, Subchapter J.

Fiber-optic equipment sold in the United States that contains a laser must meet 21 CFR, Subchapter J. However, it is not required by 21 CFR, Subchapter J to meet any standard. Many fiber-optic products that contain lasers only meet 21 CFR, Subchapter J.

21 CFR, Subchapter J encompasses all consumer products that contain a laser. In this chapter, only the areas of 21 CFR, Subchapter J typically required for fiber-optic communication and test equipment will be addressed; detailed information can be obtained from the FDA website:

`www.accessdata.fda.gov/scripts/cdrh/cfdocs/cfcfr/cfrsearch.cfm?FR=1040.10`

The laser classifications described in 21 CFR, Subchapter J are based on emission duration and emission limits for specific wavelengths. There are two wavelength windows that apply to fiber-optic communication and test equipment:

>400nm and ≤ 1400nm

>1400nm and ≤ 2500nm

CFR, Subchapter J divides laser products into classes and defines the labeling requirements based on the optical output power:

♦ Class I products do not require a label.

♦ Class IIa products require a label; however, there is no logotype for this label.

♦ Class II products require a "Caution" label affixed bearing the warning logotype "A."

♦ Class IIIa with an irradiance less than 2.5mW must have a "Caution" label affixed bearing the warning logotype "A."

♦ Class IIIa with an irradiance greater than 2.5mW, Class IIIb, and Class IV laser products must have a "Danger" label affixed bearing the warning logotype "B" similar to the labels, as shown in Figure 6.1.

FIGURE 6.1
Warning placards
for lasers

FDA Warning Logotypes

The FDA warning logotypes are available for download from the FDA website.

The classification tables for accessible emission limits for laser radiation found in 21 CFR, Subchapter J are very detailed and cover all consumer laser products. The following is a list of the hazards associated with each class. Many of the products used in fiber-optic communication systems and test equipment are Class I; however, some equipment may be Class IV.

Class I These levels of laser radiation are not considered to be hazardous.

Class IIa These levels of laser radiation are not considered to be hazardous if viewed for any period of time less than or equal to 1,000 seconds but are considered to be a chronic viewing hazard for any period of time greater than 1,000 seconds. A label is required bearing the wording "Class IIa Laser Product—Avoid Long-Term Viewing of Direct Laser Radiation."

Class II These levels of laser radiation are considered to be a chronic viewing hazard. A caution label is required that states "DO NOT STARE INTO THE BEAM" and "CLASS II LASER PRODUCT."

Class IIIa These levels of laser radiation are considered to be, depending on the irradiance, either an acute intrabeam viewing hazard or chronic viewing hazard, and an acute viewing hazard if viewed directly with optical instruments. A caution label is required with an irradiance less than or equal to 2.5mW. The caution label should state "LASER RADIATION—DO NOT STARE INTO THE BEAM OR VIEW DIRECTLY WITH OPTICAL INSTRUMENTS" and "CLASS IIIa LASER PRODUCT."

A danger label is required with an irradiance greater than 2.5mW. The danger label should state "AVOID DIRECT EYE EXPOSURE" and "CLASS IIIa LASER PRODUCT."

Class IIIb These levels of laser radiation are considered to be an acute hazard to the skin and eyes from direct radiation. A danger label is required that states "LASER RADIATION—AVOID DIRECT EXPOSURE TO BEAM" and "CLASS IIIb LASER PRODUCT."

Class IV These levels of laser radiation are considered to be an acute hazard to the skin and eyes from direct and scattered radiation. A danger label is required that states "LASER RADIATION—AVOID EYE OR SKIN EXPOSURE TO DIRECT OR SCATTERED RADIATION" and "CLASS IV LASER PRODUCT."

Intrabeam Viewing Conditions

Intrabeam viewing conditions exist when the eye is exposed to the direct or specularly reflected laser beam. A specular reflection as defined in IEC 60825-1 is a reflection from a surface that maintains angular correlation between incident and reflected beams of radiation, as with reflections from a mirror. To avoid intrabeam viewing conditions, do not directly view the beam or view reflections from the surfaces hit by the beam.

STANDARDS

Two organizations have published recognized laser safety standards: the International Electrotechnical Commission (IEC) and the American National Standards Institute (ANSI). ANSI Z136.2 addresses the safe use of optical fiber communication systems using laser diode and LED sources. IEC 60825-2 addresses the safety of optical fiber communication systems. The objective of both of these standards is to protect people from optical radiation released by an optical communication system.

These organizations have also published other laser standards. ANSI Z136.1-200 addresses the safe use of lasers in general and is not specific to optical fiber communication systems. IEC 60825-1 addresses the safety of laser products in general, not specific to optical fiber communication systems.

Test equipment or an optical fiber communication system (OFCS) containing a laser sold in the United States should have an FDA classification and the manufacturer may advertise compliance to an ANSI or IEC standard. Any OFCS or test equipment sold outside the United States may or may not have an FDA classification and may or may not comply with an ANSI or IEC standard.

OSHA Standards for OFCS Service Groups

OSHA also classifies an OFCS. Section III, Chapter 6 of the OSHA Technical Manual describes the Optical Fiber Service Group (SG) Designations. It is based on ANSI Z136.2. These SG designations relate to the potential for ocular hazards to occur only when the OFCS is being serviced. "Being serviced" can involve something as simple as removing a connector from a receptacle. The SGs outlined in the OSHA Technical Manual are described next, and you can get detailed information from the OSHA:

```
https://www.osha.gov/dts/osta/otm/otm_iii/otm_iii_6.html
```

SG1   An OFCS in this SG has a total output power that is less than the accessible emission limit (AEL) for Class I (400nW), and there is no risk of exceeding the maximum permissible irradiance (MPI) when viewing the end of a fiber with a microscope, an eye loupe, or the unaided eye.

SG2   An OFCS is in this SG only if wavelengths between 400nm and 700nm are emitted. Such lasers are potentially hazardous if viewed for more than 0.25 second.

SG3A   An OFCS in this SG is not hazardous when viewed with the unaided eye and is hazardous only when viewed with a microscope or eye loupe.

SG3B   An OFCS in this SG does not meet any of the previous criteria.

OSHA Laser Hazard Classes

If the total power for an OFCS is at or above 0.5W, it does not meet the criteria for an optical fiber SG designation and should be treated as a standard laser system. The OSHA laser hazard classes are based on ANSI Z136.1 and summarized next:

Class 1   These lasers cannot emit laser radiation at known hazard levels, typically 400nW for a continuous wave at visible wavelengths. Users of these lasers are generally exempt from radiation hazard controls during operation and maintenance, but not necessarily during service.

Class II   These lasers only emit visible wavelengths above Class I but below 1mW. These lasers can cause damage if you look directly at them for more than 0.25 second, but they have limited requirements for protection. It is assumed that the normal aversion to pain will cause anyone looking at the bright light to turn away or close their eyes before any damage can take place.

Class IIIA   These intermediate power lasers have a continuous wave output from 1.0 to 5.0mW. Directly viewing the beam can damage the eye.

Class IIIB   These moderate power lasers have a continuous wave output in the 5–500mW range (up to +27dBm) and can cause damage to the eye, even if the beam is reflected.

Class IV   These high-power lasers have a continuous wave output above 500mW, or +27dBm, and they can burn almost any living tissue they contact under any viewing condition. They also pose a fire hazard.

IEC Laser Hazard Classifications

IEC standard 60825-1 addresses the safety of laser products and is not specific to an OFCS. However, components in the OFCS may be classified by this standard. Classifications that appear in this standard are summarized here:

Class 1   These levels of laser radiation are safe under normal operating conditions, including the use of optical instruments such as a microscope or eye loupe.

Class 1M   These levels of laser radiation are safe under normal operating conditions; however, they may be hazardous when viewed with optical instruments such as a microscope or eye loupe.

Class 2   These levels of laser radiation at wavelengths between 400nm and 700nm are not hazardous if viewed for less than 0.25 second. It is assumed that the normal aversion to pain will cause anyone looking at the bright light to turn away or close his or her eyes (blink).

Class 2M   These levels of laser radiation at wavelengths between 400nm and 700nm are not hazardous if viewed for less than 0.25 second. It is assumed that the normal aversion to pain will cause anyone looking at the bright light to turn away or close his or her eyes (blink). However, they may be hazardous when viewed with optical instruments such as a microscope or eye loupe.

Class 3R   These levels of laser radiation are potentially hazardous when viewed directly.

Class 3B   These levels of laser radiation are normally hazardous when viewed directly.

Class 4   These levels of laser radiation are considered to be an acute hazard to the skin and eyes from direct beam viewing or diffuse reflections.

IEC Laser Hazard Levels for OFCS

IEC 60825-2 addresses the safety of an OFCS. This standard describes hazard levels associated with an OFCS. It also defines labeling requirements and the location type for each hazard level. There are three location types:

Unrestricted   This is a location where access to the protective housing is unrestricted. Examples of this type of access include domestic premises and any public areas.

Restricted This is a location where access to the protective housing is restricted; it is not open to the public. Examples of this type of access include secured areas within industrial and business/commercial premises not open to the public; general areas within switching centers; delimited areas not open to the public on trains, ships, or other vehicles; overhead fiber-optic cables and cable drops to a building; and optical test sets.

Controlled This is a location where access to the protective housing is controlled and accessible only to authorized personnel who have been trained in laser safety and the system they are servicing. Examples of this type of access include cable ducts, street cabinets, manholes, and dedicated and delimited areas of network operator distribution centers.

The hazard level labeling requirements for each location type that appear in this standard are summarized here:

Hazard level 1 There are no labeling requirements for unrestricted, restricted, or controlled location types.

Hazard level 1M Labels are required for unrestricted, restricted, or controlled location types.

Hazard level 2 and 2M Labels are required for unrestricted, restricted, or controlled location types.

Hazard level 3R This hazard level is not permitted in an unrestricted location type. Labels are required for restricted or controlled location types.

Hazard level 3B These hazard levels are not permitted in an unrestricted or restricted location type. A label is required for a controlled location type.

Hazard level 4 These hazard levels are not permitted in unrestricted, restricted, or controlled location types.

Laser Safety

Because most lasers used in fiber-optic systems emit IR radiation, you cannot see the beam, no matter how powerful it is. As a result, you will not be able to tell if the system is powered, especially if you are working on a piece of fiber far from the transmitter.

You should treat an optical fiber coupled to a laser with the same caution that you would treat electrical cables connected to a breaker panel. Do not assume that the system is turned off, especially if you have to use a microscope to look at the fiber end. Do not take anyone else's word that the system is off or that the fiber is uncoupled from the laser. You will have to endure the results for the rest of your life.

Unless you can be sure that the fiber is not coupled to a laser, do not look at the fiber end without some kind of protection. Use filters and protective eyewear that block out the specific wavelengths used by the lasers. Use a video microscope when available instead of a handheld optical microscope to view the endface of a connector.

Hazardous laser areas should be clearly identified with warning placards and signs stating that access is limited to personnel with proper safety gear and authorized access, as shown in Figure 6.1. Do not ignore these signs or think that they don't apply to you. They are there for your protection and for the safety of those working inside the restricted areas.

If the lab has a separate door and a hazardous laser is operating inside, the door should have interlocks to kill the laser before the door is opened. Some of these doors may have separate combination locks to prevent unauthorized entry.

Handling Fiber

In spite of optical fiber's flexibility, remember that it is glass. In short pieces, it is stiff enough to pierce your skin or eyes and cause discomfort, pain, or damage. If the pieces become airborne, you may even accidentally inhale or swallow them, risking damage to your throat or respiratory system. You can protect yourself with correct procedures and the right PPE.

When you cut, cleave, scribe, or accidentally break optical fibers, the ends can get lost easily, either by becoming airborne or by rolling along a surface. These ends can have extremely sharp edges, and if they are mishandled, they can lodge in your skin or eyes. If they are not removed immediately, the pieces can work themselves in further, increasing the risk of damage or infection. As shown in Figure 6.2, always work over a nonreflective black surface, which makes it easier to keep track of cut fiber ends. Also, keep a separate labeled container with a lid nearby for cut fiber ends.

FIGURE 6.2
A nonreflective black surface, a fiber waste container, and safety glasses can help prevent injury from fiber ends.

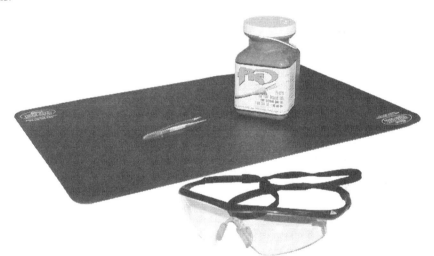

To prevent injury to your hands, always handle cut pieces of fiber with tweezers. To prevent eye injury, always wear proper eye protection. It takes only one piece of glass to damage your vision permanently. If you have been handling fiber, do not rub your eyes or put your hands near them until you have washed your hands. If you do get a piece of fiber in your skin, remove it immediately with a pair of tweezers or seek medical attention.

You may not always have the convenience of a laboratory or workshop environment for your fiber work. Work areas for splicing, building connectors, or other tasks may include basements, crawlspaces, underground vaults, an attic, or the back of a van. Don't take shortcuts just because you don't have the luxury of a full workshop at your disposal. Make sure you have an appropriate work surface and the proper tools and safety equipment before you start working.

Additional details on splicing and connectorization safety for fiber-optic cables and connectors can be found in Chapter 8, "Splicing," and Chapter 9, "Connectors."

REMOVING FIBER FROM YOUR DIET

Whenever you are cleaving fiber, you are going to wind up with ends. No matter how short or long they are, rest assured that they will be hard to spot when it's time to clean up. Your first line of defense against these hard-to-spot fibers is immediate response. In other words, as soon as you cleave, pick up and dispose of the end.

If this discipline doesn't work for you or if you find yourself with stray ends anyway, you'll still have to round them up and dispose of them properly. There are some low-tech and high-tech strategies to help you, as long as you can prepare them beforehand.

First, to prevent fiber ends from traveling beyond your work area, create a barrier around your work surface. This will keep bouncers and fliers confined to the work surface, where you can find them more easily.

If you want to make sure you've collected all of your strays or if you know that you created more ends than you have collected, use a bright light source, such as an LED flashlight. To find the fibers, place the light at the level of the work surface and radiate along the surface. The fiber ends will pick up the light and Fresnel reflection will cause them to shine brightly.

As a last, low-tech measure, double up a piece of adhesive tape and pat it all over the work surface. This will pick up any minute fiber particles that you could not spot or retrieve using other methods.

Chemicals

In your work with fiber optics, you will use several types of chemicals, including 99 percent isopropyl alcohol for cleaning components, solvents for removing adhesives and other materials, and anaerobic epoxy for making connectors.

Each of these chemicals poses a number of hazards and should be handled carefully. Each chemical that you use is accompanied by a Material Safety Data Sheet (MSDS), which provides important information on the chemical's properties, characteristics such as appearance and odor, and common uses. The MSDS also gives you information on specific hazards posed by each chemical and ways to protect yourself through specific handling procedures, protective clothing and equipment, and engineering controls such as ventilation. Finally, the MSDS describes emergency procedures, including first aid for exposure to the chemical, methods for fighting fires in the case of flammable chemicals, and cleanup procedures for spills.

Even if you think you're familiar with the chemicals you handle, take time to read the MSDS. The information you gain could help prevent an accident or save valuable time in an emergency.

Let's look at some of the most important hazards associated with the chemicals you'll be handling.

Isopropyl Alcohol

Even though alcohol is commonly used in the home and the lab, its hazards should not be ignored or taken lightly. Alcohol vapors escape into the air easily, and they can cause damage to your liver and kidneys if they are inhaled. The vapors are also highly flammable and can ignite if exposed to a spark or flame in high enough concentrations.

Alcohol can also cause irritation to your eyes, skin, and mucous membranes (nose and mouth) if it comes in direct contact with them.

Always use alcohol in a proper dispenser, shown in Figure 6.3. Store and transfer it carefully to avoid spills and excess evaporation. If you do spill any alcohol, clean it up with a dry cloth and dispose of the cloth in a container designed for flammable waste materials. As with all flammables, do not use alcohol in areas where sparks, open flames, or other heat sources will be present.

FIGURE 6.3
An alcohol dispenser helps reduce the risk of spills and fire.

It's important to remember that alcohol flames are almost invisible, and spills could lead to a mad scramble as you try to dodge flames you cannot see. Alcohol fires can be extinguished with water or a Class A fire extinguisher.

Solvents

Many solvents have similar properties that require that you handle them with great care. Like alcohol, solvents are very volatile and sometimes flammable. Their primary danger, however, is their hazard to your health.

One of the health hazards posed by solvents comes from the fact that they can cause excessive drying in your skin and mucous membranes, and the resultant cracking of the surface layers can leave you open to infection. The hazard is more serious if you inhale the solvent vapors, as they can damage your lungs and respiratory system.

Solvents can also cause organ damage if inhaled or ingested. The molecules that make up most solvents can take the place of oxygen in the bloodstream and find their way to the brain and other organs. As the organs are starved of oxygen, they can become permanently damaged. One of the first signs that this kind of damage is taking place is dizziness or a reaction similar to intoxication. If you feel these symptoms, get to fresh air immediately.

Solvents may come in glass or plastic containers. Make sure that the container you are using is properly marked for the solvent it contains. Do not leave solvent in an unmarked container.

If you are carrying solvent in a glass bottle, use a rubber cradle to carry the bottle. The cradle protects the bottle from breaking if it falls.

Never leave a solvent container open. Keep the top off just long enough to transfer the amount necessary for the job, and replace it firmly to prevent the vapors from escaping.

Anaerobic Epoxy

The two-part epoxy used for making connectors is typically used in small quantities and does not present any immediate health hazards. If you are working in an enclosed space, however, such as the back of a van or an access area, vapors from the adhesive portion can irritate your eyes, nose, and throat. If you feel any of these symptoms, get to an open space immediately. The adhesive can also irritate your skin or eyes on contact. If you get any of the adhesive on you, wash the area immediately. If any material splashes in your eyes, flush them for 15 minutes at an eyewash station or sink.

Use caution when working with the primer portion of these adhesives. With a flash point of –18° C, it is highly flammable. Do not use anaerobic epoxy where there is an open spark or flame or where heating components or elements are being used.

As with solvents, do not leave epoxy containers open any longer than necessary to dispense the amount you are using.

Site Safety

Many of the locations for fiber-optic components may be in areas that require special safety precautions. These may include construction sites, enclosed areas, locations near high-voltage power lines, or areas requiring access by ladder or scaffold.

Always follow the on-site safety requirements and observe all warning signs. Here are some general safety rules to help you.

Electrical

When fiber-optic systems run through the same area as electrical wiring or cabinets, use extreme caution with tools and ladders. One wrong move can send enough voltage through your body to kill you. Remember that electrical fields can exist beyond a cable's insulation if high voltages are present, so use wooden ladders to reduce the possibility of exposure to induced voltages. Use care with cutters and other tools to avoid accidental contact with electrical wires, and report any hazardous conditions that may exist.

Remember that high voltage causes most of its damage by making muscles seize up, including the heart and lungs. The greatest chance of damage comes when voltage passes through your heart to get the ground, such as when you touch a wire with your hand and your opposite leg provides the path to ground, or when the current passes from one hand to the other.

If you accidentally grab a live wire, the voltage may keep your hand clenched, making it nearly impossible to release the wire. If you see someone who is in this situation, do not try to pull them away with your hands. You may be caught up in the circuit as well. Instead, use a nonconducting stick, such as a wooden or fiberglass broom handle, to knock the victim away from the voltage source.

If the victim is not breathing, artificial respiration may be necessary to get the heart and lungs operating again.

Ladders

You may often find that you need a ladder to reach a work area. When choosing a ladder, make sure you select one that matches your requirements. Self-supporting ladders, such as stepladders, should be used only if the work area is not near a vertical support such as a wall and the floor beneath the work area is even and firm.

Non-self-supporting ladders, such as extension ladders, are useful when there is a firm vertical support near the work area and there is a stable, nonslip surface on which to rest the ladder. When setting up a non-self-supporting ladder, it is important to place it at an angle of 75½° to support your weight and be stable. A good rule for finding the proper angle is to divide the working height of the ladder (the length of the ladder from the feet to the top support) by 4, and place the feet of the ladder that distance from the wall. For example, if the ladder is 12 feet tall, the bottom should be 3 feet from the wall.

Be sure that the ladder you select can carry your weight along with the weight of any tools and equipment you are carrying. Read the labels and warnings on the ladder you select to make sure that it is the right one for the work you are performing.

If you are working near live electrical systems, be sure to use a nonconducting ladder of wood or fiberglass. If you are working near high heat, select an aluminum ladder to avoid scorching or melting your only means of support.

In the work area, place the ladder so that you can work comfortably without having to reach too high or too far to the side. Overreaching can cause you to lose your balance and fall, or place too much weight above or to the side of the ladder, causing the entire ladder to come down. If your work area extends beyond a comfortable reach, climb down the ladder and move it. Do not try to "walk" the ladder to a new work area.

Inspect ladders before using them. Make sure the rungs and rails are in good shape and are not split, broken, or bent. Make sure all fittings and fasteners are secure and that all locking mechanisms are working properly.

When carrying a ladder to the worksite, reach through the rail and balance it on your shoulder. Be aware of obstacles and corners as you carry it, and make sure others are aware that you have an awkward load, especially if you are walking through hallways or other limited visibility areas.

Trenches

If you are working on a fiber system in a trench, be sure the trench is properly dug and shored before entering it. Never work in a trench without someone else around, in case of a collapse.

If you have never worked in a trench before, learn the proper way to enter and exit a trench. Always use a ladder. Never jump into a trench or try to climb down the sides. You could trigger a collapse.

If you witness a collapse and others are trapped but not in immediate danger, do not try to dig them out yourself. You risk making the problem worse. Get help immediately. Special training is required to recover victims from a trench collapse, so leave it to the experts.

Emergencies

It takes only one slip-up to create an emergency. It could come from a moment of carelessness, an attempt at taking a shortcut, or ignorance of the proper procedures. Whatever the cause, the first response is always the same. Remain calm. Panic can cause even more damage and complicate matters beyond repair.

The best way to handle emergencies is to accept the fact that they will occur and be ready for them. Make sure you know what can go wrong with the materials and chemicals you handle and what you can do to minimize the damage.

Injury

Injuries can be caused by misuse of tools, fibers penetrating your skin or eyes, burns, falls, or any number of other mishaps. Make sure that you and your co-workers know the location of first-aid kits in your work area. Also, make sure you know how to reach emergency personnel. If you are on a new job site, make it a priority to familiarize yourself with emergency procedures and contact information.

Chemical Exposure

Accidental chemical exposure can result in anything from temporary discomfort to permanent injury or death. The first few seconds of an emergency involving chemical exposure can be critical in the victim's recovery.

If the chemical is splashed on the skin or in the eyes, flush the affected area with clean water immediately, using a shower or eyewash station if available. Continue flushing the area for at least 15 minutes. This washes the chemical away, but also dilutes its effects if it has been absorbed by the skin or eyes.

In case of inhalation, move the victim to fresh air immediately and call for medical attention. If a chemical has been accidentally swallowed, induce vomiting unless the chemical is corrosive and could damage the esophagus and throat as it comes back up. Use a neutralizing liquid such as milk to dilute corrosive chemicals if they have been swallowed, and seek medical attention immediately.

Fire

If a fire breaks out in your work area, it may be small enough for you to handle alone. If it is small, you can smother it with a damp cloth. If it is larger, but contained in a trash can or other enclosure, use the appropriate fire extinguisher for the material that is burning.

To use a fire extinguisher properly, remember the acronym PASS as you use the following procedure:

Pull   Pull the pin from the fire extinguisher trigger.

Aim   Aim the extinguisher at the base of the fire.

Squeeze   Squeeze the handle firmly to activate the extinguisher.

Sweep   Sweep the extinguisher discharge at the base of the fire until the flames are out.

Do not give the fire a chance to trap you. If you think that the extinguisher will not put out the fire completely and there is a risk that your exit will be cut off, leave immediately and call for help. You can do more good with a phone call than you can in a failed attempt at being a hero.

The Bottom Line

Classify a light source based on optical output power and wavelength. The OSHA Technical Manual describes the Optical Fiber Service Group (SG) Designations based on ANSI Z136.2. These SG designations relate to the potential for ocular hazards to occur only when the OFCS is being serviced. OSHA also classifies standard laser systems.

>**Master It** An OFCS does not fall into any of the SG designations defined in the OSHA Technical Manual because during servicing it is possible to be exposed to laser emissions greater than 750mW. What classification would be assigned to this OFCS based on the OSHA Technical Manual?

Identify the symptoms of exposure to solvents. Solvents can cause organ damage if inhaled or ingested. The molecules that make up most solvents can take the place of oxygen in the bloodstream and find their way to the brain and other organs. As the organs are starved of oxygen, they can become permanently damaged.

>**Master It** Your co-worker spilled a bottle of liquid several minutes ago and now appears impaired. What has your co-worker potentially been exposed to?

Calculate the proper distance the base of a ladder should be from a wall. Remember non-self-supporting ladders, such as extension ladders, require a firm vertical support near the work area and a stable, nonslip surface on which to rest the ladder. When setting up a non-self-supporting ladder, it is important to place it at an angle of 75½° to support your weight and be stable.

>**Master It** A non-self-supporting ladder is 10' in length. How far should the feet of the ladder be from the wall?

Chapter 7

Fiber-Optic Cables

So far we have studied the principles and characteristics of individual optical fibers. While they are certainly adequate for the job of carrying signals from one place to another, they are not rugged enough to withstand the rigors of handling, transportation, and installation. In addition, some installations require multiple optical fibers for sending and receiving or for routing to a number of locations.

For an optical fiber to be suitable for everyday use, it must be incorporated into cables that provide standardized fiber groupings, protection from the environment, and suitable size for handling.

In this chapter, we will describe standard and harsh environment fiber-optic cables used in many types of installations. We will detail different types of fiber-optic cables and the uses for which they were designed. We will also describe some of the basic requirements for handling and installation.

In this chapter, you will learn to:

♦ Determine the cable type from the NEC markings

♦ Identify the fiber number from the color code

♦ Determine the fiber number from the cable markings

♦ Identify the optical fiber type from the color code

♦ Determine the optical fiber type from the cable markings

♦ Determine the cable length using sequential markings

Basic Cable

You may already be familiar with cables used for electrical wiring. These cables typically consist of two or more insulated wires bundled together and surrounded by a protective outer covering called the *jacket* or *sheath*. In addition to holding the wires in place, the jacket or sheath protects the wires from the environment and damage that may occur during handling or installation.

Some of the largest cables, used for telephone transmissions, can be several inches in diameter and contain hundreds of wires, as shown in Figure 7.1. These cables are very heavy, difficult to bend, and expensive.

Photo courtesy of Roger Crider

Fiber-optic cables like the one shown in Figure 7.2 do not need to be as large as the cable shown in Figure 7.1 because the bandwidth of an optical fiber over a long distance is many times greater than the bandwidth of a wire. Greater bandwidth means fewer optical fibers are required to carry the same information the larger cable is carrying. This results in a small, low-cost fiber-optic cable that is much easier to handle and terminate than the cable shown in Figure 7.1. Chapter 15, "Fiber-Optic System Design Considerations," compares the performance of fiber-optic and copper cables.

FIGURE 7.2
Typical fiber-optic cable

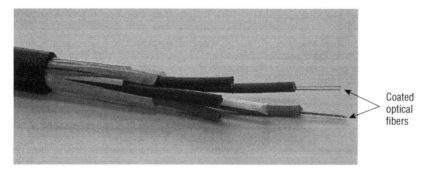

Coated optical fibers

Optical fibers are used in many different configurations and environments; manufacturers have created a wide variety of cable types to meet the needs of almost any application. The type of signal being carried and the number of optical fibers required are just two of the many considerations when selecting the right cable for an application. Other factors include the following:

♦ Tensile strength

♦ Temperature range

- ◆ Bend radius
- ◆ Flammability
- ◆ Buffer type
- ◆ Structure type
- ◆ Jacket type
- ◆ Weight
- ◆ Armor
- ◆ Crush resistance

The exact combination of these factors varies, depending on the application and operational environment. A cable installed inside an office building, for example, will be subject to less extreme temperatures than one installed outdoors or in an aircraft. Cabling installed in a manufacturing facility may be exposed to abrasive dusts, corrosive chemicals, or hotwork, such as welding, requiring special protection. Some cables may be buried underground, where they are exposed to burrowing or chewing animals, whereas others may be suspended between poles, subject to their own weight plus the weight of animals and birds who think the cables were put there for them.

Cable Components

Whether a cable contains a single optical fiber, several optical fibers, or hundreds of optical fibers, it has a basic structure in common with other cables. As shown in Figure 7.3, a typical fiber-optic cable consists of the optical fiber (made up of the core, cladding, and coating), a buffer, a strength member, and an outer protective jacket or sheath. Let's look at these components individually.

FIGURE 7.3
Fiber-optic cable
components

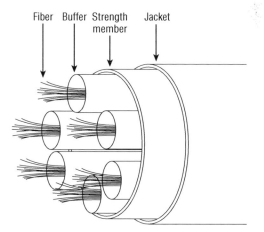

Buffer

As you learned in Chapter 4, "Optical Fiber Construction and Theory," all optical fiber has a coating. The coating is the optical fiber's first protective layer. The protective layer placed around the coating is called the buffer. There are two buffer types: *loose buffer* and *tight buffer*.

LOOSE BUFFER

Loose buffer is also referred to as loose tube buffer or loose buffer tube. A loose buffer consists of a buffer layer or tube that has an inner diameter much larger than the diameter of the coated optical fiber, as shown in Figure 7.4.

FIGURE 7.4
Loose-buffered cable has a buffer diameter greater than the fiber diameter.

Loose buffer tube   12 coated optical fibers   Loose buffer tube   12 coated optical fibers

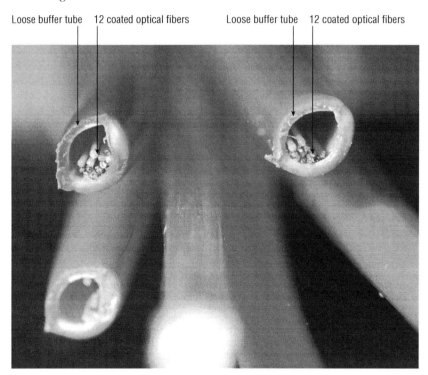

The loose buffer provides room for additional optical fiber when the cable is manufactured. This extra room also allows the optical fiber to move independently of the buffer and the rest of the cable components. This is an important factor when the cable is subjected to temperature extremes. A variety of materials are used to make up a fiber-optic cable. These materials may not expand or contract the same way as the optical fiber during temperature changes. The additional fiber in the loose buffer will allow the cable to expand (increase in length) or contract (decrease in length) without placing tensile forces on the optical fiber.

Loose-buffered cable may be single-fiber or multifiber (not to be confused with single-mode and multimode fiber), meaning that it may have one or many optical fibers running through each buffer tube. In addition, a cable may contain a number of loose buffers grouped together, as shown in Figure 7.5. In such cables, loose-buffered tubes are grouped around a central core that provides added tensile strength and a resistance to bending.

FIGURE 7.5
Loose-buffered
tubes in a cable

Central core

Loose-buffered cables are designed for indoor and outdoor applications. For outdoor applications, the buffer tubes may be filled with a gel. The gel displaces or blocks water and prevents it from penetrating or getting into the cable. A gel-filled loose-buffered cable is typically referred to as a *loose tube, gel-filled (LTGF) cable.*

Loose-buffered cables are ideal for direct burial and aerial installations. They are also very popular for indoor/outdoor applications.

TIGHT BUFFER

Tight-buffered cable is typically used in more controlled environments where the cable is not subjected to extreme temperatures or water. In short, tight-buffered cable is generally used for indoor applications rather than outdoor applications.

As shown in Figure 7.6, tight-buffered cable begins with a 250µm coated optical fiber. The buffer is typically 900µm in diameter and is applied directly to the outer coating layer of the optical fiber. In this way, it resembles a conventional insulated copper wire. The buffer may have additional strength members running around it for greater resistance to stretching.

One of the benefits of tight-buffered cable is that the buffered optical fiber can be run outside of the larger cable assembly for short distances, as shown in Figure 7.7. This makes it easier to attach connectors. The standard diameter of the buffer allows many different types of connectors to be applied.

Tight-buffered cables have some advantages over loose-buffered cables. Tight-buffered cables are generally smaller in diameter than comparable loose-buffered cables. The minimum bend radius of a tight-buffered cable is typically smaller than a comparable loose-buffered cable. These two advantages make tight-buffered cable the choice for indoor installations where the cable is routed through walls and other areas requiring tight bends.

FIGURE 7.6
Tight-buffered cable uses a buffer attached to the fiber coating.

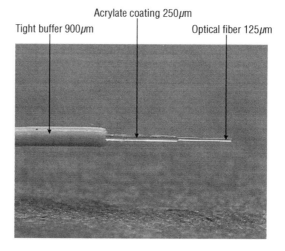

Tight buffer 900μm

Acrylate coating 250μm

Optical fiber 125μm

FIGURE 7.7
Tight-buffered optical fibers extending out of the cable assembly

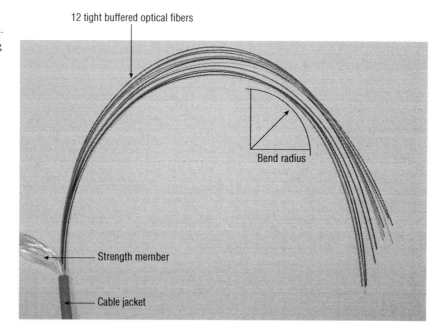

12 tight buffered optical fibers

Bend radius

Strength member

Cable jacket

Strength Members

You've already learned a bit about how strength members help increase a cable's tensile strength. The primary importance of these members is to ensure that no tensile stress is ever placed on the optical fiber. The installation of a fiber-optic cable may require a great deal of pulling, which places considerable tensile stress on the strength members. The strength members need to be able to support the installation tensile stresses and the operational tensile stresses.

Strength members may run through the center of a fiber-optic cable, or they may surround the buffers just underneath the jacket. Combinations of strength members may also be used depending on the application and the stress the cable is designed to endure. Some of the most common types of strength members are made of:

♦ Aramid yarns, usually Kevlar

♦ Fiberglass rods

♦ Steel

Aramid yarns, as shown in Figure 7.8, are useful when the entire cable must be flexible. These fine, yellowish or gold yarns are made of the same material used in high-performance sails, bulletproof vests, and fire protection gear. They have the advantages of being light, flexible, and quite strong. Aramid yarns may be used in cable subgroups if they will be bundled into larger cables.

FIGURE 7.8
Aramid (Kevlar)
yarns used as
strength members

Kevlar™ strength member

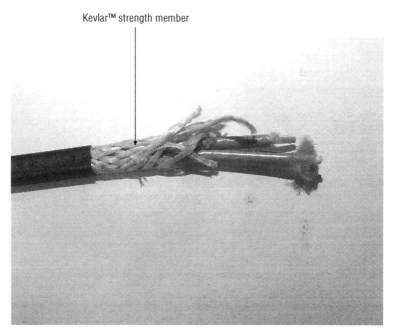

Larger cables and most loose-buffered cables typically have a central member of either fiberglass or steel added, as shown in Figure 7.9. This central member provides tensile strength and a resistance to bending. This resistance to bending prevents the cable from being over-bent and kinking the loose-buffered tubes.

Although the tensile strength of steel is greater than the tensile strength of fiberglass, many applications require a *dielectric*, or nonconductive cable. Fiberglass is nonconductive whereas steel is not. Steel running through a cable outdoors makes an excellent lightning rod.

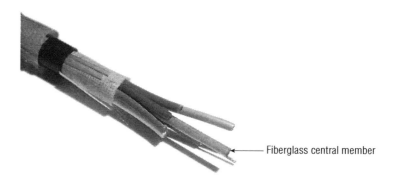

Fiberglass central member

Jacket

The jacket is the cable's outer protective layer. It protects the internal components from the outside world. The jacket may be subject to many factors such as sunlight, ice, animals, equipment accidents, and ham-handed installers. The jacket must also provide protection from abrasion, oil, corrosives, solvents, and other chemicals that could destroy the components in the cable.

Jacket materials vary depending on the application. Typical jacket materials include:

Polyvinyl chloride (PVC)   PVC is used primarily for indoor cable runs. It is fire retardant and flexible, and it's available in different grades to meet different conditions. PVC is water resistant; however, PVC does not stand up well to solvents. In addition, it loses much of its flexibility at low temperatures.

Polyethylene   This material is typically used outdoors. It offers excellent weather and sun resistance in addition to excellent water resistance and flexibility in low temperatures. Specially formulated polyethylene offers low-smoke, zero-halogen performance.

Polyvinyl difluoride (PVDF)   This material is chosen for its low-smoke and fire-retardant properties for use in cables that run through airways or *plenums* in a building. PVDF cables are not as flexible as other types, so their fire safety properties are their primary draw.

Polytetrafluoroethylene (PTFE)   This material is primarily used in aerospace fiber-optic cables. It has an extremely low coefficient of friction and can be used in environments where temperatures reach as high as 260° C.

Some cables contain multiple jackets and strength members, as shown in Figure 7.10. In such cables, the outermost jacket may be referred to as the sheath; the inner protective layers are still called jackets.

The *ripcord* is a piece of strong thread running through the cable just under the jacket or sheath as shown in Figure 7.11. When the ripcord is pulled, it splits the jacket easily to allow the fibers to be separated for connectorization or splicing. The ripcord reduces the risk incurred when cutters or knives are used to split the jacket. The ripcord will not penetrate the jackets of the individual simplex cables as a cutter or knife may when the blade depth is not accurately controlled.

FIGURE 7.10
The sheath is the outer layer of a cable with multiple jackets.

Sheath

FIGURE 7.11
The ripcord splitting the cable sheath

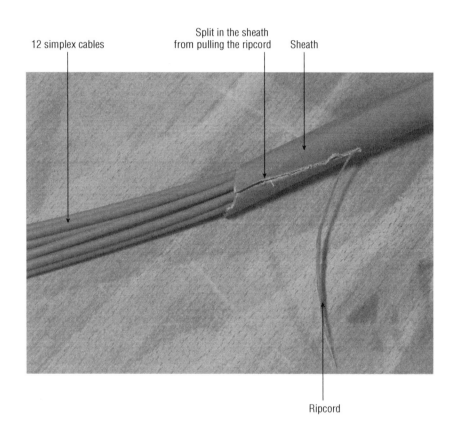

12 simplex cables

Split in the sheath from pulling the ripcord

Sheath

Ripcord

Cable Types

As uses for optical fiber have become more varied, manufacturers have begun producing cables to meet specific needs. Cable configurations vary based on the type of use, the location, and future expansion needs, and it is likely that more will be created as future applications emerge.

Bear in mind that different cable arrangements are variations on a theme. Different combinations of buffer type, strength members, and jackets can be used to create cables to meet the needs of a wide variety of industries and users.

Let's look at some of the commonly available optical fiber cables.

Cordage

The simplest types of cables are called *cordage* and are used in connections to equipment and patch panels. They are typically made into patch cords or jumpers. The major difference between cordage and cables is that cordage has only one optical fiber/buffer combination in a jacket, whereas cables may have multiple optical fibers inside a jacket or sheath.

The two common types of cordage are *simplex* and *duplex*.

SIMPLEX CORDAGE

Simplex cordage, shown in Figure 7.12, consists of a single optical fiber with a tight buffer, an aramid yarn strength member, and a jacket.

FIGURE 7.12
Simplex cordage

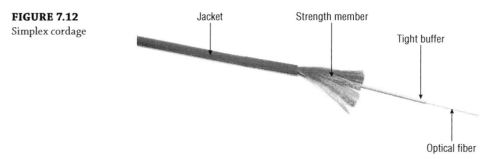

Simplex cordage gets its name from the fact that, because it is a single fiber, it is typically used for one-way, or simplex, transmission, although bidirectional communications are possible using a single fiber.

DUPLEX CORDAGE

Duplex cordage, also known as *zipcord*, is similar in appearance to household electrical cords, as you can see in Figure 7.13. Duplex cordage is a convenient way to combine two simplex cords to achieve duplex, or two-way, transmissions without individual cords getting tangled or switched around accidentally.

FIGURE 7.13
Duplex cordage

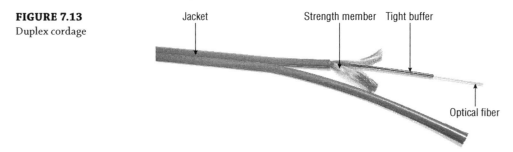

Distribution Cable

When it is necessary to run a large number of optical fibers through a building, distribution cable is often used. Distribution cable consists of multiple tight-buffered fibers bundled in a jacket with a strength member. These cables, like the one shown in Figure 7.14, may also feature a dielectric central member to increase tensile strength, resist bending, and prevent the cable from being kinked during installation.

FIGURE 7.14
Distribution cable with dielectric central member

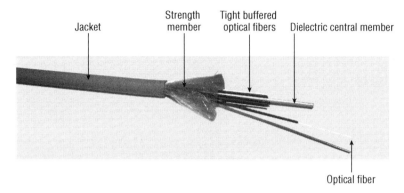

Jacket Strength member Tight buffered optical fibers Dielectric central member

Optical fiber

Distribution cables are ideal for inter-building routing. Depending on the jacket type they may be routed through plenum areas or riser shafts to telecommunications rooms, wiring closets, and workstations. The tight-buffered optical fibers are not meant to be handled much beyond the initial installation, because they do not have a strength member and jacket.

Distribution cables may carry up to 144 individual tight-buffered optical fibers, many of which may not be used immediately but allow for future expansion.

Breakout Cable

Breakout cables are used to carry optical fibers that will have direct termination to the equipment, rather than being connected to a patch panel.

Breakout cables consist of two or more simplex cables bundled with a strength member and/or central member covered with an outer sheath, as shown in Figure 7.15. These cables are ideal for routing in exposed trays or any application requiring an extra rugged cable that can be directly connected to the equipment.

FIGURE 7.15
Breakout cable containing simplex cables bundled with a central strength member

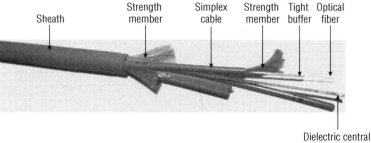

Sheath Strength member Simplex cable Strength member Tight buffer Optical fiber

Dielectric central member

Armored Cable

Armored cable can be used for indoor applications and outdoor applications. An armored cable typically has two jackets. The inner jacket is surrounded by the armor and the outer jacket or sheath surrounds the armor.

An armored cable used for outdoor applications, shown in Figure 7.16, is typically a loose tube construction designed for direct burial applications. The armor is generally a corrugated aluminum tape surrounded by an outer polyethylene jacket. This combination of outer jacket and armor protects the optical fibers from gnawing animals and the damage that can occur during direct burial installations.

FIGURE 7.16
Armored cable for outdoor applications

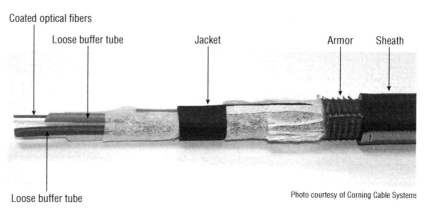

Photo courtesy of Corning Cable Systems

Armored cable used for indoor applications may feature tight-buffered or loose-buffered optical fibers, strength member(s), and an inner jacket. The inner jacket is commonly surrounded by a spirally wrapped interlocking metal tape armor. This type of armor, shown in Figure 7.17, is rugged and provides crush resistance. These cables are used in heavy traffic areas and installations that require extra protection, including protection from rodents.

FIGURE 7.17
Armored cable for indoor applications

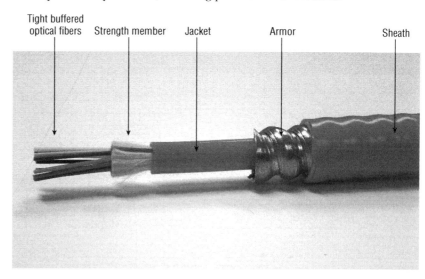

Messenger Cable

When a fiber-optic cable must be suspended between two poles or other structures, the strength members alone are typically not enough to support the weight of the cable and any additional forces that may be placed on the cable. For aerial installations a messenger wire is required. The messenger wire can be external to the fiber-optic cable or integrated into the cable.

When the messenger wire is integrated into the fiber-optic cable, the cable is typically referred to as a *messenger cable*. These cables typically feature a 0.25" stranded steel messenger wire. The messenger wire, sometimes referred to simply as the messenger, is integrated into the outer jack of the cable, as shown in Figure 7.18.

FIGURE 7.18

Messenger cable used for aerial installations

Jacket   Strength member   Tight buffered optical fibers

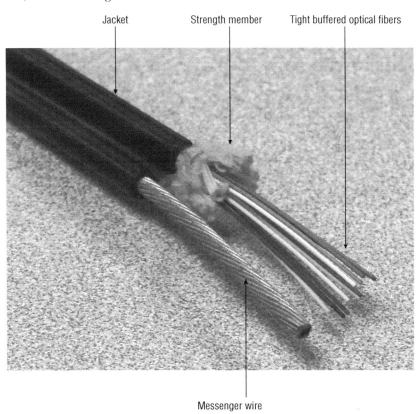

Messenger wire

Also called *figure 8 cable* for the appearance of its cross section, messenger cable greatly speeds up an aerial installation because it eliminates the need to lash the fiber-optic cable to a pre-run messenger wire.

If a messenger wire is not incorporated into the cable assembly used, the cable will have to be lashed to messenger wire. A messenger wire is a steel or aluminum wire that supports the fiber-optic cable in an aerial installation. Either way, cables in aerial installations must be able to withstand loading from high winds, ice, birds and climbing animals, and even windblown debris such as branches.

6

Ribbon Cable

Ribbon cable is a convenient solution for space and weight problems. The cable contains fiber ribbons, which are actually coated optical fibers placed side by side, encapsulated in Mylar tape (see Figure 7.19), similar to a miniature version of wire ribbons used in computer wiring. A single ribbon may contain 4, 8, or 12 optical fibers. These ribbons can be stacked up to 22 high.

FIGURE 7.19
Ribbon cables consist of parallel fibers held together with Mylar tape.

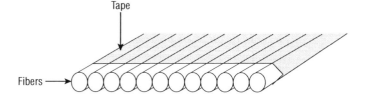

Tape

Fibers →

Because the ribbon contains only coated optical fibers, this type of cable takes up much less space than individually buffered optical fibers. As a result, ribbon cables are denser than any other cable design. They are ideal for applications where limited space is available, such as in an existing conduit that has very little room left for an additional cable.

As shown in Figure 7.20, ribbon cables come in two basic arrangements. In the *loose tube ribbon cable*, fiber ribbons are stacked on top of one another inside a loose-buffered tube. This type of arrangement can hold several hundred fibers in close quarters. The buffer, strength members, and cable jacket carry any strain while the fiber ribbons move freely inside the buffer tube.

FIGURE 7.20
Armored loose tube ribbon cable (bottom) and jacketed ribbon cable (top)

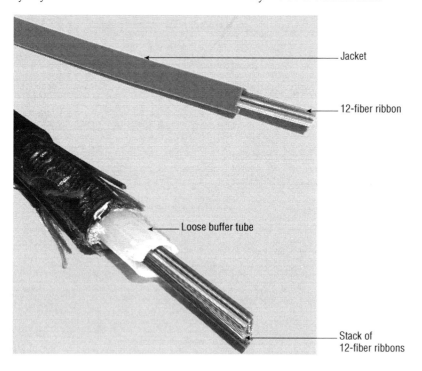

Jacket

12-fiber ribbon

Loose buffer tube

Stack of 12-fiber ribbons

The *jacketed ribbon cable* looks like a regular tight-buffered cable, but it is elongated to contain a fiber ribbon. This type of cable typically features a small amount of strength member and a ripcord to tear through the jacket.

While ribbon fiber provides definite size and weight savings, it does require special equipment and training to take advantage of those benefits. Connectors, strippers, cleavers, and fusion splicers must all be tailored to the ribbon fiber. For these reasons, ribbon fiber may not be the best solution in all situations.

Submarine Cable

Submarine cable is specially designed for carrying optical fiber underwater. Not all submarine cable is the same, however. Depending on the distance it will span and the type of service it will provide, submarine cable can take many different forms.

Submarine cable may be laid in trenches under the bottom of waterways where shipping or fishing activities threaten to snag or damage the cable, or they may be laid directly on the bottom of less-traveled waterways, or on the deep ocean floor where such activities do not penetrate. Figure 7.21 shows a multi-optical-fiber submarine cable with armor and metal strength members.

FIGURE 7.21
Submarine cable with armor and metal strength members

Jacket   Strength member   Loose buffer tube

Aerospace Cable

Aerospace cables are designed to be installed in aircraft and spacecraft. These cables are designed to operate in extreme temperature environments. In addition, they must protect the optical fiber from the shock and vibration associated with aircraft and spacecraft. While the optical fiber used in many aerospace cables may be the same optical fiber used in other cable types, the coating, buffer, and jacket are typically different.

MORE UNDERWATER HAZARDS

While fishing operations, anchoring, and other hazards pose a threat to underwater cables of all types, some fiber-optic cables running in the Grand Canary Islands encountered a special danger soon after they were laid.

The cables were part of a fiber-optic test system run by AT&T in 1985. Only three weeks after the cables were laid about ¾ of a mile below the surface, the cable stopped operating and was out of commission for a week. Technicians first thought that the cable had separated because of abrasion due to scraping across the ocean floor, but when the cable was examined more closely, it was found to have small shark teeth embedded in it. The shark attacks were repeated twice over the next several months.

Why the repeated shark bites on fiber-optic cable? Sharks are known to be drawn to prey by their electromagnetic emissions, and attacks on standard coaxial cable would have been much more likely if the cable were not shielded against electromagnetic interference. But the fiber did not carry any electricity, only light, so the sharks should have ignored it.

The answer lay in the fact that light does have an electromagnetic signature, but because the optical fiber is not subject to electromagnetic interference, no shielding was applied to the cable. The small sharks were drawn to the weak electrical fields put out by the light traveling through the fiber and in the dark water attacked it, thinking it was a meal.

STRUCTURE TYPE

A variety of aerospace fiber-optic cables are available for use in different locations on the aircraft or spacecraft. These cables can typically be separated into two structure types: tight and loose. Structure types should not be confused with buffer types. A tight-buffered optical fiber may be used in a loose structure fiber-optic cable.

ARINC Report 802, Fiber Optic Cables defines the construction requirements for loose and tight structure cables for aerospace and avionic applications. A loose structure cable allows limited movement of the buffered fiber (usually the 900µm in diameter) with respect to the outer jacket and strength member. A tight structure cable does not allow any movement of the buffered fiber with respect to the outer jacket and strength member.

APPLICATIONS

In Chapter 1, "History of Fiber Optics and Broadband Access," you learned that since the turn of the century fiber optics has helped to increase broadband download speeds by a factor of 5000. Just as the telecommunications industry is incorporating more and more optical fiber, so is the aerospace and avionics industry. The Boeing 787 Dreamliner has 110 fiber-optic links and over 1.7km of fiber-optic cable. The aerospace cable assembly shown in Figure 7.22 is similar to one that could be found on the Dreamliner.

Aerospace tight or loose structure cables can have a very wide temperature range based on the location where they are installed. Cables designed to be used inside avionics boxes or cabinets typically have a temperature range of –40° C to +85° C. Cables designed to be used between cabinets typically have a temperature range of –65° C to +125° C, whereas those cables used by the engines have a temperature range of –65° C to +260° C.

FIGURE 7.22
Aerospace fiber-optic cable assembly with LC connectors

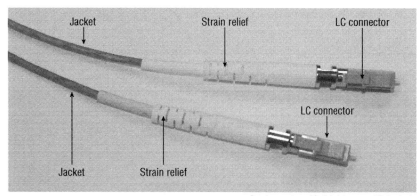

Drawing courtesy of Carlisle Interconnect Technologies

CABLE COMPONENTS

Figure 7.23 shows the components of a typical aerospace fiber-optic cable. This cable features a primary and secondary buffer. The primary buffer is the coating applied directly over the cladding of the optical fiber. Aerospace cables may use a variety of coatings depending on the temperature range required. A typical acrylate coating has a maximum operating temperature of 100° C whereas some high-temperature acrylates may perform at temperatures up to 125° C. Silicone, carbon, and polyimide coatings have greater temperature ranges than acrylate. Detailed information on optical fiber coatings can be found in Chapter 4.

FIGURE 7.23
Typical aerospace fiber-optic cable

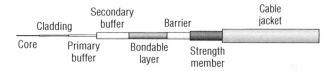

Drawing courtesy of Carlisle Interconnect Technologies

The secondary buffer shown in Figure 7.24 is polytetrafluoroethylene (PTFE). The bondable layer surrounding the secondary layer is a polyimide tape, and the barrier layer is unsealed PTFE. The strength member is braided Kevlar, and the jacket shown in Figure 7.25 is extruded FEP (fluorinated ethylene propylene). This jacket material has a service temperature range of –100° C to +200° C as well as excellent chemical and UV resistance properties; it is also flame retardant.

FIGURE 7.24
PTFE secondary buffer

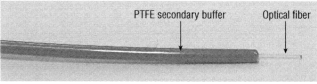

Photo courtesy of Carlisle Interconnect Technologies

Photo courtesy of Carlisle Interconnect Technologies

FIGURE 7.25
FEP jacket and braided Kevlar strength member

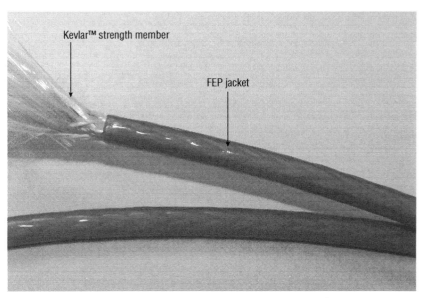

Hybrid Cable

Hybrid cable, as applied to fiber optics, combines multimode and single-mode optical fibers in one cable. Hybrid cable should not be confused with composite cable, although the terms have been used interchangeably in the past.

Composite Cable

Composite cable, as defined by the National Electrical Code (NEC), is designed to carry both optical fiber and current-carrying electrical conductors. These cables may also contain non-current-carrying conductive members such as metallic strength members, metallic vapor barriers, metallic armor, or a metallic sheath.

The composite cable shown in Figure 7.26 consists of optical fibers along with twisted-pair wiring typical of telephone wiring. This arrangement is convenient for networks that carry fiber-optic data and conventional telephone wiring to the same user. Composite cable also provides installers with a way to communicate during fiber installation and provides electrical power to remote equipment, such as repeaters, along the fiber's route.

FIGURE 7.26
Composite cable carries fiber and wiring in the same run.

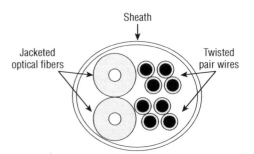

Cable Duty Specifications

The various combinations of strength members, jacket materials, and fiber arrangements are determined by the specific requirements of an installation. Among the factors considered are the amount of handling a cable will take, the amount of stress the cable must endure in normal use, and the locations where it will run.

Cable duty specifications are typically divided into two basic types:

Light-duty cables These are designed for basic protection of the fiber within and minimal handling. A good example of a light-duty fiber cable is a simplex cord, which consists of a tight-buffered fiber with a strength member and a jacket. Simplex cordage is not engineered to withstand excessive pulling forces, and the jacket is engineered for flexibility and ease of handling, not harsh environments.

Heavy-duty cables These are designed for more and rougher handling, with additional strength members and jacketing around the fiber. They are made for harder pulling during installation, and they protect the fiber within from damage in exposed or extreme environments.

Cable Termination Methods

Some fibers, such as those found in simplex and duplex cords and breakout cable, are already set up to receive connectors and can be handled easily. Others, including loose-buffered cables, must be prepared for connectors and handling with special kits.

These kits, known as *fanout kits* and *breakout kits*, are designed to adapt groups of coated fibers for connectors by separating them and adding a tight buffer (fanout kit) or a tight buffer, strength member, and jacket (breakout kit) to each one.

Fanout Kit

The fanout kit, shown in Figure 7.27, converts loose-buffered fibers into tight-buffered fibers ready for connectors. A typical fanout kit contains an enclosure sometimes called a *furcation unit*. The furcation unit attaches to the loose-buffer tube. Hollow tight-buffer tubes 900µm in diameter are applied over the optical fibers and passed into the furcation unit, which is then closed, locking the tight buffers in place on the fibers. After the fibers have the buffers applied, connectors can be attached for use in a patch panel or other protected enclosure.

Breakout Kit

The breakout kit, shown in Figure 7.28, is similar to the fanout kit in that it spreads the fibers from the loose-buffer tube through a furcation kit and provides 900µm tight buffers to be applied over the optical fiber. The breakout kit, however, is designed to allow the optical fiber to be connected directly to equipment with standard connectors.

In addition to buffer material, the breakout kit provides a 3mm diameter jacket with an aramid strength member that slips over the optical fiber. Heat-shrink tubing and epoxy is used to join the individual jackets to the cable.

FIGURE 7.27
A fanout kit adds a tight buffer to individual fibers.

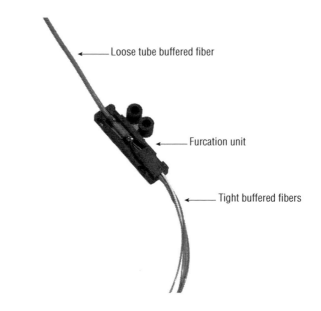

Loose tube buffered fiber

Furcation unit

Tight buffered fibers

FIGURE 7.28
A breakout kit adds a tight buffer, strength member, and a jacket to individual optical fibers.

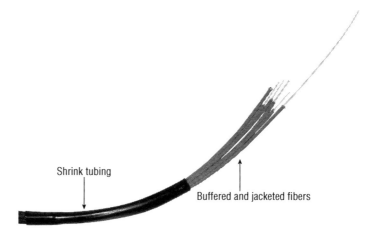

Shrink tubing

Buffered and jacketed fibers

Breakout kits and fanout kits are available to match the number of fibers in the cable being used, and some companies offer kits that can be used with ribbon cable. Different length buffer tubes are available for fanout kits.

Note that different vendors may use a variety of terms or trade names to refer to breakout kits and fanout kits.

Blown Fiber

What if you knew that you wanted to run fiber in a building, but you didn't know exactly what equipment it would be serving or what the requirements for the fiber were? One solution to this dilemma is *blown fiber*.

Blown fiber installation starts with a hollow tube about 5mm in diameter. This tube is installed just like a cable and acts as a loose tube buffer for fibers that will run through it later. The tube may also be part of a cable assembly, which may carry up to 19 tubes and is available in configurations similar to other cables, including armored, all-dielectric, plenum, and riser.

To run the fibers through the tube, you simply lead the fibers into one end of the tube and blow pressurized air through the tube. The air carries the fibers with it through the tube, similar to the way that pneumatic carriers at drive-through tellers shoot your deposits into the bank building.

Blown fiber can be used, as already described, when the fiber needs in a building are not yet known or when fiber needs to be repaired or upgraded. Blown fiber is also useful when a company cannot afford to install all of the fiber that it might potentially use. As a business grows and the new fiber can be economically justified, it can be blown in as needed.

One advantage of blown fiber, even when it is installed all at once, is that cuts or damaged sections can be repaired quickly. In case of an accidental cut from power equipment, for example, the damaged section can be cut out, a new section spliced in, and new fiber blown in. This can be a great time-savings over splicing multiple broken fibers in a cable. Blown fiber can be run 1,000' or more through a building, around curves, and uphill, in a very short time.

Blown fiber is covered in detail in Chapter 14, "Cable Installation and Hardware."

NEC Standards for Fiber-Optic Cables and Raceways

The National Electrical Code (NEC) is published by the National Fire Protection Association (NFPA). NEC Article 90 states that the purpose of the code is the practical safeguarding of persons and property from hazards arising from the use of electricity. It also states that the NEC is not intended as a design specification or an instruction manual for untrained persons. Local and state governments typically adopt it, making compliance mandatory.

It provides specific guidance for running fiber-optic cable within buildings. The NEC requires cables to be tested for fire resistance and smoke characteristics, establishes which types of cables may be run in different areas of a building, and specifies the types of raceways that can be used.

Be sure you stay current with the latest NEC guidelines. The NEC is updated every 3 years. The information is this chapter can be found in NEC Article 770, "Optical Fiber Cables and Raceways." To ensure this chapter contains the latest information a draft copy of the 2014 NEC was used for the updates.

Like the industry standards, Article 770 in the NEC has increased in page count since the turn of the century. It required approximately five pages in the 1999 NEC. That page count doubled in the draft 2014 NEC.

NEC Fiber-Optic Cable Types

The NEC recognizes three types of fiber-optic cables in Article 770:

Nonconductive   Cables containing no electrically conductive materials.

Conductive   Cables containing non-current-carrying conductive members, including strength members, armor, or sheath.

Composite   Cables containing optical fiber and current-carrying electrical conductors. These cables may also contain non-current-carrying conductive members such as metallic strength members, metallic vapor barriers, metallic armor, or a metallic sheath.

LISTED AND NONLISTED

The NEC classifies fiber-optic cables according to their electrical and fire safety characteristics and contains rules for the use of each cable type to minimize hazards. Any fiber-optic cabling run indoors must be listed as suitable for the purpose. Any outside plant nonlisted fiber-optic cabling that is brought into a structure must be terminated after no more than 15m or 50' and connectorized or spliced to a listed cable.

A listed fiber-optic cable has had its performance verified by Underwriters Laboratory (UL). The outer surface of the cable or a marker tape should be clearly marked at intervals not to exceed 40" with the following information:

- ◆ Cable type-letter designation such as OFN, OFC, OFNR, OFNP, etc.

- ◆ Manufacturer's identification

- ◆ UL symbol or the letters "UL"

Optical fiber cables installed within buildings shall be listed as being resistant to the spread of fire. Three listed types of cables are recognized in Article 770 for indoor installations: plenum cable, riser cable, and general-purpose cable.

PLENUM CABLE

Plenum cables, whether conductive or nonconductive, are suitable for use in ducts, plenums, and other space used for environmental air. These cables will have fire resistance and low smoke-producing characteristics.

RISER CABLE

Riser cables, whether conductive or nonconductive, are suitable for a vertical run in a shaft or from floor to floor. These cables will have fire-resistance characteristics capable of preventing the carrying of a fire from floor to floor.

GENERAL-PURPOSE CABLE

General-purpose cables, whether conductive or nonconductive, are resistant to the spread of fire. However, these cables are not suitable for plenum or riser applications.

Table 7.1 shows the cable types along with the markings each cable will contain. Markings should be clearly visible on the cable, as shown in Figure 7.29, and no cable should be used indoors unless it listed and contains the NEC cable type marking on it.

FIGURE 7.29
NEC cable type marking

TABLE 7.1:   NEC cable types and description

MARKING	TYPE	LOCATION	PERMITTED SUBSTITUTIONS
OFNP	Nonconductive optical fiber plenum cable	Ducts, plenums, other air spaces	None
OFCP	Conductive optical fiber plenum cable	Ducts, plenums, other air spaces	OFNP
OFNR	Nonconductive optical fiber riser cable	Risers, vertical runs	OFNP
OFCR	Conductive optical fiber riser cable	Risers, vertical runs	OFNP, OFCP, OFNR
OFNG	Nonconductive optical fiber general-purpose cable	General-purpose use except for risers and plenums	OFNP, OFNR
OFCG	Conductive optical fiber general-purpose cable	General-purpose use except for risers and plenums	OFNP, OFCP, OFNR, OFCR, OFNG, OFN
OFN	Nonconductive optical fiber general-purpose cable	General-purpose use except for risers, plenums	OFNP, OFNR
OFC	Conductive optical fiber general-purpose cable	General-purpose use except for risers, plenums	OFNP, OFCP, OFNR, OFCR, OFNG, OFN

All indoor cables must be resistant to the spread of fire, and those listed as suitable for environmental air spaces such as plenums must also have low smoke-producing characteristics. Note in the substitutions list that nonconductive cables may always be substituted for conductive cables of an equal or lower rating, but conductive cables may never be substituted for nonconductive cables. Figure 7.30 shows the order in which cables may be substituted for one another.

NOTE   The NFPA 262 describes test methods for flame travel and smoke of cables for use in air-handling spaces.

FIGURE 7.30
NEC cable substitu-
tion guide

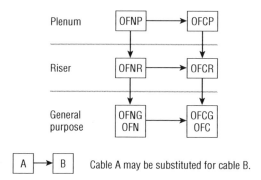

The NEC also describes the *raceways* and *cable trays* that can be used with fiber-optic cables and the conditions under which each type of cable may be used. This information is covered in Chapter 14.

Cable Markings and Codes

In addition to the NEC cable marking, optical fiber cables typically have a number of other markings and codes. Markings that appear on the jacket of the cable help identify what is inside the cable and where it originated. They may also include aids for measuring the cable length.

Inside the cable, the fiber buffers are usually color-coded with standard colors to make connections and splices easier. However, some manufacturers use numbers instead of colors.

External Markings

A cable's external markings typically consist of the manufacturer's information, including the manufacturer's name and phone number, cable part number or catalog number, the date the cable was manufactured, and occasionally the industry standards the optical fiber complies to. Information about the cable itself includes the NEC cable marking, the fiber type (multimode/single-mode or core/cladding size), and sequential cable length markings in meters or feet.

DATE, INDUSTRY STANDARDS, AND FIBER TYPE

The date the cable was manufactured can be very helpful in determining the performance standards of the optical fiber within the cable. As you learned in Chapter 5, "Optical Fiber Characteristics," there have been significant changes in the performance of multimode optical fiber since the turn of the century. The cable shown in Figure 7.31 was manufactured at the turn of the century. In addition, the cable markings identify three standards that define the optical and geometrical performance of the optical fiber. These standards are:

♦ Bellcore GR-409-CORE

♦ ISO/IEC 11801

♦ TIA/EIA-568-A

FIGURE 7.31

Duplex cordage manufactured in June 2000

The cable shown in Figure 7.31 was manufactured in June 2000. Optical Fiber Cabling Component Standard TIA/EIA-568-B.3 superseded the Commercial Building Telecommunications LAN/WAN Cabling Standard TIA/EIA-568-A in April 2000. This is an example of cable being manufactured with optical fiber that did not comply with the latest published standards. As it has been mentioned before in this book, when a standard changes, old cable is not ripped out and replaced with new. A cable manufacturer with reels of TIA/EIA-568-A optical fiber on the shelf is not going to waste that inventory just because a standard changed.

The cable shown in Figure 7.32 was manufactured in May 2008. There are no standards markings on this cable, so determining the industry standard(s) that define the optical and geometrical performance of the optical fibers in this cable is not as straightforward as with the cable in Figure 7.31. However, there is a clue: the word "laser" is printed on the jacket. This implies that the optical fiber was optimized for a laser light source. The 50μm optical fiber core size is also printed on the jacket. Combining both pieces of data indicates that the optical fiber in the cable is OM3.

[handwritten margin note: what are OM1 OM2 etc OS1]

FIGURE 7.32

Laser-optimized 6-fiber breakout cable manufactured in May 2008

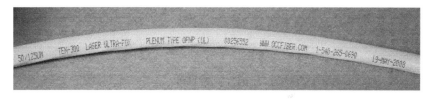

OM3 optical fiber cannot be found in TIA/EIA-568-B.3. However, it can be found in TIA-568-C.3, which was published June 2008, one month after the cable shown in Figure 7.31 was manufactured. This is an example of a cable that contains an optical fiber that exceeds the transmission performance parameters of a current published standard. As mentioned previously in this book, standards define a minimum level of performance and many manufacturers offer products that exceed the minimum performance levels. Oftentimes the working groups within standards organizations are playing catch-up, creating standards for products that currently exist and may have been in widespread use for several years or more.

Remember, you can always contact the cable manufacturer using the phone number printed on the cable to determine the optical and geometrical performance of the optical fiber.

Color Codes

As with copper wiring, optical fibers running in cables must have some way of being distinguished from one another so that they can be connected properly at each end. TIA-598-C provides color-coding schemes for premises jackets and optical fibers within a fiber-optic cable. Table 7.2 shows the color-coding scheme for individual fibers bundled in a cable.

TABLE 7.2:   TIA-598-C fiber color code

FIBER NUMBER	BASE COLOR/TRACER	FIBER NUMBER	BASE COLOR/TRACER
1	Blue	19	Red/Black
2	Orange	20	Black/Yellow
3	Green	21	Yellow/Black
4	Brown	22	Violet/Black
5	Slate	23	Rose/Black
6	White	24	Aqua/Black
7	Red	25	Blue/Double Black
8	Black	26	Orange/Double Black
9	Yellow	27	Green/Double Black
10	Violet	28	Brown/Double Black
11	Rose	29	Slate/Double Black
12	Aqua	30	White/Double Black
13	Blue/Black	31	Red/Double Black
14	Orange/Black	32	Black/Double Yellow
15	Green/Black	33	Yellow/Double Black
16	Brown/Black	34	Violet/Double Black
17	Slate/Black	35	Rose/Double Black
18	White/Black	36	Aqua/Double Black

Table 7.3 shows the color coding used on premises cable jackets to indicate the type of fiber they contain, if that is the only type of fiber they contain.

TABLE 7.3:   TIA-598-C premises cable jacket colors

FIBER TYPE	JACKET COLOR FOR NONMILITARY APPLICATIONS	JACKET COLOR FOR MILITARY APPLICATIONS
Multimode (50/125μm)	Orange	Orange
Multimode (50/125μm) Laser-optimized	Aqua	
Multimode (62.5/125μm)	Orange	Slate
Multimode (100/140μm)	Orange	Green
Single-mode (NZDS)	Yellow	Yellow
Polarized maintaining Single-mode	Blue	

Cable Numbers

The color-coding schemes for premises jackets and optical fibers within a fiber-optic cable described in TIA-598-C are not always used. For example, the jacket colors of every simplex cable shown in Figure 7.33 are the same. You can identify a specific simplex cable by locating the number that is printed on the jacket. The simplex cable number is printed every 6″ along the entire length of the cable.

FIGURE 7.33
Breakout cable with 12-simplex cables, each with printed cable numbers

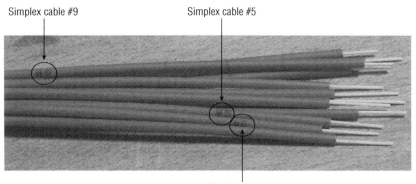

Simplex cable #9   Simplex cable #5

Simplex cable #6

With all numbering or coloring schemes, there are advantages and disadvantages. In an area that is not lit well, it can be difficult to correctly identify colors. In addition, how well color differences are perceived can vary from person to person. Numbering eliminates color perception problems.

Sequential Markings

A cable's external markings may include sequential markings, as shown in Figure 7.34. Sequential markings are numbers that appear every 2' or 1m. These markings are useful in determining how much cable is left on a reel, measuring off large runs of cable, or simply determining the length of a piece of cable without pulling out a tape measure.

FIGURE 7.34
Sequential marking in feet on a fiber-optic cable

Sequential marking in feet

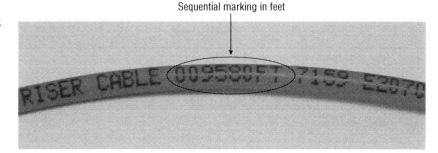

The numbers themselves indicate cable length, not optical-fiber length. To measure the length of the cable using the sequential markings, first determine the measurement standard that is being used. Next, subtract the number at the low end from the number at the high end. The difference between the two is the length.

Because some measurements dealing with fiber optics use meters rather than feet, you may need to convert any measurements in feet to the metric system. The formula for converting length is:

1 foot = 0.3048 meter

Let's practice measuring out a length of cable using sequential markings. To find the length in meters of a cable that has sequential markings of 846 at one end and 2,218 at the other end, and the markings are measuring the cable in feet, you first determine the distance between the markings using this formula:

2,218 − 846 = 1,372 feet

Now, convert feet to meters:

1,372 × 0.3048 = 418.2 meters

Remember that the length of the cable is not necessarily the length of the optical fiber inside of it. In loose tube cable, the fiber is actually slightly longer than the cable. This fact becomes important if fault location procedures tell you that there is a fault at a specific distance from the end of the fiber. That distance could be short of the measured distance on the outside of the cable since the fiber wanders inside of the buffer tube.

Bend Radius Specifications

Throughout the installation process, optical fiber's light-carrying abilities are threatened by poor handling, damage from tools or accidents, and improper installation procedures.

One of the installation hazards that can cause attenuation or damage the optical fiber is extreme bending. When an optical fiber is bent too far, the light inside no longer reflects off the boundary between the core and the cladding, but passes through to the coating.

To reduce the risk of excessive bending during installation, manufacturers specify a minimum installation and operational bend radius for their optical fiber cables. Following the manufacturer's guidelines reduces the risk of damage to the cable and optical fiber during and after installation. Chapter 14 provides detailed information on cable bend radius.

The Bottom Line

Determine the cable type from the NEC markings. Article 770 of the NEC states that optical fiber cables installed within buildings shall be listed as being resistant to the spread of fire. These cables are also required to be marked.

Master It You have been asked to install the two cables shown here in a riser space. From the markings on the cables, determine the cable types and determine if each cable can be installed in a riser space.

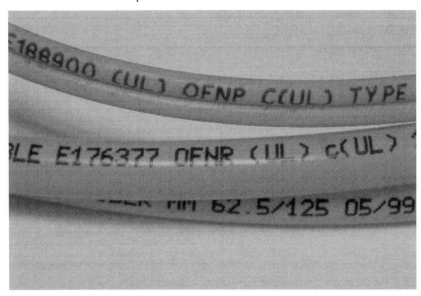

both in riser.
only plenum in plenum.

Identify the fiber number from the color code. TIA-598-C defines a color code for optical fibers within a cable assembly.

Master It You have been asked to identify the fourth and seventh tight-buffered optical fibers in the 12-fiber cable shown on the next page. There are no numbers on the tight buffers; however, each buffer is colored as defined in TIA-598-C. What are the colors of the fourth and seventh tight-buffered optical fibers?

p 140
brown
&
Rd

Determine the fiber number from the cable markings. The color-coding schemes for premises jackets and optical fibers within a fiber-optic cable described in TIA-598-C are not always used.

> **Master It** You have been asked to identify simplex cable number nine in the 12-fiber cable shown here. Describe the location of that cable.

top left.

Identify the optical fiber type from the color code. Premises cable jacket colors are defined in TIA-598-C.

table p14.

> **Master It** It is your first day on the job and your supervisor has asked you to get some laser-optimized multimode optical fiber from the back of the van. The van contains several different cables, each with a different jacket color. What is the jacket color of the laser-optimized cable?

Determine the optical fiber type from the cable markings. Premises cable jacket colors are defined in TIA-598-C. However installed cables may not comply with revision C of this standard.

Master It You have been asked to identify the optical fiber type in the cable shown here. The jacket color is slate and this cable was not intended for military applications. What type of optical fiber is used in this cable?

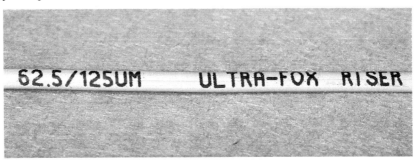

cant go by colour.

—made before standard

Determine the cable length using sequential markings. Many markings are found on a cable. Sequential markings typically appear every 2 feet or every 1m.

Master It You have been asked to determine the length of fiber-optic cable coiled up on the floor. The sequential markings for this cable are meters. The marking on one end of the cable is 0235 and the marking on the other end is 0485. How long is this cable?

250 m

Chapter 8

Splicing

In this chapter, we will discuss fiber-optic splices and examine the factors that affect splice performance. We will also describe tools and methods used to splice optical fibers as well as standards used to gauge splice performance.

In this chapter, you will learn to:

♦ Determine if the splice loss is from an intrinsic factor

♦ Determine if the splice loss is from an extrinsic factor

♦ Calculate the potential splice loss from a core diameter mismatch

♦ Troubleshoot a fusion splice

Why Splice?

Up to this point, we have not addressed how to connect or mate two optical fibers together. Nor have we discussed why we would want to do this. This chapter discusses splicing. Splicing is one way to join two optical fibers together so the light energy from one optical fiber can be transferred into another optical fiber.

A splice is a permanent connection of two optical fibers. Once the two optical fibers are joined with a splice, they cannot be taken apart and put back together as they can if you join them using connectors. A splice is typically employed for one of four reasons: to repair a damaged cable, to extend the length of a cable, to join two different cable types, or to attach a pigtail. Pigtails are covered in-depth in Chapter 9, "Connectors."

Splice Performance

How well a splice performs depends on many variables. These variables can be broken into two groups: *intrinsic factors* and *extrinsic factors*.

As you have learned in earlier chapters, optical fibers are not perfect and variations between optical fibers can affect splice performance. These variations are referred to as intrinsic factors.

The performance of a splice can also be affected by alignment and optical fiber mating issues that have nothing to do with the optical fiber. The factors that affect the alignment and/or mating of the optical fibers are referred to as extrinsic factors.

Intrinsic Factors

Even when fibers are manufactured within specified tolerances, slight variations still exist from one optical fiber to another. These variations can affect the performance of the splice even though the optical fibers are perfectly aligned when mated. The variations between two optical fibers that affect splice performance are referred to as intrinsic factors.

Let's look at the most common types of variations.

NUMERICAL APERTURE (NA) MISMATCH

A numerical aperture (NA) mismatch occurs when the NA of one optical fiber is different from the NA of the other optical fiber. If the NA of the transmitting fiber is larger than the NA of the receiving optical fiber, a loss may occur. However, a loss will not occur if the NA of the transmitting optical fiber is less than the NA receiving optical fiber.

In Chapter 5, "Optical Fiber Characteristics," we showed you how to calculate the NA of an optical fiber using the refractive indexes of the core and cladding. Recall that in order for light to be contained within a multimode optical fiber it must reflect off the boundary between the core and the cladding, rather than penetrating the boundary and refracting through the cladding. Light must enter within a specified range defined by the acceptance cone or cone of acceptance.

Light entering the core of an optical fiber from outside of the acceptance cone will either miss the core or enter at an angle that will allow it to pass through the boundary with the cladding and be lost, as shown in Figure 8.1. Light entering the core of the optical fiber at an angle greater than the acceptance angle will not propagate the length of the optical fiber. For light to propagate the length of the optical fiber, it must enter at an angle that does not exceed the acceptance angle.

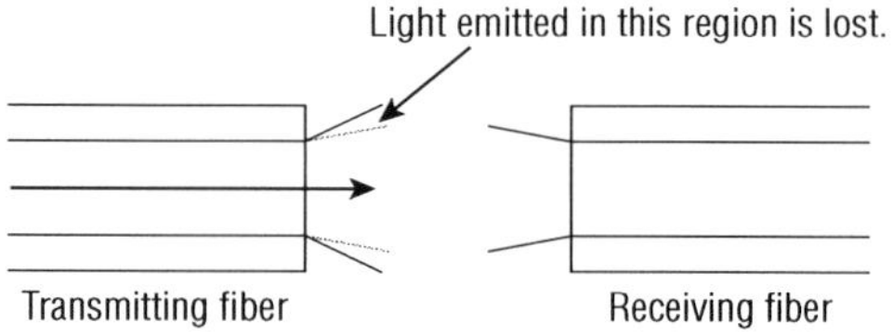

FIGURE 8.1

When NA mismatch loss occurs, the receiving optical fiber cannot gather all of the light emitted by the transmitting fiber.

Light emitted in this region is lost.

Transmitting fiber

Receiving fiber

The exact loss from an NA mismatch is difficult to calculate. Factors such as light source type, light source launch condition, optical fiber length, and bends in the optical fiber all affect the potential loss. It is possible to have an NA mismatch between two optical fibers and no loss resulting from the mismatch.

CORE DIAMETER MISMATCH

Core diameter mismatch occurs when there is a difference in the core diameters of the two optical fibers. A core diameter mismatch loss results when the core diameter of the transmitting optical fiber is greater than the core diameter of the receiving optical fiber, as shown in Figure 8.2. A loss occurs when light at the outer edge of the transmitting optical fiber core falls outside the diameter of the receiving optical fiber core. This light is lost in the cladding of the receiving optical fiber. Core diameter mismatch loss is typically only a concern with multimode optical fiber.

FIGURE 8.2

Core diameter mismatch loss is the result of the transmitting optical fiber having a larger core diameter than the receiving optical fiber.

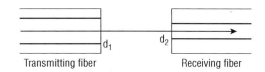

Transmitting fiber   Receiving fiber

It is not uncommon for two multimode optical fibers with different core diameters to be spliced together. As you saw in Chapter 5, ANSI/TIA-568-C.3 recognizes multimode optical fibers with 50μm cores and 62.5μm cores. We can calculate the worst-case loss percentage for a splice that joins a 50μm core and 62.5μm core using this formula:

Loss = $[(d_1)^2 - (d_2)^2] / (d_2)^2$

where d_1 is the diameter of the transmitting core and d_2 is the diameter of the receiving core.

$62.5^2 - 50^2 = 1406.25$

$1406.25/50^2 = 0.5625$ or 56.25%

Using the following formula, the decibel loss can be calculated:

dB = $10Log_{10}(P_{out} / P_{in})$

Note that the decibel formula is a ratio where the power output (P_{out}) is divided by the power input (P_{in}). In this case, the power input is the power being put into the receiving optical fiber. We know that the receiving optical fiber will not accept 56.25 percent of the light from the transmitting optical fiber. Subtracting 56.25 percent from 100 percent, as shown next, results in a difference of 43.75 percent, which can be written as 0.4375.

$100\% - 56.25\% = 43.75\%$ or 0.4375

$10Log_{10} (0.4375 / 1) = -3.59$dB

Loss = 3.59dB

The 3.59dB loss that we just calculated is the maximum loss possible from the core diameter mismatch. The actual loss may vary depending on factors such as light source type, light source launch condition, optical fiber length, and bends in the optical fiber.

MODE FIELD DIAMETER MISMATCH

Mode field diameter is only a concern in single-mode optical fibers. It describes the diameter of the light beam traveling through the core and a portion of the inner cladding. Values for the mode field diameter are defined by wavelength with longer wavelengths typically having a larger diameter than shorter wavelengths. A typical mode field diameter nominal range is 8.6 to 9.55μm at 1310nm and 8 to 11μm at 1550nm.

A mode field diameter mismatch occurs when there is a difference in the mode field diameters of two single-mode optical fibers. A mode field diameter mismatch loss results when the mode field diameter of the transmitting optical fiber is greater than the mode field diameter of the receiving optical fiber, as shown in Figure 8.3. A loss occurs when optical fiber with the smaller mode field diameter will not accept all of the light from the optical fiber with the larger mode field diameter.

FIGURE 8.3
Mode field diameter mismatch loss is the result of the transmitting optical fiber having a larger mode field diameter than the receiving optical fiber.

CLADDING DIAMETER MISMATCH

Cladding diameter mismatch occurs when the cladding diameters of the transmit and receive optical fibers are not the same. Cladding diameter mismatch loss occurs when the cores of the optical fiber are not aligned because of the cladding diameter mismatch, as shown in Figure 8.4. A cladding diameter mismatch can cause the light exiting the core of the transmitting optical fiber to enter the cladding of the receiving optical fiber. The light entering the cladding is lost, causing attenuation.

FIGURE 8.4
Cladding diameter mismatch loss results from differing cladding diameters.

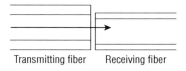

Transmitting fiber   Receiving fiber

CONCENTRICITY

Ideally, the core and cladding of an optical fiber are perfectly round and *concentric*, which means that they share a common geometric center. However, optical fibers are not perfect and there will be concentricity variations. These concentricity variations can cause the optical fiber cores to misalign, as shown in Figure 8.5, causing a loss when the light exiting the core of the transmitting optical fiber enters the core of the receiving optical fiber.

FIGURE 8.5
Off-center fiber cores cause concentricity loss.

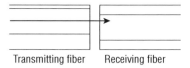

Transmitting fiber   Receiving fiber

The illustration in Figure 8.5 is greatly exaggerated to clearly show how a concentricity loss may occur. The core and cladding concentricity differences are typically less than 1μm.

NONCIRCULARITY

Just as the core and cladding of an optical fiber may not be perfectly concentric, they may also not be perfectly circular. The noncircularity of the core will cause a loss when light from the core of the transmitting optical fiber enters the cladding of the receiving optical fiber. In Figure 8.6, the core is elliptical to show how core noncircularity loss occurs.

FIGURE 8.6
Noncircularity loss takes place when the ellipticities of two optical fibers do not match exactly.

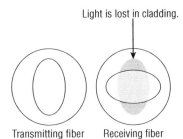

Light is lost in cladding.

Transmitting fiber   Receiving fiber

Cladding noncircularity may cause loss when it causes part of the core of the transmitting optical fiber to align with the cladding of the receiving optical fiber. Any light that enters the cladding of the receiving optical fiber will be lost, causing attenuation.

Note that the amount of loss depends on the alignment of the ellipticities of the two cores. Maximum loss occurs when the long or major axes of the cores are at right angles to one another, and minimum loss occurs when the axes of the cores are aligned. After reading about all the possible intrinsic losses, you may find it hard to believe that two optical fibers can be spliced together with virtually no loss. Optical fiber manufacturing produces optical fibers with little variation, as shown in Tables 8.1, 8.2, 8.3, and 8.4. These tables list the attributes and tolerances for different single-mode optical fibers as defined in ITU standards ITU-T G.652, ITU-T G.655, and ITU-T G.657.

TABLE 8.1: ITU-T G.652, single-mode optical fiber attributes

FIBER ATTRIBUTES, G.652.A, G.652.B, G.652.C, AND G.652.D		
Mode field diameter	Wavelength	1310nm
	Range of nominal values	8.6–9.5μm
	Tolerance	±0.6μm
Cladding diameter	Nominal	125.0μm
	Tolerance	±1μm
Core concentricity error	Maximum	0.6μm
Cladding noncircularity	Maximum	1%

TABLE 8.2: ITU-T G.655.C, non-zero-dispersion-shifted single-mode optical fiber and cable

FIBER ATTRIBUTES		
Mode field diameter	Wavelength	1550nm
	Range of nominal values	8–11μm
	Tolerance	±0.7μm
Cladding diameter	Nominal	125.0μm
	Tolerance	±1μm
Core concentricity error	Maximum	0.8μm
Cladding noncircularity	Maximum	2%

TABLE 8.3: ITU-T G.655.D, non-zero-dispersion-shifted single-mode optical fiber and cable

FIBER ATTRIBUTES		
Mode field diameter	Wavelength	1550nm
	Range of nominal values	8–11μm
	Tolerance	±0.6μm
Cladding diameter	Nominal	125.0μm
	Tolerance	±1μm
Core concentricity error	Maximum	0.6μm
Cladding noncircularity	Maximum	1%

TABLE 8.4: ITU-T G.657, bending loss-insensitive single-mode optical fiber and cable

FIBER ATTRIBUTES CATEGORIES A AND B		
Mode field diameter	Wavelength	1310nm
	Range of nominal values	8.6–9.5μm
	Tolerance	±0.4μm
Cladding diameter	Nominal	125.0μm
	Tolerance	±0.7μm
Core concentricity error	Maximum	0.5μm
Cladding noncircularity	Maximum	1%

Extrinsic Factors

Extrinsic factors that affect optical fiber splice performance are factors related to the condition of the splice itself, external to the optical fiber. In an ideal splice, the optical fibers are identical and they are aligned so that cores are perfectly centered on each other and the core axes are perpendicular to the endfaces being joined, as shown in Figure 8.7. However, there is no such thing as an ideal splice, only a real splice. In a real splice, intrinsic and extrinsic factors affect splice performance.

FIGURE 8.7
Conditions for an
ideal splice

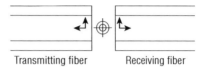

In this section, we will examine common extrinsic factors that affect splice performance. As you read about these extrinsic factors, keep in mind that many times they are caused by dirt and contamination. Microscopic particles of dirt can cause the misalignment of one or both optical fibers, creating a high-loss splice.

LATERAL MISALIGNMENT

Lateral misalignment occurs when the two optical fibers are offset, as shown in Figure 8.8. Lateral misalignment loss occurs when light from the core of the transmitting optical fiber enters the cladding of the receiving optical fiber, creating a loss. As the lateral misalignment increases, less light from the core of the transmitting optical fiber makes its way into the core of the receiving optical fiber, increasing the loss of the splice.

FIGURE 8.8
Lateral misalign-
ment of the optical
fibers

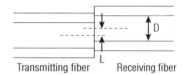

END SEPARATION

Even if the optical fibers are perfectly aligned, the splice still may still experience loss from *end separation*. End separation is simply a gap between the transmitting and receiving optical fibers, as shown in Figure 8.9.

FIGURE 8.9
End separation is a
gap between fiber
ends in a mechani-
cal splice.

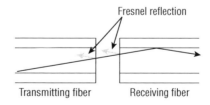

Two different types of losses can be generated from end separation. The first is through Fresnel reflection, which takes place when light passes from the higher refractive index of the core in the transmitting optical fiber into the lower refractive index of the air, and then back into the core of the receiving optical fiber. Each change in the refractive index causes a certain amount of light to be reflected and therefore lost.

One way to overcome the effects of Fresnel reflections in separated optical fibers is to use an *index matching gel*, which is a transparent gel having a refractive index close to that of the core of the optical fibers being spliced. The gel fills the gap and reduces or eliminates the Fresnel reflection. Index matching gel is typically used in all mechanical splices.

End separation also causes loss because when light exits the transmitting optical fiber it spreads out like the light from a flashlight. Some of the light leaving the core of the transmitting optical fiber may enter the cladding of the receiving optical fiber, causing a loss. How much the light spreads out depends on several variables, including the distance between the optical fibers, light source type, launch type, length and NA of the transmitting optical fiber, and the bends in the transmitting optical fiber.

ANGULAR MISALIGNMENT

If the optical fibers in a splice meet each other at an angle, a loss from *angular misalignment* may occur, as shown in Figure 8.10. The amount of loss depends on the severity of the angular misalignment and the acceptance cones of the transmitting and receiving optical fibers. Because the NA of a multimode optical fiber is greater than the NA of a single-mode optical fiber, multimode splices tolerate angular misalignment better than single-mode splices.

FIGURE 8.10
Angular misalignment results when fiber ends are not perpendicular to each other.

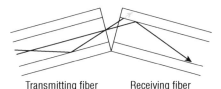

Transmitting fiber   Receiving fiber

The loss from angular misalignment occurs when light from the core of transmitting optical fiber enters the cladding of the receiving optical fiber or enters the core of receiving optical fiber at an angle exceeding the acceptance angle. Light entering the core of the receiving optical fiber at an angle exceeding the acceptance angle may not propagate the length of the receiving optical fiber.

Angular misalignment also prevents the endfaces from contacting each other, resulting in Fresnel reflections exactly like those described in the "End Separation" section earlier in this chapter. Again, this is why index-matching gel is used with every mechanical splice.

Splicing Safety

As discussed in Chapter 6, "Safety," you are responsible for your own safety as well as for the safety of your co-workers. It is up to you to know and incorporate safe work practices when splicing optical fibers. Remember, good work habits can help you prevent accidents and spot potential problems in time to correct them.

Here are some general safety rules that are discussed in detail in Chapter 6.

♦ Keep a clean workspace.

♦ Observe your surroundings.

♦ Use tools for the job they were designed to perform.

♦ Do not eat or drink in the work area.

♦ Report problems or injuries immediately.

♦ Know how to reach emergency personnel.

♦ Put your emergency contact information in your cell phone.

Splicing Hazards

Let's look at some of the hazards directly related to splicing.

LIGHT SOURCES

Most lasers used in fiber-optic systems emit infrared radiation; you cannot see the beam, no matter how powerful it is. As a result, you will not be able to tell if the system is powered, especially if you are working on a piece of fiber far from the transmitter. Always treat an optical fiber as if it is coupled to a laser unless you can be sure that the fiber is not coupled to a laser.

HANDLING FIBER

Optical fiber may be flexible; however, it is rigid enough to pierce your skin or eyes and cause discomfort, pain, or damage. If it becomes airborne, you may even accidentally inhale or swallow it, risking damage to your throat or respiratory system. Protect yourself with proper procedures and the right personal protective equipment.

When you cut, cleave, or accidentally break optical fibers, the ends can get lost easily, either by becoming airborne or by rolling along a surface. Whenever possible, work over a nonreflective black mat; it will enable you to more easily identify fiber fragments. In addition, keep track of cut fiber ends and dispose of them in a separate labeled container.

To prevent eye injury, always wear proper eye protection. To prevent injury to your hands, always handle cut pieces of fiber with tweezers. If you have been handling fiber, do not rub your eyes or put your hands near them until you have washed your hands. If you do get a piece of fiber in your skin, remove it immediately with a pair of tweezers or seek medical attention.

CHEMICALS

In your work with fiber optics, you will use various types of chemicals; each poses a number of hazards and should be handled carefully. Every chemical that you use should be accompanied by a material safety data sheet (MSDS), which provides important information on the chemical's properties and characteristics such as appearance, odor, and common uses. The MSDS also gives you information on specific hazards posed by each chemical and ways to protect yourself through specific handling procedures, protective clothing and equipment, and engineering controls such as ventilation. Finally, the MSDS describes emergency procedures, including first aid for exposure to the chemical, methods for fighting fires in the case of flammable chemicals, and cleanup procedures for spills.

Even if you think you're familiar with the chemicals you handle, take time to read the MSDS. The information you gain could help prevent an accident or save valuable time in an emergency.

ELECTRICAL

Unlike mechanical slicing, fusion splicing uses a high-voltage electric arc between two electrodes to heat and melt the fiber ends as shown in Figure 8.11. Touching the electrodes while the arc is present can cause severe electrical shock. Remember that high voltage causes most of its damage by making muscles seize up, including the heart and lungs.

FIGURE 8.11
Close-up of a high-voltage electric arc melting the fiber ends during the splicing process

Photo courtesy of Aurora Optics

Fusion splicers like the one shown in Figure 8.12 have a cover that prevents access to the splicing area during the splicing process. This device typically has a safety interlock that interrupts the splicing process if lifted. Do not attempt to gain access to the splicing area until after the arc on the fusion splicer display shown in Figure 8.13 is extinguished. The arc is extinguished when it no longer appears on the display, as shown in Figure 8.14.

FIGURE 8.12
Automatic fusion
splicer with the
cover closed over
the splicing area

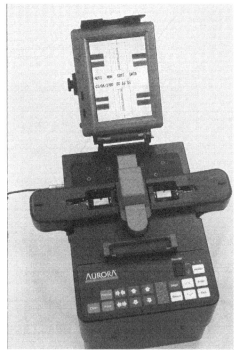

Photo courtesy of Aurora Optics

FIGURE 8.13
Electric arc melt-
ing the fiber ends
during the splicing
process shown on
the fusion splicer
display

Photo courtesy of Aurora Optics

FIGURE 8.14
Fusion splicer display after the arc is extinguished

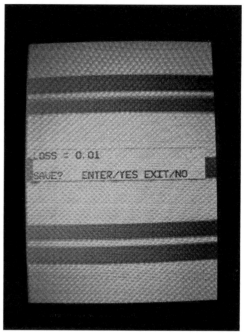

Photo courtesy of Aurora Optics

EXPLOSION

Prior to fusion splicing, ensure you are not in a potentially explosive atmosphere. Remember the arc generated during the splicing process is essentially a continuous spark capable of igniting flammable vapors such as those produced by an open container of isopropyl alcohol or the refueling of a vehicle. The ignition of these vapors will cause an explosion.

If you are using isopropyl alcohol during fusion splicing, keep it a safe distance from the fusion splicer and work in a well-ventilated area. Do not initiate the fusing process in the presence of a high concentration of isopropyl alcohol vapors.

In Chapter 7, "Fiber-Optic Cables," you learned that the aerospace and avionics industries are incorporating more optical fiber in aircraft. This is evident with the Boeing 787 Dreamliner, which has 110 fiber-optic links and over 1.7km of fiber-optic cable. As more commercial and military aircraft install optical fiber, there will be a need for fusion splicing.

A fueled aircraft is a hazardous environment for fusion splicing and special precautions must be taken prior to splicing. Only a fusion splicer like the one shown in Figure 8.15, developed specifically for explosion-proof operation, can be used in hazardous environments such as near a fueled aircraft.

FIGURE 8.15
Fusion splicer developed specifically for use in hazardous environments

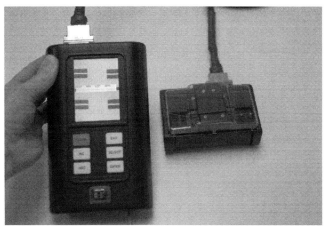

Photo courtesy of Aurora Optics

Splicing Equipment

The goal of any splice is to join two optical fiber endfaces together with as little loss as possible. This can be accomplished through mechanical or fusion splicing. This section describes the equipment required to perform a mechanical or fusion splice.

Cleaning Materials

One of the keys to performing a successful fiber-optic splice is cleanliness. The optical fiber needs to be properly cleaned prior to cleaving and splicing. In addition, the splicing area should be kept as clean as possible. As discussed earlier in the chapter, a microscopic piece of dirt could cause lateral displacement, end separation, or angular misalignment.

Prior to starting any splice, ensure you have lint-free wipes designed for optical fiber; a solvent designed to clean and remove the coating residue from the optical fiber; and if you will be fusion splicing, swabs to clean the V-groove. Lint-free wipes like those shown in Figures 25.16 and 25.17 are engineered to lift away oils, grime, and dust from the surface of the optical fiber.

When handling a lint-free wipe, be sure to only touch one side of the wipe because the wipe will absorb the oil from your skin. Use the side of the wipe that you did not touch for cleaning the optical fiber. You do not want to contaminate the optical fiber during the cleaning process. Cleaning of the optical fiber is covered in depth later in the "Splicing Procedures" section.

When you select an optical fiber cleaning solvent, ensure it has been engineered to clean optical fibers. The goal is to select a solvent that leaves little or no residue on the surface of the optical fiber. If you choose isopropyl alcohol, make sure that it is virtually free of water and has been filtered to remove impurities, like the isopropyl alcohol shown in Figure 8.18. Do not use the isopropyl alcohol commonly available at your local pharmacy; this alcohol contains a large percentage of water and will leave a residue on the optical fiber.

FIGURE 8.16
A container of lint-free wipes engineered for cleaning optical fibers

Photo courtesy of MicroCare

FIGURE 8.17
Packages of lint-free wipes engineered for cleaning optical fibers

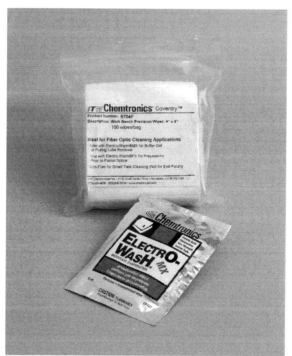

Photo courtesy of ITW Chemtronics

FIGURE 8.18
Isopropyl alcohol
for cleaning optical
fibers

Photo courtesy of MicroCare

Isopropyl alcohol poses a problem for air travel because of its flammability. If the splicing job requires air travel, you will not be able to bring isopropyl alcohol on the aircraft. For jobs requiring air travel, you need to select a cleaning fluid that is approved for air travel. Remember from Chapter 6 that isopropyl alcohol vapors are highly flammable and can ignite if exposed to a spark or flame in high enough concentrations.

Cleaning fluids like the one shown in Figure 8.19 are approved for air travel and engineered specifically to clean the optical fibers. These cleaning fluids typically contain little or no isopropyl alcohol and have been filtered to remove microscopic contaminates.

FIGURE 8.19
Optical fiber clean-
ing fluid approved
for air travel

Photo courtesy of MicroCare

As will be discussed later in this chapter, many fusion splicers have a precision V-groove that holds the optical fiber. Dirt or contamination in a V-groove can cause the optical fibers to misalign resulting in a high-loss fusion splice. When the V-grooves become contaminated, they can be cleaned using a fiber-optic swab, as shown in Figure 8.20.

FIGURE 8.20
Cleaning the V-groove with a fiber-optic swab

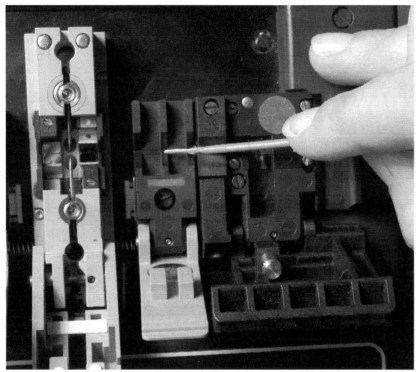

Photo courtesy of ITW Chemtronics

MicroCare, ITW Chemtronics, Corning Cable Systems, and AFL Telecommunications manufacture some of the most commonly used cleaning materials.

MicroCare www.microcare.com

ITW Chemtronics www.chemtronics.com

Corning www.corningcablesystems.com

AFL Telecommunications www.afltele.com

Cleavers

A cleaver is required for all splice types. Cleavers are available for single optical fibers or both single and ribbon fibers. Cleavers that support both single and ribbon fibers have removable adapters designed specifically for single fiber or ribbon fiber cleaving. An adapter is not inserted in the cleaver shown in Figure 8.21. A single fiber adapter is inserted in the cleaver shown in Figure 8.22.

FIGURE 8.21
Single or ribbon fiber cleaver without the adapter

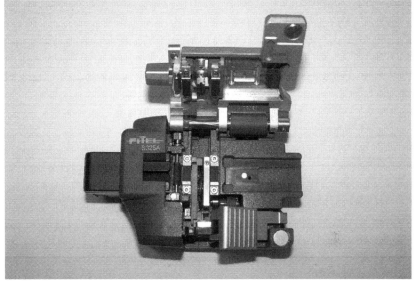

Photo courtesy of KITCO Fiber Optics

FIGURE 8.22
Single or ribbon fiber cleaver with the single fiber adapter inserted

Photo courtesy of KITCO Fiber Optics

Cleaving is typically a two-step process, and it takes place after the optical fiber has been properly prepared and cleaned. The first step in the process is scoring the optical fiber. This is accomplished with a scoring blade. The scoring blade should be held perpendicular to the optical fiber, as shown in Figure 8.23. When the cleaver is operated, the scoring blade lightly contacts the optical fiber, creating a small surface flaw.

FIGURE 8.23
Scoring blade ready
to score the optical
fiber

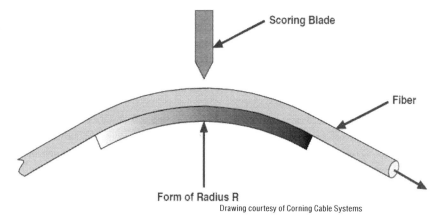

Drawing courtesy of Corning Cable Systems

After the optical fiber is scored, the next step is to bend the optical fiber. The optical fiber is bent slightly, as shown in Figure 8.23. Slightly bending the optical fiber places tensile stress on the surface of the optical fiber where it has been scored. This causes the small surface flaw to open and split the fiber.

After the optical fiber is cleaved, the endface should be perpendicular to the optical fiber without any surface defects and the cleave angle should not exceed 1.0°. Cleave angles can be evaluated only with a microscope. This is typically only done when a fusion splice is performed. Many fusion splicers will measure the cleave angle of the optical fiber prior to splicing.

Figure 8.24 is a photograph of a cleaved optical fiber with defects. In an ideal cleave, the mirror area would cover the entire endface. However, in this example the mirror area represents only a small portion of the endface. The rest of the endface looks like it was hacked with a sharp object many times. These irregular defects are commonly referred to as *hackles*.

FIGURE 8.24
Cleaved endface
with hackles

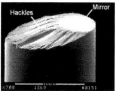

Photo courtesy of Corning Cable Systems

Mechanical Splice

Many manufacturers offer mechanical splices. Mechanical splices like the ones shown in Figures 8.25 and 8.26 are typically permanent. They align the cleaved optical fibers and hold them in place. Index matching gel inside the mechanical splice reduces or eliminates Fresnel reflections.

Mechanical splices can be used for both multimode and single-mode optical fiber. Mechanical splices do not outperform fusion splices. However, they will outperform mated connector pairs. The key advantage to a mechanical splice over a fusion splice is the low cost of the equipment required to perform the mechanical splice.

FIGURE 8.25
3M Fibrlok II universal mechanical splice

FIGURE 8.26
Corning CamSplice mechanical splice

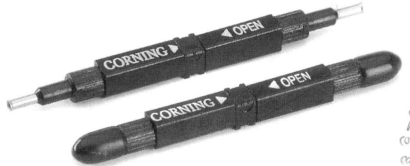

Photo courtesy of Corning Cable Systems

The assembly tool that holds the optical fibers and the mechanical splice is relatively inexpensive when compared to the price of a fusion splicer. However, the actual mechanical splices shown in Figures 8.25 and 8.26 cost considerably more than the protective sleeve required for a fusion splice.

SPECIALTY MECHANICAL SPLICE

Up to this point, we have only discussed splicing optical fibers that have been cleaved so the endface is perpendicular to the optical fiber. However, angled mechanical splices that offer better *return loss* performance than traditional mechanical splices are being introduced. Return loss is covered in detail in Chapter 12, "Passive Components and Multiplexers."

Angled mechanical splices require a special cleaver, angle cleaved fiber holders, and a special mechanical splice. The special cleaver like the one shown in Figure 8.27 places a 7° angle on the fiber endface while the angle cleaved fiber holders maintain the orientation of the cleave. Maintaining the orientation of the cleaved fibers is critical so that they align properly inside the mechanical splice.

FIGURE 8.27
3M fiber optic angle cleaver

Photo courtesy of 3M

The angled mechanical splice is an alternative to fusion splicing for analog video transmission, which typically requires lower return loss than a digital transmission. The 7° angle on the fiber endface reflects light into the cladding where it is absorbed. This reduces the amount of light energy that could be reflected back toward the light source, minimizing the back reflection.

Like the traditional mechanical splice, the angled mechanical splice is filled with index matching gel. The combination of the 7° angle and the index matching gel create a mechanical spice that meets or exceeds the return loss requirements for analog video transmission.

3M and Corning Cable Systems manufacture some of the most commonly used mechanical splices.

3M   www.3m.com

Corning   www.corningcablesystems.com

Fusion Splice

As you learned earlier in this chapter, fusion splicing uses a high-voltage electric arc between two electrodes to heat and melt the fiber ends as shown in Figure 8.11. This creates a permanent splice that is very fragile and must be protected from the outside environment and bending. This is typically accomplished using heat-shrink tubing and a small metal rod commonly referred to as a protective sleeve.

Prior to performing the splice, the optical fiber is passed through the center of the protective sleeve and the sleeve is positioned so it will not interfere with the fusion splicing process. After the fusion splice is successfully completed, the optical fiber is removed from the fusion splicer and the protective sleeve is positioned directly over the slice area.

The protective sleeve is then placed in an electric oven that is typically built into the fusion splicer. The oven heats the tubing, shrinking it around the fusion splice. The end result looks like the protected fusion splice shown in Figure 8.28.

FIGURE 8.28
A fusion splice covered with heat shrink held rigid with a metal rod

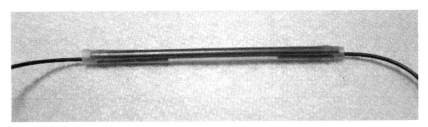

Fusion splicers are more expensive than the assembly tool required for a mechanical splice. However, they provide the lowest-loss splice possible. In addition, fusion splices do not produce Fresnel reflections.

Fusion splicing is the most accurate and durable method for joining two optical fibers. After the optical fibers are stripped and cleaved, they are placed into the fusion splicer, where the optical fibers are aligned between two electrodes, as shown in Figure 8.29. Alignment may be accomplished several ways depending on the type of fusion splicer used.

FIGURE 8.29
Optical fibers aligned between the fusion splicer electrodes

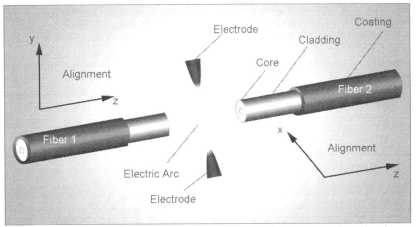

Drawing courtesy of Corning Cable Systems

Several alignment techniques are used to align the optical fibers. A common alignment technique is to use a fixed V-groove, as shown in Figure 8.30. Each optical fiber is placed in a precision-machined V-groove. The V-grooves align the cladding of the optical fibers. The fusion splicer moves the optical fibers axially into the electric arc, melting the optical fibers together.

FIGURE 8.30

Optical fibers in a V-groove aligned between the fusion splicer electrodes

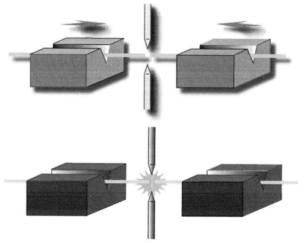

Drawing courtesy of Corning Cable Systems

Fusion splicers with this type of alignment system have the lowest cost and are typically the smallest and lightest. However, they align the cladding of the optical fiber instead of the core and can create higher loss splices than fusion splicers that align the cores of the optical fibers.

A fusion splicer that aligns the cores of the optical fibers will produce the lowest-loss splice. However, aligning the cores of the optical fibers is more difficult than aligning two optical fibers placed in V-grooves. To align the cores, the fusion splicer must be able to detect the cores and have the ability to move the optical fibers on each axis.

Several patented technologies are available to detect the cores of optical fibers. Because they are patented, they will not be discussed in detail in this chapter.

Fusion splicers on the market today typically feature a display that allows you to see the optical fibers on two different axes. The cameras in the fusion splicer magnify the optical fibers so that the endfaces can be evaluated. You cannot look at an optical fiber with the naked eye and evaluate the endface. Without magnification, optical fiber with a perfect cleave will look just like an optical fiber that has been broken. Figure 8.31 is a photograph of a fusion splicer display showing a two-axis view of properly cleaved and aligned optical fibers. Shown in Figure 8.32 is a two-axis view of two broken optical fibers. This photograph clearly shows that a broken optical fiber can have a jagged endface.

Most fusion splicers available today also have the ability to approximate the loss for the splice after it is completed. Fusion splicers are also available that have the ability to measure the loss for the splice after it is completed. Figure 8.33 is a photograph of a fusion splicer display showing the estimated splice loss. The splice loss is displayed on the bottom of the screen, and different axis views of the spliced optical fiber are shown at the top of the screen. Note that the spliced optical fibers appear as a single optical fiber with no flaws.

FIGURE 8.31
Properly cleaved
and aligned optical
fibers viewed on the
display of a fusion
splicer

Photo courtesy of Aurora Optics

FIGURE 8.32
Optical fibers with
jagged endfaces
viewed on the
display of a fusion
splicer

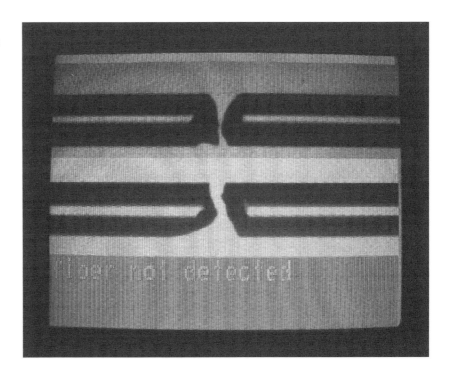

FIGURE 8.33
Fusion splicer
displaying the esti-
mated splice loss

You can find out more about specific features of fusion splicers through their manufacturers' websites. Some of the major manufacturers of fusion splicers are as follows:

Corning www.corningcablesystems.com

Aurora Optics www.aurora-optics.com

AFL Telecommunications www.afltele.com

Fitel www.fitel.com

Next let's look at the procedures used with each type of splicer.

Splicing Procedures

This section will familiarize you with the basic procedures used for mechanical or fusion splicing. While manufacturers' models will vary, many of the same principles and requirements apply. Be sure to read and follow the directions for your particular equipment carefully.

WARNING Whenever handling optical fiber, remember to follow the safety precautions described in Chapter 6.

Mechanical Splicing Procedure

To prepare for mechanical splicing, make sure that the work area is clean, dry, and well lit. Do all your work over a fiber-optic mat and place any scrap optical fibers in their proper container, as described in Chapter 6. Assemble the following tools before you begin:

♦   Mechanical splice assembly tool

- Mechanical splice

- Buffer and coating removal tool

- Optical fiber cleaning fluid

- Lint-free wipes

- Cleaver

Once your materials are assembled, clean the cleaver as directed by the manufacturer, and then proceed with the following steps:

1. Remove the 3M Fibrlok II mechanical splice from its protective packaging and load the splice into the assembly tool by pressing firmly at the ends of the splice. Do not depress the raised section on the mechanical splice.

2. Strip approximately 3cm of buffer and/or coating from the optical fiber using a stripper, as shown in Figure 8.34.

FIGURE 8.34
Stripping the buffer and coating from an optical fiber

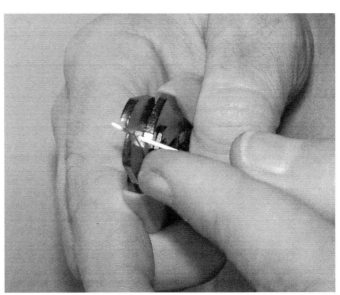

3. Clean the optical fiber by pulling the fiber through a lint-free wipe soaked in optical fiber cleaning fluid, as shown in Figure 8.35.

4. Place the optical fiber in the cleaver, as shown in Figure 8.36, to the length specified by the mechanical splice manufacturer.

FIGURE 8.35
Clean the optical fiber by pulling it through the lint-free wipe soaked in optical fiber cleaning fluid.

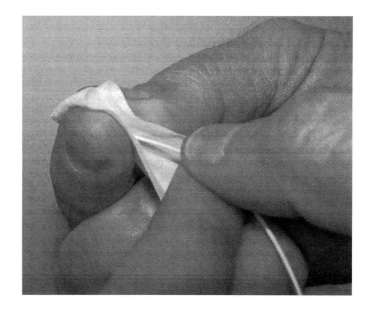

FIGURE 8.36
Place the optical fiber in the cleaver at the length specified by the manufacturer.

Photo courtesy of KITCO Fiber Optics

5. Cleave the optical fiber.

6. If a gauge is provided, check the cleave length with the gauge. If the optical fiber has touched the gauge, repeat step 3.

7. Repeat steps 2 through 6 for the other fiber end to be spliced.

8. Push one cleaved optical fiber into one end of the mechanical splice until it stops moving.

9. Push the other cleaved optical fiber into the other end of the mechanical splice until it stops moving.

10. Place both optical fibers in the clamping mechanisms on the opposite sides of the splicing tool, forming a modified loop in the optical fiber, as shown in Figure 8.37.

FIGURE 8.37
Mechanical splice
tool properly set
up to perform a
mechanical splice

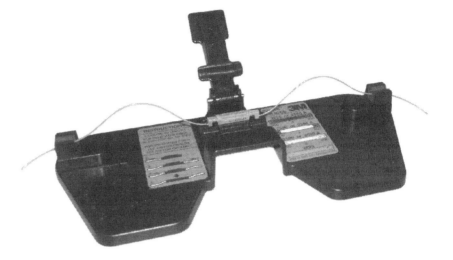

11. Pivot the splicing tool handle down until it contacts the top of the splice; then squeeze the tool handle to complete the assembly and lock the spliced ends in place.

12. Remove the optical fibers from the clamping mechanisms and lift the mechanical splice from the tool.

MECHANICAL SPLICING TROUBLESHOOTING

Following the mechanical splicing procedures outlined earlier typically results in a low-loss splice that outperforms the attenuation requirements for inside or outside plant splices. However, occasionally you will encounter a splice that does not meet the attenuation requirement. As you learned earlier, how well a splice performs depends on many variables. These variables can be broken into two groups: intrinsic factors and extrinsic factors.

Intrinsic factors are the result of slight variations from one optical fiber to another. These variations can affect the performance of the splice even though the optical fibers are perfectly aligned when mated. However, the optical fibers available today have very little variation, which means that a small percentage of the splice loss may be from intrinsic factors.

Extrinsic factors are related to the condition of the splice itself, external to the optical fiber. Many times, they are caused by dirt and contamination. Microscopic particles of dirt can cause the misalignment of one or both optical fibers, creating a high-loss splice.

Many mechanical splices can be opened or unlocked so they can be used more than one time in case the initial splice does not meet the attenuation or back reflection requirements. Next we will examine some of the common causes for a high-loss mechanical splice.

Cleave Angle

As you've learned, the endface of a properly cleaved optical fiber should be perpendicular to the optical fiber without any surface defects and the cleave angle should not exceed 1.0°. Cleave angles can be evaluated only with a microscope. Unfortunately, mechanical splicing tools are not equipped with a 2-axis microscope as most fusion splicers are. If you suspect a bad cleave, perform the following steps:

1. Open or unlock the splice and remove the optical fibers.

2. Prepare the optical fibers for cleaving adhering to the manufacturer's instructions.

3. Ensure the optical fiber and cleaver is clean.

4. Re-cleave both optical fibers.

5. Verify the cleave dimensions as stated in the manufacturer's instructions.

6. Reassemble the mechanical splice as described in the manufacturer's instructions.

Cleave Length

Cleave length is just as important as cleave angle. Each mechanical splice manufacturer defines how much bare optical fiber should be exposed after the cleaving process is completed. This dimension is typically provided in millimeters with maximum and minimum values. Exposing too much optical fiber or not enough optical fiber can create a high-loss mechanical splice. If you suspect improper cleave length, follow these steps:

1. Open or unlock the splice and remove the optical fibers.

2. Prepare the optical fibers for cleaving adhering to the manufacturer's instructions.

3. Ensure the optical fiber and cleaver is clean.

4. Re-cleave both optical fibers.

5. Verify the cleave dimensions as stated in the manufacturer's instructions.

6. Reassemble the mechanical splice as described in the manufacturer's instructions.

Contamination

Contamination on the optical fiber or cleaver that is invisible to the unaided eye can cause a mechanical splice to exceed attenuation and back reflection requirements. Because you cannot see the contamination, make sure that you are using the types of cleaning products and the cleaning process required by the splice and cleaver manufacturer or described in this chapter. Never let a cleaned optical fiber touch any surface after cleaving. If you have any cleaning related questions, you can always reach out to a cleaning product manufacturer such as MicroCare or ITW Chemtronics for help.

MicroCare www.microcare.com

ITW Chemtronics www.chemtronics.com

CAN'T AFFORD A FUSION SPLICER?

If you want the joy of fusion splicing without the purchase price, you can rent or lease a fusion splicer. Companies such as Fiber Instrument Sales, Inc. (`www.fiberinstrumentsales.com`) provide rental arrangements for fusion splicers and other high-end fiber equipment if you know you won't need it on a long-term or recurring basis.

Rental rates typically run $^1/_{10}$ to $^1/_{12}$ of the purchase price of a new fusion splicer per month. Bear in mind, though, that the rental period starts the day that the splicer is shipped to you and ends the day that it is received back at the rental office. In other words, you're going to pay for the days that it is in transit.

If you still want to buy a fusion splicer but don't want to pay full price, consider a preowned or reconditioned unit. These are occasionally available through distributors and still have years of life left in them.

Fusion Splicing Procedure

The equipment required for fusion splicing is far more complex and expensive than the equipment required for mechanical splicing. However, many of the steps required to perform a fusion splice are identical to the steps required to perform a mechanical splice.

Many fusion splicers contain a feature that automatically positions the fiber ends in proper relationship with each other and with the electrodes for the best possible splice. All that is required of the operator is to prepare the fibers properly and place them in the fusion splicer as outlined by the manufacturer.

The fusion-splicing procedures outlined in this chapter describe the typical steps required to fusion-splice single optical fibers and ribbon optical fibers. These steps are not product specific and modifications may be required depending on the fusion splicing equipment being used. Always review the manufacturer's operating documentation prior to performing a splice.

To prepare for fusion splicing, as with mechanical splicing, make sure that the work area is clean, dry, and well lit. Do all your work over a fiber-optic mat and place any scrap optical fibers in their proper container, as described in Chapter 6. Assemble the following tools before you begin:

- Fusion splicer
- Buffer and/or coating removal tool
- Optical fiber cleaning fluid
- Lint-free wipes
- Cleaver
- Heat-shrink protective covering

Once your materials are assembled, clean the cleaver and the fusion splicer as directed by the manufacturer, and then proceed with the following steps. First, we'll look at single optical fiber fusion splicing:

1. Power on the fusion splicer and select the appropriate splicing program or profile for the optical fiber you will be splicing.

2. Slide the protective heat-shrink tubing over one optical fiber end and move it far enough up the optical fiber to place it out of the way.

3. Strip approximately 3cm of buffer and/or coating from the optical fiber using a stripper, as shown in Figure 8.38.

FIGURE 8.38
Stripping the buffer and coating from an optical fiber

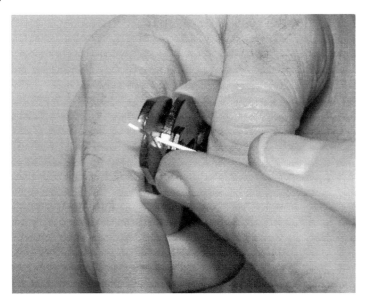

4. Clean the optical fiber by pulling the fiber through a lint-free wipe soaked in optical fiber cleaning fluid, as shown in Figure 8.39.

FIGURE 8.39
Clean the optical fiber by pulling it through the lint-free wipe soaked in optical fiber cleaning fluid.

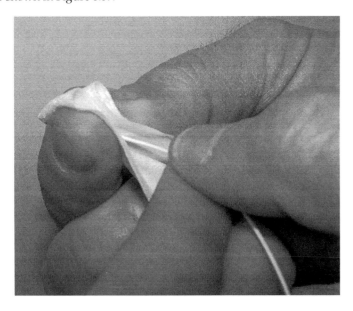

5. Place the optical fiber in the cleaver, as shown in Figure 8.40, to the length specified by the mechanical splice manufacturer.

FIGURE 8.40
Place the optical fiber in the cleaver at the length specified by the manufacturer.

Photo courtesy of KITCO Fiber Optics

6. Cleave the optical fiber.

7. Place the optical fiber in the fusion splicer following the manufacturer's instructions. Position the endface of the optical fiber between the electrodes.

8. Repeat steps 3 through 7 for the other fiber end to be spliced. The properly placed fibers should be slightly separated between the electrodes, as shown in Figure 8.41.

FIGURE 8.41
Properly placed fibers almost touch each other between the electrodes.

Photo courtesy of KITCO Fiber Optics

9. Close the electrode cover.

10. Begin the fusion-splicing process.

11. Carefully remove the splice and position the heat-shrink tubing from step 2 over it. Place the splice and tubing in the heat-shrink oven to seal and protect the splice.

Next we'll look at ribbon fiber fusion splicing:

1. Power on the fusion splicer and select the appropriate splicing program or profile for the optical fiber you will be splicing.

2. Slide the protective heat-shrink tubing over one ribbon fiber end and move it far enough up the optical fiber to place it out of the way.

3. Strip approximately 3cm of Mylar tape and coating from the ribbon fiber, as shown in Figure 8.42.

FIGURE 8.42
Ribbon fiber stripper using heat to remove the Mylar tape and coating

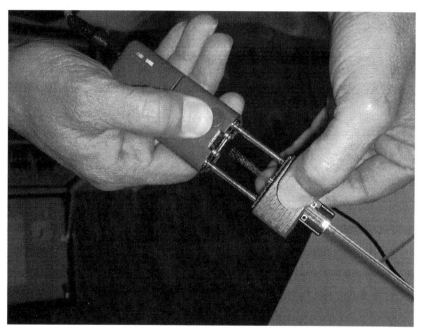

Photo courtesy of MicroCare

4. Clean the ribbon fiber by pulling the fiber through a lint-free wipe soaked in optical fiber cleaning fluid, as shown in Figure 8.43.

FIGURE 8.43
Ribbon fiber being pulled through a lint-free wipe soaked in optical cleaning fluid

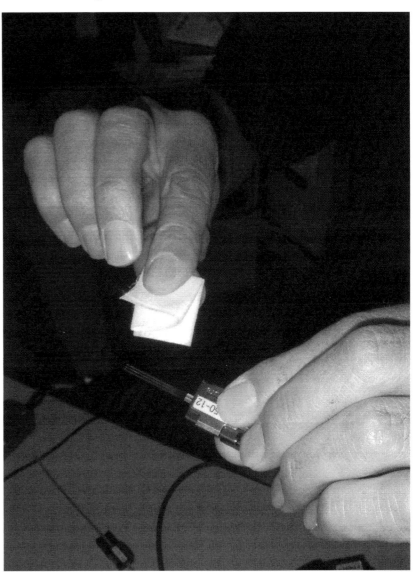

Photo courtesy of MicroCare

5. Place the ribbon fiber in the cleaver, as shown in Figure 8.44, to the length specified by the manufacturer of splicer you are using.

FIGURE 8.44
Ribbon fiber placed in the cleaver for cleaving

Photo courtesy of MicroCare

6. Cleave the ribbon fiber.

7. Place the ribbon fiber in the fusion splicer, as shown in Figure 8.45. Position the ribbon fiber as described in the manufacturer's instructions.

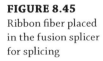

FIGURE 8.45
Ribbon fiber placed in the fusion splicer for splicing

Photo courtesy of MicroCare

8. Lock the ribbon fiber in place as described in the manufacturer's instructions.

9. Repeat steps 3 through 8 for the other fiber end to be spliced.

10. Close the electrode cover.

11. Begin the fusion-splicing process.

12. Carefully remove the splice and position the heat-shrink tubing from step 2 over it. Place the splice and tubing in the heat-shrink oven to seal and protect the splice.

FUSION SPLICING TROUBLESHOOTING

Following the fusion splicing procedures outlined earlier typically results in a low-loss splice that outperforms the attenuation requirements for inside or outside plant splices. However, occasionally you will encounter a splice that exceeds maximum attenuation requirements. As you've learned, how well a splice performs depends on many variables. As with mechanical splicing, these variables can be broken into two groups: intrinsic factors and extrinsic factors.

Intrinsic factors are the result of slight variations from one optical fiber to another. These variations can affect the performance of the splice even though the optical fibers are perfectly aligned when mated. However, the optical fibers available today have very little variation, which means that a small percentage of the splice loss may be from intrinsic factors.

Extrinsic factors are related to the condition of the splice itself, external to the optical fiber. Many times, they are caused by dirt and contamination. Microscopic particles of dirt can cause the misalignment of one or both optical fibers, creating a high-loss splice.

Next we will examine some of the common causes for a high-loss fusion splice.

Cleave Angle

As discussed earlier, the endface of a properly cleaved optical fiber should be perpendicular to the optical fiber without any surface defects and the cleave angle should not exceed $1.0°$. Cleave angles can be evaluated only with a microscope. Fortunately many fusion splicers are equipped with a 2-axis microscope that allows you to observe the cleave and the measured cleave angle values. If you or the fusion splicer detects a bad cleave, as shown in Figure 8.46, perform the following steps:

1. Remove the optical fibers from the fusion splicer.

FIGURE 8.46
Improperly cleaved optical fibers viewed on the display of a fusion splicer

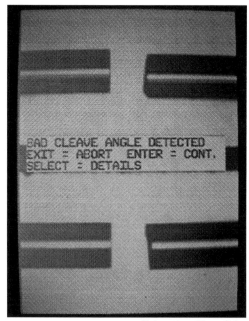

Photo courtesy of Aurora Optics

2. Prepare the optical fibers for cleaving adhering to the manufacturer's instructions.

3. Ensure the optical fiber and cleaver is clean.

4. Re-cleave both optical fibers.

5. Verify the cleave dimensions as stated in the manufacturer's instructions.

6. Place the optical fibers in the fusion splicer following the manufacturer's instructions. Position the endface of the optical fibers between the electrodes.

7. Perform the fusion splice.

Cleave Length

Cleave length is just as important as cleave angle. Each fusion splice manufacturer defines how much bare optical fiber should be exposed after the cleaving process is completed. This dimension is typically provided in millimeters with maximum and minimum values. Exposing too much optical fiber or not enough optical fiber can create a high-loss fusion splice. If you suspect improper cleave length, follow these steps:

1. Remove the optical fibers from the fusion splicer.

2. Prepare the optical fibers for cleaving, adhering to the manufacturer's instructions.

3. Ensure the optical fiber and cleaver is clean.

4. Re-cleave both optical fibers.

5. Verify the cleave dimensions as stated in the manufacturer's instructions.

6. Place the optical fibers in the fusion splicer, following the manufacturer's instructions. Position the endface of the optical fibers between the electrodes.

7. Perform the fusion splice.

Bubbles

Dirt or entrapped air may create a bubble or bubbles, resulting in a high-loss fusion splice like the one shown in Figure 8.47. To prevent bubbles:

1. Ensure you are using the types of cleaning products and the cleaning processes required by the fusion splicer and cleaver manufacturer or described in this chapter.

2. Verify that you have selected the appropriate splicing program or profile for the optical fiber you are splicing.

FIGURE 8.47
Bubbles in a high-loss fusion splice viewed on the display of a fusion splicer

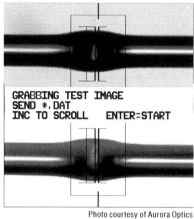

Photo courtesy of Aurora Optics

Necking

Necking is a term used to describe a fusion splice where the diameter of the fused optical fiber is smaller near the electrodes than it was prior to being fused, as shown in Figure 8.48. Necking is typically the result of too much heat during the prefuse, causing glass to be transported away from the splice area and thus producing a high-loss splice. To prevent necking:

1. Ensure you are using the types of cleaning products and the cleaning processes required by the fusion splicer and cleaver manufacturer or described in this chapter.

2. Verify that you have selected the appropriate splicing program or profile for the optical fiber you are splicing.

3. Ensure the optical fibers are properly cleaved.

FIGURE 8.48
A high-loss fusion splice caused by necking viewed on the display of a fusion splicer

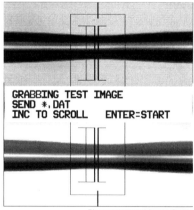

Photo courtesy of Aurora Optics

Contamination

Contamination on the optical fiber or cleaver that is invisible to the unaided eye can cause fusion splices to exceed attenuation requirements. Because you cannot see the contamination, make sure that you are using the types of cleaning products and the cleaning process required by the fusion splicer and cleaver manufacturer or described in this chapter. Never let a cleaned optical fiber touch any surface after cleaving. If you have any cleaning-related questions, you can always reach out to a cleaning product manufacturer such as MicroCare or ITW Chemtronics for help.

MicroCare www.microcare.com

ITW Chemtronics www.chemtronics.com

Splice Requirements

As with connectors, standards specify a maximum permissible loss for splices, depending on their location. For inside plant splices, ANSI/TIA-568-C.3 states that optical fiber splices, whether fusion or mechanical, will have a maximum attenuation of 0.3dB.

For outside plant splices, ANSI/TIA-758-B states that the splice insertion loss shall not exceed 0.1dB mean, with a maximum of 0.3dB, as measured with an optical time-domain reflectometer (described in Chapter 16, "Test Equipment and Link/Cable Testing").

FUSION OR MECHANICAL?

It may seem like an easy choice between mechanical and fusion splicing: fusion splices are superior to mechanical splices because they essentially make one optical fiber out of two. The cost per fusion splice is virtually nil, whereas mechanical splices typically cost from $7 to $25 per splice. In short, if you have a fusion splicer, there is really no need to even look at mechanical splicing.

So why bother with mechanical splices at all?

The one key factor that favors mechanical splicing is the initial setup cost. Mechanical splicing and fusion splicing both require a cleaver, hand tools, and cleaning supplies. However, only a mechanical splice assembly tool is required to assemble the mechanical splice. This tool typically costs around $100 whereas many fusion splicers cost more than $10,000.

If your application requires only a few splices, mechanical splicing is the most economical way to go. However, if your application requires many splices, fusion splicing may be the most economical way to go.

The Bottom Line

Determine if the splice loss is from an intrinsic factor.   Even when fibers are manufactured within specified tolerances, there are still slight variations from one optical fiber to another. These variations can affect the performance of the splice even though the optical fibers are perfectly aligned when mated.

Master It   Two optical fibers are about to be fusion spliced together, as shown here. Determine what type of problem exists with the optical fibers about to be spliced.

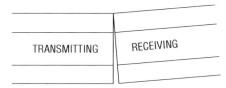

Determine if the splice loss is from an extrinsic factor.   Extrinsic factors that affect optical fiber splice performance are factors related to the condition of the splice itself, external to the optical fiber.

Master It   Two optical fibers are about to be fusion spliced together, as shown here. Determine what type of problem exists with the optical fibers about to be spliced.

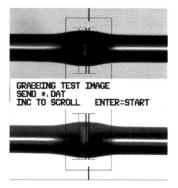

Calculate the potential splice loss from a core diameter mismatch.   A core diameter mismatch loss results when the core diameter of the transmitting optical fiber is greater than the core diameter of the receiving optical fiber.

Master It   Calculate the worst-case loss for a splice where the transmitting optical fiber has a core diameter of 51µm and the receiving optical fiber has a core diameter of 49.5µm.

Troubleshoot a fusion splice.   How well a splice performs depends on many variables. These variables can be broken into two groups: intrinsic factors and extrinsic factors.

Master It   Two optical fibers have been fused together as shown here. Determine the type of problem and potential causes.

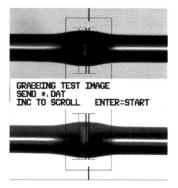

Chapter 9

Connectors

The ideal traveling environment for a pulse of light is an unbroken optical fiber. At some point, however, that optical fiber must connect to a piece of equipment or join another optical fiber in order to extend its length or change the type of cable being used.

One of the most common methods for *terminating* an optical fiber, or making its end useful, is to use a connector. A connector is a device that supports the end of the optical fiber while allowing it to be quickly and reliably joined to equipment or other optical fibers.

Connectors are often used instead of splices to join two optical fibers together because they allow the optical fibers to be disconnected and reconnected easily. Splices, on the other hand, are permanent connections between two optical fibers. Connectors can be useful when network assignments must be changed, when equipment must be removed/replaced, or when expansion is anticipated.

This chapter describes common connectors used in optical fiber termination. It investigates the factors that affect connector performance and methods used to improve performance. The chapter also discusses methods used to install connectors so they meet industry performance standards.

In this chapter, you will learn to:

♦ Evaluate connector endface polishing

♦ Evaluate connector endface cleaning

♦ Evaluate connector endface geometry

♦ Identify an optical fiber type from the color of the connector strain relief

The Fiber-Optic Connector

The job of a fiber-optic connector is to couple an optical fiber end mechanically to a piece of equipment or to another optical fiber so that the cores line up accurately and produce the smallest amount of loss. Inherent in this requirement is the need for the connector to protect the fiber from repeated handling during connection and disconnection, align the fiber end precisely with its counterpart in the interconnection, and prevent strain on the fiber itself. The strength members in the fiber-optic cable normally provide strain relief, but once the fiber itself is attached to the connector, the connector must perform that job.

Although there are several types of connectors recognized by industry standards, they all contain common components, shown in Figure 9.1.

FIGURE 9.1

Fiber-optic connec-
tor components

S T

S C

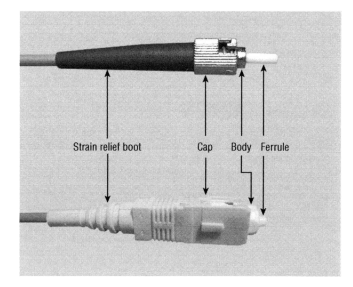

Strain relief boot   Cap   Body   Ferrule

Ferrule

Beginning at the working end of the connector, the *ferrule* holds the fiber in place. The ferrule must hold the fiber exactly centered in its endface for the best possible connection, so its construction is critical. Not only must the hole for the optical fiber be accurately placed; it must also be sized precisely to receive the exact diameter of the optical fiber cladding. The optical fiber's coating is stripped away prior to inserting the optical fiber into the ferrule.

Ferrules typically are made of metal, ceramic, or plastic with a selection of hole or bore diameters ranging from slightly larger than the optical fiber diameter to slightly smaller to allow for minute variations in the manufactured optical fiber cladding diameters. For example, for a 125µm optical fiber, ferrules might be available with hole sizes ranging from 124µm to 127µm. — *why* Because the ferrule must align the optical fiber end precisely, it must meet several important criteria:

♦ The ferrule must be strong enough to withstand many cycles of connection and disconnection without bending, cracking, or breaking.

♦ The ferrule must maintain dimensional stability to ensure proper alignment of the optical fiber.

♦ The ferrule must be of the right shape and have material properties to ensure a low-loss interconnection.

Ceramic materials such as aluminum oxide and zirconium oxide are among the best materials for ferrules, offering the best combination of characteristics. They are hard enough to protect the fiber end, and their *coefficient of thermal expansion*, the measure of how much a material expands and contracts with temperature changes, is about the same as the optical fiber itself.

Metal ferrules, typically made of stainless steel, are stronger than ceramic, but they are less dimensionally stable. Plastic ferrules are less expensive than metal or ceramic, but they are neither as strong nor as stable as the other materials.

When a connector is assembled, which we'll describe in detail later in this chapter, the optical fiber is typically epoxied into the ferrule with the end protruding slightly beyond the endface of the ferrule. The optical fiber end is later trimmed and polished with the ferrule endface for a precise fit.

The ferrule fits inside the next component, the body of the connector. The body, which can be either metal or plastic, holds the ferrule, the optical fiber, and the cable in place and transfers any strain placed on the connector to the cable rather than the optical fiber.

Cap

The cap, sometimes called the coupling nut, fits over the body of the connector and provides a means to secure the connector, as shown in Figure 9.2. The cap can be a locking mechanism, a threaded ring, or a snap attachment, depending on the type of connector.

FIGURE 9.2
Connector cap
types

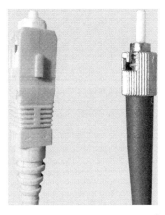

The strength member of the fiber-optic cable is secured to the connector. This allows all the tensile stresses placed on the connector to be handled by the cable's strength member instead of the optical fiber. There are several ways to accomplish this; crimping a ring or band around the strength member, securing it to the connector body, is a very popular method. Figure 9.3 shows the band of an LC connector securely attaching the strength member to the connector body.

FIGURE 9.3
Crimped band of
an LC connector
securely attaching
the cable strength
member to the con-
nector body

Body

Depending on the connector type, the cable jacket may or may not be securely fastened to the connector body. Sometimes a band or ring is crimped around the cable jacket to secure it to the connector body, as shown in Figure 9.4. Other times a piece of shrink tubing is used to secure the jacket to the connector body, as you can see in Figure 9.5.

FIGURE 9.4
Crimped band of an SC connector securely attaching the cable jacket to the connector body

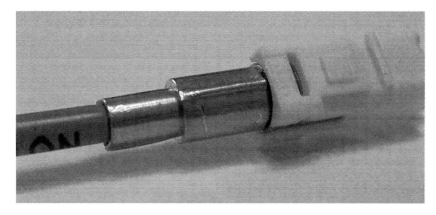

FIGURE 9.5
Shrink tubing attaching the cable jacket to an LC connector body

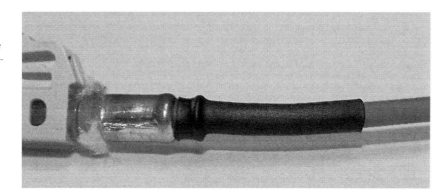

Strain Relief

A strain relief, or boot, is typically placed over the cable jacket and secured to the connector body. The boot is generally made of an elastic material that slides over the connector body. Friction between the boot and the connector body holds the boot in place. The boot prevents the cable from being pulled at too great an angle against the connector, as shown in Figure 9.6.

Fiber-optic connectors, commonly referred to as *plugs*, mate with other connectors or receptacles. The ferrule of the connector is the plug, and when two connectors mate, their ferrules are aligned with a sleeve that is typically called a *mating sleeve* or an *alignment sleeve*. Figure 9.7 is a photograph of an LC alignment sleeve that mates two LC connectors.

FIGURE 9.6
Boot or strain relief
preventing the
fiber-optic cable
from being over
bent

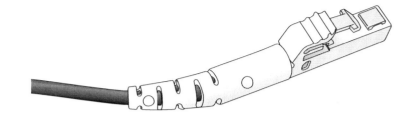

FIGURE 9.7
LC mating or align-
ment sleeve with
connectors

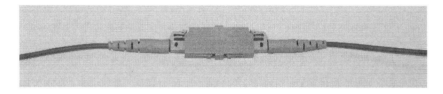

Connectors also mate with equipment or device receptacles. The device may be a transmitter or receiver, as discussed in Chapter 10, "Fiber-Optic Light Sources and Transmitters," and
Chapter 11, "Fiber-Optic Detectors and Receivers." Or it may be a passive device like one of
those described in Chapter 12, "Passive Components and Multiplexers." The receptacle is similar to one-half of a mating sleeve. Figure 9.8 shows an ST mating sleeve and Figure 9.9 shows ST
transceiver receptacles. Looking at these two photographs, you can see how closely they match.
You can also see a slot at the receiving end of the receptacle and both ends of the sleeve. This slot
is for the key on the connector body.

FIGURE 9.8
ST mating or align-
ment sleeve without
connectors

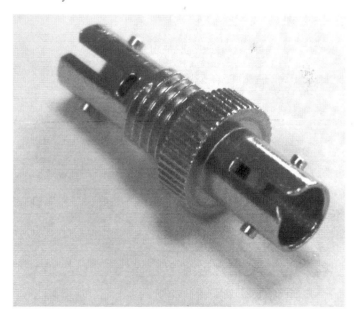

NOTE The terms *alignment sleeve*, *mating sleeve*, and *adapter* are interchangeable. ANSI/TIA-568-C defines an adapter as a mechanical device designed to align and join two optical fiber connectors (plugs) to form an optical connection. However, sometimes the term *adapter* is used when referring to a device that mates two different connector types such as an SC and ST. Every adapter contains an alignment sleeve that aligns the connector ferrules or plugs.

FIGURE 9.9
ST transceiver
receptacles

The connector body can take many forms; however, most connector body types fall into one of four categories:

♦ Round

♦ Square

♦ Rectangular

♦ D-shaped

The connector body may or may not have a key to orient the connector. The key ensures the connector only goes into the receptacle or alignment sleeve one way, and it prevents the ferrule from rotating and damaging the optical fiber when it makes contact with another optical fiber. Connectors without a key cannot make contact with the optical fiber of another connector when they are mated. Contact and noncontact connectors are discussed in detail later in this chapter.

The ST connector is an example of a round body type contact connector; the key typically has a pin-like shape, as shown in Figure 9.10, that must be aligned with a slot in the receptacle or the alignment sleeve, as shown in Figures 9.8 and 9.9 earlier. If the key is not aligned, you will not be able to seat the connector and a high-loss interconnection will result.

The LC connector is an example of a square body type with a rectangular key at the top of the connector as shown in Figure 9.11. The MTP connector shown in Figure 9.12 is an example of a rectangle body type with a rectangular key at the top.

The SC connectors shown in Figure 9.13 are an example of a D-shaped body that serves as the key to align it properly as it is inserted into the cap during assembly. The cap is rectangular with a rectangular key on the top that ensures it is not plugged in upside down.

FIGURE 9.10
ST connector with key visible

FIGURE 9.11
LC connector with key visible

FIGURE 9.12
MTP connector with key visible

FIGURE 9.13
SC connectors with
the D-shaped body
visible

The key design in each of these connectors ensures that they can be inserted into a receptacle or alignment sleeve in only one way. All of these connectors are discussed in detail in the "Connector Types" section later in this chapter.

2-

Connection Performance

As described earlier, the connector mates with either another connector or the equipment. The performance of two connectors mated together, also referred to as a mated connector pair, can be measured. This is described in detail in Chapter 16, "Test Equipment and Link/Cable Testing."

The performance of a connector mating with a piece of equipment such as a transceiver or passive optical device requires specialized test equipment that is described in detail in Chapter 16. Because of the specialized equipment required to perform these measurements, they typically are not performed in the field.

Connection performance describes the performance of an interconnection. In this section, we are going to examine the factors that affect the performance of an interconnection. This interconnection may consist of two connectors and a mating sleeve or a connector and a receptacle.

Connection performance is dependent on several factors, including intrinsic factors, extrinsic factors, geometry, endface finish, and cleanliness. Intrinsic factors, extrinsic factors, and cleanliness apply to the connector, mating sleeve, and the receptacle. Endface finish and geometry apply only to the connector. Endface finish and the cleaning of the connector endface are discussed at the end of the chapter after connector assembly. Cleaning and inspection of the mating sleeve and receptacle is discussed in detail in Chapter 17, "Troubleshooting and Restoration."

Intrinsic Factors

As discussed in detail in Chapter 8, "Splicing," even when optical fibers are manufactured within specified tolerances, there are still slight variations from one optical fiber to another. The same is true for the mating sleeve and receptacle. Variations in these components can affect the performance of the connection.

Let's briefly review the most common types of intrinsic variations:

NA mismatch   A numerical aperture (NA) mismatch occurs when the NA of one optical fiber is different from the NA of the other optical fiber. A loss may occur if the NA of the transmitting optical fiber is larger than the NA of the receiving optical fiber.

Core diameter mismatch   Core diameter mismatch occurs when there is a difference in the core diameters of the two optical fibers. A loss may occur when the core diameter of the transmitting optical fiber is greater than the core diameter of the receiving optical fiber.

Mode field diameter mismatch   A mode field diameter mismatch occurs when there is a difference in the mode field diameters of two single-mode optical fibers. A loss may occur when the mode field diameter of the transmitting optical fiber is greater than the mode field diameter of the receiving optical fiber.

Cladding diameter mismatch   Cladding diameter mismatch occurs when the cladding diameters of the two optical fibers are not the same. A loss may occur when the cores of the optical fibers are not aligned because of the cladding diameter mismatch.

Ferrule bore diameter mismatch   Ferrule bore diameter mismatch occurs when the bore or hole diameter of the transmit ferrule is different from the bore diameter of the receive ferrule. A loss may occur if the bore diameter mismatch prevents the cores of the optical fibers from properly aligning.

Mating sleeve diameter mismatch   Mating sleeve diameter mismatch occurs when the diameters of the transmit side of the mating sleeve and the receive side are not the same. A loss may occur when the cores of the optical fibers are not aligned because of the mating sleeve diameter mismatch.

Optical fiber concentricity   The core and cladding of an optical fiber should be round and concentric, which means that they share a common geometric center. Concentricity variations can cause the optical fiber cores to misalign, creating a loss when the light exiting the core of the transmitting optical fiber enters the cladding of the receiving optical fiber.

Ferrule concentricity   Concentricity variations in the ferrule can also cause a loss. The ferrule and the bore or hole through the ferrule that accepts the optical fiber should have the same geometric center just like the core and cladding of the optical fiber. Variations in this geometric center can cause the optical fiber cores to misalign when two connectors are mated, creating a loss when light from the core of the transmit optical fiber enters the cladding of the receive optical fiber.

Mating sleeve concentricity   Concentricity variations in the transmit side and receive side of the mating sleeve can also cause a loss. The transmit and receive sides of the mating sleeve should have the same geometric center just like the core and cladding of the optical fiber. Variations in the geometric center of the two sides of the sleeve can cause the optical fiber cores to misalign when two connectors are mated, which may create a loss when light from the core of the transmit optical fiber enters the cladding of the receive optical fiber.

Optical fiber noncircularity   In an ideal optical fiber, the core and the cladding are perfectly circular. Noncircularity of the core or cladding can cause the cores of the transmit and receive fibers to misalign. This misalignment will cause a loss when light from the core of the transmitting optical fiber enters the cladding of the receiving optical fiber.

Ferrule noncircularity   Just as the core and cladding of an optical fiber may not be perfectly circular, the ferrule and the bore through the ferrule may also not be perfectly circular. The noncircularity of the ferrule or the bore through the ferrule may cause the core of the transmit and receive optical fibers to misalign. A loss will occur when light from the core of the transmitting optical fiber enters the cladding of the receiving optical fiber.

Mating sleeve noncircularity   Noncircularity in either side of the mating sleeve may prevent the transmit and receive ferrules from properly aligning, which may cause the cores of the optical fibers to misalign. A loss will occur if the misalignment from the noncircularity causes light from the core of the transmitting optical fiber to enter the cladding of the receiving optical fiber.

Extrinsic Factors

Extrinsic factors that affect connection performance are factors related to the condition of the interconnection itself or factors that affect the interconnection. An ideal interconnection would have identical optical fibers, ferrules, and mating sleeve ends. These components would allow the cores of the transmitting and receiving optical fibers to align perfectly, resulting in a low-loss interconnection. However, there is no such thing as a perfect interconnection, only an imperfect interconnection. In a real interconnection, intrinsic and extrinsic factors affect connection performance.

Extrinsic factors and cleanliness go hand in hand. Dirt or contamination can cause any or all of the extrinsic factors described in this chapter. This section briefly reviews the common extrinsic factors that affect connection performance. Cleaning is covered later in the chapter and extrinsic factors are covered in detail in Chapter 8.

Lateral misalignment   Lateral misalignment occurs when the two optical fibers are offset laterally. A loss occurs when the lateral misalignment causes light from the core of the transmitting optical fiber to enter the cladding of the receiving optical fiber. As the lateral misalignment increases, the loss increases because less light from the core of the transmitting optical fiber makes its way into the core of the receiving optical fiber.

In an interconnection, lateral misalignment can be caused by intrinsic variations in the optical fiber, ferrule, or mating sleeve. It may also be the result of dirt, contamination, or wear.

End separation   End separation is simply an unintended air gap between the endfaces of the transmitting and receiving optical fibers. The air gap will cause a loss and produce a Fresnel reflection. End separation may be caused by dirt, contamination, an improperly polished connector endface, or wear. However, it is more often the result of not fully inserting and securing the connector in the receptacle or mating sleeve.

Figure 9.14 shows two ST connectors in a mating sleeve. The connector on the right has been inserted into the mating sleeve; however, it has not been latched like the connector on the left of the mating sleeve. This happens often and will cause end separation.

FIGURE 9.14
ST connectors in a mating or alignment sleeve: the connector on the left is latched and the connector on the right is not latched.

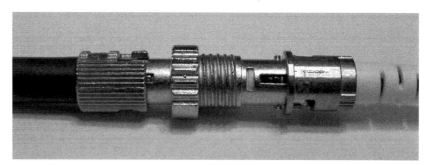

Angular misalignment Angular misalignment occurs when the optical fibers in an interconnection meet each other at an angle. Loss from angular misalignment occurs when light from the core of transmit optical fiber enters the cladding of the receive optical fiber or enters the core of receive optical fiber at an angle exceeding the acceptance angle. Light entering the core of the receive optical fiber at an angle exceeding the acceptance angle typically does not propagate the length of the receive optical fiber.

As with end separation and lateral misalignment, angular misalignment may be caused by dirt, contamination, or wear.

Geometry

Geometry refers to the shape of the ferrule endface. It seems at first that the ideal shape would be a flat surface that would mate with the flat surface of another ferrule, but this geometry actually presents the most potential problems. Endface geometry can be broken into four categories:

♦ Flat

♦ Curved

♦ Angled

♦ Lensed

This section provides an overview of each of these geometries.

FLAT

Flat endfaces like the one shown in Figure 9.15 will always have some polishing irregularities; they will never be perfectly flat across. When two precision flat metal pieces such as the head and cylinder block of a car engine are brought together, a gasket is required. The gasket fills the small voids between the head and engine block, compensating for the irregularities of the metal surfaces. Without the gasket, the car would not run properly.

The slightest variation in the flat endface surface will keep the fiber ends far enough apart to cause end separation loss due to Fresnel reflection. In addition, some of the reflected light may return down the core of the optical fiber to the transmitter as *return reflection* or *back reflection*, which can interfere with the operation of the light source. Light interference is discussed in detail in Chapter 10.

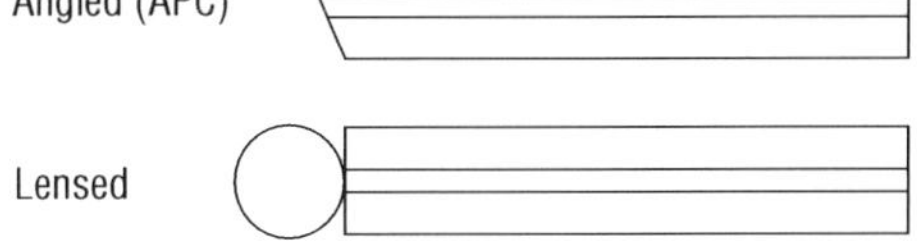

FIGURE 9.15
Endface geometry
configurations

Flat

Curved (PC)

Angled (APC)

Lensed

CURVED

To ensure *physical contact* (PC) between optical fiber ends, the best endface geometry is a convex curve. This curve, or *PC finish*, shown in Figure 9.15, ensures that the highest feature or *apex* on the endface is the center of the optical fiber end. When the optical fiber ends are in direct physical contact, the light behaves as if the connected ends are a continuous piece of optical fiber and passes through with very little loss and back reflection.

ANGLED

Another method of reducing return reflection is with an *angled PC (APC) finish*. This type of finish, also shown in Figure 9.15, puts an angle of about 8° on the endface, with the intent that it will mate with a similarly angled endface when properly aligned. With the endface and the fiber end angled, any light that is reflected is sent into the cladding and absorbed by the coating, rather than traveling back through the core of the optical fiber toward the light source.

LENSED

The *lensed ferrule* endface is more commonly known as the as the *expanded beam*. This endface is not a new concept; it has been in use for over three decades. However, until recently it was not very well known.

Unlike the other geometries that have been discussed, the lensed geometry literally has a lens at the end of the ferrule, as shown in Figure 9.15. One of two different lens types can be used: the collimating lens and the imaging lens. Most expanded beam connectors feature the collimating lens.

When the light traveling through the optical fiber enters the transmit lens, it is expanded and *collimated*, as shown in Figure 9.16. The collimated light travels through the air gap between the two lenses and enters the receive lens. The receive lens collects the collimated light and focuses it to the point where it is within the cone of acceptance of the receiving optical fiber, as you can see in Figure 9.16.

FIGURE 9.16

The basic components and operation of an expanded beam interconnection

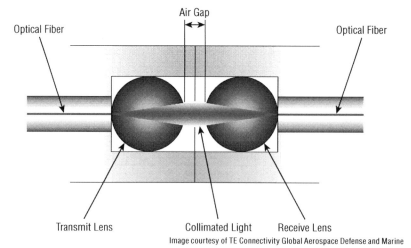

Image courtesy of TE Connectivity Global Aerospace Defense and Marine

Expanded beam offers some advantages over the other geometries that make it ideal for harsh environments such as those associated with avionics, aerospace, space, mining, and drilling. As of this writing, SAE International is preparing to publish their first expanded beam standard AIR6112, A Guideline for Aerospace Platform Fiber Optic Expanded Beam Interconnect Technology. This standard discusses the operational advantages expanded beam has over physical contact optical interfaces, such as the following:

♦   Greater tolerance to particulate contamination

♦   Ease of cleaning

♦   Protection of the optical fiber core

♦   Non-physically contacting interface

♦   Lower mating forces

Interferometer

An interferometer like the one shown in Figure 9.17 is a device that uses light to take precision measurements of the fiber-optic connector endface. The measurements are used to generate a three-dimensional (3D) map of the surface and quantitatively define three critical parameters: *radius of curvature*, *apex offset*, and *fiber undercut or protrusion*. These critical parameters are required by Telcordia GR-326 to evaluate connector endface geometry for single-mode connectors and jumper assemblies. Telcordia GR-326 is an industry standard that defines generic requirements for single-mode connectors and jumpers.

FIGURE 9.17
Interferom-
eter with a con-
nector inserted for
evaluation

Photo courtesy of PROMET International Inc.

The interferometer shown in Figure 9.17 also has the capability to perform a two-dimensional (2D) optical analysis as you would with an inspection microscope. Two-dimensional analysis is discussed in detail in the "Endface Cleaning" section later in this chapter.

THREE-DIMENSIONAL (3D) ANALYSIS

When two connectors with a PC finish are brought together in a mating sleeve, springs are used to provide compressive forces that force the endfaces to touch under pressure. These compressive forces maintain glass on glass contact, ideally producing a low-loss and low-reflection interconnection. The geometry of the connector endface is critical to ensure a high-performance interconnection and can be evaluated only with a 3D analysis from an interferometer.

In addition to being able to quantitatively define the radius of curvature, apex offset, and fiber undercut or protrusion, 3D analysis can be used to evaluate surface roughness and contamination. The surface of both the ferrule and optical fiber appear very rough in Figure 9.18, whereas the surface of the ferrule in Figure 9.19 is smooth, although the optical fiber is scratched in several places. The endface shown in Figure 9.20 features a smooth ferrule and optical fiber, but both surfaces are covered with debris.

FIGURE 9.18
3D endface image of surface roughness on the ferrule and the optical fiber

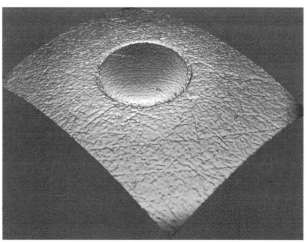

Image courtesy of PROMET International Inc.

FIGURE 9.19
3D endface image of an optical fiber with scratches and a smooth ferrule

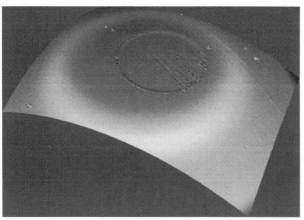

Image courtesy of PROMET International Inc.

FIGURE 9.20
3D endface image
of a smooth ferrule
and optical fiber
both covered with
debris

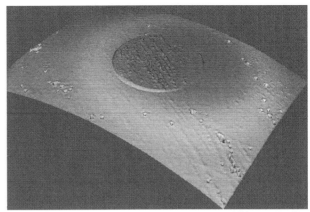

Image courtesy of PROMET International Inc.

Radius of curvature   The radius of curvature describes the roundness of the connector end-face, and it is measured from the center axis of the connector ferrule, as shown in Figure 9.21. The minimum and maximum values for the radius of curvature as defined in Telcordia GR-326 are 7mm and 22mm. Values below or above this range increase the risk of optical fiber damage. They also increase the possibility of reflection and insertion loss.

FIGURE 9.21
Connector radius of
curvature

what dimension does this relate to ?

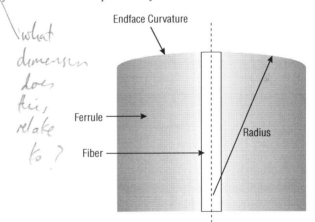

Drawing courtesy of Corning Cable Systems

Apex offset   The apex of the connector endface is the highest point of the rounded endface. Ideally, the apex would be in the center of the optical fiber. Not having the apex in the center of the optical fiber can lead to the lack of physical contact between the two optical fiber cores in the mating sleeve. Apex offset is the displacement between the apex of the endface and the center of the optical fiber core, as shown in Figure 9.22. Telcordia GR-326 states that apex offset cannot exceed 50μm.

FIGURE 9.22
Connector apex
offset

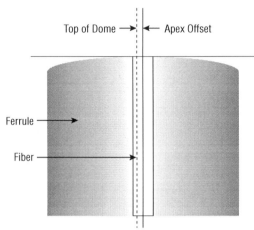

Drawing courtesy of Corning Cable Systems

The image captured by an interferometer in Figure 9.23 uses grayscale shading to display the apex. The darkest shade of gray is the apex. This shaded section should be centered directly over the core of the optical fiber; however, it is offset by roughly the radius of the optical fiber, or 62.5μm. The apex in this image is directly over the epoxy ring, exceeding the Telcordia GR-326 maximum of 50μm.

FIGURE 9.23
3D endface image
showing an apex
offset

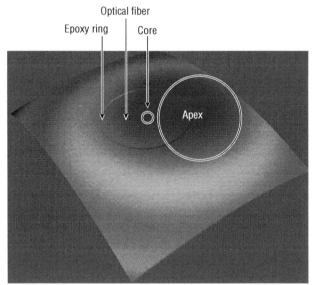

Image courtesy of PROMET International Inc.

Fiber undercut or protrusion   Optical fiber undercut or protrusion describes the distance the optical fiber endface is below or above the rounded connector endface, as shown in Figure 9.24. When the optical fiber protrudes too far from the ferrule endface, as shown in Figure 9.25, the compressive forces on the optical fiber increase and so does the chance that the optical fiber will be damaged. When the optical fiber sits too far below the ferrule endface, as shown in Figure 9.26, an air gap will result, increasing the loss and reflection. Telcordia GR-326 states that an undercut value cannot exceed ±55nm.

FIGURE 9.24
Connector undercut

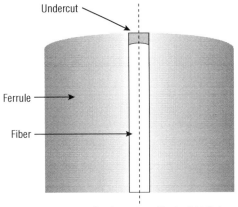

Drawing courtesy of Corning Cable Systems

FIGURE 9.25
3D endface image showing excessive fiber protrusion

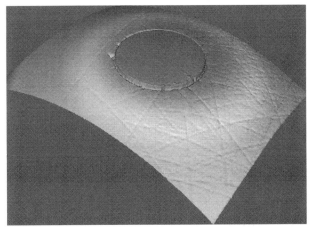

Image courtesy of PROMET International Inc.

how/if are these 3D images used in practice? fluke

FIGURE 9.26
3D endface image
showing fiber
undercut

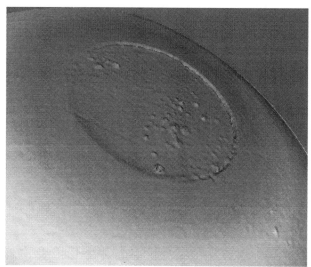

Image courtesy of PROMET International Inc.

A connector endface is like a snowflake—no two are exactly alike. It is for this reason that Telcordia GR-326 provides ranges for critical geometric values. An interferometer can typically display information several different ways to help with the 2D and 3D evaluation of the endface. The manner in which the information is displayed on the monitor and in a report will vary from manufacturer to manufacturer.

The computer screen capture shown in Figure 9.27 features side-by-side 3D and 2D images and analysis of the connector endface. A metrology report below the 3D image provides geometric measurement data and a pass/fail indication. The measurement data includes radius of curvature, fiber height, apex offset, fiber diameter, ferrule angle, key angle, and epoxy width. A defects report below the 2D image displays the number of chips, particles, scratches, total defects, and a pass/fail indication. This connector has a fiber height greater than 55nm and therefore fails to meet the Telcordia GR-326 requirement.

Connector Types

The purpose of all fiber-optic connectors is the same—to align a terminated optical fiber end with another terminated optical fiber end or a piece of equipment and hold it firmly in place. The manner in which this has been accomplished, however, has continued to evolve as technology improves and as requirements change. As a result, manufacturers have created many types of connectors. Most of the connectors are still in use, though some have become more common whereas others exist only in older or legacy systems.

Connectors can be broken into two categories: single-fiber connectors and multifiber connectors. Whether the endfaces make contact or do not make contact when mated defines them as a contact or noncontact connector.

FIGURE 9.27
Interferometer screen capture with 3D and 2D views, along with metrology data and defect analysis

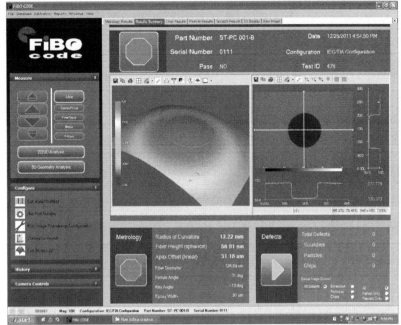

Image courtesy of PROMET International Inc.

Single-fiber connectors are designed for only one optical fiber, but they may be used to break out multiple optical fibers from a cable for individual termination at a patch panel or piece of equipment. Multifiber connectors are available in many configurations and may be used to connect paired fibers in duplex systems, or to connect up to 72 optical fibers in a single ferrule array.

The majority of the connectors covered in this chapter are contact type connectors. The endfaces of contact type connectors, regardless of the endface geometry, physically touch when mated. However, the endfaces of noncontact type connectors, regardless of their geometry, do not physically touch when mated. These types of interconnections will always have an air gap.

Because there are so many types of connectors, the Telecommunications Industry Association (TIA) has not attempted to recognize or create standards for all of the available types. In the ANSI/TIA-568-C standard, however, they recognize performance standards for connectors, which will be described at the end of this chapter.

Let's look at the various types of connectors.

Single-Fiber Contact Connectors

Single-fiber contact connectors have a wide variety of connection methods. Some, including the earliest types, are engaged by pushing and twisting or by using a threaded sleeve to draw the connector tight. However, many of the newer form factors are square or rectangular snap-in connectors. These are inserted with a simple push that engages a locking mechanism.

SC CONNECTORS

The SC (subscriber connector), shown in Figure 9.28, is among the most widely used connectors. Originally developed by Nippon Telephone and Telegraph (NTT), the SC has a standard-sized 2.5mm ferrule and a snap-in connection, which was created as an alternative to connectors that required turning or twisting to keep them in place. In addition, SC connectors can be installed in only one orientation, making them suitable for an APC endface.

FIGURE 9.28
The SC connector is one of the most common types in use.

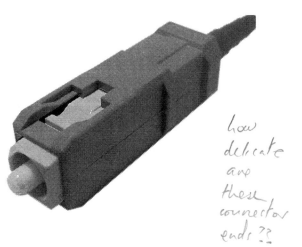

ST CONNECTORS

The ST (straight tip) connector, shown in Figure 9.29, was developed by AT&T as a variation on a design used with copper coaxial cables. This connector has a metal connector cap that must be twisted to lock into place. The ST is considered a legacy connector, as it has been around for quite some time and can still be found in many installations today.

FIGURE 9.29
An ST connector must have its connector cap twisted to lock in place.

ST

FC CONNECTORS

An FC (face contact) connector is a rugged metal connector with a screw-on connector cap and a 2.5mm ferrule. Like the SC connector, the FC, shown in Figure 9.30, is used in connections where proper polarization must be maintained. Because the connector is cylindrical, it must be aligned with a built-in key. Note, however, that there are several different variations for the size of the key, meaning that the connector must be properly matched with its adapter or receptacle.

FIGURE 9.30
FC connector with screw-on connector cap

Photo courtesy of Corning Cable Systems

LC CONNECTORS

The LC connector, shown in Figure 9.31, is a *small form factor* connector. Developed by Lucent Technologies, this snap-in connector is considered to be a smaller version of the SC connector and is sometimes referred to as a mini-SC. The small form factor connector has a 1.25mm ferrule, half the diameter of an SC connector ferrule.

FIGURE 9.31
The LC connector is a small form factor version of the SC connector.

LC

The smaller form factor allows two LC connectors to fit into roughly the same size space as a single SC connector. Figure 9.32 shows two LC connectors on one side of an SC mating sleeve and an SC connector plugged into the other side of the mating sleeve. The two LC connectors require no more room than a single SC.

FIGURE 9.32
SC mating sleeve with an SC connector on the left side and two LC connectors on the right side

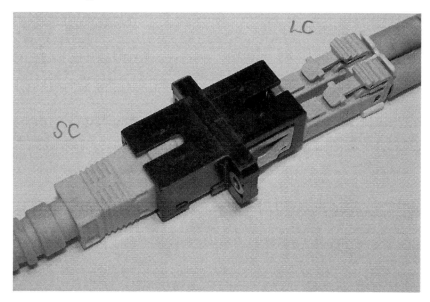

[handwritten note:] is the data on one fibre now going into 2 ???

D4 CONNECTORS

The D4 connector is an older-style connector with a 2.5mm ferrule and a threaded connector cap that must be screwed on to secure the connector. It is similar in function to the FC connector, but its profile is slightly smaller, allowing it to be used in smaller spaces than the FC connector. D4 connectors are keyed and the optical fibers make contact when mated.

MINI-BNC CONNECTORS

The mini-BNC, shown in Figure 9.33, is similar in appearance to its counterpart made for copper coaxial cables. The mini-BNC (short for either Bayonet Nut Connector or Bayonet Neill-Concelman, the inventor of the original BNC) is a metal twist-lock connector with a 2.5mm ferrule. The mini-BNC is considered a legacy connector, so though it may never be installed in new a system, you may find it on older installations.

FIGURE 9.33
The mini-BNC connector resembles the BNC connector used in RF applications.

9.2 Single-Fiber Noncontact Connectors

Single-fiber noncontact connectors are typically engaged by pushing and twisting or by using a threaded sleeve to draw the connector tight.

SMA Connectors

The SMA connector was developed by Amphenol from its line of microwave connectors known as SubMiniature A. The connector has a 3mm stainless steel ferrule and a connector cap that is threaded on the inside. Because it was originally developed before the invention of single-mode optical fiber, the SMA connector does not provide as precise a connection as more recent designs. In addition, they do not have an alignment key and can rotate during the mating process. To prevent damage to the optical fiber, they are noncontact. SMA connectors can still be found in military applications and in the delivery of high-power laser light.

Biconic Connectors

The biconic connector shown in Figure 9.34 may feature a stainless steel or ceramic cone-shaped ferrule that helps in aligning fiber ends during connection. Biconic connectors do not have an alignment key and can rotate during the mating process. It is for this reason that they are noncontact.

The biconic is threaded on the outside for screw-in placement, and although they are considered legacy connectors, they are still used in military applications. The biconic connector ferrule is also incorporated into the tactical fiber-optic cable assembly (TFOCA) that is discussed later in this chapter.

FIGURE 9.34
The biconic connector has a cone-shaped ferrule.

9.3 Multiple-Fiber Contact Connectors

Multiple-fiber contact connectors are mostly designed to support duplex operations and ribbon fibers, allowing a greater number of connections in a smaller space. Some duplex connectors resemble larger or ganged versions of single-fiber connectors.

FDDI Connectors

The FDDI connector was developed for use in Fiber-Distributed Data Interface (FDDI) networks. This plastic duplex connector is also called FDDI MIC for *medium interface connector*, a connector that is used to link electronics and fiber transmission systems. It features protective shields around the two 2.5mm ferrules to prevent damage. The connector locks into place with latches on the sides and can be keyed to ensure proper orientation of the two fiber ends.

ESCON CONNECTORS

The ESCON connector was developed for IBM's ESCON (Enterprise Systems Connection) architecture in the early 1990s. It is similar to the FDDI, but it has a shroud that retracts from the ferrules when the connector is engaged.

SC DUPLEX CONNECTORS

The SC duplex connector, shown in Figure 9.35, is actually two single SC connectors joined with a plastic clip. This arrangement can be extended into even more plugs, if necessary, but is most commonly applied in the duplex configuration.

FIGURE 9.35
The SC duplex is made of two single SC connectors joined together with a clip.

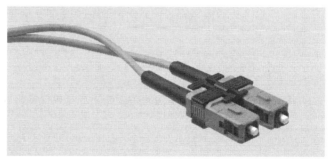

Photo courtesy of Norfolk Wire & Electronics

LC DUPLEX CONNECTORS

The LC duplex connector, shown in Figure 9.36, is actually two single LC connectors joined with a plastic clip. The clip keeps the transmit and receive optical fibers for a single channel together.

FIGURE 9.36
The LC duplex is made of two single LC connectors joined together with a clip.

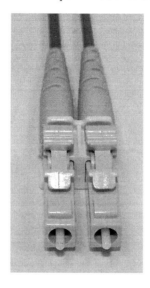

MPO Connectors

The MPO (Multifiber Push On) connector, shown in Figure 9.37, is built on the MT ferrule. The MT (mechanical transfer) ferrule is rectangular and the endface measures 6.4mm × 2.5mm. It is designed to hold from 4 to 72 fibers. The MT ferrule is ideally suited for use with ribbon fiber because the optical fibers are arranged in rows of 4, 8, or 12 optical fibers. The optical fibers are spaced 0.25mm apart, as shown in the 12-fiber ferrule in Figure 9.38. The MT ferrule can support up to six rows of 12 optical fibers.

FIGURE 9.37
The MPO connector packs 4 to 72 fibers into a single ferrule.

FIGURE 9.38
12-fiber MT ferrule with 0.25mm spacing

The MT ferrule can accept two 0.7mm in diameter precision-machined guide pins that maintain the alignment necessary for connecting 4 to 72 fibers at once. These guide pins can be arranged as necessary between the mating connectors depending on the way they will be used. The 12-fiber ribbon patch cord shown in Figure 9.39 features one MT ferrule with guide pins and one without.

The MT ferrule is available with a flat or angled endface. The 8-degree angled endface is typically used for single-mode optical fiber to reduce back reflections. The MT ferrule endface cannot be polished with a radius because the optical fibers in the center of the ferrule would extend further out than the optical fibers away from the center. This would create a large air gap between the outside fibers when mated with another connector.

FIGURE 9.39
12-fiber ribbon patch cord featuring one MT ferrule with guide pins and one without

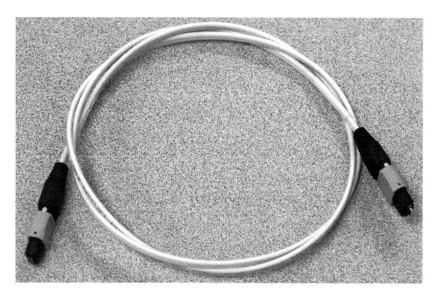

NOTE The MPO connector has a plastic body that is spring-loaded like many other fiber-optic connectors to keep the connectors together. International standard IEC 61754-7 defines the standard interface dimensions for the different types of MPO connectors available.

Single ferrule connectors such as the MPO that contain multiple optical fibers arranged in a row or in rows and columns are also known as array connectors.

MTP CONNECTORS

The US Conec MTP connector shown in Figure 9.40 is built around the MT ferrule. It is designed as a high-performance version of the MPO and will interconnect with MPO connectors. Details on the performance differences between MPO and MTP connectors are explained on the US Conec website at www.usconec.com.

FIGURE 9.40
MTP connector with alignment pins

MT-RJ CONNECTORS

The MT-RJ connector, shown in Figure 9.41, was developed to emulate the functionality of the RJ-45 modular connector, which is used in most office networking. It provides a snap-in, duplex connection in a housing that resembles the familiar modular network plug. In fact, the connector is designed to fit in the same physical opening as the RJ-45, so many networking fixtures and wall plates can be retained after the hardware inside them has been upgraded for fiber.

FIGURE 9.41
The MT-RJ connector is designed to emulate the RJ-45 modular plug in fit and ease of use.

The MT-RJ's single ferrule holds two fibers in a housing smaller than most single-fiber connectors, so it is attractive in many applications where size matters. The connector is built like a mechanical splice, so the fibers are inserted into the ferrule and the cam is rotated to hold the fibers in place without the need for epoxy.

MT-RJ connectors may be used as most connectors are, with a sleeve that joins two connectors together, as shown in Figure 9.42, or in a plug-and-receptacle arrangement with a transceiver.

FIGURE 9.42
Two MT-RJ connectors and a mating sleeve

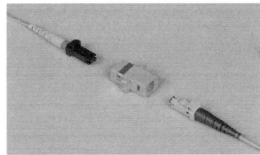

Photo courtesy of Corning Cable Systems

Many other connector types are in use, and more are being added to support new applications and higher bandwidths in a variety of environments. Table 9.1 gives you a quick reference to the connectors we've discussed and their key features.

TABLE 9.1:   Quick reference for fiber-optic connectors

CONNECTOR	NO. OF FIBERS	CONNECTION	SHAPE	FERRULE SIZE
Single-fiber				
SC	1	Push-in	Rectangular/ D-shaped	2.5mm
ST	1	Twist-lock	Round	2.5mm
FC	1	Screw-on	Round	2.5mm
LC	1	Push-in	Square	1.25mm
D4 (DIN)	1	Screw-on	Round	2.5mm
SMA	1	Screw-on	Round	3mm
Biconic	1	Screw-on	Round	Conical
Mini-BNC	1	Twist-lock	Round	2.5mm
Multifiber				
FDDI	2	Push-in	Rectangular	(2) 2.5mm
ESCON	2	Push-in	Rectangular	(2) 2.5mm
SC Duplex	2	Push-in	Rectangular	(2) 2.5mm
MPO	4 to 74	Push-in	Rectangular	6.4 mm × 2.5 mm
US Conec MTP®	4 to 72	Push-in	Rectangular	6.4 mm × 2.5 mm
MT-RJ	2	Push-in	Rectangular	5 mm × 2.5 mm

PIGTAILS

Pigtails are fiber ends with connectors factory-attached for future splicing into a system, as shown in Figure 9.43. Typically, a pigtail starts as a manufactured patch cord or jumper with a connector at each end. You can then cut the jumper in half and have two pigtails ready to splice.

Pigtails are available with a variety of connectors, depending on your needs. These products have advantages over field terminations because the connectors are factory-installed and polished to exacting standards. This can save time over attaching a connector to the end of an optical fiber, and it typically produces a better connection. They are especially useful in situations where many connectors have to be added to cables in a relatively short time, or in a location where it is easier to make a splice than it is to add a connector.

On the downside, pigtails require hardware to protect the splice and an investment in mechanical or fusion splicing equipment.

FIGURE 9.43
Use a pigtail to
splice a connector
to an existing fiber.

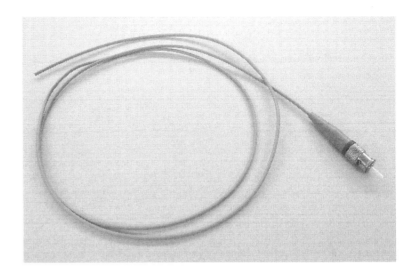

SPECIALIZED HARSH ENVIRONMENT CONNECTORS

While standard commercial connectors are used in most networks and communications links, fiber-optic connectors are also available in specialized configurations for harsh environments such as those associated with avionics, aerospace, space, mining, and underground drilling. They feature durable cases, heavy-duty connector bodies, and multiple-fiber connections, such as the TFOCA III connector shown in Figure 9.44. This connector is often found in military applications where heavy use and destructive environments are common.

FIGURE 9.44
This 24-channel
TFOCA III con-
nector is a special-
ized connector for
harsh environment
applications.

Photo courtesy of MicroCare

The connectors shown in Figure 9.45 are typically used in aerospace applications. These are military qualified MIL-DTL-38999 four-optical-fiber connectors cabled for an aerospace application.

FIGURE 9.45
Two MIL-DTL-38999 connectors cabled for an aerospace application

Photo courtesy of Carlisle Interconnect Technologies

NOTE The vocabulary associated with harsh environment connectors contains some terms that are typically not used when working with the other types of connectors described in the text. A contact describes the ferrule and all the other physical components required to complete the mechanical assembly. A single contact may be referred to as a *terminus* and multiple contacts may be referred to as *termini*. A terminus can be a socket or a pin. The socket terminus would typically have an alignment sleeve.

As mentioned earlier in this chapter, expanded beam offers some advantages over the other geometries that make it ideal for harsh environments. It is for this reason that SAE International created AIR6112, A Guideline for Aerospace Platform Fiber Optic Expanded Beam Interconnect Technology. Shown in Figure 9.46 is a detailed drawing of a lensed socket terminus and a lensed pin terminus. The actual disassembled socket and pin lensed termini represented in that drawing are shown in Figure 9.47. These termini can be used in multi-contact connectors like the ones shown in Figure 9.48.

FIGURE 9.46
Detailed drawing of a socket and pin lensed termini

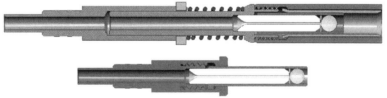

Photo courtesy of TE Connectivity Global Aerospace Defense and Marine

FIGURE 9.47
Photograph of a dis-
assembled socket
and pin lensed
termini

Photo courtesy of TE Connectivity Global Aerospace Defense and Marine

FIGURE 9.48
Multi-contact
expanded beam
connectors broken
out to individual LC
connectors

Photo courtesy of TE Connectivity Global Aerospace Defense and Marine

Connector Termination

One of the most exacting jobs you will encounter in working with fiber optics is connector termination. If you do the job properly, an optical fiber the size of a human hair will mate with
another optical fiber or a piece of hardware and transfer a large percentage of the light passing
through it. If you do the job wrong, the light could stop at the connector and go no further, or
so little of the light could transfer that it would essentially be useless. With those encouraging
words urging us forward, let's look at three ways in which an optical fiber can be terminated
with a connector.

NOTE There are many precision mechanical tools used for fabricating connections, but the
actual process is largely an art that must be practiced. Do not make the mistake of making your
first connection in a real installation; practice before every installation.

4.1. ## Tools

Regardless of the method you choose to terminate the optical fiber, tools will be required. In this section, we discuss three termination methods. Each of these methods requires some of the tools discussed in this section.

4.1.1.
Shears   Even though most cabling materials are fairly easy to cut, the aramid or Kevlar strength member is strong, fibrous, and loose fitting, making it difficult to cut. You will need a good pair of sharp shears like the pair shown in Figure 9.49 to cut through the strength member quickly and cleanly.

FIGURE 9.49
Shears designed
to cut the Kevlar
strength member

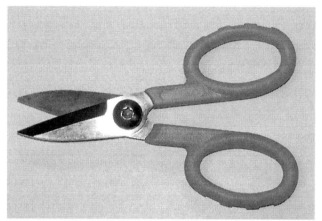

Photo courtesy of KITCO Fiber Optics

4.1.2
Stripper   A stripper is used to remove the outer jacket, tight buffer, and coating so that only the optical fiber itself is exposed. Some strippers are designed to only remove the tight buffer and coating, like the one shown in Figure 9.50. Other strippers may be designed to remove the jacket, tight buffer, and coating, like the one shown in Figure 9.51. Some are adjustable, like the one shown in Figure 9.52, and can be used to remove the jacket, tight buffer, or coating. However, they need to be adjusted for each application.

FIGURE 9.50
No-Nik stripper
designed to remove
only the tight buffer
and coating

FIGURE 9.51
Stripper designed to remove the jacket, tight buffer, and coating

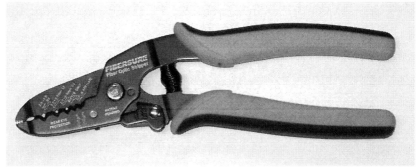

FIGURE 9.52
Adjustable stripper that can be set to remove the jacket, tight buffer, or coating

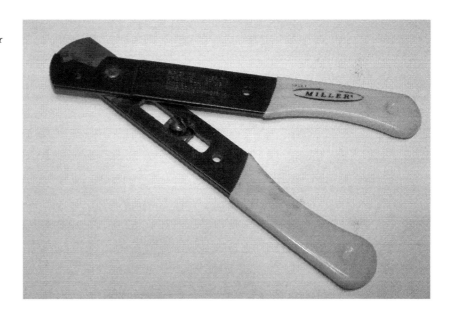

4.1.3 **Scribe** The scribe, shown in Figure 9.53, is used for precision work in removing the fiber end once it has been adhered inside the ferrule. To use it, you only need to score or nick one side of the optical fiber lightly. This breaks the surface of the cladding, which provides the fiber's tensile strength. You'll need some practice with a scribe to get the right results, since it requires just the right touch to keep from damaging or breaking the optical fiber.

FIGURE 9.53
Scribe used to score or nick the optical fiber

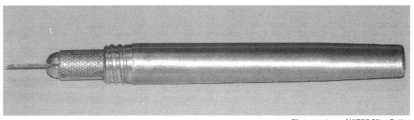

4.14

Polishing fixture (puck)   The polishing fixture, or puck as it is typically referred to, is shown in Figure 9.54. The puck is used to ensure that the ferrule stays perpendicular to the polishing film during the polishing process; see Figure 9.55. Even the slightest variation in the polishing angle can affect the endface finish of the connector, which in turn could impact the performance of the connector.

FIGURE 9.54
Polishing fixture (puck) for a 2.5mm ferrule

Photo courtesy of KITCO Fiber Optics

FIGURE 9.55
Polishing fixture (puck) holding an LC connector perpendicular to the polishing film

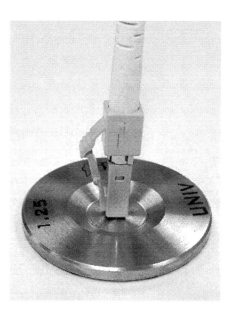

4.15.

Lint-free wipes   Lint-free wipes like those shown in Figure 9.56 are engineered to lift away oils, grime, and dust from the surface of the optical fiber. When handling a lint-free wipe, only touch one side of the wipe because the wipe will absorb the oil from your skin. Use the side of the wipe that you did not touch for cleaning the optical fiber. You do not want to contaminate the optical fiber during the cleaning process.

4.16.

Cleaning fluid   Cleaning fluids like those shown in Figure 9.57 are approved for air travel and engineered specifically to clean the optical fibers without leaving a residue. These cleaning fluids typically contain little or no isopropyl alcohol and have been filtered to remove microscopic contaminates.

FIGURE 9.56
Lint-free wipes
engineered for
cleaning optical
fibers

Photo courtesy of MicroCare

FIGURE 9.57
Optical fiber clean-
ing fluids approved
for air travel

Photo courtesy of MicroCare

4.1.7   **Alcohol**   Isopropyl alcohol is used to remove the residual coating from the optical fiber prior to termination. While it is widely used, it does leave a residue and you may want to consider a cleaning fluid that does not leave a residue. If you choose isopropyl alcohol, make sure that it is virtually free of water and has been filtered to remove impurities, like the isopropyl alcohol shown in Figure 9.58. Do not use the isopropyl alcohol commonly available at your local pharmacy; this alcohol contains a large percentage of water and will leave a residue on the optical fiber. While you are using the isopropyl alcohol, ensure it is kept covered when not in use. Doing so will not only prevent spills; it will also prevent the isopropyl alcohol from absorbing the moisture in the atmosphere and help to minimize contamination.

FIGURE 9.58
Isopropyl alcohol
for cleaning optical
fibers

Photo courtesy of MicroCare

↑ 18.

Cleaver   A cleaver is only required for prepolished connectors. Cleavers like the one shown in Figure 9.59 score and break the optical fiber, leaving a perpendicular finish. Cleaving is typically a two-step process and it takes place after the optical fiber has been properly prepared and cleaned. Cleaving is described in detail in Chapter 8.

FIGURE 9.59
Optical fiber cleaver

Photo courtesy of KITCO Fiber Optics

4.1.9.

Nonreflective mat   Always work over a nonreflective black surface like the one shown in Figure 9.60. The nonreflective black mat makes it easier to keep track of cut optical fiber ends. Keep in mind that the small fiber pieces are sharper and smaller than a hypodermic needle and can painlessly penetrate your skin.

FIGURE 9.60
A nonreflective black mat makes it easier to see the bare optical fiber.

4.1.10

Optical fiber disposal container   Always use a labeled container with a lid like the one shown in Figure 9.61 to dispose of bare optical fibers.

FIGURE 9.61
Container designed and labeled for optical fiber waste

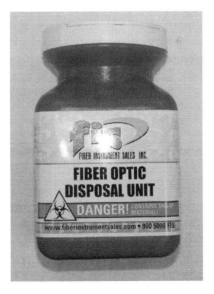

Safety glasses   To prevent eye injury, always wear proper eye protection like the safety glasses shown in Figure 9.62.

FIGURE 9.62
Safety glasses
designed to prevent
eye injuries

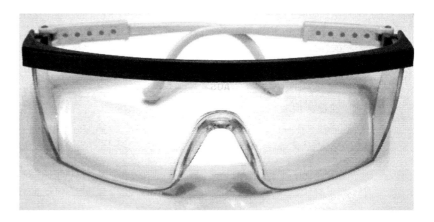

4.1.11. **Tweezers**   To prevent injury to your hands from the optical fiber, you may want to handle pieces of bare optical fiber with tweezers, like the pair shown in Figure 9.63.

FIGURE 9.63
Tweezers can be
used to pick up the
pieces of optical
fiber.

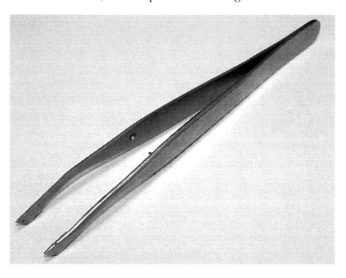

4.1.12 **Crimper**   The crimper like the one shown in Figure 9.64 is typically used to secure the strength member and jacket to the connector.

4.1.13 **Eye loupe**   An eye loupe like the one shown in Figure 9.65 provides moderate magnification and is used to view the end of the connector ferrule after the scribing process.

4.1.14 **Rubber pad**   The rubber pad like the one shown in Figure 9.66 is placed under the polishing film. It provides a cushioned surface for PC connectors.

FIGURE 9.64
Crimpers are used to secure the strength member and jacket to the connector.

Photo courtesy of KITCO Fiber Optics

FIGURE 9.65
An eye loupe provides moderate magnification for viewing the endface of the connector ferrule.

Photo courtesy of KITCO Fiber Optics

FIGURE 9.66
The rubber pad provides a cushioned surface for the polishing film.

Photo courtesy of KITCO Fiber Optics

Cure adapter The cure adapter like the one shown in Figure 9.67 is placed over the ferrule of the connector. It helps transfer heat from the curing oven and protects the optical fiber during the curing process.

FIGURE 9.67
The cure adapter slides over the connector ferrule and helps transfer heat from the curing oven to the connector.

Photo courtesy of KITCO Fiber Optics

Curing oven The curing oven is only required for epoxies that need to be heated to a specific temperature to cure. The curing oven like the one shown in Figure 9.68 accepts the cure adapter.

FIGURE 9.68
Curing oven used to heat the connector and cure the epoxy

Photo courtesy of W.R. Systems

4.2 # Epoxy

This section of the chapter discusses oven-cured and anaerobic epoxies. There are advantages and disadvantages to each type. The epoxy should be chosen based on the application. This section provides a general overview on epoxy. You should always consult the epoxy data sheet for specific information. If you have specific questions about the type of epoxy to use for your application, consult the manufacturer of the connector or epoxy.

OVEN-CURED EPOXY

Oven-cured epoxy is probably the best type you can use, but it is also the most cumbersome. The epoxy itself consists of a resin and a hardener, which must be mixed in the right proportions. Once it is mixed, it has a limited *pot life* before it begins to harden, and any epoxy that is unused must be discarded. On the other hand, once a batch is mixed, several connectors can be assembled at the same time, so it is useful in making a large number of terminations at one location.

Once the connector has been assembled, you have to insert it into a specially built oven to cure the epoxy and then let the assembly cool before you can begin polishing it. The oven will hold a number of connectors, so as each one is built it can be inserted for curing.

While oven-cured epoxy is time-consuming and equipment-intensive, it produces a hard, fully cured bead around the base of the optical fiber, as shown in Figure 9.69. This bead reduces the risk that the fiber will break inside the ferrule during cleaving, which could ruin the connector.

FIGURE 9.69
The bead of oven-cured epoxy protects the fiber during cleaving.

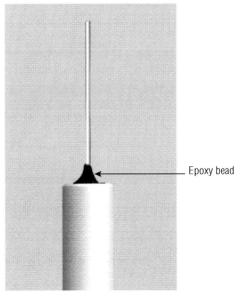

Epoxy bead

Photo courtesy of W.R. Systems

ANAEROBIC EPOXY

Anaerobic epoxy is typically used where there is no power available or when time is at a premium. While the adhesive still comes in two parts, each part is applied separately to the fiber and the ferrule. When the two are joined, the epoxy hardens and cures in about 10 seconds.

Although this may seem to be an ideal solution for all situations, there are some drawbacks to anaerobic epoxy. For one or two connectors, there is a significant time-savings. However, because of its quick hardening time, anaerobic epoxy does not lend itself to be used in batches as oven-cured epoxy does. You typically prepare and work on only one connector at a time. In addition, this epoxy does not form a hard bead the way oven-cured epoxy does, so greater skill is required for cleaving and polishing the optical fiber. Finally, this epoxy is not approved for air travel.

4.3. Abrasives

Abrasives are used to polish the optical fiber end. Depending on the application, different types of abrasives are needed. Essentially, they are like very fine sandpaper, but the abrasive material is adhered or fixed to a Mylar film backing instead of paper. Two popular abrasive materials are aluminum oxide and diamond.

ALUMINUM OXIDE

Aluminum oxide is an abrasive that is harder than the optical fiber but softer than the ceramic ferrule. Because of this, it will polish down the optical fiber but not the ferrule. While this abrasive prevents the person polishing the connector from altering the endface geometry of the ferrule, it does not produce a very attractive finish when compared to diamond abrasive, and it chips the optical fiber at the epoxy ring. (The epoxy ring is the area between the optical fiber and the ferrule where the epoxy rests.) This abrasive will produce an endface similar to the one shown in Figure 9.70.

FIGURE 9.70
Multimode end-
face polished with
aluminum oxide
polishing film

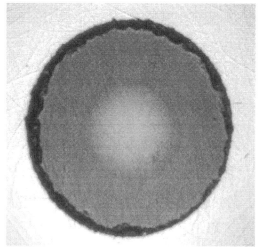

Photo courtesy of W.R. Systems

While the cosmetic finish of an endface polished with aluminum oxide may not look as attractive as an endface polished with diamond, like the one shown in Figure 9.71, aluminum oxide polishing films are significantly less expensive than diamond. The cosmetic finish does not always tell the whole story when it comes to connector performance. The endface shown in Figure 9.70 will offer insertion loss performance that exceeds the ANSI/TIA-568-C.3 requirements for multimode interconnections.

FIGURE 9.71
A single-mode endface that was polished with diamond polishing film and is absent of scratches or chips

Photo courtesy of MicroCare

DIAMOND

Diamond is an abrasive that is harder than both the optical fiber and the ceramic ferrule. Because of this, it will polish down the optical fiber and the ceramic ferrule. This abrasive does not prevent the person polishing the connector from altering the ferrule's endface geometry. However, it does produce an attractive finish and does not chip the optical fiber at the epoxy ring. This abrasive will produce an endface like the one shown in Figure 9.71.

Although the cosmetic finish of an endface polished with diamond may be very attractive, its geometry may also be altered. When polishing with a diamond abrasive, you have to be careful not to overpolish. As discussed earlier in this chapter, an interferometer can be used to evaluate the geometry of the connector endface. However, an interferometer is expensive and not always available.

You can use the inspection microscope to determine if you are polishing the ferrule endface and possibly changing its geometry. When you look at Figure 9.70, which shows a connector endface polished with aluminum oxide, you can see some texture markings on the ferrule. The same type of texture marking exists on the unpolished connector ferrule shown in Figure 9.72. This clearly shows that the aluminum oxide abrasive did not polish down the ferrule endface. However, when you look at the endface polished with a diamond abrasive shown in Figure 9.71, you see there are no texture markings on the ferrule because they have been removed with the diamond abrasive.

The risk associated with polishing the ferrule is that you may alter the shape of the ferrule endface. When you buy a connector, the endface meets the geometric specifications you selected. Polishing the connector with a diamond abrasive may alter those specifications. It is easier to control the geometry with a polishing machine than with hand polishing.

FIGURE 9.72
An unpolished
connector endface
without an optical
fiber

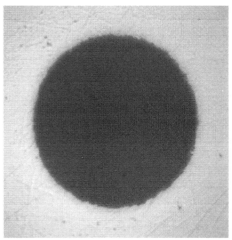

Photo courtesy of W.R. Systems

Hand Polishing

Some of the procedures involved may vary, but the main task is the same: you must insert a bare optical fiber into a hole in the ferrule so that it sticks out slightly beyond the ferrule endface and secure it there; then you must *score*, or *cleave*, the excess optical fiber and polish the ferrule endface and the optical fiber together to achieve the proper profile and finish.

Although this task sounds simple, it requires some careful planning and execution. Remember that the hole in the ferrule is almost microscopic, since it must be only slightly larger than the optical fiber itself.

There are many different fiber-optic connectors on the market from a variety of manufacturers. Each manufacturer has its own polishing process. This section discusses two basic connector assembly and polishing processes using oven-cured epoxy and anaerobic epoxy. The connectors will be terminating simplex cordage.

Assembling the Connector

Assemble your tools before you start so you don't have to look for them in the middle of the process. It may seem like a fussy detail, but your work will proceed more smoothly and efficiently if you have prepared all of your tools and materials and laid them out where you can reach them easily when you need them.

CABLE AND OPTICAL FIBER PREPARATION

The following steps describe how to prepare simplex cordage for connectorization:

1. Cut the cable 2″ longer than you'll need.

2. Install the strain relief boot on the cable end, small diameter first. Important: The strain relief boot will not fit over the connector, so it must be placed on the cable first.

3. Install the crimp sleeve if one was provided with the connector.

4. Locate the appropriate strip chart from the connector manufacturer.

5. Mark the cable jacket as shown on the strip chart, cut the jacket with the stripper, and then twist the jacket and pull it away from the cable to remove it.

6. Pull the strength member back; it will be cut to length in the last step.

7. Strip away the tight-buffer and coating to the length shown on the strip chart in ¼″ increments to reduce the chance of breaking the optical fiber.

8. Clean the fiber with a lint-free wipe soaked in optical fiber cleaning fluid; this will remove the coating residue so that the epoxy in the ferrule can bond with the cladding surface.

9. Using the shears, cut the strength member to the length shown on the strip chart, and then distribute it evenly around the tight buffer.

Oven-Cured Epoxy Application, Connector Assembly, and Polishing

The following procedures apply to oven-cured epoxies. Remember these are general procedures and you should always refer to the epoxy and connector manufacturer data sheets for detailed information.

Oven-Cured Epoxy Application and Connector Assembly

This procedure describes how to prepare the epoxy, inject it into the connector, and assemble the connector for polishing:

1. Turn on the curing oven and set the temperature as recommended by the epoxy manufacturer.

2. Mix the epoxy according to the manufacturer's instructions.

3. Secure the needle to the syringe. Load the syringe with epoxy and ensure that the syringe is free of air pockets.

4. Insert the needle tip into the back of the connector until it bottoms out against the ferrule. Maintain pressure and slowly inject the epoxy until a bead appears on the end of the ferrule tip. Continue to inject epoxy until the bead covers about a third of the ferrule diameter end.

5. Release the pressure on the plunger, wait 5 seconds, and then remove the needle.

6. If required, place the cure adapter over the connector ferrule.

7. Gently feed the optical fiber through the connector, as shown in Figure 9.73, making sure the strength member fans back as you push the optical fiber forward.

FIGURE 9.73
Feed the fiber through the connector.

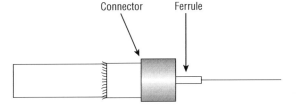

8. Crimp the connector using the crimper as described in the manufacturer's data sheet.

9. Place the connector in the curing oven and cure the epoxy according to the manufacturers cure schedule.

10. After the epoxy has cured, carefully remove the connector and allow it to cool. Then proceed to polishing the connector, described in the next section.

Oven-Cured Epoxy Connector Polishing

This procedure describes how to polish the endface of the connector after the epoxy has cured and the connector has cooled. This polishing process is only for PC finishes. The same process can be applied for a flat finish by substituting a glass plate for the rubber pad.

The following procedure is a general polishing procedure. You should always refer to the connector manufacturer data sheet for detailed information.

1. Score the fiber with the scribe where it exits the epoxy bead, and then gently pull the optical fiber, lifting it from the connector endface. A small optical fiber nub should be protruding from the connector endface. You can use the eye loupe to view the connector endface.

2. Polish down or de-burr the fiber end by holding the connector so that it is facing up. Arch a piece of 5μm polishing film over it. Lightly rub the film in a circle over the connector until you no longer feel the fiber end "grab" the film. Then stop.

3. Place a clean piece of 5μm aluminum oxide polishing film over the rubber pad.

4. Insert the connector into the puck and place the puck on the polishing film.

5. Slowly move the puck in a figure-8 pattern over the polishing film to begin removing the epoxy bead. After about five or ten figure-8 strokes, you should start to feel less resistance as the epoxy bead polishes down. Stop polishing before you have removed all of the epoxy.

6. Remove the 5μm polishing film and place a 1μm polishing film on the rubber pad. Slowly move the puck in a figure-8 pattern over the polishing film to begin removing what is left of the epoxy bead. After about five or ten figure-8 strokes, the polishing puck should glide smoothly over the polishing film. Stop polishing and go to the next step for single-mode applications.

7. For single-mode applications, remove the 1μm polishing film and place a 0.3 or 0.1μm diamond polishing film on the rubber pad. Slowly move the puck in a figure-8 pattern over the polishing film and stop after ten figure-8 strokes. Polishing is complete.

NOTE An aluminum oxide abrasive works well for most non-laser-optimized multimode applications; however, a diamond abrasive is recommended for single-mode and laser-optimized multimode applications. Do not overpolish when using a diamond abrasive.

ANAEROBIC EPOXY APPLICATION, CONNECTOR ASSEMBLY, AND POLISHING

The following procedure applies to anaerobic epoxies. Remember this is a general procedure and you should always refer to the epoxy and connector manufacturer data sheets for detailed information.

Anaerobic Epoxy Application and Connector Assembly

This procedure describes how to inject the adhesive into the connector, apply the primer to the optical fiber, and assemble the connector:

1. Place a needle on top of the anaerobic adhesive and insert the needle tip into the back of the connector until it bottoms out against the ferrule. Maintain pressure and slowly inject the epoxy until a bead appears on the end of the ferrule tip. Continue to inject epoxy until the bead covers about a third of the ferrule diameter end.

2. Open the primer. Using the small brush on the cap of the primer, coat the bottom two-thirds of the exposed optical fiber with the primer. Ensure that you place the lid back on the primer immediately after use; the primer evaporates quickly and will melt the nonreflective mat if you spill it on the mat.

3. Gently feed the optical fiber through the connector, as shown in Figure 9.74, making sure the strength member fans back as you push the optical fiber forward. Wait 10 seconds for the epoxy to cure before going to the next step.

FIGURE 9.74
Feed the fiber through the connector.

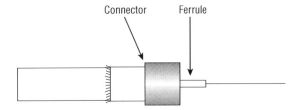

Connector   Ferrule

WARNING   Do not pull back on the optical fiber; doing so may break the optical fiber.

4. Crimp the connector as described in the manufacturer's data sheet, and then proceed to the next section.

Anaerobic Epoxy Connector Polishing

This procedure describes how to polish the endface of the connector after the anaerobic epoxy has cured. This polishing process is only for PC finishes. The same process can be applied for a flat finish by substituting a glass plate for the rubber pad.

The following procedure is a general polishing procedure. You should always refer to the connector manufacturer data sheet for detailed information.

NOTE   The anaerobic epoxy bead is not hard like the oven-cured epoxy bead and does not offer any support to the optical fiber.

1. Score the optical fiber with the scribe where it exits the connector endface, and then gently pull the optical fiber, lifting it from the connector endface. A small optical fiber nub should be protruding from the connector endface. You can use the eye loupe to view the connector endface.

2. Polish down or de-burr the optical fiber end by holding the connector so that it is facing up. Arch a piece of 5μm polishing film over it. Lightly rub the film in a circle over the connector until you no longer feel the fiber end "grab" the film. Then stop.

3. Place a clean piece of 1μm polishing film on the rubber pad.

NOTE An aluminum oxide abrasive works well for most non-laser-optimized multimode applications; however, a diamond abrasive is recommended for single-mode and laser-optimized multimode applications. Do not overpolish when using a diamond abrasive.

4. Insert the ferrule into the puck and place the puck on the polishing film.

5. Slowly move the puck in a figure-8 pattern over the polishing film to begin polishing down the optical fiber. After about five or ten figure-8 strokes, you should start to feel less resistance and the puck should glide across the polishing film. Stop polishing and go to the next step for single-mode applications.

6. For single-mode applications, remove the 1μm polishing film and place a 0.3 or 0.1μm diamond polishing film on the rubber pad. Slowly move the puck in a figure-8 pattern over the polishing film and stop after ten figure-8 strokes. Polishing is complete.

Machine Polishing

Machine polishing is typically performed at the factory. Depending on the polishing machine, one or more connectors can be polished at a time. Machine polishing produces the best connector endface and produces more consistent results than hand polishing. Most single-mode connectors are polished with a machine.

The same steps required to assemble a connector for hand polishing are also required to assemble a connector for machine polishing. The polishing machine is substituted for the hand polishing. Connectors are inserted into the polishing machine after they have been cleaved and de-burred. When the polishing process is complete, remove the connectors and inspect them.

Pre-polished Connectors

Pre-polished connectors have been around for many years and do not require hand or machine polishing. A pre-polished connector is similar to a mechanical splice because a cleaved piece of optical fiber is inserted into the rear of the connector and mated with a small piece of optical fiber that was bonded to the connector ferrule at the factory. The gap between the two optical fibers is filled with index matching gel to reduce Fresnel reflections. The small piece of optical fiber bonded to the connector ferrule has a cleaved perpendicular finish on the side mating with the other optical fiber. The other end of the optical fiber is polished with a machine to produce an exacting endface.

Pre-polished connectors are available for single or multiple fiber applications. When using a pre-polished connector, you must ensure that the optical fiber in the connector is the same size and type of optical fiber you are terminating. You cannot terminate a multimode optical fiber that has a 50μm core with a pre-polished connector that has a 62.5μm core, and vice versa. You should not terminate a non-laser-optimized multimode optical fiber that has a 50μm core with a pre-polished connector that has a laser-optimized 50μm core, and vice versa.

Three popular brands of pre-polished connectors are the UniCam from Corning Cable Systems, the OptiCam from Panduit, and the Light Crimp Plus from TE Connectivity. Both the UniCam and OptiCam feature a cam-type locking mechanism that holds the optical fiber firmly in place throughout the life of the connector. If a problem occurs during termination, the cam-type locking mechanism can be unlocked to remove the optical fiber and reattempt the termination.

Shown in Figure 9.75 is a UniCam 50µm laser-optimized multimode LC connector and in Figure 9.76 is a single-mode SC connector. Shown in Figure 9.77 is an OptiCam 62.5µm multimode SC connector and in Figure 9.78 is a single-mode LC connector. The Light Crimp Plus multimode SC connector kit is shown in Figure 9.79 and the single-mode LC is shown in Figure 9.80.

FIGURE 9.75
Corning Cable Systems' UniCam 50µm laser-optimized multimode LC connector

Photo courtesy of Corning Cable Systems

FIGURE 9.76
Corning Cable Systems' UniCam single-mode SC connector

Photo courtesy of Corning Cable Systems

FIGURE 9.77
Panduit OptiCam 62.5µm multimode SC connector

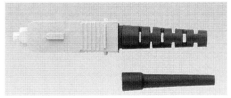

Photo courtesy of Panduit

FIGURE 9.78
Panduit OptiCam
single-mode LC
connector

Photo courtesy of Panduit

FIGURE 9.79
TE Connectivity
LightCrimp Plus
multimode SC con-
nector kit

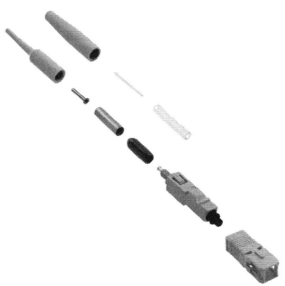

Photo courtesy of TE Connectivity Global Aerospace Defense and Marine

FIGURE 9.80
TE Connectivity
LightCrimp Plus
single-mode LC
connector kit

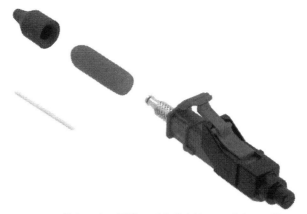

Photo courtesy of TE Connectivity Global Aerospace Defense and Marine

All three manufacturers offer many different versions of these connectors and kits with the required tools for assembly. They have trained personnel who can answer any questions you may have about your application. Shown in Figure 9.81 is the UniCam LANscape kit; Figure 9.82 shows the Panduit OptiCam kit, and Figure 9.83 shows the LightCrimp Plus kit.

FIGURE 9.81
Corning Cable
Systems' UniCam
LANscape kit

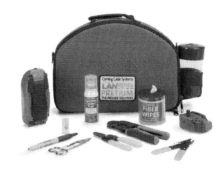

Photo courtesy of Corning Cable Systems

FIGURE 9.82
Panduit
OptiCam kit

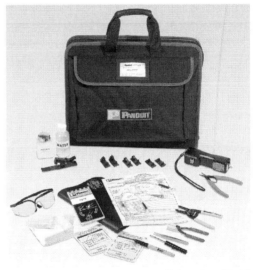

Photo courtesy of Panduit

FIGURE 9.83
TE Connectivity
LightCrimp Plus kit

Photo courtesy of TE Connectivity Global Aerospace Defense and Marine

Cleaning and Inspection

Good cleaning and inspection skills are essential for low-loss and low back reflection interconnections. Often the cause of a problem is contamination, which can be discovered through inspection and repaired by cleaning. Contamination can take many forms, and there is no spot on a fiber-optic connector that contamination cannot find.

Until recently, cleaning and inspection focused on the endface. However, that school of thought has changed and when working in fiber optics you need to be aware that contamination anywhere on the ferrule or alignment sleeve can affect the performance of an interconnection. Also keep in mind that contamination can migrate, which means it can move from one location to another both during and after the cleaning process. It is very easy to pull contamination from the side of the ferrule onto the endface during cleaning.

There are many ways that contamination can migrate or move from one location on the ferrule to another. Some of the ways are more obvious than others. A typical ferrule is made of ceramic; both ceramic and optical fiber are dielectrics. A dielectric is another name for an insulator.

As a dielectric, both the ferrule and optical fiber are capable of holding an electrical charge. This electrical charge may be referred to as an electrostatic charge that is commonly known as static electricity. Because the ferrule and optical fiber can hold a charge, they are capable of attracting particles (contamination) with an opposite charge. It is for this reason that most fiber-optic cleaning fluids are engineered to dissipate those charges.

Endface Cleaning

After a connector is polished or before it is inserted into a receptacle or mating sleeve, the end-face should be cleaned and inspected. Mating a clean connector with a dirty connector will result in two dirty connectors. Figure 9.84 shows what a formerly clean connector endface looks like after it has been mated with a connector that has skin oil on the endface. You can tell that this connector had been mated because the oil is distributed in a crater-like pattern.

FIGURE 9.84
Skin oil distributed in a crater-like pattern, the result of mating a clean and dirty connector

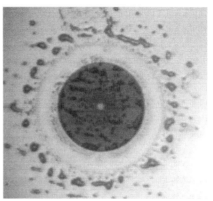

Photo courtesy of MicroCare

Cleaning the endface of a connector is a simple process; however, it does require cleaning products engineered for that purpose. You need only two basic products: lint-free wipes and a cleaning fluid engineered to clean the optical fiber endface.

In August 2012, SAE International published AIR6031, Fiber Optic Cleaning. That standard states that manual cleaning can be done using a wet-dry process or a dry wipe process. This section will describe dry and wet-dry cleaning techniques.

DRY CLEANING TECHNIQUES

When a cleaning fluid is not required to clean a connector endface, the connector endface can be cleaned with a dry, lint-free wipe. Multiple manufacturers offer lint-free wipes developed specifically for fiber-optic cleaning applications. They are typically packaged in different ways, as shown in Figure 9.85, for different applications or environments. These wipes have been engineered to lift oils, dirt, and contamination from the endface of the connector.

The foil packaged wipes shown in Figure 9.85 are sealed to prevent the wipe from being contaminated prior to use. This wipe is glued to the foil on one side; that way, you can handle the wipe without getting oil and dirt from your hand on the wipe. To clean the endface with this wipe, open the package and place the wipe in the palm of your hand, as shown in Figure 9.86, and move the connector in a straight line from top to bottom several times, each time starting in a different location. Be sure to lift the connector from the wipe before it reaches the end of the wipe. If the connector comes in contact with your skin or hair, it will pick up the oil, as shown in Figure 9.87.

FIGURE 9.85
Lint-free wipes
packaged in a round
container, box,
and sealed in a foil
package

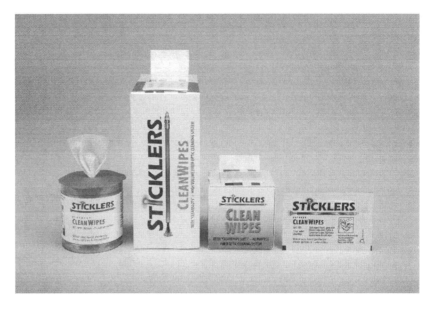

Lint-free wipes are also available in round containers, as shown in Figure 9.85. These are the same wipes found in the sealed foil package. When using these wipes, be careful to touch only one side of the wipe with your hand. Only use the side that did not touch your hand to clean the connector endface.

FIGURE 9.86
Cleaning the con-
nector endface with
a lint-free wipe that
was sealed in a foil
package

Photo courtesy of MicroCare

FIGURE 9.87
Oil on the endface
of a connector

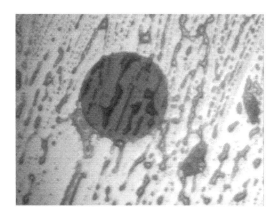

To clean the endface with this wipe, place the wipe in the palm of your hand, as shown in Figure 9.88, and move the connector in a straight line from top to bottom several times, each time starting in a different location. Be sure to lift the connector from the wipe before it reaches the end of the wipe. If the connector comes in contact with your skin or hair, it will pick up the oil, as you saw in Figure 9.87.

FIGURE 9.88
Lint-free wipe prop-
erly placed in the
palm of a hand for
cleaning a connec-
tor endface

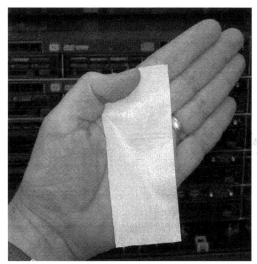

Photo courtesy of MicroCare

Lint-free wipes are also available in boxes, also shown in Figure 9.85. These are the same wipes found in the sealed foil package; however, depending on the manufacturer, they may be packaged so the wipe can be placed over a slotted stencil that prevents the endface from touching sections of the wipe that were contaminated during the cleaning process. When using these wipes, be careful to touch only one side of the wipe with your hand while positioning it over the stencil.

To clean the endface with this wipe, place the endface at the top of a slot, as shown in Figure 9.89, and move the connector in a straight line from top to bottom several times, using a different slot each time. Be sure to lift the connector from the wipe before it reaches the end of the wipe. If the connector comes in contact with your skin, it will pick up the oil, as you saw in Figure 9.87.

FIGURE 9.89
Connector properly
placed in the slot
for cleaning the
endface

Photo courtesy of MicroCare

WET-DRY CLEANING TECHNIQUES

If cleaning the endface of the connector with a dry, lint-free wipe does not remove the contamination, a wet-dry clean technique is required. As discussed earlier, several manufacturers offer lint-free wipes developed specifically for fiber-optic cleaning applications packaged in different ways for different applications or environments, as shown earlier in Figure 9.85. These wipes have been engineered to lift oils, dirt, and contamination from the endface of the connector.

The foil packaged wipes shown in Figure 9.85 are sealed to prevent the wipe from being contaminated prior to use. To clean the endface with this wipe, open the package and wet the wipe with an optical fiber cleaning fluid. Place the wipe in the palm of your hand, as shown in Figure 9.90, and move the connector in a straight line from top to bottom several times, each time starting in a different location. Be sure to lift the connector from the wipe before it reaches the end of the wipe. If the connector comes in contact with your skin or hair, it will pick up the oil, as you saw in Figure 9.87.

Lint-free wipes are also available in round containers, as shown in Figure 9.85. These are the same wipes found in the sealed foil package. When using these wipes, be careful to touch only one side of the wipe with your hand. Use only the side that did not touch your hand to clean the connector endface.

To clean the endface with this wipe, wet the wipe with an optical fiber cleaning fluid. To do this, fold the wipe over, place it in the cleaning fluid dispenser, and pump the dispenser, as shown in Figure 9.91. Place the wipe in the palm of your hand, as shown in Figure 9.92, and move the connector in a straight line from top to bottom several times, each time starting in a different location. Be sure to lift the connector from the wipe before it reaches the end of the wipe. If the connector comes in contact with your skin or hair, it will pick up the oil, as you saw in Figure 9.87.

FIGURE 9.90
Cleaning the con-
nector endface
with a wet-dry
lint-free wipe that
was sealed in a foil
package

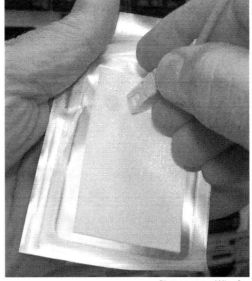

Photo courtesy of MicroCare

FIGURE 9.91
Wetting the lint-
free wipe with a
cleaning fluid

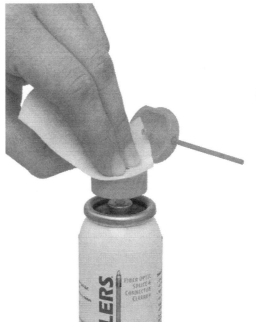

Photo courtesy of MicroCare

FIGURE 9.92
Wet-dry lint-free wipe properly placed in the palm of a hand for cleaning a connector endface

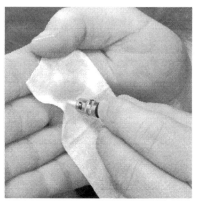

Photo courtesy of MicroCare

Lint-free wipes are also available in boxes, as shown in Figure 9.85. These are the same wipes found in the sealed foil package; however, depending on the manufacturer they may be packaged without a stencil. When using these wipes, be careful to touch only one side of the wipe with your hand while positioning it over the stencil.

To clean the endface with this wet-dry wipe, moisten the wipe, as shown in Figure 9.93. Place the endface at the top, as shown in Figure 9.94, and move the connector in a straight line from top to bottom several times, using a different location each time. Be sure to lift the connector from the wipe before it reaches the end of the wipe. If the connector comes in contact with your skin, it will pick up the oil, as you saw in Figure 9.87.

FIGURE 9.93
Wetting the lint-free wipe with a cleaning fluid

Photo courtesy of ITW Chemtronics

FIGURE 9.94
Connector properly placed on the lint-free wipe for cleaning the endface

Photo courtesy of ITW Chemtronics

A connector that has been cleaned properly will be free of any dirt or contamination, as shown in Figure 9.95. Only after the connector has been cleaned and inspected should it be mated with another connector or receptacle.

FIGURE 9.95
Properly cleaned single-mode connector endface

Photo courtesy of MicroCare

Special kits developed just for cleaning fiber-optic components are available from several manufacturers, among them the ones shown in Figures 26.96 and 26.97.

FIGURE 9.96
Sticklers fiber-optic
cleaning kit

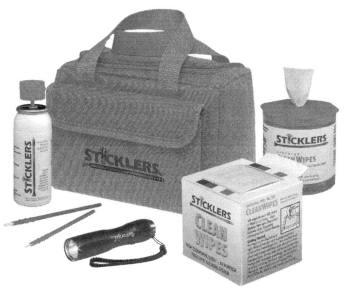

Photo courtesy of MicroCare

FIGURE 9.97
ITW Chemtronics
complete fiber-optic
cleaning kit

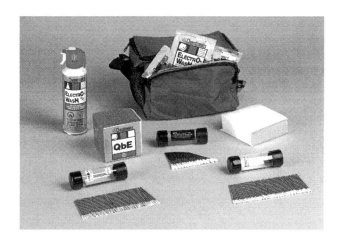

MicroCare, ITW Chemtronics, Corning Cable Systems, and AFL Telecommunications manufacture some of the most commonly used cleaning materials.

MicroCare www.microcare.com

ITW Chemtronics www.chemtronics.com

Corning www.corningcablesystems.com

AFL Telecommunications www.afltele.com

5.2 **Endface Inspection**

IEC 61300-3-35 is a standard that provides methods to quantitatively evaluate the endface quality of a polished fiber-optic connector. This standard breaks the endface into four zones: core, cladding, adhesive or epoxy ring, and contact zone, as shown in Figure 9.98. As of this writing, this standard was under revision. It is unknown at this time if an additional zone will be included in the next revision. In this book, however, zone five, the side of the ferrule, will be discussed. In addition, this section will discuss the tools required to inspect the connector endface and examine common defects and their causes.

FIGURE 9.98

The four endface zones

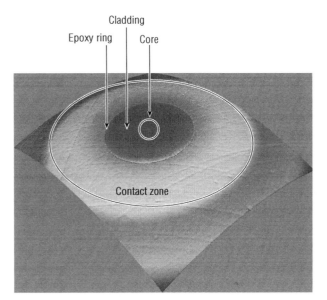

Epoxy ring Cladding Core

Contact zone

INSPECTION MICROSCOPES

To evaluate the endface of a fiber-optic connector, you need a microscope designed for this application. This type of microscope is typically referred to as an *inspection microscope*. Many different inspection microscopes are available, and they can be grouped into two basic categories: optical microscopes and video microscopes. Video microscopes can be further broken into two groups: bench top and video probe.

An optical inspection microscope allows the direct viewing of the connector endface.

WARNING Care should be taken to ensure that there is no light source on the other end of the connector endface being viewed. As you have learned in this book, fiber-optic light sources are infrared and are not visible to the naked eye. Viewing a connector endface that is radiating infrared light with an optical microscope can temporarily or permanently damage your eye.

Optical inspection microscopes are typically handheld and are available with magnification levels from 100X to 400X. Three different light sources are also available to illuminate the endface: incandescent lamp, LED, and laser. The 400X handheld inspection microscope shown in Figure 9.99 uses a laser to illuminate the connector endface, and the 400X model shown in Figure 9.100 uses an LED.

FIGURE 9.99
400X microscope
with a laser for
illumination

£60

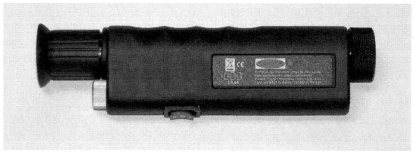

FIGURE 9.100
400X microscope
with an LED for
illumination and
an LC connec-
tor inserted for
inspection

The key advantage to the LED light source is long battery life, and the LED should never
need replacing. The laser light source requires more power than the LED; therefore battery life
is reduced. However, the laser does make it easier to identify light scratches and defects in the
optical fiber endface, and it too should never need replacing. Incandescent lamps are typically
found in lower-cost 100X microscopes like the one shown in Figure 9.101.

FIGURE 9.101
100X microscope
with an incan-
descent lamp for
illumination

Video inspection microscopes allow you to view the connector endface indirectly. This eliminates any possibility of your eye being damaged from infrared light while inspecting the endface with the microscope. Figure 9.102 is a photograph of a 200X video probe microscope.

FIGURE 9.102

200X video probe microscope

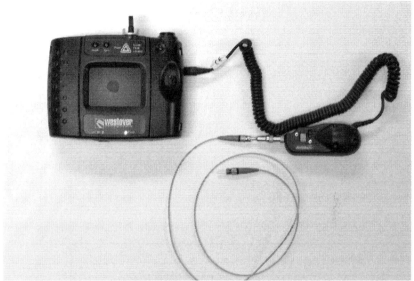

Photo courtesy of KITCO Fiber Optics

Typically, a multimode connector can be evaluated with a 100X microscope, whereas a single-mode connector requires a minimum of 200X magnification. A 400X microscope, of course, works great for both multimode and single-mode.

Higher magnification microscopes allow a close inspection of the core, cladding, epoxy ring, and contact zones. However, they will not show as much of the endface as a lower power microscope and may miss contamination.

Contaminated endfaces at different magnifications are shown in Figures 9.103 and 9.104. Less endface is visible in the higher magnification image shown in Figure 9.103 when compared to the lower magnification image shown in Figure 9.104.

FIGURE 9.103

Alcohol residue contaminated endface with higher power magnification

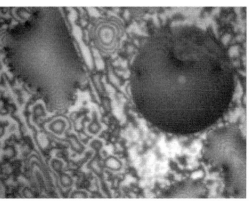

Photo courtesy of MicroCare

FIGURE 9.104
Alcohol residue con-
taminated endface
with lower power
magnification

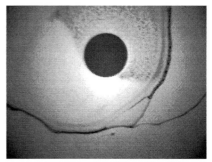

Photo courtesy of MicroCare

As mentioned earlier in this chapter, zone five includes the side of the ferrule and contamination on the side of the ferrule can affect the performance of the interconnection. It is not uncommon to find cured epoxy on the side of the ferrule. Cured epoxy will not wipe off, and because it
is on the side of the ferrule, it will not be visible with an inspection microscope.

Examining the side of the ferrule requires little or no magnification. It is a good practice to
perform this as part of your inspection routine. An eye loupe like the one shown in Figure 9.105
provides moderate magnification and is helpful in locating contamination on the ferrule.

FIGURE 9.105
An eye loupe pro-
vides moderate
magnification for
viewing the endface
and ferrule the
connector.

Photo courtesy of KITCO Fiber Optics

EXAMPLES OF ENDFACE INSPECTION

This section will examine different endface finishes. Some finishes will be a hand polish with
an aluminum oxide abrasive, and some will be a machine polish with a diamond abrasive. Each
endface will be evaluated with a video microscope. As mentioned earlier in the chapter this
type of evaluation is a 2D evaluation that cannot quantitatively define radius of curvature, apex
offset, fiber undercut, fiber protrusion, the depth of scratches, or the height of debris as can be
done with an interferometer.

Good Quality, Clean Endfaces

The multimode endface shown in Figure 9.106 was hand polished with a 5μm aluminum oxide abrasive to remove the epoxy, followed with a 1μm aluminum oxide. The light scratches on the core and the cladding are typical of a 1μm aluminum oxide abrasive. The chipping and pitting around the epoxy ring is also typical for this abrasive. This endface is an example of a good hand polish with an aluminum oxide abrasive.

FIGURE 9.106
Good hand polish with an aluminum oxide abrasive viewed at 400X

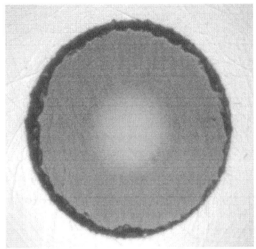

Photo courtesy of W.R. Systems

The single-mode endface shown in Figure 9.107 was machine polished. The final polishing step used a 0.1μm diamond abrasive. This endface is cosmetically perfect; there are no visible scratches on the optical fiber and no chipping or pitting near the epoxy ring. This endface is an example of a good machine polish with a diamond abrasive.

FIGURE 9.107
Good machine polish with a diamond abrasive viewed at 400X

Photo courtesy of MicroCare

Good Quality, Dirty Endfaces

The multimode endface shown in Figure 9.108 was hand polished with a 5μm aluminum oxide abrasive to remove the epoxy, followed by a 1μm aluminum oxide. The light scratches on the core and the cladding are typical of a 1μm aluminum oxide abrasive. Note that the endface is covered with skin oil and needs to be cleaned.

FIGURE 9.108
Skin oil covering a good hand polish with an aluminum oxide abrasive viewed at 400X

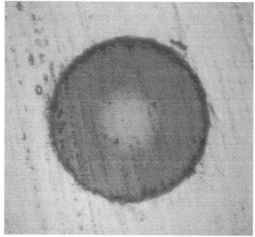

Photo courtesy of W.R. Systems

The single-mode endface shown in Figure 9.109 was machine polished. The final polishing step used a 0.1μm diamond abrasive. This endface is cosmetically perfect; however, it is covered with hair oil. You can typically tell the difference between hair oil and skin oil by the size of the oil deposits. Skin oil as shown in Figure 9.108 typically leaves smaller deposits than hair oil, and the oil tends to look it was applied in streaks. This connector needs to be cleaned.

FIGURE 9.109
Hair oil covering a good single-mode machine polish with a diamond abrasive viewed at 400X

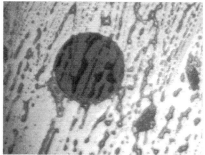

Photo courtesy of MicroCare

The single-mode endface shown in Figure 9.110 was machine polished. The final polishing step used a 0.1μm diamond abrasive. This endface is cosmetically perfect; however, it was cleaned with a shirttail instead of a lint-free wipe. Debris from the shirttail was deposited on the endface of the connector. This connector needs to be cleaned with a lint-free wipe.

FIGURE 9.110
Debris from a
shirttail covering a
good single-mode
machine polish with
a diamond abrasive
viewed at 200X

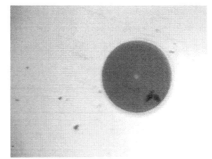

Photo courtesy of MicroCare

Broken or Damaged Endfaces

It is impossible to tell how a connector was broken or damaged if it is inspected only after polishing is complete. If damage happens more than once, stop after every step and inspect the endface to determine which step is causing the problem.

The multimode endface shown in Figure 9.111 was hand polished with a 5μm aluminum oxide abrasive to remove the epoxy, followed by a 1μm aluminum oxide. Somewhere in the polishing process, the optical fiber was broken and a majority of the break extended below the surface of the endface.

FIGURE 9.111
Broken multimode
endface viewed at
400X

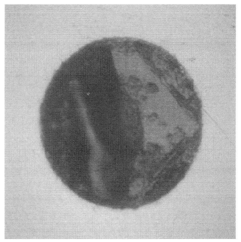

Photo courtesy of W.R. Systems

The multimode endface shown in Figure 9.112 was hand polished with a 5μm aluminum oxide abrasive to remove the epoxy, followed by a 1μm aluminum oxide. Somewhere in the polishing process, the optical fiber was broken; however, only part of the break extended below the surface of the endface.

FIGURE 9.112
Damaged mul-
timode endface
viewed at 400X

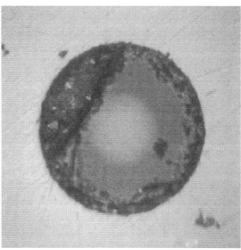

Photo courtesy of W.R. Systems

Connector Performance

If you have assembled, polished, cleaned, and inspected the connector properly, there is a good chance that your connector will provide a low-loss interconnection. Appendix A of ANSI/TIA-568-C.3 defines fiber-optic connector performance. The maximum insertion loss allowed by ANSI/TIA-568-C.3 for a multimode or single-mode mated connector pair is 0.75dB.

Return loss as defined in ANSI/TIA-568-C.3 is the ratio of the power of the outgoing optical signal to the power of the reflected signal. This ratio is expressed in dB. The minimum return loss for a multimode interconnection is 20dB. This means that light energy from the light source traveling through the interconnection is 20dB greater than the light energy being reflected back by the interconnection toward the source.

The minimum return loss for a single-mode interconnection is 26dB unless that interconnection is part of a broadband analog video (CATV) application. Interconnections used in CATV applications must have a minimum return loss of 55dB. Optical return loss testing is discussed in Chapter 16.

Connector Color Code

Multimode and single-mode connectors and adapters can be identified using the color code in section 5.2.3 of ANSI/TIA-568-C.3. The color code in Table 9.2 should be used unless color-coding is used for another purpose. This table lists the strain relief and adapter housing color for different optical fiber types.

TABLE 9.2: Multimode and single-mode connector and adapter identification

OPTICAL FIBER TYPE OR CONNECTOR TYPE	STRAIN RELIEF AND ADAPTER HOUSING COLOR
850nm laser-optimized 50/125μm optical fiber	Aqua
50/125μm optical fiber	Black
62.5/125μm optical fiber	Beige
Single-mode optical fiber	Blue
Angled contact ferrule single-mode connectors	Green

The Bottom Line

Evaluate connector endface polishing. To evaluate the endface of a fiber-optic connector, you need a microscope designed for this application.

Master It You are inspecting the multimode endface shown here with a 200X inspection microscope. Is there anything wrong with this connector endface?

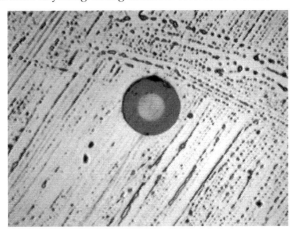

Evaluate connector endface cleaning. A connector that has been cleaned properly will be free of any dirt or contamination.

Master It You are inspecting the connector endfaces shown in these two graphics. Which of these connectors has oil on the endface?

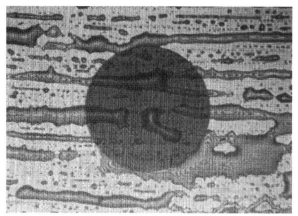

Photo courtesy of MicroCare

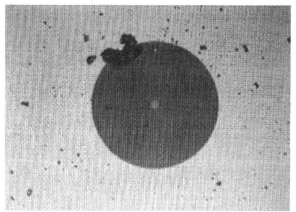

Photo courtesy of MicroCare

Evaluate connector endface geometry. The key measurements provided by the interferometer are radius of curvature, apex offset, and fiber undercut or protrusion. These critical parameters are required by Telcordia GR-326 to evaluate connector endface geometry.

    Master It You are evaluating a machine polished endface with the interferometer. The radius of curvature measured by the interferometer is 25mm. Does this endface radius of curvature fall within the range specified by Telcordia GR-326?

Identify an optical fiber type from the color of the connector strain relief. Multimode and single-mode connectors and adapters can be identified using the color code in ANSI/TIA-568-C.3.

    Master It You need to mate two 850nm laser-optimized 50/125μm fiber-optic cables together with a jumper at a patch panel. What color strain reliefs would the correct jumper have?

Chapter 10

Fiber-Optic Light Sources and Transmitters

As discussed in Chapter 1, "History of Fiber Optics and Broadband Access," the idea of transmitting information with light is not new—only the technology that makes it easily possible. Like optical fiber, light source technology has improved rapidly over the decades. These technological advances have greatly increased data transmission rates and reduced costs. Fiber-optic transmitters are available to support every standardized network with a variety of connector choices.

This chapter discusses current fiber-optic light source and transmitter technology as it applies to common telecommunication network standards, industrial control systems, and general-purpose systems. The performance standards that we discuss in this chapter do not represent the highest levels achievable in the lab or on the test bench. They represent commonly available parts for standardized networks. All performance values were obtained from manufacturers' data sheets or networking standards.

In this chapter, you will learn to:

- ♦ Determine the minimum optical output power for an LED transmitter
- ♦ Determine the maximum optical output power for an LED transmitter
- ♦ Determine the minimum optical output power for a laser transmitter
- ♦ Determine the maximum optical output power for a laser transmitter

Semiconductor Light Sources

The light sources used in fiber-optic communication systems are far different from the light sources used to illuminate your home or office. Fiber-optic light sources must be able to turn on and off millions to billions of times per second while projecting a near-microscopic beam of light into an optical fiber. On top of this performance, they must be reasonably priced, highly reliable, easy to use, and available in a small package.

Semiconductor light source technology has made all this possible. Today's fiber-optic communication systems use *light-emitting diodes (LEDs)* and *laser diodes* (from this point forward, the laser diode will be referred to as the *laser*) exclusively. These semiconductor light sources are packaged to support virtually every fiber-optic communication system imaginable.

LED Sources

A basic LED light source is a semiconductor diode with a *p* region and an *n* region. When the LED is *forward biased* (a positive voltage is applied to the *p* region and a negative voltage to the *n* region), current flows through the LED. As current flows through the LED, the junction where the *p* and *n* regions meet emits random photons. This process is referred to as *spontaneous emission*. Figure 10.1 shows a forward-biased LED in a basic electric circuit.

FIGURE 10.1
Forward-biased
LED

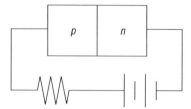

Photons emitted from the junction where the *p* and *n* regions meet are not in phase, nor are they launched in the same direction. These out-of-phase photons are called *incoherent light*. This incoherent light cannot be focused so that each photon traverses down the optical fiber. Because of this, only a small percentage of the photons emitted will be coupled into the optical fiber. Figure 10.2 shows the out-of-phase photons being spontaneously emitted from the LED.

FIGURE 10.2
Radiating forward-
biased LED

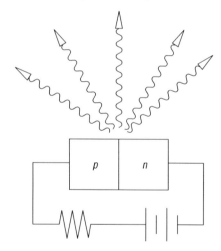

Two types of LEDs, the *surface-emitting LED* and the *edge-emitting LED*, are commonly used in fiber-optic communication systems. Surface-emitting LEDs are a *homojunction structure*, which means that a single semiconductor material is used to form the *pn* junction. Incoherent photons radiate from all points along the *pn* junction, as shown in Figure 10.3.

Edge-emitting LEDs are a *heterojunction structure*, which means that the *pn* junction is formed from similar materials with different refractive indexes. The different refractive indexes are used to guide the light and create a directional output. Light is emitted through an etched opening in the edge of the LED, as shown in Figure 10.4.

FIGURE 10.3
Radiating surface-
emitting LED

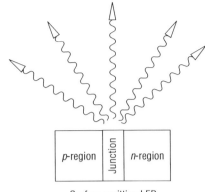

Surface-emitting LED

FIGURE 10.4
Radiating edge-
emitting LED

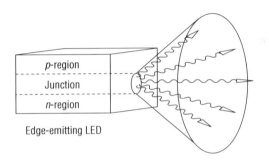

Edge-emitting LED

Laser Sources

The term *laser* is actually an acronym that stands for *light amplification by stimulated emission of radiation*. Like the LED, the laser is a semiconductor diode with a *p* region and an *n* region. Unlike the LED, the laser has an optical cavity that contains the emitted photons with reflecting mirrors on each end of the diode. One of the reflecting mirrors is only partially reflective. This mirror allows some of the photons to escape the optical cavity, as shown in Figure 10.5.

FIGURE 10.5
Radiating laser

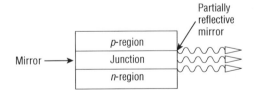

Every photon that escapes the optical cavity is a duplicate of the first photon to escape. These photons have the same wavelength, phase relationship, and direction as the first photon. This process of generating light energy is called *stimulated emission*. The photons radiated from the laser have a fixed relationship that is referred to as *coherent light* or coherent radiation. Figure 10.5 shows the in-phase light waves emitted from the laser.

There are three families of lasers used in fiber-optic communication systems: *Fabry-Pérot, distributed feedback (DFB),* and the *vertical-cavity surface-emitting laser (VCSEL)*. Each laser family has unique performance characteristics that will be discussed in this chapter and is designed to support a specific telecommunication application.

2. Light Source Performance Characteristics

This section will compare the performance characteristics of the LED and laser light sources. The performance of a light source can be judged in several areas: output pattern, wavelength, spectral width, output power, and modulation speed. These performance areas determine the type of optical fiber that the source can be coupled to, transmission distance, and data rate. Without a doubt, the laser is the hands-down winner when it comes to ultra-high-speed long-distance data transmission. However, many applications require only a fraction of the performance that a laser offers; for these applications, an LED is used.

2.1 Output Pattern

numerical aperture

The LED and laser semiconductors used in fiber-optic light sources are packaged to couple as much light as possible into the core of the optical fiber. The output pattern or NA of the light source directly relates to the energy coupled into the core of the optical fiber. The LED has a wide output pattern compared to a laser and does not couple all its light energy into the core of a multimode optical fiber. The output pattern of a laser light source is very narrow, allowing a majority of the light energy to be coupled into the core of a single-mode or multimode optical fiber.

LED Output Pattern

Unpackaged surface-emitting and edge-emitting LEDs have wide output patterns, as shown earlier in Figures 10.3 and 10.4. An LED light source must be assembled in a package to mate with a specific connector type and optical fiber. However, occasionally there may be a requirement for an LED to be packaged as a *pigtail*. A pigtail is a short length of tight-buffered optical fiber permanently bonded to the light source package, as shown in Figure 10.6.

To couple as much light as possible into the core of the optical fiber, the manufacturer typically mounts a micro lens in the shape of a sphere directly on top of the LED. The manufacturer may also use an additional lens to further control the output pattern. Figure 10.7 shows a surface-emitting LED with a series of lenses directing light into the optical fiber.

Even the best lens can't direct all the light energy into the core of the optical fiber. The typical LED light source overfills the optical fiber, allowing light energy to enter the cladding and the core at angles exceeding the NA. These high-order or marginal modes are attenuated over a short distance. Light energy that enters the core exceeding the acceptance angle will not propagate to the end of the optical fiber. This light energy will refract into the cladding. Figure 10.8 shows a typical LED source overfilling an optical fiber, which is illustrated by the light spot that the source projects.

FIGURE 10.6
Pigtailed light
source

FIGURE 10.7
Packaged surface-
emitting LED

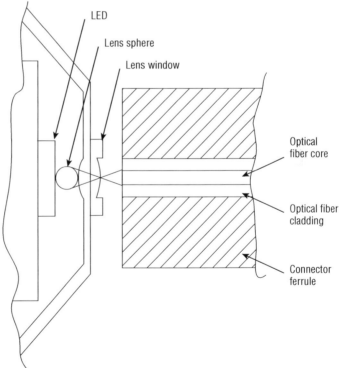

FIGURE 10.8
Laser and LED spot
sizes

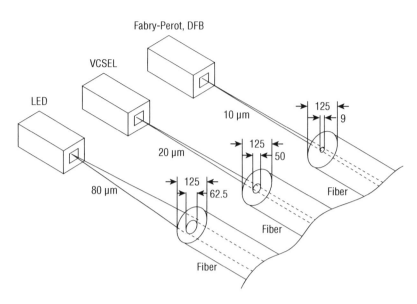

The process of an LED transmitter overfilling a multimode optical fiber is referred to as an *overfilled launch (OFL)*. An OFL is used to define minimum effective modal bandwidth length product of a multimode optical fiber. Multimode OFL optical fiber bandwidth requirements are defined in ANSI/TIA-568-C.3 and ISO/IEC 11801, as shown in Table 10.1.

TABLE 10.1:   Characteristics of ANSI/TIA-568-C.3 and ISO/IEC 11801-recognized optical fibers

OPTICAL FIBER AND CABLE TYPE	WAVELENGTH (nm)	MAXIMUM ATTENUATION (dB/km)	MINIMUM OVERFILLED MODAL BANDWIDTH-LENGTH PRODUCT (MHz · km)	MINIMUM EFFECTIVE MODAL BANDWIDTH-LENGTH PRODUCT (MHz · km)
62.5/125µm Multimode OM1 Type A1b TIA 492AAAA	850 1300	3.5 1.5	200 500	Not required Not required
50/125µm Multimode OM2 TIA 492AAAB Type A1a.1	850 1300	3.5 1.5	500 500	Not required Not required

TABLE 10.1: Characteristics of ANSI/TIA-568-C.3 and ISO/IEC 11801-recognized optical fibers *(continued)*

OPTICAL FIBER AND CABLE TYPE	WAVELENGTH (nm)	MAXIMUM ATTENUATION (dB/km)	MINIMUM OVERFILLED MODAL BANDWIDTH-LENGTH PRODUCT (MHz · km)	MINIMUM EFFECTIVE MODAL BANDWIDTH-LENGTH PRODUCT (MHz · km)
850nm laser-optimized 50/125µm Multimode OM3 Type A1a.2 TIA 492AAAC	850 1300	3.5 1.5	1500 500	2000 Not required
850nm laser-optimized 50/125µm Multimode OM4 Type A1a.3 TIA 492AAAD	850 1300	3.5 1.5	3500 500	4700 Not required

LASER OUTPUT PATTERN

Unlike the LED, the laser light source has a narrow output pattern or NA. Like the LED, the laser must be packaged to align the source with the optical fiber and couple as much energy as possible into the core. Lasers can be packaged for a specific connector or bonded to a pigtail.

Fabry-Pérot and DFB laser light sources are designed for either multimode or single-mode applications. VCSELs are currently designed only for multimode optical fiber applications. The VCSEL emits a wider output pattern and larger spot size than the Fabry-Pérot or DFB lasers. The output of the VCSEL fills only the center of the core of the multimode optical fiber, as shown earlier in Figure 10.8.

Fabry-Pérot and DFB lasers are designed for single-mode and multimode optical fiber applications. The output pattern and spot size of these lasers for single-mode applications are narrower and smaller than the VCSEL. Figure 10.8 shows the typical output pattern of the Fabry-Pérot, DFB, and VCSEL laser.

As bandwidth requirements increase, LED transmitters are being replaced with laser transmitters. However, the multimode optical fiber is not being replaced with single-mode. A laser transmitter, unlike an LED transmitter, does not overfill the multimode optical fiber. Power from a laser transmitter is distributed in the narrow center region of the multimode optical fiber core. This is very different from an LED transmitter, where power is distributed to 100 percent of the core.

Multimode optical fiber has a greater bandwidth potential with a laser transmitter than an LED transmitter. The launch from a laser transmitter results in far fewer modes than the overfilled launch from an LED transmitter. Fewer modes mean less modal dispersion, resulting in greater potential bandwidth.

While all of the laser types—Fabry-Pérot, DFB, and VCSEL—are used with multimode optical fiber, only the VCSEL is used for 850nm applications. The VCSEL is not currently available at a wavelength of 1300nm. For 1300nm or 1310nm applications, the Fabry-Pérot or DFB type is used.

Because the launch conditions for the VCSEL transmitter are different from those for the LED, ANSI/TIA-568-C.3 now defines the bandwidth of a multimode optical fiber both with an OFL and with a *restricted mode launch (RML)*. The minimum effective modal bandwidth length product of a multimode optical fiber at 850nm is defined with a restricted mode launch condition. RML applies only to laser-optimized multimode optical fiber, as shown in Table 10.1.

Source Wavelengths

The performance of a fiber-optic communication system is dependent on many factors. One key factor is the wavelength of the light source. Short wavelengths (650nm through 850nm) have more modes and greater attenuation than longer wavelengths (1300nm through 1600nm). The wavelength of the light source can determine system bandwidth and transmission distance.

LED WAVELENGTHS

LED light sources are manufactured from various semiconductor materials. The output wavelength of the LED depends on the semiconductor material that it's manufactured from. Table 10.2 breaks down the semiconductor materials and their associated wavelengths. LED wavelengths are chosen for fiber-optic communication systems based on their application. Visible wavelengths (650nm) are typically used in short-distance, low-data-rate systems with large-core-diameter optical fiber. These systems are typically found in industrial control systems. Infrared wavelengths (820nm, 850nm, and 1300nm) are typically used for longer-distance, higher-data-rate systems with smaller-core optical fiber.

TABLE 10.2:   LED semiconductor materials

WAVELENGTH	SEMICONDUCTOR MATERIALS
650nm	Aluminum (Al), gallium (Ga), arsenic (As)
820nm	Aluminum (Al), gallium (Ga), arsenic (As)
850nm	Aluminum (Al), gallium (Ga), arsenic (As)
1300nm	Indium (In), gallium (Ga), arsenic (As), phosphorus (P)

LASER WAVELENGTHS

Just like the LED, a laser light source output wavelength is dependent on the semiconductor material that it's manufactured from, as shown in Table 10.3. As you learned in Chapter 5, "Optical Fiber Characteristics," single-mode optical fiber performance can be defined at 1310nm, 1550nm, or 1625nm. Fabry-Pérot and DFB lasers can be manufactured to operate at or near these wavelengths.

TABLE 10.3:   Laser semiconductor materials

WAVELENGTH	SEMICONDUCTOR MATERIALS
850nm	Aluminum (Al), gallium (Ga), arsenic (As)
1310nm	Indium (In), gallium (Ga), arsenic (As), phosphorus (P)
1550nm	Indium (In), gallium (Ga), arsenic (As), phosphorus (P)

NOTE   As VCSEL technology improves, it will begin to replace the Fabry-Pérot for 1310nm and 1550nm applications.

The DFB laser is unique because it can be tuned to a specific wavelength. This allows many lasers to operate around a center wavelength in a multichannel system. Multichannel systems will be discussed in greater detail in Chapter 12, "Passive Components and Multiplexers."

The DFB laser is virtually a single-wavelength laser and has the least dispersion of laser fiber-optic sources. Because of this, the DFB is the choice for long-distance, high-speed applications. The 1550nm DFB laser can support a 2.5Gbps data rate at a transmission distance of 80km.

Source Spectral Output

The spectral output of a fiber-optic light source can have a significant impact on the bandwidth of a fiber-optic communication system. Different wavelengths of light travel at different speeds through an optical fiber. Because of this, a light pulse made up of more than one wavelength will disperse as each wavelength travels at a different velocity down the optical fiber. This dispersal limits the bandwidth of the system. To achieve the highest bandwidth possible, a fiber-optic light source must output the narrowest range of wavelengths possible. The range of wavelengths that a light source outputs is described by its *spectral width*.

LED SPECTRAL WIDTH

The spectral width of an LED light source is much wider than the spectral width of a laser. LED spectral width is typically described by the term Full Width, Half Maximum (FWHM). When the spectral width of an LED is displayed graphically, FWHM is measured at one-half the maximum intensity across the full width of the curve, as shown in Figure 10.9.

FIGURE 10.9
Typical FWHM
spectral width of
650nm, 850nm,
and 1300nm LEDs

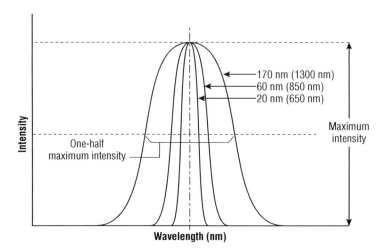

LED spectral widths vary tremendously. A short wavelength 650nm visible LED may have a spectral width as narrow as 20nm. The spectral width of a long wavelength 1300nm infrared LED may be as wide as 170nm. Because an LED is used only for multimode applications, the wide spectral width has no effect on the bandwidth of the communication system. Figure 10.9 compares the spectral width of various LED wavelengths.

LASER SPECTRAL WIDTH

Laser light sources offer the highest level of performance for a fiber-optic communication system. Unlike the LED, lasers output a very narrow spectrum of light. The manufacturer may describe the spectral output width of a laser several ways. These include FWHM, Root Mean Squared (RMS), and 20dB below peak spectral density.

All laser light sources output a narrow spectrum of light. However, there are significant differences in the spectral widths of each laser family. The spectral width of the Fabry-Pérot laser may span as much as 5nm, causing dispersion problems in high-speed or long transmission distance systems. VCSELs have shorter cavities than Fabry-Pérot lasers, which reduce their spectral output width. The DFB laser is virtually a one-frequency laser with spectral widths less than 0.1nm.

The DFB laser achieves this narrow spectral output width by using a series of corrugated ridges on the semiconductor substrate to diffract unwanted wavelengths back into the laser cavity. These corrugated ridges form an internal diffraction grating, which is discussed in detail in Chapter 12. The laser manufacturer can set the desired output wavelength with the spacing of the corrugated ridges on the internal diffraction grating. The DFB laser is used in the highest performance systems. Figure 10.10 compares the spectral widths of the three laser families.

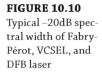

FIGURE 10.10

Typical –20dB spectral width of Fabry-Pérot, VCSEL, and DFB laser

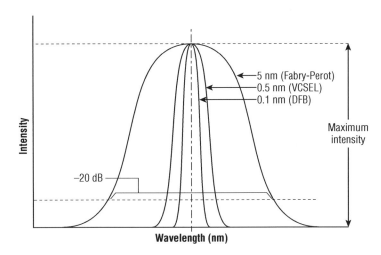

2.4 Source Output Power

The output power of the light sources used in fiber-optic communication systems varies dramatically depending on the application. LED light sources are typically designed to support transmission distances up to 2km whereas laser light sources may support transmission distances well in excess of 80km. Laser optical power output levels can exceed LED optical output power levels by more than 20dB for premises local area network (LAN) and storage area network (SAN) applications. This number increases to 50dB or greater for outside-plant fiber-optic communication systems that incorporate automatic reduction to reduce emissions to a lower hazard level. These systems may operate with power levels up to 2.6 watts at a 1550nm wavelength.

LED Output Power

Today's LED technology seems to be changing daily. High-power LED light sources are beginning to replace many incandescent light sources. As of this writing, single-lamp 20-watt LED white light flashlights are available with beams so powerful that viewing them directly is like viewing a flashbulb. However, the LED light sources used in fiber-optic communication output considerably less power than that flashlight.

The optical output power of a typical LED light source used in fiber-optic communications is 10dB or more below the lowest power laser light source. Output power is typically expressed by the manufacturer at the Beginning of Life (BOL) and the End of Life (EOL). An LED light source will lose some output power over its usable lifetime. The industry convention allows for a degradation of 1.5dB from BOL to EOL.

LED light sources couple only a small portion of their light energy into the core of the optical fiber. The amount of energy coupled into the optical fiber depends on the optical fiber core size and NA. The larger the core size and greater the NA, the more energy coupled into the core.

Off-the-shelf LED light sources are available to support a wide range of glass, plastic, and PCS optical fiber. Most LED light sources used in telecommunication applications are designed to work with either 50/125µm or 62.5/125µm optical fiber. The light energy coupled into the core of a 50/125µm optical fiber with an NA of 0.20 is approximately 3.5dB less than the energy coupled into the core of a 62.5/125µm optical fiber with an NA of 0.275. Because the same LED light source may be used with different sizes of optical fiber, the manufacturer typically measures the output power at the end of one meter of optical fiber with the cladding modes removed. The cladding modes can be removed by a mode filter, which is discussed in detail in Chapter 16, "Test Equipment and Link/Cable Testing."

LASER OUTPUT POWER

Laser light sources offer the highest output power available for fiber-optic communication systems. The higher output power of the laser allows greater transmission distances than with an LED. Laser output power varies depending on the application. VCSELs used in multimode applications have the lowest output power. Fabry-Pérot and DFB lasers used in single-mode applications have the greatest output power.

Like the LED, the output power of the laser will diminish over its usable lifetime. Laser manufacturers typically provide BOL and EOL minimum optical power levels. The industry convention for lasers allows for a 1.5dB reduction in power from BOL to EOL. Optical output power levels are normally expressed as the amount of light coupled into a one-meter optical fiber.

Source Modulation Speed

There are many advantages to using optical fiber as a communications medium. High bandwidth over long distances is one of them. What limits the bandwidth of today's fiber-optic communication systems is not the optical fiber but the light source. The modulation speed of a light source is just one factor that can limit the performance of a fiber-optic communication system.

LED MODULATION SPEED

Today's LED light sources can be modulated to support data rates greater than 400Mbps. However, most LED light sources are designed to support network standards that do not require a data rate that high. An example of this is the IEEE 802.3 Ethernet standard, which establishes network data rates at 10Mbps, 100Mbps, 1Gbps, and 10Gbps. The LED can easily support the 10Mbps and 100Mbps data. However, current LED technology does not support the 1Gbps or greater data rate.

LASER MODULATION SPEED

Laser light sources are constantly evolving. Laser manufacturers have been able to modulate all three-laser families at data rates up to 10Gbps. Today laser transmitters that support 1Gbps, 2.5Gbps, or 10Gbps are very affordable and are being integrated into more and more LANs and SANs. Laser light sources may also be used in low-date-rate applications such as 10Mbps or 100Mbps when transmission distances greater than 2,000 meters are required.

Transmitter Performance Characteristics

Up to this point, we have discussed only the characteristics of the unpacked and packaged LED and laser light source with no mention of the electronics required to drive these light sources. The incredible data rates we've discussed in the previous text would not be possible without integrated electronics packaged into the transmitter.

For most users, the LED or laser transmitter is a black box. The manufacturer has neatly integrated everything required to convert the electrical input signal into light energy in the smallest package possible. However, it is important to be able to interpret the manufacturer data sheet and verify the optical output of the transmitter.

LED Transmitter Performance Characteristics

LED transmitters are designed to support multimode optical fibers with core sizes ranging from as small as 50μm to as large as 1mm. LED transmitters are directly modulated. (Direct modulation is when the drive current through the LED is varied.) The output power of the LED is directly proportional to the current flow through the LED, as shown in Figure 10.11. In a digital application, the drive current is switched on and off. In an analog application, the drive current is varied.

FIGURE 10.11
LED and laser optical output power versus drive current

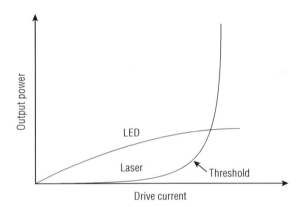

A typical LED transmitter contains an electrical subassembly, optical subassembly, and receptacle, as shown in Figure 10.12. The electrical subassembly amplifies the electrical input signal with the driver integrated circuit (IC). The driver IC provides the current to drive the LED in the optical subassembly. The optical subassembly mates with the receptacle to direct light into the optical fiber.

FIGURE 10.12
Block diagram of a typical LED transmitter

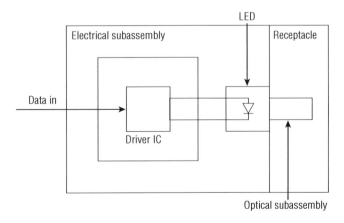

Because LED fiber-optic links are simplex, they require a transmit optical fiber and a receive optical fiber. Typically, the LED transmitter is packaged with the receiver section. Packaging the transmitter and receiver together reduces the overall space required, simplifies circuit board design, and reduces cost. Figure 10.13 is a photograph of a 1300nm transceiver with an ST receptacle.

FIGURE 10.13
A 100Mbps 1300nm LED transceiver with an ST receptacle

NOTE The term *transceiver* describes a device in which the fiber-optic transmitter and receiver are combined into a single package. Standalone transmitters and receivers are still available; however, the vast majority of standardized networks use transceivers.

The performance characteristics of the LED transmitter are typically broken up into four groups: recommended operating conditions, electrical characteristics, optical characteristics, and data rate. The recommended operating conditions describe maximum and minimum temperature and voltage ranges that the device can operate in without damage. Table 10.4 shows the typical recommended operating conditions for a 1300nm 100Mbps LED transmitter.

TABLE 10.4: LED transmitter recommended operating conditions

PARAMETER	SYMBOL	MIN.	TYP.	MAX.	UNIT
Ambient Operating Temperature	T_A	0		70	$^{\circ}C$
Supply Voltage	V_{CC}	4.75		5.25	V
Data Input Voltage—Low	$V_{IL}-V_{CC}$	–1.810		–1.475	V

TABLE 10.4: LED transmitter recommended operating conditions *(continued)*

PARAMETER	SYMBOL	MIN.	TYP.	MAX.	UNIT
Data Input Voltage—High	$V_{IH}-V_{CC}$	−1.165		−0.880	V
Data and Signal Detect Output Load	R_L		50		Ω

The electrical characteristics of the LED transmitter describe the supply current requirements, the data input requirements, and the power dissipated by the device. Table 10.5 shows the typical electrical characteristics for a 1300nm 100Mbps LED transmitter.

TABLE 10.5: LED transmitter electrical characteristics

PARAMETER	SYMBOL	MIN.	TYP.	MAX.	UNIT
Supply Current	I_{CC}		145	185	mA
Power Dissipation	P_{DISS}		0.76	0.97	W
Data Input Current—Low	I_{IL}	−350	0		μA
Data Input Current—High	I_{IH}		14	350	μA

The optical characteristics of the LED transmitter at a minimum include output power, center wavelength, and spectral width. Table 10.6 shows the typical optical characteristics for a 1300nm 100Mbps LED transmitter.

TABLE 10.6: LED transmitter optical characteristics

PARAMETER		SYMBOL	MIN.	TYP.	MAX.	UNIT
Optical Output Power	BOL	P_0	−19	−16.8	−14	dBm avg.
62.5/125μm, NA = 0.275 Fiber	EOL		−20			
Optical Output Power	BOL	P_0	−22.5	−20.3	−14	dBm avg.
50/125μm, NA = 0.20 Fiber	EOL		−23.5			
Optical Extinction Ratio				0.001	0.03	%
				−50	−35	dB
Optical Output Power at Logic "0" State		P_0("0")			−45	dBm avg.

TABLE 10.6: LED transmitter optical characteristics *(continued)*

PARAMETER	SYMBOL	MIN.	TYP.	MAX.	UNIT
Center Wavelength	λ_c	1270	1308	1380	nm
Spectral Width—FWHM	$\Delta\lambda$		137	170	nm
Optical Rise Time	t_r	0.6	1.0	3.0	ns
Optical Fall Time	t_f	0.6	2.1	3.0	ns
Duty Cycle Distortion Contributed by the Transmitter	DCD		0.02	0.6	ns_{P-P}
Data-Dependent Jitter Contributed by the Transmitter	DDJ		0.02	0.6	ns_{P-P}

As you can see in Table 10.6, there is a minimum, maximum, and typical value for the optical output power of an LED transmitter. These values are defined for 62.5/125µm optical fiber and for 50/125µm optical fiber. As you learned in Chapter 5, the NA of a 62.5/125µm optical fiber is typically 0.275 and the NA of a 50/125µm optical fiber is 0.20.

The amount of optical output power coupled into a multimode optical fiber depends on the core diameter and the NA of the fiber. As you saw earlier in this chapter, the same LED transmitter will typically couple 3.5dB more power into the 62.5/125µm optical fiber than into the 50/125µm. The 3.5dB difference is shown in Table 10.6. There is a 3.5dB difference between the minimum optical output power for the 62.5/125µm and 50/125µm optical fiber.

Earlier in this chapter, we stated that light sources age, and as they age, their output power decreases. The industry convention allows for 1.5dB of aging. However, the difference between the BOL value and EOL for the LED transmitter described in Table 10.6 is only 1dB. The difference between BOL and EOL output power will vary from manufacturer to manufacturer. Always obtain this data from the manufacturer's data sheet.

As you will learn in Chapter 15, "Fiber-Optic System Design Considerations," the optical output power of a fiber-optic transmitter is required to create a power budget for a fiber-optic link. The output power of most fiber-optic transmitters is very close to the typical value. However, the output power can vary as defined by the minimum and maximum values in Table 10.6.

To find the minimum optical output power for a specific optical fiber, locate the Optical Output Power row for that optical fiber in Table 10.6. In the Min. column, you will see a BOL and an EOL value. The higher number is the BOL value, and the lower number is the EOL value.

To find the maximum optical output power for a specific optical fiber, locate the Optical Output Power row for that optical fiber in Table 10.6. The number in the Max. column is the value for the maximum optical output power. The optical output of the fiber-optic transmitter should never exceed this value. There is no BOL or EOL value for the maximum optical output power.

LED Transmitter Applications

LED transmitters are often designed to support one or more network standards. A 100Mbps 1300nm transmitter could support 100Base-FX applications or 155Mbps ATM applications. Table 10.7 lists the data rates, wavelengths, and optical fiber types for various LED transmitters used in IEEE 802.3 Ethernet communication systems.

TABLE 10.7:   LED transmitters for Ethernet applications

NETWORK	DATA RATE	WAVELENGTH AND MEDIA
10BASE-FL	10Mbps	850nm, multimode fiber
100BASE-FX	100Mbps	1300nm, multimode fiber
10/100BASE-SX	10/100Mbps	850nm, multimode fiber

Laser Transmitter Performance Characteristics

Laser transmitters are designed to support either single-mode optical fiber systems or multi-mode optical fiber systems. The single-mode transmitter is designed to interface with 8–11µm optical fiber cores. Until recently, all single-mode transmitters were from the Fabry-Pérot or DFB families. However, manufacturers keep improving VCSEL technology and the VCSEL is beginning to emerge as an alternative 1300nm single-mode optical fiber transmitter.

PROBLEMS ASSOCIATED WITH CORE DIAMETER MISMATCH

LED transmitters are typically designed to support different core size multimode optical fiber. For example, a 100BASE-FX transmitter is designed to operate with 50/125µm or 62.5/125µm optical fiber, as shown in Table 10.6. If the LED transmitter receptacle is designed to accept an SC connector, any SC connector regardless of optical fiber size or type can be plugged into the transmitter.

So what happens when a 62.5/125µm patch cord from the LED transmitter is mated with a 50µm horizontal cable? Back reflections and attenuation. As you may remember from the "LED Output Power" section, we mentioned that an LED light source couples roughly 3.5dB more light energy into a 62.5µm core than a 50µm core. So when you take the 62.5µm core from the transmitter and mate it with the 50µm core horizontal cable, there is roughly a 3.5dB loss at that connection. As mentioned in Chapter 2, "Principles of Fiber-Optic Transmission," a loss of 3dB means that half of the power is lost. So if less than half the photons from a light pulse get coupled into the horizontal cable, where do the rest of the photons go?

Well, some of the photons will be absorbed by the coating of the 50µm optical fiber. Others will be reflected back into the cladding of the 62.5µm optical fiber and absorbed by the coating. However, some of the photons will be reflected back into the core of the 62.5µm optical fiber. Many of those photons will travel back to the transmitter, where they will be reflected off the transmitter window or lens assembly back into the core. These photons will travel the length of the optical fiber to the receiver. There is a good possibility that the receiver will detect these photons and convert their light energy into electrical energy. Now the circuit decoding the electrical energy is being bombarded with good and bad data that prevents the system from operating.

The fix for this problem is simple: use a 50µm patch cord. The bottom line is that you should pay careful attention to core diameter to prevent core diameter mismatch. You can't look at only the color of the patch cord to determine the core diameter. There are orange patch cords with a 62.5µm core and there are orange patch cords with a 50µm core. You must read the cable markings from the manufacturer to find out the actual core diameter.

Like the LED transmitter, the laser transmitter can be broken into three sections: the electrical subassembly, optical subassembly, and receptacle or pigtail, as shown in Figure 10.14. The electrical subassembly amplifies the input signal and provides the drive current for the laser. Unlike the LED, the laser emits very few photons until the drive current passes a threshold level, as shown earlier in Figure 10.11. A feedback circuit that monitors laser output intensity typically controls the drive current provided to the laser to ensure a constant output level.

FIGURE 10.14
Block diagram
of a typical laser
transmitter

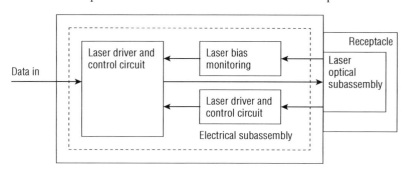

The optical subassembly contains a photodiode that receives part of the energy output by the laser. The electrical output of the photodiode is monitored in the electrical subassembly to help control drive current. The optical subassembly may mate with an external modulator, receptacle, or pigtail.

Laser transmitters may or may not be packaged with the receiver section. How the transmitter is packaged depends on the application. Laser transmitters operating up to 2.5Gbps typically do not require coolers and are often packaged with a receiver section. Laser transmitters operating at 10Gbps or greater produce enough heat to require an integrated cooler to help the laser maintain a constant temperature. Temperature variations will affect the laser's threshold current, output power, and frequency.

The VCSEL is used in multimode transmitters. Currently VCSEL multimode transmitters support only 850nm operation with 50/125μm or 62.5/125μm optical fiber. The VCSEL can support data rates as high as 10Gbps through multimode optical fiber.

The VCSEL, Fabry-Pérot, or DFB laser is capable of direct modulation up to 10Gbps. To achieve data rates greater than 10Gbps, the laser must typically be indirectly modulated. Indirect modulation does not vary the drive current as direct modulation does. Maintaining a constant drive current prevents the output wavelength from changing as the electron density in the semiconductor material changes. This is commonly referred to as *laser chirp,* which causes dispersion and limits bandwidth over long transmission distances.

Indirect modulation requires a constant drive current to the laser. The laser light source outputs a continuous wave (CW) that is modulated by an external device. The external modulation device is typically integrated into the laser transmitter package. Indirect modulation of a laser allows higher modulation rates (up to 40Gbps or greater) and less dispersion than direct modulation. Figure 10.15 shows a functional block diagram of direct and indirect modulation.

The performance characteristics of the laser transmitter are typically broken up into four groups: recommended operating conditions, electrical characteristics, optical characteristics, and data rate. The recommended operating conditions describe the maximum and minimum temperatures and voltage ranges that the device can operate in without damage. Table 10.8 shows the typical recommended operating conditions for a 1300nm, 2.5Gbps laser transmitter.

FIGURE 10.15

Block diagram of direct and indirect modulation

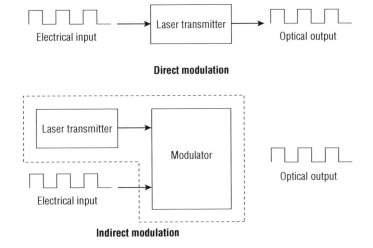

Direct modulation

Indirect modulation

TABLE 10.8: Laser transmitter recommended operating conditions

PARAMETER	SYMBOL	MIN.	TYP.	MAX.	UNIT
Ambient Operating Temperature	T_A	0		+70	°C
Supply Voltage	V_{CC}	3.1		3.5	V
Power Supply Rejection	PSR		100		mVPP
Transmitter Differential Input Voltage	V_D	0.3		2.4	V
Data Output Load	R_{DL}		50		Ω
TTL Signal Detect Output Current—Low	I_{OL}			1.0	mA
TTL Signal Detect Output Current—High	I_{OH}	−400			μA
Transmit Disable Input Voltage—Low	T_{DIS}			0.6	V
Transmit Disable Input Voltage—High	T_{DIS}	2.2			V
Transmit Disable Assert Time	T_{ASSERT}			10	μS
Transmit Disable Deassert Time	$T_{DEASSERT}$			50	μS

The electrical characteristics of the laser transmitter describe the supply current requirements, the data input requirements, and the power dissipated by the device. Table 10.9 shows the typical electrical characteristics for a 1300nm, 2.5Gbps laser transmitter.

TABLE 10.9: Laser transmitter electrical characteristics

PARAMETER	SYMBOL	MIN.	TYP.	MAX.	UNIT
Supply Current	I_{CCR}		115	140	mA
Power Dissipation	P_{DISS}		0.38	0.49	W
Data Output Voltage Swing (single-ended)	$V_{OH} - V_{OL}$	575		930	mV
Data Output Rise Time	t_r		125	150	ps
Data Output Fall Time	t_f		125	150	ps
Signal Detect Output Voltage—Low	V_{OL}			0.8	V
Signal Detect Output Voltage—High	V_{OH}	2.0			V
Signal Detect Assert Time (OFF to ON)	AS_{MAX}			100	μs
Signal Detect Deassert Time (ON to OFF)	ANS_{MAX}			100	μs

The optical characteristics of the laser transmitter at a minimum include output power, center wavelength, spectral width, and back-reflection sensitivity. Excessive back reflections can interfere with the operation of the laser. Table 10.10 shows the typical optical characteristics for a 1300nm, 2.5Gbps laser transmitter.

TABLE 10.10: Laser transmitter optical characteristics

PARAMETER	SYMBOL	MIN.	TYP.	MAX.	UNIT
Optical Output Power 9μm SMF	P_{OUT}	−10	−6	−3	dBm
Center Wavelength	λc	1260		1360	nm
Spectral Width—rms	$\Delta\lambda$		1.8	4	nm rms
Optical Rise Time	t_r		30	70	ps
Optical Fall Time	t_f		150	225	ps

TABLE 10.10: Laser transmitter optical characteristics *(continued)*

PARAMETER	SYMBOL	MIN.	TYP.	MAX.	UNIT
Extinction Ratio	E_R	8.2	12		dB
Optical Output Eye	Compliant with eye mask Telecordia GR-253-CORE				
Back-Reflection Sensitivity				−8.5	dB
Jitter Generation	pk to pk			70	mUI
	RMS			7	mUI

As you can see in Table 10.10, there is a minimum, maximum, and typical value for the optical output power of a laser transmitter. These values are defined for a single-mode optical fiber with a core diameter of 9μm. However, the NA is not stated as it was in the LED transmitter data sheet. Recall from Chapter 5 that single-mode optical fibers have the smallest NA and narrowest acceptance cone. A typical NA for a single-mode optical fiber is 0.14.

Like the LED, the laser ages and its output power decreases over time. In Table 10.10, there is no BOL or EOL value. If the manufacturer does not provide these values, use the industry convention of 1.5dB and assume that the values stated for optical output power are the BOL values.

To find the minimum optical output power, locate the Optical Output Power row in Table 10.10. In the Min. column, there is a value. Since BOL and EOL are not stated, assume this value is the BOL value. To approximate the EOL value, subtract 1.5dB.

To find the maximum optical output power, locate the Optical Output Power row in Table 10.10. The number in the Max. column is the value for the maximum optical output power. The optical output of the fiber-optic transmitter should never exceed this value. There is no BOL or EOL value for the maximum optical output power.

Laser Transmitter Applications

Laser transmitters are available for serial transmission applications and parallel transmission applications. A typical transmitter may support multiple data rates and multiple network standards such as IEEE 802.3 and Fibre Channel. IEEE 802.3 is an international standard for LANs and metropolitan area networks (MANs). Fibre Channel is the foundation for over 90 percent of all SAN installations globally.

SERIAL LASER TRANSMITTERS

A serial laser transmitter is designed to be used with network standards such as IEEE 802.3 or Fibre Channel. It transmits data in a serial format at a single wavelength over a single optical fiber. These transmitters can be designed to support a broad range of data rates over multimode or single-mode optical fiber. Tables 10.11 and 10.12 list the data rates, wavelengths, and optical fiber types for various serial laser transmitters used in IEEE 802.3 Ethernet communication systems and Fibre Channel communication systems.

TABLE 10.11: Serial fiber laser transmitters for Ethernet applications

NETWORK	DATA RATE	WAVELENGTH AND MEDIA
100BASE-LX	100Mbps	1310nm FP, single-mode fiber
1000BASE-SX	1Gbps	850nm VCSEL, multimode fiber
1000BASE-LX	1Gbps	1310nm FP, multimode or single-mode fiber
10GBASE-S	10Gbps	850nm VCSEL, multimode fiber
10GBASE-L	10Gbps	1310nm DFB, single-mode

TABLE 10.12: Serial fiber laser transmitters for Fibre Channel applications

NETWORK	DATA RATE	WAVELENGTH AND MEDIA
100-MX-SN-I	1062Mbaud	850nm VCSEL, multimode fiber
100-SM-LC-L	1062Mbaud	1310 FB, single-mode fiber
200-MX-SN-I	2125Mbaud	850nm VCSEL, multimode
200-SM-LC-L	2125Mbaud	1310 FP, single-mode
1200-MX-SN-I	10512Mbaud	850nm VCSEL, multimode fiber
1200-SM-LL-L	10512Mbaud	1310nm DFB, single-mode

PARALLEL OPTIC LASER TRANSMITTERS

Up to this point in the chapter only single-fiber transmitters that transmit serial data at a single wavelength have been discussed. These transmitters have limitations to the data rates they can achieve, and that limit as defined by IEEE 802.3 and Fibre Channel is approximately 10Gbps. To achieve higher data rates, you can take one of two approaches. One approach is to combine the output from multiple transmitters, each transmitting at a different wavelength, into a single optical fiber; this is called *wavelength division multiplexing* (WDM) and it is covered in Chapter 12, "Passive Components and Multiplexers." WDM is ideal for high-data-rate systems over long distances using single-mode optical fiber.

Another approach to achieving high data rates over short distances using multimode optical fiber is parallel optics. With parallel optics, multiple transmitters each operating at the same wavelength are used. The difference between parallel optics and WDM is that parallel optics uses one optical fiber for each transmitter vice, combining outputs of multiple transmitters into a single optical fiber. Since each transmitter is dedicated to a single optic fiber, all the transmitters can operate at the same wavelength.

The light source of choice for parallel optics is the VCSEL. VCSEL technology has matured to where multiple 10Gbps transmitters can be manufactured in the form of an array that physically aligns with the fibers in an MT ferrule. As you learned in Chapter 9, "Connectors," the optical fibers in an MT ferrule are spaced 0.25mm apart, as shown in Figure 10.16.

FIGURE 10.16
MT ferrule optical fiber spacing is 0.25mm.

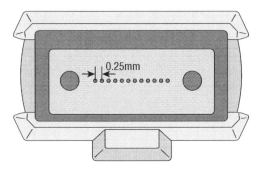

IEEE Standard 802.3-2012, Section 6 introduces 40Gbps and 100Gbps Ethernet. It defines the maximum transmission distances for each data rate using single-mode optical fiber, multimode optical fiber, or copper cabling for the physical layer. When multimode OM3 or OM4 optical fiber is used for the physical layer, multiple transmitters operating in parallel are required, as described in Table 10.13. Achieving a 40Gbps data rate requires four 10Gbps transmitters operating in parallel. A 100Gbps data rate is achieved using ten 10Gbps transmitters operating in parallel.

As of this writing, the maximum transmission distance is 100m for systems using OM3 optical fiber and 150m for systems using OM4 optical fiber.

TABLE 10.13:   Parallel optic laser transmitters for Ethernet applications

NETWORK	DATA RATE	DATA RATE PER CHANNEL	NUMBER OF TRANSMITTERS	WAVELENGTH AND MEDIA
40GBASE-SR4	40Gbps	10Gbps	4	850nm VCSEL, multimode fiber
100GBASE-SR10	100Gbps	10Gbps	10	850nm VCSEL, multimode fiber

To support parallel operation, the transmitters must be physically arranged to accept an MPO female plug like the one shown in Figure 10.17. For 40GBASE-SR-4 networks, the transmitters use the four leftmost optical fibers or optical transmit lanes and the receivers use the four rightmost optical fibers or receive lanes, as shown in Figure 10.18. The four optical fibers in the center are not used. This arrangement allows for 40Gbps full duplex operation using a single 12-fiber ribbon.

FIGURE 10.17
MPO female plug
with 12 optical
fibers

FIGURE 10.18
40GBASE-SR4
optical lane
assignments

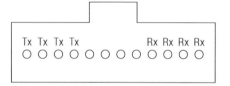

For 100GBASE-SR10 networks the optical lanes can be configured for a single-receptacle using Option A shown in Figure 10.19; this approach is recommended in IEEE 802.3. They can also be configured using the two-receptacle Option B shown in Figure 10.20 or Option C shown in Figure 10.21; these options are not recommended and considered alternatives.

FIGURE 10.19
100GBASE-SR10
Option A optical
lane assignments

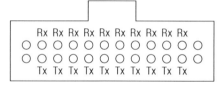

FIGURE 10.20
100GBASE-SR10
Option B optical
lane assignments

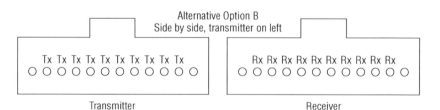

FIGURE 10.21
100GBASE-SR10
Option C optical
lane assignments

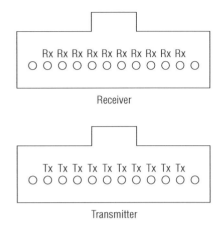

Higher Power Transmitters

5.5

LANs and SANs are not limited to premises applications. These networks are being used in spacecraft, aircraft, on the battlefield, and in numerous other harsh environments. In many of the applications, the number of interconnections required to connect the transmitter to the receiver is typically greater than a premises-based system. In addition, the connectors may experience greater loss than is typically experienced in a premises-based system.

To compensate for the greater number of interconnections and the increased loss at each interconnection, a higher power transmitter may need to be selected. For example, the transceiver shown in Figure 10.22 will support a 1000BASE-SX LAN. The manufacturer offers it with a standard output power or a higher output power for applications with a greater number of interconnections or higher loss interconnections.

FIGURE 10.22
Higher-power
850nm VCSEL
multi-rate
transceiver

Photo courtesy of COTSWORKS

IEEE 802.3 states that the minimum average launch power for a 1000BASE-SX transmitter is –9.5dBm. The transceiver shown in Figure 10.22 can be purchased to meet this requirement, or a higher power version can be purchased that has a minimum average launch power of –5dBm. The guaranteed extra 4dB compensates for the additional losses caused by a greater number of interconnections and/or higher loss interconnections while still meeting the Class I eye safety requirements. You can obtain additional information about higher power transmitters at www.cotsworks.com.

4. ## Light Source Safety

The light sources used in fiber-optic communication systems typically operate at low power levels. Unlike an incandescent lamp, their light energy is distributed over a narrow spectrum, typically the infrared spectrum. This narrow spectrum light energy is often invisible, and it can be a danger to the fiber-optic installer or fiber-optic technician. The level of danger depends on the classification of the light source and the amount of energy coupled into the optical fiber. Laser light sources pose the greatest risk because of their narrow spectral width coherent light.

4.1 ### Classifications

We have learned that fiber-optic light sources are very different from the lightbulbs used to illuminate our home or office. We shop for lightbulbs by their wattage. The wattage shown on the package of the lightbulb tells us how much energy the bulb requires to operate. A 60-watt incandescent lightbulb does not output 60 watts of light energy; it requires 60 watts of power to operate. Much of the energy required to operate the bulb is actually emitted as heat, not visible light. We all know how hot a 60-watt incandescent lightbulb can get.

That same 60-watt lightbulb emits a very broad spectrum of visible light. In other words, the bulb emits many different wavelengths of visible light that combine together to create white light. Each wavelength of light radiating from the bulb represents only a small fraction of the overall light energy output by the bulb. Unlike the laser, the visible light that radiates from the bulb is incoherent. Because the broad visible spectrum of light radiating from the bulb is incoherent and the light energy is spread out over many wavelengths, the light from the bulb will not damage your eye. The coherent light from the laser and its lack of divergence is what can cause eye damage. An ordinary incandescent lightbulb is not an eye hazard like a laser. However, this doesn't mean that you should stare at 60-watt light bulbs.

Fiber-optic light sources are classified by their ability to cause damage to your eye. As you learned in Chapter 6, "Safety," there are standards and federal regulations that classify the light sources used in fiber-optic transmitters. Every laser transmitter sold in the United States must be classified as defined in 21CFR, Chapter 1, Subchapter J. Only these classifications are covered in this chapter, additional information can be found in Chapter 6. Listed here are the hazards associated with each class:

Class I   These levels of laser radiation are not considered to be hazardous.

Class IIa   These levels of laser radiation are not considered to be hazardous if viewed for any period of time less than or equal to 1,000 seconds, but are considered to be a chronic viewing hazard for any period of time greater than 1,000 seconds.

Class II   These levels of laser radiation are considered to be a chronic viewing hazard.

Class IIIa These levels of laser radiation are considered to be, depending on the irradiance, either an acute intrabeam viewing hazard or chronic viewing hazard, and an acute viewing hazard if viewed directly with optical instruments.

Class IIIb These levels of laser radiation are considered to be an acute hazard to the skin and eyes from direct radiation.

Class IV These levels of laser radiation are considered to be an acute hazard to the skin and eyes from direct and scattered radiation.

4.2 Safe Handling Precautions

There are some fundamental precautions that you should always observe when working with a fiber-optic communication system.

♦ Assume that the fiber-optic cable assembly you are handling is energized.

♦ Never directly view the end of an optical fiber or the endface of a fiber-optic connector without verifying that the optical fiber is dark.

♦ View the endface of a fiber-optic connector from at least 6″ away when testing continuity or using a visual fault locator.

The Bottom Line

Determine the minimum optical output power for an LED transmitter. The minimum optical output power of an LED transmitter should be defined in the manufacturer's data sheet.

Master It Refer to the following table to determine the BOL and EOL values for the minimum optical output power of a 50/125μm optical fiber.

PARAMETER		SYMBOL	MIN.	TYP.	MAX.	UNIT
Optical Output Power	BOL	P_0	−22	−19.8	−17	dBm avg.
62.5/125μm, NA = 0.275 Fiber	EOL		−23			
Optical Output Power	BOL	P_0	−25.5	−23.3	−17	dBm avg.
50/125μm, NA = 0.20 Fiber	EOL		−26.5			
Optical Extinction Ratio				0.001	0.03	%
				−50	−35	dB
Optical Output Power at Logic "0" State		P_0("0")			−45	dBm avg.
Center Wavelength		λ_c	1270	1308	1380	nm
Spectral Width—FWHM		$\Delta\lambda$		137	170	nm
Optical Rise Time		t_r	0.6	1.0	3.0	ns

PARAMETER		SYMBOL	MIN.	TYP.	MAX.	UNIT
Optical Fall Time		t_f	0.6	2.1	3.0	ns
Duty Cycle Distortion Contributed by the Transmitter		DCD		0.02	0.6	ns$_{p\text{-}p}$
Data-Dependent Jitter Contributed by the Transmitter		DDJ		0.02	0.6	ns$_{p\text{-}p}$

Determine the maximum optical output power for an LED transmitter.   The manufacturer's data sheet should contain the maximum value for the optical output power of an LED transmitter.

> **Master It**   Refer to the following table to determine the maximum optical output power of a 62.5/125μm optical fiber.

PARAMETER			SYMBOL	MIN.	TYP.	MAX.	UNIT
Optical Output Power	BOL	P_0		−25	−21.8	−19	dBm avg.
62.5/125μm, NA = 0.275 Fiber	EOL			−25			
Optical Output Power	BOL	P_0		−27.5	−25.3	−19	dBm avg.
50/125μm, NA = 0.20 Fiber	EOL			−28.5			
Optical Extinction Ratio					0.001	0.03	%
					−50	−35	dB
Optical Output Power at Logic "0" State		P_0 ("0")				−45	dBm avg.
Center Wavelength		λ_c		1270	1308	1380	nm
Spectral Width—FWHM		$\Delta\lambda$			137	170	nm
Optical Rise Time		t_r		0.6	1.0	3.0	ns
Optical Fall Time		t_f		0.6	2.1	3.0	ns
Duty Cycle Distortion Contributed by the Transmitter		DCD			0.02	0.6	ns$_{p\text{-}p}$
Data-Dependent Jitter Contributed by the Transmitter		DDJ			0.02	0.6	ns$_{p\text{-}p}$

Determine the minimum optical output power for a laser transmitter.   The manufacturer's data sheet should contain the minimum, maximum, and typical value for the optical output power of a laser transmitter.

Master It   Refer to the following table to determine the BOL and EOL values for the minimum optical output power.

PARAMETER	SYMBOL	MIN.	TYP.	MAX.	UNIT
Optical Output Power 9μm SMF	P_{OUT}	−6	0	+3	dBm
Center Wavelength	λ_c	1260		1360	nm
Spectral Width—rms	$\Delta\lambda$		1.8	4	nm rms
Optical Rise Time	t_r		30	70	ps
Optical Fall Time	t_f		150	225	ps
Extinction Ratio	E_R	8.2	12		dB
Optical Output Eye	Compliant with eye mask Telecordia GR-253-CORE				
Back-Reflection Sensitivity				−8.5	dB
Jitter Generation	pk to pk			70	mUI
	RMS			7	mUI

Determine the maximum optical output power for a laser transmitter.   The manufacturer's data sheet should contain the maximum value for the optical output power of a laser transmitter.

Master It   Refer to the following table to determine the maximum optical output power.

PARAMETER	SYMBOL	MIN.	TYP.	MAX.	UNIT
Optical Output Power 9μm SMF	P_{OUT}	−12	−8	−5	dBm
Center Wavelength	λ_c	1260		1360	nm
Spectral Width—rms	$\Delta\lambda$		1.8	4	nm rms
Optical Rise Time	t_r		30	70	ps
Optical Fall Time	t_f		150	225	ps
Extinction Ratio	E_R	8.2	12		dB

PARAMETER	SYMBOL	MIN.	TYP.	MAX.	UNIT
Optical Output Eye	Compliant with eye mask Telecordia GR-253-CORE				
Back-Reflection Sensitivity				−8.5	dB
Jitter Generation	pk to pk			70	mUI
	RMS			7	mUI

Chapter 11

Fiber-Optic Detectors and Receivers

In Chapter 1, "History of Fiber Optics and Broadband Access," we introduced you to the fiber-optic receiver. The job of the receiver is to take light energy from the optical fiber and convert it to electrical energy. In this chapter, we will explain the basic components that make up the fiber-optic receiver, starting with the photodiode. You will learn about the effects of optical input power on the performance of the receiver and the performance characteristics of LED and laser receivers.

In this chapter, you will learn to:

♦ Calculate the dynamic range for an LED receiver

♦ Calculate the dynamic range for a laser receiver

Photodiode Fundamentals

A *photodiode* in a fiber-optic receiver is like the tire on your car. The photodiode is where the rubber meets the road. Light energy from the optical fiber stops at the photodiode. It's the job of the photodiode to convert the light energy received from the optical fiber into electrical energy. Just as there are different performance-level tires that you can put on your car, so there are different performance-level photodiodes that can be incorporated into a receiver. This section of the chapter discusses the fundamentals of basic photodiode operation and the various types of photodiodes that may be used in a receiver.

The best way to imagine a photodiode is to think about a *solar cell*. Most of us have seen the exhibits at museums where a solar cell or a group of solar cells powers a small boat or car with the light energy from a lightbulb. Maybe you own a solar charger for your boat battery or have decorative outdoor lighting that uses solar cells to recharge the batteries.

The solar cell takes the light energy it receives and converts it into electrical energy. In other words, the photons absorbed by the solar cell cause electrons to flow within the solar cell. This electron flow is called a *current*. The current from the solar cell flowing through the motor of the small boat or car causes the motor to rotate. The more current, the faster the motor rotates and the faster the boat or car travels.

Like the LED you learned about in Chapter 10, "Fiber-Optic Light Sources and Transmitters," the basic photodiode is a semiconductor diode with a *p* region and an *n* region, as shown in Figure 11.1. Photons absorbed by the photodiode excite electrons within the photodiode in a process called *intrinsic absorption*. When stimulated with an outside bias voltage, these electrons produce a current flow through the photodiode and the external circuit providing the bias voltage.

FIGURE 11.1
PN photodiode

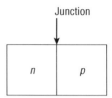

The *PN photodiode* is *reverse biased* when used in an electrical circuit. This is the opposite of how the LED is used in an electrical circuit. Reverse bias means that the *n* region of the photodiode is connected to a positive electrical potential and the *p* region is connected to a negative electrical potential.

In an electrical circuit, as shown in Figure 11.2, light that is absorbed by the photodiode produces current flow through the entire external circuit. As current flows through the *resistor*, it produces a voltage drop across the resistor. This voltage drop is input to an amplifier for amplification.

FIGURE 11.2
PN photodiode in
an electrical circuit

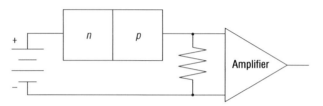

You may be wondering why the output of the photodiode needs to be amplified since there is no amplification used with a solar cell. The solar cell is typically supplied with photons by a powerful light source such as the sun or a very bright lamp. The photodiode used in a fiber-optic receiver gets all of its light energy from the optical fiber connected to it. As you have learned in this book, the core of an optical fiber is extremely small and carries very little light energy in comparison to the energy that a solar cell receives. A photodiode that is stimulated by the light energy in an optical fiber does not produce a great amount of electrical current flow. This is why amplification is required. It's also a reason why different photodiodes have been developed for various fiber-optic receiver applications.

Other Types of Photodiode

The basic photodiode is called a PN photodiode, but there are two other types: the PIN photodiode and the avalanche photodiode. This section describes how they work.

PIN Photodiode

The *PIN photodiode* works like a PN photodiode; however, it is manufactured to offer better performance. The better performance comes in the form of improved efficiency and greater speed. Improved efficiency means that it has a better photon-to-carrier conversion ratio. If the same amount of light energy hit a PN photodiode and a PIN photodiode, the PIN photodiode would generate more current flow through an external circuit.

Greater speed means that the diode can turn on and off faster. Remember that in fiber optics the light pulses being sent by the transmitter happen at a very fast rate. The photodiode needs to be able to stop and start electron flow fast enough to keep up with the incoming light pulses.

The PIN photodiode shown in Figure 11.3 is constructed a little differently than the PN photodiode. An *intrinsic* layer is used to separate the *p* region and the *n* region. This creates a large depletion region that absorbs the photons with improved efficiency when compared to the PN photodiode.

FIGURE 11.3
PIN photodiode

Avalanche Photodiode

The *avalanche photodiode (APD)* works just as its name suggests. On a snow-covered mountain, a small vibration can trigger an avalanche of snow. With the APD, a small bundle of photons can trigger an avalanche of electrons. The APD accomplishes this through a process called photomultiplication.

The APD is constructed with one more *p* region than the PIN photodiode, as shown in Figure 11.4. When the APD is biased very close to its breakdown voltage, it acts like an amplifier with a multiplication factor, or *gain*. An APD with a multiplication factor of 50 sets free on the average 50 electrons for each photon absorbed. The free electrons produce current flow through the electrical circuit connected to the APD. The APD is typically used in receivers that operate with lower optical input power levels than those associated with PIN diodes. These receivers typically have data rates ≤1Gbps.

FIGURE 11.4
Avalanche
photodiode

Photodiode Responsivity, Efficiency, and Speed

This section describes the factors determining how responsive a photodiode is.

Responsivity

The *responsivity* of the photodiode describes how well the photodiode converts a wavelength or a range of wavelengths of optical energy into electrical current. It's the ratio of the photodiode electrical output current to its optical input power. The greater the responsivity, the greater the electrical current output for a given amount of optical input power.

Responsivity is described in amperes/watt (A/W). However, a photodiode in a fiber-optic receiver will never generate an ampere of electrical current. That's not to say that a photodiode *couldn't* be built to generate an ampere of electrical current, but remember that a photodiode in a fiber-optic receiver receives a very small amount of light energy. A typical LED receiver works well with an optical input power as low as one microwatt. One microwatt is one millionth of a watt.

The overall responsivity of a photodiode depends on three factors:

♦ Semiconductor material makeup

♦ Wavelength

♦ Diode construction

Photodiodes are constructed for specific wavelengths. Some semiconductor materials perform better at longer wavelengths and some perform better at shorter wavelengths. Silicon photodiodes, for example, perform best in the visible and short infrared wavelengths. Germanium and indium gallium arsenic (InGaAs) photodiodes perform best at long infrared wavelengths, as shown in Figure 11.5. Diode construction also plays a large role in responsivity. The responsivity of an APD photodiode may be 100 or more times greater than a PIN photodiode.

FIGURE 11.5
Photodiode
semiconductor
responsivity

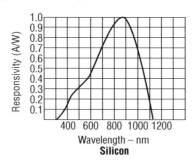

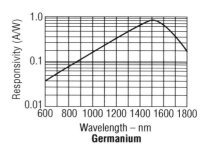

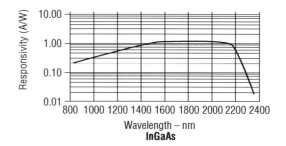

Quantum Efficiency

The responsivity of a photodiode depends on its quantum efficiency. Quantum efficiency describes how efficiently a photodiode converts light energy into electrical energy—that is, photons into free electrons. It is typically expressed as a percentage. A quantum efficiency of 47 percent means that for every 100 photons absorbed, 47 current-generating electrons will be created.

Switching Speed

It has been stressed many times in this book that fiber-optic systems transmit and receive millions or billions of light pulses per second. When data is being moved at this rate, there is little time for the photodiode to switch on and off. Switching speed depends on physical size, construction, and electrical biasing. Photodiodes in fiber-optic receivers are small in size and biased to produce the fastest possible switching time.

4. Fiber-Optic Receiver

Many different fiber-optic receiver designs are in use today. The complexity of these receivers varies by application. The job of the receiver is to take the light energy from the optical fiber and convert it into electrical energy. The output of the receiver is designed to interface with electronics that handle the information after the light-to-electricity conversion. A typical receiver can be broken into three subassemblies: *receptacle*, optical subassembly, and electrical subassembly, as shown in Figure 11.6.

FIGURE 11.6
Block diagram of a typical LED receiver

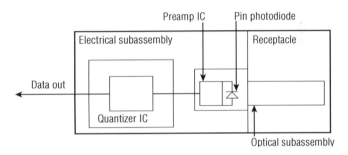

Receptacle

Regardless of the receiver packaging, the receptacle is the part of the fiber-optic receiver that accepts the connector. Its job is to provide alignment for the ferrule or terminus so that the optical fiber is perpendicular to the photodiode in the optical subassembly.

Optical Subassembly

The *optical subassembly* guides the light energy from the optical fiber to the photodiode. Depending on the receiver packaging, a window or terminus endface in the optical subassembly makes contact with the optical fiber endface. The window may be shaped like a lens to focus the light energy onto the photodiode, as shown in Figure 11.7, or a lens may be placed between the window and the photodiode.

In many receivers, the photodiode and preamplifier are housed in the optical subassembly. Sometimes the optical subassembly is referred to as the photodiode preamplifier subassembly. As we learned earlier in this chapter, the photodiode used in a receiver does not produce a substantial electrical signal. Mounting the preamplifier as physically close as possible to the photodiode allows the preamplifier to receive the strongest electrical signal.

FIGURE 11.7
Photodiode and
window

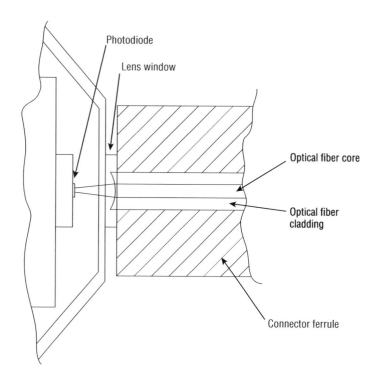

Electrical Subassembly

The electrical subassembly is typically built on a multilayer printed circuit board. This printed circuit contains the *quantizer IC* and other passive components required to complete the electrical circuit. The photodiode and preamplifier contained in the optical subassembly are electrically connected to the printed circuit board. A metal shield, shown in Figure 11.8, is typically placed over the electrical subassembly. The shield serves two purposes; it protects the electrical subassembly from external electromagnetic radiation, and it reduces the amount of electromagnetic radiation from the electrical subassembly.

FIGURE 11.8
Transceiver module
with EMI shield
exposed

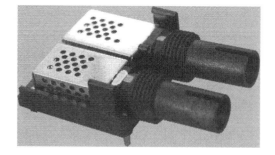

When the receiver is incorporated into a cavity insert, the metal housing acts as a shield. The metal receptacle also provides additional shielding.

The photodiode converts the light pulses from the optical fiber to electrical current pulses, as discussed earlier in this chapter. The *transimpedance amplifier*, or preamplifier, amplifies the electrical current pulses from the photodiode and outputs voltage pulses to the quantizer IC. The *limiting amplifier* in the quantizer IC amplifies the voltage pulses and provides a binary decision. It determines whether the electrical pulses received represent a binary 1 or a binary 0.

The quantizer IC also measures the received optical energy. It sets the signal detect line when there is adequate signal strength to convert the light energy into electrical energy. This prevents the electronics from trying to decode a weak signal or noise.

Receiver Performance Characteristics

This section examines several key performance characteristics of a fiber-optic receiver. It generalizes dynamic range and operating wavelength. The specific performance characteristics of LED and laser receivers are covered in detail. The data used in this section was extracted from manufacturer datasheets and represents typical performance characteristics of readily available fiber-optic receivers.

Dynamic Range

Fiber-optic receivers are limited in the amount of optical input power they can receive. Too much optical input power will saturate the photodiode. Saturating the photodiode prevents the photodiode from turning off after the light pulse has been absorbed. This can cause electrical output pulses of the receiver to overlap, creating a *bit error*.

The dynamic range of the receiver is measured in decibels. It is the difference between the maximum and minimum optical input power that the receiver can accept. If the maximum optical input power is –14dBm and the minimum optical input power is –32dBm, the dynamic range would be the difference between the two values, or 18dB.

The receiver will generate a minimum number of errors when the optical input power is kept within the minimum and maximum values. All receivers generate errors. Error generation is typically described by the receiver's *bit error rate* (BER). A typical receiver may have a BER of one error in a billion to one error in a trillion. Typically this would be written as 10^{-9} or 10^{-12}, respectively.

Operating Wavelength

Fiber-optic receivers are designed to operate within a range of wavelengths. A typical 1300nm receiver may have an operating wavelength range from 1270nm to 1380nm. This is because fiber-optic transmitters output optical energy within a wavelength range. The receiver must be able to accept a 1300nm transmitter that has a center wavelength of 1275nm or 1375nm. A receiver designed for 1300nm may not perform well or may not perform at all when connected to an 850nm or 1550nm transmitter because the responsivity of the photodiode changes over a range of wavelengths.

LED Receiver Performance Characteristics

The performance characteristics of the LED receiver are typically broken up into four groups:

♦ Recommended operating conditions

♦ Electrical characteristics

♦ Optical characteristics

♦ Data rate

The recommended operating conditions describe the maximum and minimum temperature and voltage ranges that the device can operate in without damage. Table 11.1 shows the typical recommended operating conditions for a 1300nm 100Mbps LED receiver.

TABLE 11.1:   LED receiver recommended operating conditions

PARAMETER	SYMBOL	MIN.	TYP.	MAX.	UNIT
Ambient operating temperature	T_A	0		70	°C
Supply voltage	V_{CC}	4.75		5.25	V
Data input voltage—low	$V_{IL} - V_{CC}$	–1.810		–1.475	V
Data input voltage—high	$V_{IH} - V_{CC}$	–1.165		–0.880	V
Data and signal detect output load	R_L		50		Ω

The electrical characteristics of the LED receiver describe the supply current requirements, data output voltages, signal detect output voltages, rise/fall times, and power dissipated by the device. Table 11.2 shows the typical electrical characteristics for a 1300nm 100Mbps LED receiver.

TABLE 11.2:   LED receiver electrical characteristics

PARAMETER	SYMBOL	MIN.	TYP.	MAX.	UNIT
Supply current	I_{CC}		82	145	mA
Power dissipation	P_{DISS}		0.3	0.5	W
Data output voltage—low	$V_{OL} - V_{CC}$	–1.840		–1.620	V
Data output voltage—high	$V_{OH} - V_{CC}$	–1.045		–0.880	V
Data output rise time	t_r	0.35		2.2	ns
Data output fall time	t_f	0.35		2.2	ns

TABLE 11.2:   LED receiver electrical characteristics   *(continued)*

PARAMETER	SYMBOL	MIN.	TYP.	MAX.	UNIT
Signal detect output voltage—low	$V_{OL} - V_{CC}$	−1.840		−1.620	V
Signal detect output voltage—high	$V_{OH} - V_{CC}$	−1.045		−0.880	V
Signal detect output rise time	t_r	0.35		2.2	ns
Signal detect output fall time	t_f	0.35		2.2	ns

The optical characteristics of the LED receiver at a minimum include minimum optical input power, maximum optical input power, and operating wavelength. Table 11.3 shows the typical optical characteristics for a 1300nm 100Mbps LED receiver.

TABLE 11.3:   LED receiver optical characteristics

PARAMETER	SYMBOL	MIN.	TYP.	MAX.	UNIT
Optical input power minimum at window edge	$P_{IN\,Min.}$ (W)		−33.5	−31	dBm avg.
Optical input power maximum	$P_{IN\,Max.}$	−14	−11.8		dBm avg.
Operating wavelength	λ	1270		1380	nm

If the LED receiver optical characteristics are known, the dynamic range for the receiver can be calculated. The dynamic range is the difference between the minimum value for the maximum optical input power and the maximum value for the minimum optical input. The minimum value for the maximum optical input power in Table 11.3 is −14dBm. The maximum value for the minimum optical input power is −31dBm. (Remember that this information should be obtained from the manufacturer's datasheet.)

To calculate the dynamic range, subtract the minimum value for the maximum optical input power ($P_{IN\,Max}$) from the maximum value for the minimum optical input power ($P_{IN\,Min}$), as shown in the following formula:

$P_{IN\,Max} - P_{IN\,Min}$ = Dynamic range

−14dBm − −31dBm = 17dB

Dynamic range = 17dB

LED Receiver Applications

LED receivers are typically designed to support one or more network standards. A 100Mbps 1300nm receiver could support a 100Base-FX Ethernet application or a 100Mbps ATM application. Table 11.4 lists the data rates, wavelengths, and optical fiber type for various LED receivers used in IEEE 802.3 Ethernet communication systems.

TABLE 11.4: LED receivers for Ethernet applications

NETWORK	DATA RATE	WAVELENGTH AND MEDIA
10Base-FL	10Mbps	850nm, multimode fiber
100Base-FX	100Mbps	1300nm, multimode fiber
100Base-SX	10/100Mbps	850nm, multimode fiber

Laser Receiver Performance Characteristics

The performance characteristics of the laser receiver, like the LED receiver, are typically broken up into four groups:

♦ Recommended operating conditions

♦ Electrical characteristics

♦ Optical characteristics

♦ Data rate

The recommended operating conditions describe maximum and minimum temperature and voltage ranges that the device can operate in without damage. Table 11.5 shows the typical recommended operating conditions for a 1300nm 2.5Gbps laser transmitter.

TABLE 11.5: Laser receiver recommended operating conditions

PARAMETER	SYMBOL	MIN.	TYP.	MAX.	UNIT
Ambient operating temperature	T_A	0		+70	°C
Supply voltage	V_{CC}	3.1		3.5	V
Power supply rejection	PSR		100		mV_{P-P}
Transmitter differential input voltage	V_D	0.3		2.4	V
Data output load	R_{OL}		50		Ω
TTL signal detect output current—low	I_{OL}			1.0	mA
TTL signal detect output current—high	I_{OH}	−400			μa

TABLE 11.25:   Laser receiver recommended operating conditions   *(continued)*

PARAMETER	SYMBOL	MIN.	TYP.	MAX.	UNIT
Transmit disable input voltage—low	T_{DIS}			0.6	V
Transmit disable input voltage—high	T_{DIS}	2.2			V

The electrical characteristics of the laser receiver describe the supply current requirements, data output characteristics, and power dissipated by the device. Table 11.6 shows the typical electrical characteristics for a 1300nm 2.5Gbps laser receiver.

TABLE 11.6:   Laser receiver electrical characteristics

PARAMETER	SYMBOL	MIN.	TYP.	MAX.	UNIT
Supply current	I_{CC}		115	140	mA
Power dissipation	P_{DISS}		0.38	0.49	W
Data output voltage swing (single-ended)	$V_{OH} - V_{OL}$	575		930	mV
Data output rise time	t_r		125	150	ps
Data output fall time	t_f		125	150	ps
Signal detect output voltage—low	V_{OL}			0.8	V
Signal detect output voltage—high	V_{OH}	2.0			V
Signal detect assert time (OFF to ON)	AS_{MAX}			100	µS
Signal detect deassert time (ON to OFF)	ANS_{MAX}			100	µS

The optical characteristics of the laser receiver at a minimum include minimum optical input power, maximum optical input power, center wavelength, spectral width, and reflectance. *Reflectance* is the ratio of reflected power to incident power, where incident power is the light energy exiting the optical fiber into the receiver and reflected power is the light energy reflected from the receiver into the core of the optical fiber traveling toward the light source. Table 11.7 shows the typical optical characteristics for a 1300nm 2.5Gbps laser receiver.

TABLE 11.7: Laser receiver optical characteristics

PARAMETER	SYMBOL	MIN.	TYP.	MAX.	UNIT
Receiver sensitivity	$P_{IN\ Min}$		−23	−19	dBm avg.
Receiver overload	$P_{IN\ Max}$	−3	+1		dBm avg.
Input operating wavelength	λ	1260		1570	nm
Signal detect—asserted	P_A		−27.3	−19.5	dBm avg.
Signal detect—deasserted	P_D	−35	−28.7		dBm avg.
Signal detect—hysteresis	P_H	0.5	1.4	4	dB
Reflectance			−35	−27	dB

If the laser receiver optical characteristics are known, the dynamic range for the receiver can be calculated. As with the LED receiver, the dynamic range for the laser receiver is the difference between the minimum value for the maximum optical input power and the maximum value for the minimum optical input. The minimum value for the maximum optical input power in Table 11.7 is −3dBm. The maximum value for the minimum optical input power is −19dBm. (Remember that this information should be obtained from the manufacturer's datasheet.)

To calculate the dynamic range, subtract the minimum value for the maximum optical input power ($P_{IN\ Max}$) from the maximum value for the minimum optical input power ($P_{IN\ Min}$) as shown in this formula:

$P_{IN\ Max} - P_{IN\ Min}$ = dynamic range

−3dBm − −19 dBm = 16dB

Dynamic range = 16dB

Laser Receiver Applications

Laser receivers are available for serial transmission applications and parallel transmission applications. A typical receiver may support multiple data rates and multiple network standards such as IEEE 802.3 and Fibre Channel. IEEE 802.3 is an international standard for LANs and metropolitan area networks (MANs). Fibre Channel is the foundation for over 90 percent of all SAN installations globally.

SERIAL LASER RECEIVERS

A serial laser receiver is designed to be used with network standards such as IEEE 802.3 or Fibre Channel. It transmits data in a serial format at a single wavelength over a single optical fiber. These receivers can be designed to support a broad range of data rates over multimode or single-mode optical fiber. Tables 11.8 and 11.9 list the data rates, wavelengths, and optical fiber types for various serial laser receivers used in IEEE 802.3 Ethernet communication systems and Fibre Channel communication systems.

TABLE 11.8:   Serial fiber laser receivers for Ethernet applications

NETWORK	DATA RATE	WAVELENGTH AND MEDIA
100BASE-LX	100Mbps	1310nm FP, single-mode fiber
1000BASE-SX	1Gbps	850nm VCSEL, multimode fiber
1000BASE-LX	1Gbps	1310nm FP, multimode or single-mode fiber
10GBASE-S	10Gbps	850nm VCSEL, multimode fiber
10GBASE-L	10Gbps	1310nm DFB, single-mode

TABLE 11.9:   Serial fiber laser receivers for Fibre Channel applications

NETWORK	DATA RATE	WAVELENGTH AND MEDIA
100-MX-SN-I	1062Mbaud	850nm VCSEL, multimode fiber
100-SM-LC-L	1062Mbaud	1310 FB, single-mode fiber
200-MX-SN-I	2125Mbaud	850nm VCSEL, multimode
200-SM-LC-L	2125Mbaud	1310 FP, single-mode
1200-MX-SN-I	10512Mbaud	850nm VCSEL, multimode fiber
1200-SM-LL-L	10512Mbaud	1310nm DFB, single-mode

PARALLEL OPTIC LASER RECEIVERS

Up to this point in the chapter only single-fiber receivers operating at a single wavelength have been discussed. These receivers have limitations to the data rates they can achieve and that limit as defined by IEEE 802.3 and Fibre Channel is approximately 10Gbps. To achieve higher data rates, you can take one of two approaches. One is to combine the output from multiple transmitters, each transmitting at a different wavelength, into a single optical fiber; this technique is called wavelength division multiplexing (WDM) and it is covered in Chapter 12, "Passive Components and Multiplexers." WDM is ideal for high data rates systems over long distances using single-mode optical fiber.

Another approach to achieving high data rates over short distances using multimode optical fiber is parallel optics. With parallel optics, multiple transmitters and receivers, each operating at the same wavelength, are used. The difference between parallel optics and WDM is that the parallel optics approach uses one optical fiber for each transmitter and receiver pair instead of combining the outputs of multiple transmitters into a single optical fiber and then splitting out each wavelength to a separate receiver. Because each transmitter and receiver pair is dedicated to a single optical fiber, all the transmitters can operate at the same wavelength.

The light source of choice for parallel optics is the VCSEL. VCSEL technology has matured to where multiple 10Gbps transmitters can be manufactured in the form of an array that physically aligns with the fibers in an MT ferrule. As you learned in Chapter 9, the optical fibers in an MT ferrule are spaced 0.25mm apart, as shown in Figure 11.9.

FIGURE 11.9
The MT ferrule
optical fiber spacing
is 0.25mm.

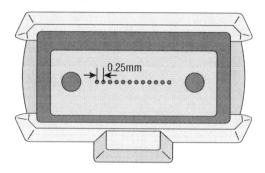

IEEE Standard 802.3-2012, Section 6 introduces 40Gbps and 100Gbps Ethernet. It defines the maximum transmission distances for each data rate using single-mode optical fiber, multimode optical fiber, or copper cabling for the physical layer. When multimode OM3 or OM4 optical fiber is used for the physical layer, multiple transmitter and receiver pairs operating in parallel are required, as described in Table 11.10. Achieving a 40Gbps data rate requires four 10Gbps transmitter and receiver pairs operating in parallel. A 100Gbps data rate is achieved using ten 10Gbps transmitter and receiver pairs operating in parallel.

As of this writing, the maximum transmission distances for these systems using OM3 optical fiber is 100 meters and 150 meters with OM4 optical fiber.

TABLE 11.10:   Parallel optic laser receivers for Ethernet applications

NETWORK	DATA RATE	DATA RATE PER CHANNEL	NUMBER OF TRANSMITTERS	WAVELENGTH AND MEDIA
40GBASE-SR4	40Gbps	10Gbps	4	850nm VCSEL, multimode fiber
100GBASE-SR10	100Gbps	10Gbps	10	850nm VCSEL, multimode fiber

To support parallel operation, the transmitters and receivers must be physically arranged to accept an MPO female plug like the one shown in Figure 11.10. For 40GBASE-SR-4 networks, the transmitters use the four leftmost optical fibers or optical transmit lanes and the receivers use the four rightmost optical fibers or receive lanes, as shown in Figure 11.11. The four optical fibers in the center are not used. This arrangement allows for 40Gbps full-duplex operation using a single 12-fiber ribbon.

FIGURE 11.10
MPO female plug
with 12 optical
fibers

FIGURE 11.11
40GBASE-SR4
optical lane
assignments

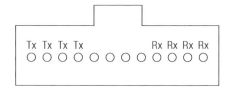

For 100GBASE-SR10 networks, the optical lanes can be configured for a single-receptacle using Option A, shown in Figure 11.12; this approach is recommended in IEEE 802.3. They can also be configured using the two-receptacle Option B, shown in Figure 11.13, or Option C, shown in Figure 11.14. These options are not recommended and are considered alternatives.

FIGURE 11.12
100GBASE-SR10
Option A optical
lane assignments

FIGURE 11.13
100GBASE-SR10
Option B optical
lane assignments

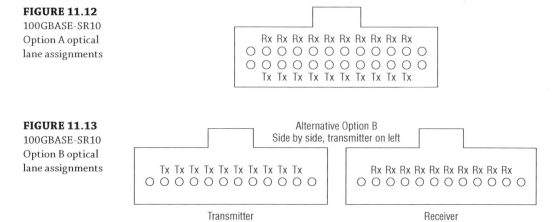

FIGURE 11.14
100GBASE-SR10
Option C optical
lane assignments

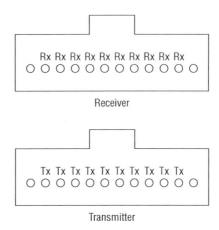

Receiver

Transmitter

IS IT THE COMPUTER HARDWARE OR SOFTWARE?

We were troubleshooting a communication problem between a piece of rack-mounted comput-ing equipment and a router. The equipment and the router were communicating over 50/125µm multimode optical fiber. Everything would work for a while, and then the equipment and router would stop communicating.

Typically, when things like this happen, the programmers blame the failure on the hardware and we hardware engineers blame the failure on the software. Because we had worked together on the design of the network switch in the computing equipment, we immediately became involved when the communication failure occurred. The next phase was troubleshooting to prove that the hardware was not the problem.

During the troubleshooting process, someone questioned whether the receiver on the router was receiving any light energy from the switch's transmitter. Because the two pieces of equip-ment communicate at a wavelength of 1300nm, the light is not visible. To quickly answer that question, we used a power meter to measure the optical output power from the transmitters on both pieces of equipment. We used a mode filter on the 1 meter jumper from the transmitter to the power meter.

We compared the measurements we obtained to the manufacturer's optical characteristics for both the transmitter and the receiver. The optical output power for each transmitter was within specifications. About the same time that we determined the hardware was functioning properly, the programmers made some minor changes and the problem was resolved.

Being able to measure the transmitter's optical output power and determine that it's within the acceptable range for the receiver is a fundamental skill that will be needed in various trouble-shooting scenarios.

Transceivers

As you learned in Chapter 10, fiber-optic receivers are typically packaged with the transmitter. Together, the receiver and transmitter form a *transceiver*. Engineers are finding more and more applications for fiber optics in various types of environments. Because of this, fiber-optic transceivers are being developed in many different form factors or packages.

Form Factors

In this section, we will look at transceiver form factors that were created as part of a multi-source agreement (MSA) and form factors where the design is driven by the application.

An ad hoc committee made up of representatives from various transceiver manufacturers typically creates the transceiver form factors developed under an MSA. These committees produce operational requirement specifications that are used by manufacturers, system integrators, and suppliers of that specific form factor transceiver. These specifications are typically submitted to standards bodies such as the Electronic Industries Association (EIA) or the Accredited Standards Committee (ASC) to be adopted as separate standards, referenced in standards, or incorporated into a standard.

SFF Committee Form Factors

The SFF Committee is an ad hoc group that was formed in 1990. Initially the goal of this committee was to define the mechanical envelopes for disk drives being developed for computers and other small products. In 1992, this committee broadened its charter to include pin outs for interface applications. The transceivers shown in Figure 11.15 are examples of the 1×9 package style. These transceivers have one row of nine pins. The position and function of each pin is defined in an MSA.

FIGURE 11.15
1×9 package style
LED transceivers
soldered to circuit
board

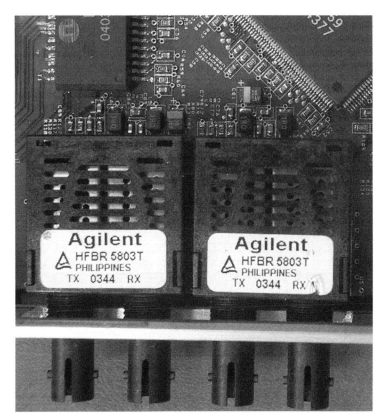

The specifications developed by the SFF Committee typically include:

♦ Electrical interfaces that describe the electrical connector, recommended printed circuit board layout requirements, and pinouts for:

  ♦ Data control

  ♦ Status

  ♦ Configuration

  ♦ Test signals

♦ Management interfaces such as specific multidata rate and multiprotocol implementations

♦ Optical interfaces that include the following:

  ♦ Receptacle

  ♦ Mating optical fiber

  ♦ Mating connector

  ♦ Recommended breakout cable assembly

♦ Mechanical requirements, including:

  ♦ Package dimensions

  ♦ Latching details

  ♦ Mating optical connector

  ♦ Mating electrical connector

  ♦ Panel cutout dimensions

  ♦ Labeling

  ♦ Host circuit board layout

♦ Thermal requirements such as:

  ♦ Case temperatures

  ♦ Ambient air temperatures

♦ Electromagnetic interference (EMI) recommendations such as:

  ♦ Component shielding

  ♦ Packaging shielding

  ♦ Panel cutout sealing

♦ Electrostatic discharge (ESD) requirements

Popular form factors developed by the SFF Committee include the small form factor pluggable (SFP), enhanced small form factor pluggable (SFP+), and the quad small form factor pluggable (QSFP+).

SFP transceivers are used in applications with data rates ranging from 155Mbps up to 2.5Gbps. These transceivers contain a printed circuit board like the one shown in Figure 11.16 that mates with the SFP electrical connector on the host equipment printed circuit board. The transceiver printed circuit board is designed so that as it is inserted into the host electrical connector, the ground trace is the first to make contact with the electrical connector, followed by the power trace, and the signal traces. This electrical mating sequence allows the SFP module to be removed or inserted without powering down the host equipment. This may be referred to as hot swappable, hot swapping, hot pluggable, or hot plugging.

FIGURE 11.16
SFP transceiver
with the printed
circuit board
exposed

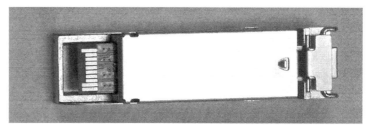

Photo courtesy of Electronic Manufacturers' Agents, Inc.

SFP+ transceivers are approximately the same physical size as the SFP transceivers. Like the SFP transceivers, they are hot pluggable. These transceivers support 10Gbps data rates. They are available with an 850nm VCSEL for use with multimode optical fiber. They are also available with either a 1310nm or 1550nm light source for use with single-mode optical fiber over distances as great as 80km.

QSFP+ are slightly larger than the SFP or SFP+ transceivers. These transceivers feature four independent full-duplex channels, each capable of supporting 10Gbps or greater data rates. These transceivers are designed for 12-fiber MPO connectors like the one shown in Figure 11.10. They are used for 40Gbps applications such as 40GBASE-SR4.

Additional information on the SFF Committee and the various form factors may be obtained from their website, `www.sffcommittee.com/ie/`.

Application-Driven Form Factors

Engineers are continually finding new applications for fiber optics. Some of these applications may require the transceiver to operate in harsh environments where shock, vibration, and temperature requirements drive the transceiver form factor. These types of transceivers are typically used for avionics, aerospace, or military applications. Four form factors employed in these applications are optical inserts, integrated bulkhead receptacles, rugged RJ, and rugged chip-scale pluggable (RCP).

CAVITY INSERTS

Cavity inserts like those shown in Figure 11.17 are designed to be inserted into a standardized connector such as an ARINC 600. They would be inserted into the receptacle instead of a terminus. As of this writing, these inserts were available only as a transmitter or a receiver. Two inserts would be required to form a transceiver and achieve full-duplex operation.

FIGURE 11.17
Size 8 cavity inserts with Gigabit Ethernet receiver

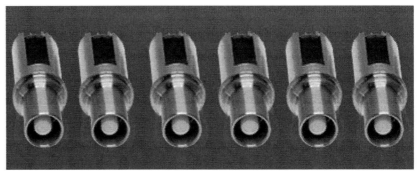

Photo courtesy of Moog Protokraft

The transmitter cavity insert contains the electrical and optical subassemblies. The electrical subassembly contains the electronics required to drive and monitor the laser light source. The optical subassembly guides the light from the laser to the endface of the terminus or ferrule.

The receiver cavity insert contains the electrical and optical subassemblies. The receiver electrical subassembly contains the photodiode, preamplifier or transimpedance amplifier, and the postamplifier or limiting amplifier. The optical subassembly guides the light from the terminus endface to the photodiode.

INTEGRATED BULKHEAD RECEPTACLES

Integrated bulkhead receptacles like the one shown in Figure 11.18 have the transmitter and receiver electrical and optical subassemblies packaged within a harsh environment receptacle such as an MIL-DTL-38999. This receptacle features two transmitters and two receivers or two transceivers. Integrating the optical and electrical subassemblies for these transceivers into the harsh environment receptacle eliminates an interconnection and the losses associated with that interconnection.

Before the development of integrated bulkhead receptacles, optical fibers were required to guide the light to and from the bulkhead receptacle to the transceiver within the enclosure. This configuration requires more space within the enclosure than the integrated bulkhead receptacle. Additional optical loss is incurred because there is an interconnection at the bulkhead receptacle and the transceiver. Integrating the transceiver into the receptacle eliminates one interconnection loss, frees up space within the enclosure, and eliminates the short lengths of optical fiber from the bulkhead receptacle to the transceiver.

NOTE Bulkhead receptacles typically mount to a panel or side of an enclosure such as an avionics box.

Additional information on the cavity inserts or integrated bulkhead receptacles can be obtained from the Moog Protokraft website: www.protokraft.com.

FIGURE 11.18
Transceiver bulk-
head receptacle
soldered to the cir-
cuit board

Photo courtesy of Moog Protokraft

RUGGED RJ

The rugged RJ transceiver shown in Figure 11.19 accepts LC connectors. It was designed so
the exterior mechanical dimensions are very close to those of RJ-45 copper connectors with a
front extension. To ensure reliable operation in harsh shock and vibration environments, two
screws are used to secure the transceiver to the circuit board after it has been soldered to the
circuit board. The threaded screw holes are on the bottom of the transceiver and not visible in
Figure 11.19. This transceiver features a broad operational temperature range that permits operation in cold or hot temperatures. In addition, the electronic components can be conformal coated
for reliable operation in high moisture–level environments.

RUGGED CHIPSCALE PLUGGABLE (RCP)

The rugged RCP transceiver shown in Figure 11.20 accepts ARINC 801 termini. It supports four
independent optical channels at data rates up to 11Gbps. Like the rugged RJ, screws are used
to secure the transceiver to the circuit board, ensuring reliable operation in a harsh shock and
vibration environment. The threaded screw holes are on the bottom of the transceiver and are
not visible in Figure 11.20.

To simplify manufacturing and facilitate quick swap-outs in the field, the rugged RCP transceiver has a separable electrical connector on the bottom that mates with the receptacle on the
circuit board. The transceiver is installed onto the circuit board by snapping the electrical connector into the receptacle and securing it to the circuit board with two screws. This transceiver
also features a broad operational temperature range that permits operation in cold or hot temperatures. In addition, the electronic components can be conformal coated for reliable operation
in high moisture–level environments.

FIGURE 11.19
Rugged RJ
transceiver

Photo courtesy of COTSWORKS

FIGURE 11.20
RCP transceiver

Photo courtesy of COTSWORKS

Additional information on the rugged RJ and the RCP can be obtained from the COTSWORKS website: www.cotsworks.com.

Transceiver Health Monitoring

Transceiver health monitoring, also referred to as built-in test (BIT), can provide real-time information about the physical, electrical, and optical characteristics of the transceiver. Two organizations developing specifications related to transceiver health monitoring are the SFF Committee and SAE International's AS-3 Fiber Optics and Applied Photonics Committee. In April 2013, AS-3 began work on a new optical performance monitoring system standard, AIR6552, that outlines the sensors and basic architecture required to detect, localize, and isolate impairments as well as assist in failure prediction of the physical layer of a fiber-optic network. This fiber-optic network end-to-end data link evaluation system standard is also known as NEEDLES.

Since NEEDLES is a multisensor architecture, it has a main document, AIR6552, and several slash sheets. Each slash sheet addresses a particular type of sensor. As of this writing, the main document and three slash sheets are under development.

Slash sheet AIR6552/3 addresses transceiver health monitoring. Transceiver health monitoring can provide real-time information about the physical, electrical, and optical characteristics of the transceiver. This document establishes methods to obtain, store, and access data about the health of a fiber-optic network using commercially available sensors located in or near the transceiver.

NEEDLES is a relatively new project when compared to work that has been published by the SFF Committee. The SFF Committee SFF-8472 specification for the diagnostic monitoring interface for optical transceivers was initially published in April 2001, and it has been revised many times since it was first published. This document defines the enhanced digital diagnostic monitoring interface for optical transceivers. It allows real-time access to device operating parameters over a two-wire serial data bus. These parameters include:

♦ Internally measured transceiver temperature

♦ Internally measured transceiver supply voltage

♦ Measured transmitter bias current in mA

♦ Measured transmitter output power in mW

♦ Measured received optical power in mW

NOTE Slash sheets are typically a family of documents where the main document, such as AIR6552, will refer to an overall/general aspect of the document and then the slash sheet document(s) will refer to specific items within that overall document.

ADVANCED BUILT-IN TEST TRANSCEIVERS

Advanced built-in test transceivers offer real-time access to device operating parameters such as optical power levels, voltage levels, and laser characteristics. However, they also feature a unique optical time domain reflectometer (OTDR) function that can be used to identify a fault or a break in an optical fiber. OTDR theory, operation, testing, and trace analysis is covered in Chapter 16, "Test Equipment and Link/Cable Testing."

The two advanced built-in test transceivers shown in Figure 11.21 each feature four 12.5Gbps transmitter and receiver pairs operating in parallel. These transceivers are extremely small and designed to operate in harsh environments. Figure 11.21 shows the bottom of one transceiver and the top of another. The 12-fiber ribbon and integrated connector are held in place with two screws, ensuring a secure interconnection.

FIGURE 11.21
Advanced built-in
test transceivers

Photo courtesy of Ultra Communications

Additional information on this advanced built-in test transceiver can be obtained from the Ultra Communications website: www.ultracomm-inc.com.

The Bottom Line

Calculate the dynamic range for an LED receiver. The dynamic range of the LED receiver is the difference between the minimum value for the maximum optical input power and the maximum value for the minimum optical input.

Master It Refer to the following table and calculate the dynamic range for the LED receiver.

PARAMETER	SYMBOL	MIN.	TYP.	MAX.	UNIT
Optical input power minimum at window edge	$P_{IN\,Min.}$ (W)		−30.5	−29	dBm avg.
Optical input power maximum	$P_{IN\,Max.}$	−17	−13.8		dBm avg.
Operating wavelength	λ	1270		1380	nm

Calculate the dynamic range for a laser receiver. The dynamic range of the laser receiver is the difference between the minimum value for the maximum optical input power and the maximum value for the minimum optical input.

Master It Refer to the following table and calculate the dynamic range for the laser receiver.

PARAMETER	SYMBOL	MIN.	TYP.	MAX.	UNIT
Receiver sensitivity	$P_{IN\,Min}$		−20	−18	dBm avg.
Receiver overload	$P_{IN\,Max}$	−3	+3		dBm avg.
Input operating wavelength	λ	1260		1570	nm
Signal detect—asserted	P_A		−24.3	−16.5	dBm avg.
Signal detect—deasserted	P_D	−32	−25.7		dBm avg.
Signal detect—hysteresis	P_H	0.5	1.4	4	dB
Reflectance			−35	−27	dB

Chapter 12

Passive Components and Multiplexers

The objective of this chapter is to help you gain an understanding of fiber-optic passive components and *multiplexers*. This chapter covers not only particular devices and their applications, but also the reasons why the components are chosen and when they should be used.

Fiber-optic passive components and multiplexers are found in applications that require the transmission, combining, or distribution of optical signals. Some of the optical devices covered in this chapter are *couplers, power taps, switches, attenuators, isolators, amplifiers,* and *filters.* Multiplexers and their associated processes, in particular *wavelength division multiplexing* and *dense wavelength division multiplexing,* are also examined.

In this chapter, you will learn to:

♦ Calculate the output power of a real tee coupler port

♦ Calculate the output power of a real star coupler

♦ Calculate attenuator values

Standards

There are many standards for passive components and multiplexers. Passive components and multiplexers play a key role in bringing fiber optics to your home or business. They are the core of a *passive optical network (PON),* and the application of many of the devices in this chapter is covered in Chapter 13, "Passive Optical Networks."

This chapter provides an overview of many passive components and their operation. For each of these components, one or more standards define performance and testing requirements. The International Electrotechnical Commission (IEC) has developed many of these standards.

IEC has two technical subcommittees—86B and 86C—under Technical Committee 86. The subcommittees focus on standards for many of the components in this chapter. Subcommittee 86B focuses on fiber-optic interconnecting devices and passive components. These components include attenuators, switches, wavelength multiplexers/demultiplexers, couplers, and isolators. This subcommittee has developed numerous performance and testing standards for interconnecting devices and passive components.

IEC Subcommittee 86C focuses on fiber-optic systems and active devices, including terminology, test and measurement methods, functional interfaces, and mechanical, optical, environmental, and electrical requirements. The documents developed by this subcommittee help to ensure the interoperability of components and systems, which leads to reliable system performance.

SAE International's Aerospace Information Report AIR5667—titled Fiber Optic Wavelength Division Multiplexed (WDM) Single-mode Interconnect and Component Standards Mapping for Aerospace Platform Applications – Device Level Specification—is an excellent resource for all designers developing WDM networks. It provides an overview of WDM architectures, environmental requirements, telecommunications WDM standards, and individual component parameters. Additional information on this standard is available on SAE International's website: www.sae.org.

2. Parameters

This chapter discusses several passive devices and some of the common parameters that apply to each device. When working with passive components, you should have a basic understanding of common parameters. Some parameters that you need to be familiar with are optical fiber type, connector type, *center wavelength, bandwidth, insertion loss, excess loss, polarization-dependent loss (PDL), return loss, crosstalk, uniformity, power handling,* and *operating temperature.*

Connector types and optical fiber types   Many passive devices are available with receptacles or pigtails. The pigtails may or may not be terminated with a connector. If the device is available with a receptacle or connector, the type of receptacle or connector needs to be specified when ordered. You should also note the type of optical fiber used by the manufacturer of the device to ensure it is compatible with the optical fiber used for your application.

Center wavelength and bandwidth   Passive devices also have a center wavelength and bandwidth or bandpass. The center wavelength is the nominal operating wavelength of the passive device. The bandwidth and bandpass are the range of wavelengths over which the manufacturer guarantees the performance of the device. Some manufacturers will list an operating wavelength range instead.

Types of loss   Insertion loss, excess loss, PDL, and return loss are all measured in decibels.

Insertion loss   This is the optical power loss caused by the insertion of a component into the fiber-optic system. When working with passive devices, you need to be aware of the insertion loss for the device and the insertion loss for an interconnection. Insertion loss as stated by the manufacturer typically takes into account all other losses, including excess and PDL. Insertion loss is the most useful parameter when designing a system.

Excess loss   This may or may not be defined by the manufacturer. Excess loss, associated with optical couplers, is the amount of light lost in the coupler in excess of the light lost from splitting the signal. In other words, when a coupler splits a signal, the sum of the power at the output ports does not equal the power at the input port; some optical energy is lost in the coupler. Excess loss is the amount of optical energy lost in the coupler. This loss is typically measured at the specified center wavelength for the device.

Polarization-dependent loss (PDL)   This is only a concern for single-mode passive devices. It is often the smallest value loss, and it varies as the polarization state of the propagating light wave changes. Manufacturers typically provide a range for PDL or define a not-to-exceed number.

Return loss   When a passive device is inserted, some of the optical energy from the source is going to be reflected back toward the source. This is typically described as return loss, or reflection loss. Return loss is expressed in decibels and is the negative quotient of the power received divided by the power transmitted. Return loss is covered in detail in Chapter 16, "Test Equipment and Link/Cable Testing."

Crosstalk Crosstalk in an optical device describes the amount of light energy that leaks from one optical conductor to another. Crosstalk is not a concern in a device where there is a single input and multiple outputs. However, it is a concern with a device that has multiple inputs and a single output, such as an optical switch. Crosstalk is expressed in decibels, where the value defines the difference between the optical power of one conductor and the amount of leakage into another conductor. In an optical switch with a minimum crosstalk of 60dB, there is a 60dB difference between the optical power of one conductor and the amount of light that leaked from that conductor into another conductor.

Uniformity Uniformity is expressed in decibels. It is a measure of how evenly optical power is distributed within the device. For example, if a device is splitting an optical signal evenly into four outputs, how much those outputs could vary from one another is defined by uniformity. Uniformity is typically defined over the operating wavelength range for the device.

Power handing Power handling describes the maximum optical power at which the device can operate while meeting all the performance specifications defined by the manufacturer. Power handling may be defined in milliwatts or decibels, where 0dBm is equal to 1mW.

Operating temperature Operating temperature describes the range of temperatures that the device is designed to operate in. This can vary significantly between devices, because some devices are only intended for indoor applications whereas others may be used outdoors or in other harsh environments.

Couplers

In many applications, it may not be possible to have a design of many point-to-point connections. In these cases, optical couplers are used. A fiber-optic coupler is a device that combines or splits optical signals. A coupling device may combine two or more optical signals into a single output, or the coupler may be used to take a single optical input and distribute it to two or more separate outputs. Figure 12.1 is an example of a basic four-port coupler.

FIGURE 12.1
Four-port coupler

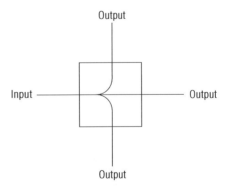

Many couplers are designed bidirectionally, which enables the same coupler to be used to combine signals or split signals. A coupler being used to split a signal may be referred to as a splitter.

Couplers are available with a wide range of input and output ports. A basic coupler may have only one input port and two output ports. Today's technology supports couplers with up to 64 input and 64 output ports, as shown in Figure 12.2.

FIGURE 12.21
28-port coupler

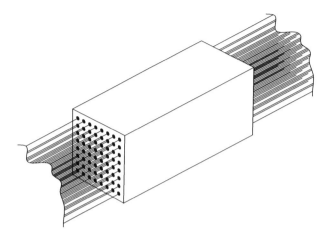

There are many different types of couplers, and the number of input and output ports is dependent on the intended usage. Some of the types of *optical couplers* are *optical combiners, Y couplers, star couplers, tee couplers, tree couplers,* and *optical splitters*; in this chapter, we will focus on the tee coupler and the star coupler.

The Tee Coupler

A *tee coupler* is a three-port optical coupling device that has one input port and two output ports, as shown in Figure 12.3.

FIGURE 12.3
Tee coupler

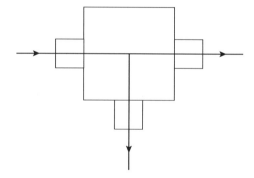

The tee coupler is a passive device that splits the optical power from the input port into two output ports. The tee coupler is in essence an optical splitter. The uniqueness of the tee coupler is that this type of coupler typically distributes most of the optical input power to one output and only a small amount of power to the secondary output. Note that when the outputs are evenly distributed, the coupler is called a Y coupler. The tee coupler is also referred to as an optical tap, due to the nature of the device. A majority of the power continues forward, but a portion of the signal (determined by the splitting ratio) is tapped to be used for an output port.

DISTRIBUTION AND LOSS OF OUTPUT POWER IN A TEE COUPLER

The tee coupler is a 1×2 coupler, meaning that it has one input port (or connection) and two output ports. As previously stated, the optical output power of the two output ports is typically not evenly distributed. Common splitting ratios are 90:10, 80:20, 70:30, 60:40, and 50:50 (a Y coupler). Not all manufacturers follow the convention of placing the larger value to the left of the colon and the smaller to the right. Some manufacturers simply reverse this and place the smaller value to the left of the colon and the larger to the right.

A typical use for a tee coupler would be to supply optical signals to a bus type network of inline terminals. Assuming ideal conditions and a 90:10 split on the tee coupler, the first terminal would receive 10 percent of the optical signal and 90 percent of the optical signal would go forward to the next tee coupler, and so on, as shown in Figure 12.4.

FIGURE 12.4

Ideal tee couplers in a bus type network

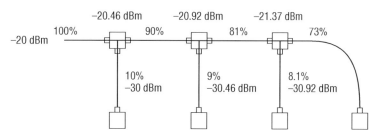

It's easy to see how the optical signal power levels decrease from one terminal to the next. Keep in mind that we did not account for any losses; this is an ideal system. If we were to take losses into account, the results (shown in Figure 12.5) would be much different.

FIGURE 12.5

Real tee couplers in a bus type network

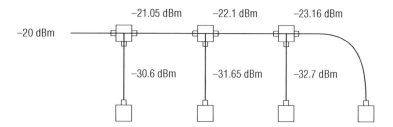

Now let's look at some of the common losses associated with couplers and their effects on the network of interconnections. To be realistic, we have to take into account the insertion loss for the device and the insertion loss for each individual interconnection, whether a splice or a mated connector pair.

METHODS OF DESCRIBING INSERTION LOSS

The insertion loss for the coupler may be described two ways. One manufacturer may provide an insertion loss value for a device and you will have to calculate the power at each output port based on the splitting ratio. However, another manufacturer may provide the insertion loss combined with the loss from the splitting ratio. In this section, you will learn how to calculate the power at an output port using the insertion loss values provided by the manufacturer.

In Figure 12.5, the insertion loss for each coupler and each interconnection is identical; each interconnection had an insertion loss of 0.3dB and each coupler had an insertion loss of 0.3dB. These losses add up and affect the optical output power of each port. Each coupler adds additional insertion losses; these losses reduce the number of taps a real tee coupler can support when compared to an ideal tee coupler. Figure 12.6 graphically compares the performance differences between ideal tee couplers and real tee couplers.

FIGURE 12.6
Comparison of
ideal and real tee
couplers

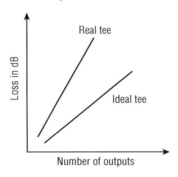

As shown by the previous example, taking into account the losses of the device and the interconnections can make a large difference in anticipated optical output power. Remember that actual coupler losses can vary from manufacturer to manufacturer. Always refer to the manufacturer's datasheet for performance information.

CALCULATING THE OUTPUT POWER AT THE PORTS OF A COUPLER

The decibel rules of thumb described in Chapter 2, "Principles of Fiber-Optic Transmission," can be used to determine the output power at the ports of a coupler.

Calculating from the Insertion Loss Value

To determine the output power at both output ports of a tee coupler with a 50:50 splitting ratio using the decibel rules of thumb, we will assume that the insertion loss for each interconnection is the same and that the manufacturer provided the insertion loss only for the coupler and did not take into account the loss for the splitting ratio. In this case, the total loss per port is the sum of the insertion losses for the interconnections (LI) and the insertion loss for the device (LC) as shown in this formula:

$LI_1 + LI_2 = LI_T$

$LI_T + LC = Total\ Loss$

For this example, let's assume that the insertion loss for each interconnection is 0.3dB and the insertion loss for the coupler is 0.5dB. The total coupler loss per port would be 1.1dB, as shown here:

0.3dB + 0.3dB = 0.6dB

0.6dB + 0.5dB = 1.1dB

Remember from earlier in the chapter that uniformity is a measure of how evenly optical power is distributed within the device. In the next step, we will calculate the optical power at each output port without taking into account uniformity. To find the optical output power at each port, we must know the input power.

Assume that the input power to the coupler is –20dBm. The first step is to subtract the total coupler losses per port (1.1dB) from the input power (–20dBm). As shown here, the remaining power available to the ports is –21.1dBm:

–20dBm – 1.1dB = –21.1dBm

The second step is to split the remaining power using the decibel rules of thumb covered in Chapter 2.

Because this coupler has a 50:50 splitting ratio, each output port should receive 50 percent of the energy. A loss of 50 percent is a change of 3dB. The output power at each port will be 3dB less than the power remaining after all the insertion losses have been accounted for, as shown here:

–21.1dBm – 3dB = –24.1dBm

Each output port should have a power output of –24.1dBm. However, couplers are not perfect and the amount of power actually available will vary depending on the uniformity. Assume that the uniformity for this coupler is 0.2dB. This means that the difference between the highest and lowest insertion loss for the coupler output ports will not exceed 0.2dB within the bandpass. In this example, the difference between the output power in each port could vary as much as 0.2dB.

Calculating with the Insertion Losses

In the next example, the manufacturer provided the insertion loss combined with the loss from the splitting ratio. This tee coupler has a splitting ratio of 80:20. The decibel rules of thumb will not be required to calculate the power available at each output port because the manufacturer states the insertion loss in decibels as 1.5/7.6. This means that the power at the 80 percent port will be 1.5dB less than the input power to the coupler and the power at the 20 percent port will be 7.6dB less than the input power to the coupler.

If the input power to the coupler is –20dBm, as in the previous example, the power at each port can be calculated by subtracting the insertion loss for each port defined by the manufacturer and subtracting the interconnection insertion losses from –20dBm. Assume that insertion loss for each interconnection is 0.25dB. As shown here, the output power at the 80 percent port is –22.0dBm and the output power at the 20 percent port is –28.1dBm.

–20dBm – 1.5dB – 0.25dB – 0.25dB = –22dBm

–20dBm – 7.6dB – 0.25dB – 0.25dB = –28.1dBm

The Star Coupler

The star coupler is used in applications that require multiple ports—input and/or output. The star coupler will distribute optical power equally from one or more input ports to two or more output ports. Figure 12.7 shows a basic star coupler with four input ports and four output ports. Star couplers are available in 1×64 up to 64×64 configurations.

FIGURE 12.7
Eight-port star
coupler

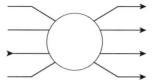

A special version of the star coupler, called a *tree coupler*, is used when there is one input port and multiple output ports or when there are multiple input ports and one output port.

Star couplers are frequently used in network applications when there are a large number of output terminals. In our tee coupler example, we had to account for interconnection insertion loss and coupler insertion loss at each tee connection. However, with the star coupler there is only one coupler insertion loss regardless of the number of ports. With only one coupler insertion loss, a multiple port star coupler is more efficient than a series of tee couplers, as shown in Figures 12.4 and 12.5. So the larger the network is, the more efficient the star coupler becomes.

Two types of star couplers are commonly used: the reflective star and the transmissive star. Couplers are typically considered to be a black box—that is, only the manufacturer knows what's inside. However, many star couplers are made of fused optical fibers.

The reflective star, shown in Figure 12.8, is defined as a coupler that distributes optical power to all input and output ports when a signal is applied to only one port.

FIGURE 12.8
Fused reflective star
coupler

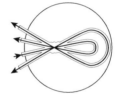

The transmissive star, shown in Figure 12.9, is defined as a coupler that distributes optical power to all the output ports when a signal is applied to any of the input ports.

FIGURE 12.9
Fused transmissive
star coupler

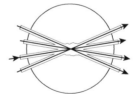

CALCULATING THE LOSSES WITH A STAR COUPLER

In our previous example, we looked at the losses in a network of four terminals with tee couplers. Let's compare the losses for the same number of terminals using a star coupler. Figure 12.10 shows the power delivered to each terminal of a four-port star coupler using the same values as in the previous example, 0.3dB interconnection insertion losses and 0.3dB coupler insertion loss (due to only one coupler). This example shows a tree coupler with four output ports and one input port, similar to the sequential tee-coupled workstation network shown in Figure 12.5.

FIGURE 12.10

Four-output port
tree coupler

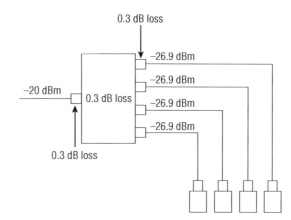

Let's determine the output power at each output port of a five-output port star coupler using the decibel rules of thumb. (Note that this exercise does not use the coupler shown in Figure 12.10.)

In this example, we will assume that the insertion loss for each interconnection is the same. Unlike with the tee coupler, manufacturers typically provide the insertion loss only for the coupler and do not take into account the loss for the splitting ratio. In this case, the total loss per port is the sum of the insertion losses for the interconnections (LI) and the insertion loss for the coupler (LC) as shown here:

$$LI_1 + LI_2 = LI_T$$

$$LI_T + LC = \text{Total Loss}$$

For this example, let's assume that the insertion loss for each interconnection is 0.3dB and the insertion loss for the coupler is 1.2dB. This coupler has five output ports and one input port. The total coupler loss per port would be 0.6dB, as shown here:

0.3dB + .03dB = 0.6dB

or

0.3dB × 2 = .06dB

0.6dB + 1.2dB = 1.8dB

Assume that the input power to the coupler is –7dBm. The first step is to account for the star coupler insertion losses and the interconnection losses. This is done by subtracting 1.8dB from the input power of –7dBm. The remaining power available to the ports is –8.8dBm, as shown here:

–7dBm – 1.8dB = –8.8dBm

The second step is to split the remaining power using the decibel rules of thumb covered in Chapter 2.

Each output port will receive 20 percent of the energy because the energy is distributed evenly between the output ports. A loss of 80 percent is a change of 7dB. The output power at each port will equal –8.8dBm minus 7dB. Each output port will have a power output of –15.8dBm, as shown here:

–8.8dBm – 7dB = –15.8dBm

Remember that couplers are not perfect and the amount of power actually available will vary depending on the uniformity. If we assume that the uniformity for this coupler is 0.3dB, the difference between the highest and lowest insertion loss for the coupler output ports will not exceed 0.3dB within the bandpass.

ADVANTAGES OF STAR COUPLER COMPARED TO TEE COUPLER

The key advantage to the star coupler is that there is only one insertion loss caused by the coupler. The only remaining insertion losses are from the interconnections. The advantage of the star coupler becomes very apparent as the number of ports increases. A simple loss-comparison graph, as shown in Figure 12.11, can reveal the significance in the number of terminals versus loss for the tee and star couplers.

FIGURE 12.11
Real tee coupler vs. real star coupler comparison graph

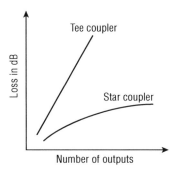

A star coupler has another advantage over a series of tee couplers. If one of the tee couplers in the series is disconnected, none of the other terminals down the line will receive an optical signal. However, disconnecting a terminal from the star coupler will not impact the operation of the other terminals.

Inline Power Tap

An inline power tap like the one shown in Figure 12.12 is an optically passive device with an input optical fiber, output optical fiber, and integrated photodiode. These devices can be manufactured for single-mode or multimode optical fiber applications. However, they are primarily used in single-mode networks. Applications for the power tap include:

♦ Monitoring transmitter output power

♦ Monitoring receiver input power

♦ Monitoring back reflections

FIGURE 12.12
Inline power tap

Photo courtesy of EigenLight Corporation

Power taps can also be integrated into control loops that monitor and control the amount of amplification when using erbium doped fiber amplifiers (EDFAs) or Raman amplification. Both are covered in detail later in this chapter. As of this writing, AIR6552/1, Inline Optical Power Monitoring, was under development within SAE International's AS-3 Fiber Optics and Applied Photonics Committee. This document will provide detailed information about the inline power tap.

An inline power tap is similar to a tee coupler where there is one input and two outputs. However, unlike the tee coupler, one of the outputs from the inline power tap is directed toward a photodiode and the light from this output never exits the tap, as shown in Figure 12.13. Most of the light entering the tap exits through the output optical fiber.

FIGURE 12.13
Inline power tap
functional diagram

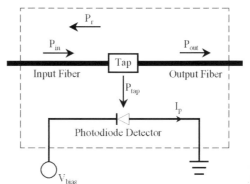

Drawing courtesy of EigenLight Corporation

Like the tee coupler, the power tap can be designed to direct different percentages of light toward the photodiode. Power taps that direct more light toward the photodiode will have a greater insertion loss than power taps that direct less light. However, power taps that direct less light toward the photodiode will accept higher input power levels than taps that direct more light.

Single-mode power taps like the one shown in Figure 12.12 have high *directivity*. Directivity is the sensitivity of forward-directed light relative to backward-directed light. This means that the power tap is more sensitive to light traveling from the input fiber to the output fiber.

The power tap's high directivity makes it an excellent device to measure forward or reverse power in an optical fiber. To measure forward power, the tap is installed so the output of the transmitter enters the power tap on the input fiber, as shown in Figure 12.13. Reverse power is measured by turning the tap around so the output of the transmitter enters the tap on the output fiber. How the power tap is used for inline power monitoring is covered in Chapter 16.

Optical Switches

The next device we will be looking at is an optical switch. The fiber-optic switch can be a mechanical, optomechanical, or electronic device that opens or closes an optical circuit. The switch can be used to complete or break an optical path. Passive fiber-optic switches will route an optical signal without electro-optical or optoelectrical conversion. However, a passive optical switch can use an electromechanical device to physically position the switch. An optical switch can have one or more input ports and two or more output ports. Figure 12.14 shows a basic optical switch with one input port and four output ports.

FIGURE 12.14
Basic optical switch

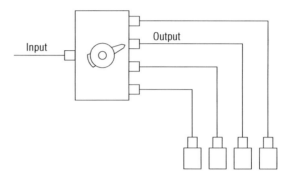

As with any other type of switch, the optical switch has many uses, depending on the complexity of the design. In essence, the switch is the control for making, breaking, or changing the connections within an optical circuit. This definition can be expanded to incorporate the concept of the switch as the control that interconnects or transfers connections from one optical circuit to another.

It is important to be aware of the basic switch parameters for an optical switch. Some of the performance parameters to consider are the number of input and output ports (required size of the switch), optical fiber type, connector type, center wavelength, bandwidth, losses, crosstalk, switching speed, durability (number of switching cycles), power handling, and repeatability (the amount of change in output power each time the switch changes state).

Optomechanical Switch

An *optomechanical* switch redirects an optical signal by moving fiber or bulk optic elements by means of mechanical devices. These types of switches are typically stepper motor driven. The stepper will move a mirror that directs the light from the input to the desired output, as shown in Figure 12.15. Although optomechanical switches are inherently slow due to the actual physical movement of the optical elements, their reliability, moderate insertion losses, and minimal crosstalk make them a widely deployed type of switch.

The optomechanical switch works on the premise that the input and output light beams are collimated and "matched" within the switching device—the beams are moved within the device to ensure the switched connection from the input to the output. The optomechanical switch can be physically larger than alternative switches, but many micromechanical fiber-optic switches are becoming available.

FIGURE 12.15
Optomechanical
switch

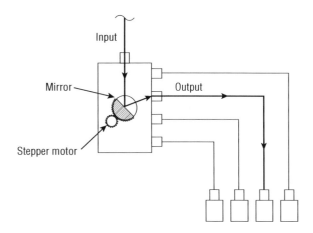

Thermo-Optic

The *thermo-optic* switch is based on waveguide theory and uses waveguides made in polymers or silica. In other words, this switch uses the thermal/refractive index properties of the device's material. The principle of this switch relies on the altering of the waveguide's refractive index due to a temperature change.

The temperature change can be accomplished in many ways, but generally the device is heated by using a resistive heater, which has the effect of slowing down light in one of the paths. The device then combines the light in the two paths in a constructive or destructive effect, making it possible to attenuate or switch the signal. This type of switch is inherently slow due to the time it takes to heat the waveguide. It's like a burner on an electric stove: it takes a while to heat up and a while to cool down.

This type of device typically has less optical loss than the optomechanical switch. Thermo-optic switches are attractive for several reasons: they work well in low optical power applications, are small in size, and have the potential to be integrated with a number of devices based on silicon wafer theory.

Electro-Optic

Electro-optic refers to a variety of phenomena that occur when an electromagnetic wave in the optical spectrum travels through a material under the stress of an electric field. An electro-optic switch is based on the changing of the refractive index of a waveguide by using an electric field. This device is semiconductor based and therefore boasts high speed and low optical power loss similar to that of the thermo-optic devices. This device is still in the research stage; however, the technology is rapidly advancing.

In summary, optical switches can be used in a variety of applications, large and small. The use of a fiber-optic switch allows data to be routed where and when it is needed.

6.

Optical Attenuators

An *optical attenuator* is a passive device that is used to reduce the power level of an optical signal. The attenuator circuit will allow a known source of power to be reduced by a predetermined factor, which is usually expressed as decibels. Optical attenuators are generally used in single-mode long-haul applications to prevent optical overload at the receiver.

Optical attenuators typically come in two forms of packaging. The bulkhead optical attenuator shown in Figure 12.16 can be plugged into the receiver receptacle. The inline attenuator resembles a patch cord and is typically used between the patch panel and the receiver.

FIGURE 12.16
Bulkhead optical
attenuator

Principles of Optical Attenuators

Optical attenuators use several principles in order to accomplish the desired power reduction. Attenuators may use the gap-loss, absorptive, or reflective technique to achieve the desired signal loss. The types of attenuators generally used are fixed, stepwise variable, and continuously variable.

GAP-LOSS PRINCIPLE

The principle of gap-loss is used in optical attenuators to reduce the optical power level by inserting the device in the fiber path using an inline configuration. Gap-loss attenuators are used to prevent the saturation of the receiver and are placed close to the transmitter. Gap-loss attenuators use a longitudinal gap between two optical fibers so that the optical signal passed from one optical fiber to another is attenuated. This principle allows the light from the transmitting optical fiber to spread out as it leaves the optical fiber. When the light gets to the receiving optical fiber, some of the light will be lost in the cladding because of gap and the spreading that has occurred. The gap-loss principle is shown in Figure 12.17.

FIGURE 12.17
Gap-loss principle
attenuator

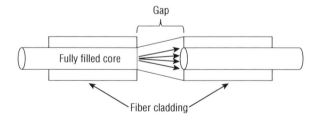

The gap-loss attenuator will only induce an accurate reduction of power when placed directly after the transmitter. These attenuators are very sensitive to modal distribution ahead of the transmitter, which is another reason for keeping the device close to the transmitter to keep the loss at the desired level. The farther away the gap-loss attenuator is placed from the transmitter, the less effective the attenuator is, and the desired loss will not be obtained. To attenuate a signal farther down the fiber path, you should employ an optical attenuator that uses absorptive or reflective techniques.

Keep in mind that the air gap will produce a Fresnel reflection, which could cause a problem for the transmitter.

USING A BULKHEAD ATTENUATOR TO TEST RECEIVER SENSITIVITY

In Chapter 10, "Fiber-Optic Light Sources and Transmitters," you learned about the output parameters of a fiber-optic transmitter, and in Chapter 11, "Fiber-Optic Detectors and Receivers," you learned about the input parameters of a fiber-optic receiver. In Chapter 16, you will learn how to apply this information to analyze the performance of a fiber-optic link. Knowing how to test the sensitivity of a fiber-optic receiver is an important skill.

A fiber-optic receiver provides optimal performance when the optical input power is within a certain range. But how do you test the receiver to see if it will provide optimal performance at the lowest optical input powers? One way is to use optical attenuators, such as bulkhead attenuators. Typically, only a couple of values are required to complete your testing. This process involves three steps:

1. Measure the optical output power of the fiber-optic transmitter with the power meter. Remember from Chapters 10 and 11 that industry standards define transmitter optical output power and receiver optical input power for a particular network standard. If you are testing a 100BaseFX receiver, you should be using a 100BaseFX transmitter. The optical output power of the transmitter should be within the range defined by the manufacturer's datasheet.

2. Connect the transmitter to the receiver and verify proper operation at the maximum optical output power that the transmitter can provide. You need to test the receiver at the minimum optical input power that the receiver can accept while still providing optimal performance. To do this, you need to obtain the lowest optical input power level value from the manufacturer's datasheet.

3. Calculate the attenuation level required for the test. Let's say that the transmitter's optical output power is –17dBm and the minimum optical power level for the receiver is –33dBm. The difference between –17dBm and –33dBm is 16dB. You would use a 16dB bulkhead attenuator at the input of the receiver and retest the receiver. If the receiver still operates properly, it's within specifications.

ABSORPTIVE PRINCIPLE

The absorptive principle, or absorption, accounts for a percentage of power loss in optical fiber. This loss is realized because of imperfections in the optical fiber that absorb optical energy and convert it to heat. (See Chapter 5, "Optical Fiber Characteristics," for a detailed discussion of the subject.) This principle can be employed in the design of an optical attenuator to insert a known reduction of power.

The absorptive principle uses the material in the optical path to absorb optical energy. The principle is simple, but it can be an effective way to reduce the power being transmitted and/or received. Figure 12.18 shows the principle of the absorption of light.

FIGURE 12.18

Absorptive principle attenuator

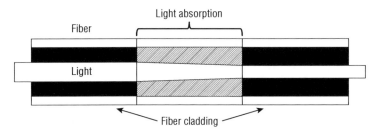

REFLECTIVE PRINCIPLE

The reflective principle, or scattering, accounts for the majority of power loss in optical fiber and again is due to imperfections in the optical fiber, which in this case cause the signal to scatter. This topic is also discussed in detail in Chapter 5. The scattered light causes interference in the optical fiber, thereby reducing the amount of transmitted and/or received light. This principle can be employed in the planned attenuation of a signal. The material used in the attenuator is manufactured to reflect a known quantity of the signal, thus allowing only the desired portion of the signal to be propagated. This principle is illustrated in Figure 12.19.

FIGURE 12.19

Reflective principle attenuator

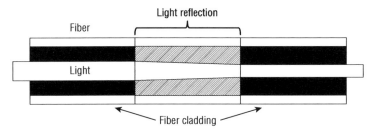

Now that we have looked at the principles behind the attenuator theories, we will discuss some of the types of attenuators. We will examine fixed, stepwise variable, and continuously variable attenuators and when they should be used.

Types of Attenuators

The types of attenuators generally used are fixed, stepwise variable, and continuously variable.

FIXED ATTENUATORS

Fixed attenuators are designed to have an unchanging level of attenuation. They can theoretically be designed to provide any amount of attenuation that is desired. The output signal is attenuated relative to the input signal. Fixed attenuators are typically used for single-mode applications.

STEPWISE VARIABLE ATTENUATORS

A stepwise variable attenuator is a device that changes the attenuation of the signal in known steps such as 0.1dB, 0.5dB, or 1dB. The stepwise attenuator may be used in applications dealing with multiple optical power sources—for example, if there are three inputs available, there may be a need to attenuate the signal at a different level for each of the inputs.

Conversely, the stepwise attenuator may also be used in situations where the input signal is steady, yet the output requirements change depending on the device that the signal is output to.

The stepwise attenuator should be used in applications where the inputs, outputs, and operational configurations are known.

CONTINUOUSLY VARIABLE ATTENUATORS

A continuously variable attenuator is an attenuator that can be changed on demand. These attenuators generally have a device in place that allows the attenuation of the signal to change as required. A continuously variable attenuator is used in uncontrolled environments where the input characteristics and/or output needs continually change. This allows the operator to adjust the attenuator to accommodate the changes required quickly and precisely without any interruption to the circuit.

Calculating the Attenuation Value

In summary, there are many types of attenuators and many principles on which they work. The key to choosing the appropriate one is to understand the theory on which each operates and the application that the attenuator will be applied to. Of course, you also need to be able to determine the attenuator value in decibels required for your application.

In this example let's assume that the maximum optical input power a fiber-optic receiver can operate with is −6dBm. If the input power exceeds this power level, the receiver will be overloaded. The transmitter, which is located 10km from the receiver, has an output power of 3dBm. The loss for the 10km of optical fiber, including interconnections, is 5dB.

To calculate the minimum attenuation required to prevent the receiver from being overloaded, we need to subtract all the known losses from the output power of the transmitter as shown here:

Transmitter power (TP) = 3dBm

Receiver maximum optical input power (MP) = −6dBm

Total losses (TL) = 5dB

Minimum attenuation required = MP + TL − TP

−6dBm + 5dB − 3dBm = −4dB

At a minimum, a 4dB attenuator is required. However, an attenuator with a larger value could be used as long as it did not overattenuate the signal. Refer to Chapter 11 to determine the dynamic range of a receiver.

7. Optical Isolator

Many laser-based transmitters and optical amplifiers use an optical isolator because the components that make up the optical circuit are not perfect. Connectors and other types of optical devices on the output of the transmitter may cause reflection, absorption, or scattering of the optical signal. These effects on the light beam may cause light energy to be reflected back at the source and interfere with source operation. To reduce the effects of the interference, you must use an *optical isolator*.

The optical isolator consists of elements that will permit only forward transmission of the light; it does not allow any reflections to return to the transmitter or amplifier. There are a variety of optical isolator types, such as polarized (dependent and independent), composite, and magnetic.

Polarized Optical Isolator

As mentioned, the polarized optical isolator transmits light in one direction only. This is accomplished by using the polarization axis of the linearly polarized light. The incident light is transformed to linearly polarized light by traveling through the first polarizer. The light then goes through a Faraday rotator; this takes the linearly polarized light and rotates the polarization 45 degrees, then the light passes through the exit polarizer. The exit polarizer is oriented at the same 45 degrees relative to the first polarizer as the Faraday rotator is. With this technique, the light is passed through the second polarizer without any attenuation. This technique allows the light to propagate forward with no changes, but any light traveling backward is extinguished entirely.

The loss of backward-traveling light occurs because when the backward light passes through the second polarizer, it is shifted again by 45 degrees. The light then passes through the rotator and again is rotated by 45 degrees in the same direction as the initial tilt. So when the light reaches the first polarizer, it is polarized at 90 degrees. And when light is polarized by 90 degrees, it will be "shut out." Figure 12.20 shows the forward-transmitted light in a dependent polarized optical isolator.

FIGURE 12.20
Forward-transmitted light through a polarized optical isolator

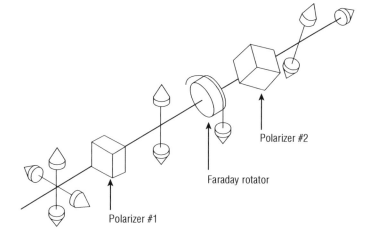

Polarizer #2

Faraday rotator

Polarizer #1

Figure 12.21 shows the backward-traveling light in a dependent polarized optical isolator.

FIGURE 12.21
Reverse-transmitted light through a polarized optical isolator

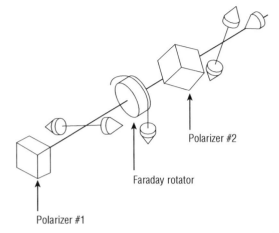

Polarizer #2

Faraday rotator

Polarizer #1

It should be noted that these figures depict the dependent type of polarized optical isolator. There is also an independent polarized optical isolator. The independent device allows all polarized light to pass through, not just the light polarized in a specific direction. The principle of operation is roughly the same as the dependent type, just slightly more complicated. The independent optical isolators are frequently used in optical fiber amplifiers.

Magnetic Optical Isolator

Magnetic optical isolators are another name for polarized isolators. The magnetic portion of any isolator is of great importance. As mentioned, there is a Faraday rotator in the optical isolators. The Faraday rotator is a rod composed of a magnetic crystal having a Faraday effect and operated in a very strong magnetic field. The Faraday rotator ensures that the polarized light is in the correctly polarized plane, thus ensuring that there will be no power loss. Figure 12.22 shows a basic magnetic optical isolator.

FIGURE 12.22
Magnetic optical isolator

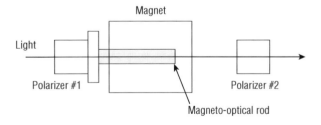

Magnet

Light

Polarizer #1

Polarizer #2

Magneto-optical rod

In summary, optical isolators are used to ensure stabilization of laser transmitters and amplifiers as well as to maintain good transmission performance.

8.

Wavelength Division Multiplexing

Wavelength division multiplexing (WDM) is the combining of different optical wavelengths into one optical fiber. This combining or coupling of the wavelengths can be very useful in increasing the bandwidth of a fiber-optic system. WDM multiplexers are used in pairs: one at the beginning of the fiber to couple the different wavelength inputs and one at the end of the fiber to decouple the wavelengths. Figure 12.23 shows a simple WDM system composed of multiple light sources, a multiplexer or combiner that combines the wavelengths into one optical fiber, and a *demultiplexer* or splitter that separates the wavelengths to their respective receivers.

FIGURE 12.23

Simple WDM system

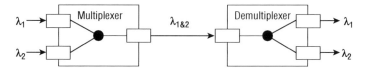

WDM multiplexer and demultiplexer classifications are defined in ITU-T G.671, Transmission Characteristics of Optical Components and Subsystems. A wide WDM (WWDM) device has channel wavelength spacing greater than or equal to 50nm. WWDM devices typically separate a channel in one conventional transmission window such as 1310nm from another such as 1550nm. These types of devices are typically referred to as wideband or crossband. Figure 12.24 shows a basic wideband or crossband WDM system.

FIGURE 12.24

Basic wideband or crossband WDM system

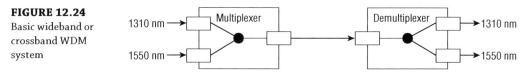

A coarse WDM (CWDM) device has channel spacing less than 50nm but greater than 1000Ghz. However, these types of devices typically operate with 20nm spacing as defined in ITU-T G.694.2, Spectral Grids for WDM Applications: CWDM: Wavelength Grid. Table 12.1 lists the nominal central wavelengths for 20nm spacing. Note that the lower and upper wavelengths defined in Table 12.1 do not represent the end of either spectrum; these wavelengths are illustrative. ITU-T G.694.2 does not define the minimum or maximum spectral wavelengths.

TABLE 12.1:   CWDM nominal central wavelengths, 20nm spacing

WAVELENGTHS IN nm
1271
1291
1311
1331

TABLE 12.1:   CWDM nominal central wavelengths, 20nm spacing   *(continued)*

WAVELENGTHS IN nm
1351
1371
1391
1411
1431
1451
1471
1491
1511
1531
1551
1531
1551
1571
1591
1611

Source: ITU-T G.694.2

A dense WDM (DWDM) device has channel spacing less than or equal to 1000GHz. These types of devices typically operate with 100GHz, 50GHz, 25GHz, or 12.5GHz spacing as defined in ITU-T G.694.1 Spectral Grids for WDM Applications: DWDM Frequency Grid. Unlike the CWDM standard, this standard uses frequency rather than wavelength to define the DWDM grid.

In Chapter 3, "Basic Principles of Light," you learned how to calculate wavelength from a given frequency and velocity using the following equation:

$$\lambda = v \div f$$

In that chapter, the speed of light in a vacuum was approximated at 300,000km/s. This approximated value is easy to remember and works very well for many applications. However, when working with DWDM frequencies you cannot approximate the speed of light in a vacuum; you must use 299,792.458km/s to ensure the correct wavelength is calculated.

The DWDM spectral grid is based on a nominal central frequency of 193.1THz. To calculate the nominal central wavelength of the DWDM spectral grid you can use the following equation:

$\lambda = 299,792.458\text{km/s} \div 193.1\text{THz}$

$\lambda = 1552.5244\text{nm}$

The nominal central wavelength of the DWDM spectral grid is 1552.5244nm. Channel spacing is anchored to this wavelength or frequency of 193.1THz. Channels above 193.1THz will have a shorter wavelength and channels below will have a longer wavelength. As with the CWDM standard, ITU-T G.694.1 does not define the lower and upper wavelengths. DWDM systems may operate in the O-, E-, S-, C-, L-, or U-bands. The nominal wavelength range of each band is defined in Table 12.2.

TABLE 12.2:   DWDM bands

BAND	NOMINAL WAVELENGTH RANGE IN (NM)
O	1260 to 1360
E	1360 to 1460
S	1460 to 1530
C	1530 to 1565
L	1565 to 1625
U	1625 to 1675

A DWDM system with a channel spacing of 1000GHz or approximately 8nm is typically referred to as a narrowband WDM system. Table 12.3 shows the wavelength and frequency data for a narrowband WDM system.

TABLE 12.3:   Narrowband WDM channel spacing

λ (NM)	F (THZ)
1531.90	195.7
1539.77	194.7
1547.72	193.7
1555.75	192.7

Figure 12.25 shows a basic narrowband WDM system.

FIGURE 12.25
Basic narrowband
WDM system

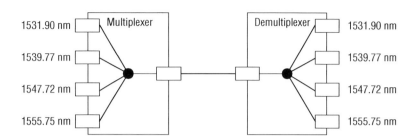

Although ITU-T G.694.1 defines 100GHz, 50GHz, 25GHz, and 12.5GHz spacing, it does not imply that multiplexers and laser transmitters are readily available to support 25GHz or 12.5GHz side-by-side channel spacing. While 50GHz multiplexers supporting up to 96 side-by-side channels were commercially available at the time this text was written, 100 GHz multiplexers were more commonly used. Table 12.4 lists the wavelengths and frequencies for 11 channels above and below the DWDM nominal central frequency of 193.1THz.

TABLE 12.4:   DWDM nominal central frequencies

NOMINAL CENTRAL FREQUENCIES (THZ) FOR SPACING OF:				APPROXIMATE NOMINAL CENTRAL WAVELENGTHS (NM)
12.5GHz	**25GHz**	**50GHz**	**100GHz**	
193.2375	—	—	—	1551.4197
193.2250	193.225	—	—	1551.5200
193.2125	—			1551.6204
193.2000	193.200	193.20	193.2	1551.7208
193.1875	—	—	—	1551.8212
193.1750	193.175	—	—	1551.9216
193.1625	—	—	—	1552.0220
193.1500	193.150	193.15	—	1552.1225
193.1375	—	—	—	1552.2229
193.1250	193.125	—	—	1552.3234
193.1125	—	—	—	1552.4239
193.1000	193.100	193.10	193.1	1552.5244
193.0875	—	—	—	1552.6249
193.0750	193.075	—	—	1552.7254

TABLE 12.4: DWDM nominal central frequencies *(continued)*

NOMINAL CENTRAL FREQUENCIES (THZ) FOR SPACING OF:				APPROXIMATE NOMINAL CENTRAL WAVELENGTHS (NM)
193.0625	—	—	—	1552.8259
193.0500	193.050	193.05	—	1552.9265
193.0375	—	—	—	1553.0270
193.0250	193.025			1553.1276
193.0125	—	—	—	1553.2282
193.0000	193.000	193.00	193.0	1553.3288
192.9875	—	—	—	1553.4294
192.9750	192.975	—	—	1553.5300
192.9625	—	—	—	1553.6307

Source: ITU-T G.694.1

As shown in Table 12.4, the closer the channels are spaced together, the higher the number of channels that can be inserted into a band. It is important to note that as the spacing or the width of each channel decreases, the smaller the spectral width becomes. This is relevant because the wavelength must be stable or sustainable long enough not to drift into an adjacent channel.

Now let's look at a different view of channel spacing. Figure 12.26 shows a four-channel narrowband DWDM spectrum using DFB laser transmitters with a spectral width of 1.0nm measured at –20dB.

FIGURE 12.26
Four-channel narrowband DWDM
spectrum

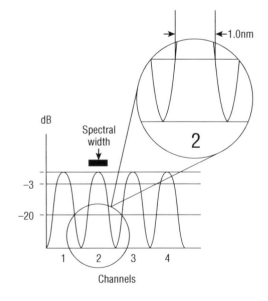

Figure 12.27 shows a 32-channel DWDM spectrum using distributed feedback (DFB) laser transmitters with a spectral width of 0.3nm measured at –20dB.

FIGURE 12.27
32-channel DWDM
spectrum

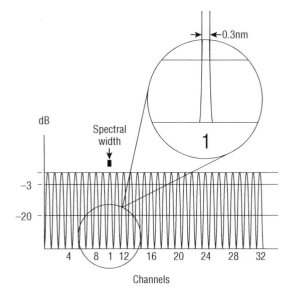

You can quickly see that as the channel spacing decreases, the laser transmitter spectral width must also decrease. To achieve 50GHz channel spacing, the laser transmitter spectral width needs to be very narrow.

Besides having a very narrow spectral width, the laser transmitter cannot drift—it must output the same wavelength at all times. If the laser transmitter's output wavelength changes even a few tenths of a nanometer, it could drift into the next channel and cause interference problems.

There are different configurations of WDM multiplexers. Everything we have covered up to this point describes a unidirectional WDM system. The unidirectional WDM multiplexers are configured so that the multiplexer connects only to optical transmitters or receivers. In other words, they allow the light to travel in only one direction and they provide only simplex communication over a single optical fiber. Therefore, full-duplex communications require two optical fibers.

A WDM multiplexer that is designed to connect with both transmitters and receivers is called bidirectional; in essence, the multiplexer is designed for optical transmission in both directions using only one optical fiber. Two channels will support one full-duplex communication link. Figure 12.28 shows two bidirectional WDM multiplexers communicating over a single optical fiber.

FIGURE 12.28
Two-channel bidi-
rectional WDM
system

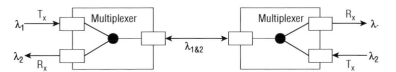

As with any other device that is added to a fiber-optic network, certain factors must be considered. As mentioned earlier in the chapter, losses are a factor that must be taken into account. When using WDM multiplexers, remember that the greater the number of channels, the greater the insertion losses. Other specifications to keep in mind when using WDM multiplexers are isolation, PMD, and the spectral bandwidth.

In summary, WDM multiplexers are widely used devices. They provide a way to use the enormous bandwidth capacity of optical fiber without the expense of using the fastest laser transmitters and receivers. Just think about it: An eight-channel WDM system using directly modulated 2.5Gbps laser transmitters carries twice as much data as a single indirectly modulated 10Gbps laser transmitter. WDM systems allow designers to combine modest performance parts and create an ultra-performance system. WDM systems deliver outstanding bang for the buck!

Optical Amplifier

As optical signals travel through an optical fiber, they are attenuated. In long-haul applications, the signal is attenuated to the point where re-amplification is required. Traditionally, a device commonly referred to as a repeater accomplished this re-amplification.

A repeater is basically a receiver and transmitter combined in one package. The receiver converts the incoming optical energy into electrical energy. The electrical output of the receiver drives the electrical input of the transmitter. The optical output of the transmitter represents an amplified version of the optical input signal plus noise.

The technology available today eliminates the need for repeaters. Optical amplifiers are now used instead of repeaters. An optical amplifier amplifies the signal directly without the need for optical-to-electric and electric-to-optical conversion. There are several different techniques with which to amplify an optical signal: erbium doped fiber amplifiers, semiconductor optical amplifiers, and Raman amplification, all of which use a technique called laser pumping.

Erbium doped fiber amplifiers (EDFAs) *Erbium doped fiber amplifiers (EDFAs)* are generally used for very long fiber links such as undersea cabling. The EDFAs use a fiber that has been treated or "doped" with erbium, and this is used as the amplification medium. The pump lasers operate at wavelengths below the wavelengths that are to be amplified. The doped fiber is energized with the laser pump. As the optical signal is passed through this doped fiber, the erbium atoms transfer their energy to the signal, thereby increasing the energy or the strength of the signal as it passes. With this technique, it is common for the signal to be up to 50 times or 17dB stronger leaving the EDFA than it was when it entered.

An example of an EDFA is shown in Figure 12.29. EDFAs may also be used in series to further increase the gain of the signal. Two EDFAs used in series may increase the input signal as much as 34dB.

FIGURE 12.29
Erbium doped fiber amplifier

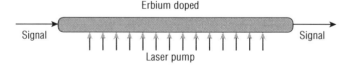

Semiconductor optical amplifiers (SOAs)   *Semiconductor optical amplifiers (SOAs)* use a technique similar to that of EDFAs but without doping the optical fiber. Unlike the EDFA, which is energized with a laser pump, the SOA is energized with electrical current. The SOAs use an optical waveguide and a direct bandgap semiconductor that is basically a Fabry-Pérot laser to inject light energy into the signal, as shown in Figure 12.30. This technique, however, does not offer the high amplification that the EDFAs do. SOAs are typically used in shorter fiber links such as *metropolitan area networks (MANs)*.

FIGURE 12.30

Semiconductor optical amplifier

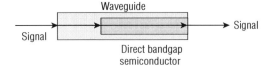

One problem with SOAs is that the gain can be hard to control. By using the semiconductor technique and a waveguide, the signal may deplete the gain of a signal at another wavelength. This can introduce crosstalk among channels by allowing the signal at one wavelength to modulate another.

Raman amplification   *Raman amplification* is a method that uses pump lasers to donate energy to the signal for amplification. However, unlike EDFAs, this technique does not use doped fiber, just a high-powered pumping laser, as shown in Figure 12.31. The laser is operated at wavelengths 60nm to 100nm below the desired wavelength of the signal. The laser signal energy and the photons of the transmitted signal are coupled, thereby increasing the signal strength.

FIGURE 12.31

Raman amplification

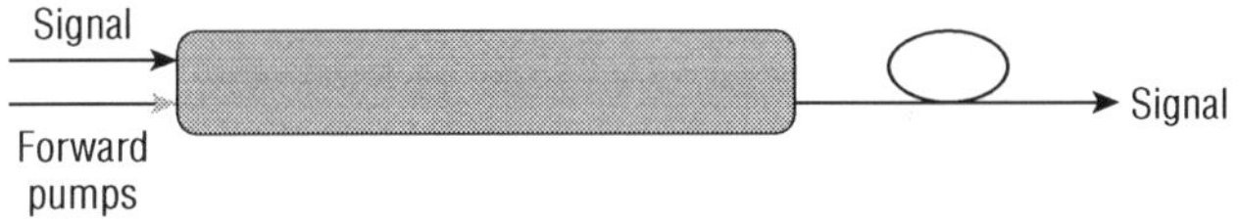

One problem with SOAs Raman amplification does not amplify as much as the EDFAs, but it does have an advantage in that it generates much less noise.

These techniques can be combined to take advantage of their amplification characteristics. In some cases, Raman and EDF amplifiers are combined in long-haul fiber links to ensure high amplification and decreased noise levels.

In summary, each amplification technique has advantages and disadvantages. Remember to keep in mind the amplification that the amplifier is being used in. For example, if a signal needed amplification but noise was an issue, a Raman amplifier would most likely be the best choice. If the signal needed to be amplified by just a small amount, the SOA might be best.

All of these amplification methods have one big advantage: optical amplifiers will amplify all signals on a fiber at the same time. Thus, it is possible to simultaneously amplify multiple wavelengths. But it is important to keep in mind that the power levels must be monitored carefully because the amplifiers can become saturated, thereby causing incorrect operation.

10.

Optical Filter

An optical filter is a device that selectively permits transmission or blocks a range of wavelengths. Optical filters are typically bandpass or band-reject.

A bandpass optical filter allows a certain range of optical wavelengths to pass and attenuates the rest, as shown in Figure 12.32.

FIGURE 12.32
Optical bandpass
filter response

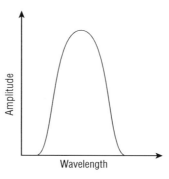

A band-reject optical filter attenuates a band of optical wavelengths and allows the others to pass, as shown in Figure 12.33.

FIGURE 12.33
Optical band-reject
filter response

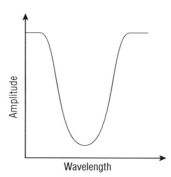

An example of a basic optical filter would be the optical filter used on a traffic light. A typical traffic light contains three optical filters: one red, one yellow, and one green. The bulb behind each optical filter is the same and emits a wide range of visible wavelengths. The optical filters allow only a certain range of wavelengths to pass, creating the red, yellow, or green light.

Bandpass optical filters are designed to transmit a specific waveband or wavelength range. A wideband optical filter may allow wavelengths plus and minus 20nm off the center frequency to pass. This type of optical filter would be used when signals are separated by several hundred nanometers, such as with a 1310nm and 1550nm source.

A narrowband optical filter allows only a very narrow range of optical energy to pass, as shown in Figure 12.34. The bandwidth of a narrowband optical filter may be less than one nanometer. The narrowband optical filter would be used in a DWDM application to reject adjacent optical channels.

FIGURE 12.34

Narrowband optical
filter response

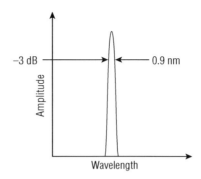

Each of these optical filter types is simple in theory yet is a vital part of some fiber-optic systems. As stated, an optical filter is a device that selects the wavelengths it will allow to pass and will reject the others.

The Bottom Line

Calculate the output power of a real tee coupler port.   Remember that manufacturers may or may not provide the insertion loss combined with the splitting ratio.

> **Master It**   The manufacturer provided the insertion loss combined with the loss from the splitting ratio for a 90:10 tee coupler with an insertion loss in decibels of 0.63/11. Calculate the power available at each output port.

Calculate the output power of a real star coupler.   Remember that unlike the tee coupler, a manufacturer typically only provides the insertion loss for the coupler and does not take into account the loss for the splitting ratio.

> **Master It**   Calculate the output power at each port of a star coupler with one input port and 10 output ports. Take into account the interconnection insertion losses. Assume the insertion loss for the coupler is 1.7dB and each interconnection insertion loss is 0.25dB. The input power to the coupler is –5dBm.

Calculate attenuator values.   Remember that if the maximum optical input power for a fiber-optic receiver is exceeded, the receiver will not operate properly.

> **Master It**   Calculate the minimum attenuation required to prevent the receiver from being overloaded. Assume the transmitter power is –3dBm, the total losses are 6dB, and the maximum optical input power the receiver can tolerate is –14dBm.

Chapter 13

Passive Optical Networks

Today we have access to more information than ever before. We live in a digital world and bandwidth is what makes a digital world happen. Aging copper networks are being taxed by residential and business customers. However, passive optical networks (PONs) such as fiber-to-the-home (FTTH) are increasingly being deployed to meet the current and future bandwidth needs that the aging copper networks cannot support.

The fastest-growing global broadband technology today is FTTH. According to ABI Research's "Broadband Subscribers" market report, FTTH will serve 19 percent of subscribers worldwide by the end of 2013, an increase of 23 percent since the end of 2012. With this type of growth, telecommunications service providers are moving quickly to maximize the number of services that can be offered to a residential or commercial customer.

As you learned in Chapter 1, "History of Fiber Optics and Broadband Access," at the turn of the century broadband speed in the United States was 200kbps and only 4.4 percent of the households in America had a broadband connection to their home. In 2013, however, the basic broadband speed was defined as 3 million bits per second (Mbps) downstream and 768 thousand bits per second (kbps) upstream. While the basic broadband speed may support only a 3Mbps download, more than 94 percent of the homes in America exceeded 10Mbps, 75 percent exceeded 50Mbps, 47 percent exceeded 100Mbps, and more than 3 percent exceeded 1Gbps or greater.

According to the "Four Years of Broadband Growth" report published in June 2013 by the United States Office of Science and Technology Policy and The National Economic Council, every broadband technology type supports the basic broadband speed of 3Mbps. This includes cable, digital subscriber line (DSL), wireless, copper, and optical fiber. However, for broadband data rates of 1Gbps or higher, optical fiber is used for 99.99 percent of the subscribers. These types of data rates are not supported by cable, DSL, or wireless, and copper is used for only one in 10,000 subscribers.

This chapter discusses the fundamentals of a PON, including fiber-to-the-home, fiber-to-the-building, fiber-to-the-curb, and fiber-to-the-node. It also introduces outside plant components, standards, active equipment, and radio frequency over fiber.

In this chapter, you will learn to:

♦ Identify the different PON configurations

♦ Identify the cables used in a PON

♦ Identify the different access points in a PON

Passive and Active Network Fundamentals

There are many types of networks carrying many different types of information. However, all these individual networks can be placed in one of two categories: passive or active. A *passive network* does not use electrically powered equipment or components (excluding the transmitter) to get the signal from one place to another. An *active network* uses electrically powered equipment or components in addition to the transmitter to get the signal from one place to another.

Passive Copper Network

There are various types of passive copper networks. However, the one virtually everyone is familiar with is their home cable TV network. In a copper cable TV network, the cable provider supplies the signal to the home over a coaxial cable.

In the most basic network, the cable enters the home and is routed to a single television. However, few homes have a single television. For homes with multiple televisions, the signal from the cable provider must be split for each television to receive the signal. The splitting is usually accomplished with an inexpensive device commonly referred to as a *splitter*. The splitter requires no electrical power. It will typically have a single input and may have two, three, four, or more outputs. Figure 13.1 is an example of a splitter that has a single input and four outputs. An individual cable is routed from the splitter to each television.

FIGURE 13.1
Splitter with one input and four outputs

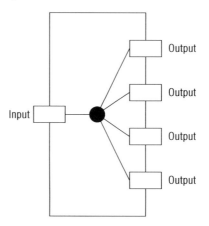

One of the problems with this type of network is the loss of signal strength. As the signal from the cable provider is split and routed to multiple televisions, the signal strength to each television is reduced. Adding too many televisions can reduce the signal strength to the point where none of the televisions receives adequate signal strength to operate properly. When this happens, it is time to look at installing an active cable TV network.

Active Copper Network

Just as there are various types of passive copper networks, there are also various types of active copper networks. The previous section focused on a passive home cable TV network and pointed out that you can connect only a limited number of televisions to this type of network.

To have adequate signal strength for multiple televisions—for example, one in each room—an active network is required.

In one example of an active home cable TV network, one cable enters the home and is routed to a distribution amplifier. The distribution amplifier boosts or amplifies and splits the signal from the cable provider. Each output of the distribution amplifier has a signal strength approximately equal to the signal strength on the input cable from the cable provider. An individual cable is routed from the distribution amplifier to each television.

This type of active network overcomes the signal strength problem associated with a passive network. However, it does add a level of complexity and requires power. If the distribution amplifier were to fail, all the televisions would lose their signals. The same would be true if the distribution amplifier were accidentally unplugged: every television in the house would be without a signal.

Passive Optical Network

There are many variations of passive optical networks (PONs). One variation is very similar to the passive cable TV network previously described. However, the coaxial cable is replaced with optical fiber. In Chapter 12, "Passive Components and Multiplexers," you learned about many different passive devices that are available to support various types of physical network topologies. Couplers are the core of any PON. A coupler may combine two or more optical signals into a single output, or the coupler may take a single optical input and distribute it to two or more separate outputs.

Figure 13.2 is an example of a seven-port coupler. The coupler is splitting a single input signal into six outputs.

FIGURE 13.2
Seven-port coupler splitting a single input into six outputs

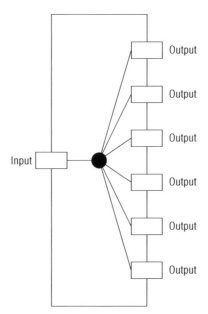

Many couplers are designed for bidirectional operation, which enables the same coupler to be used either to combine signals or to split signals. In a bidirectional coupler, therefore, each port can be either an input or an output. However, for a PON application, a coupler being used to split a signal may be referred to as a splitter.

In a PON, the input to the coupler in Figure 13.2 would typically be split equally between the six outputs. Data from the transmitter going into the coupler would be sent to each output just as the signal from the cable TV provider is sent to each TV in the passive copper network. Although each output will carry the same information as the input, the signal strength will be reduced based on the number of outputs, as shown in the ideal splitter in Figure 13.3.

FIGURE 13.3

Ideal seven-port splitter showing the reduction in signal strength at each output after the split

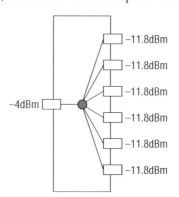

There is a finite limit on the number of outputs for a PON application; typically, the limit is 32. However, some applications may support more. Increasing the number of outputs may require an increase in transmitter output power, an increase in the sensitivity of the receiver, or a reduction in the overall length of the fiber-optic link. For most applications, transmitter output power is limited by eye safety requirements defined in federal regulations and international standards. These regulations and standards are discussed in detail in Chapter 6, "Safety."

Active Optical Network

An active optical network is very similar to the active home cable TV network previously described. One optical fiber connects to a hub instead of a distribution amplifier. The hub rebroadcasts the data to each individual user. A separate fiber-optic cable is routed from the hub to each individual user.

This type of active network overcomes the signal strength problem associated with a passive network. However, it does add a level of complexity and requires power. If the hub were to fail, all the users would lose access to incoming data. The same would be true if the hub lost power: data would stop flowing.

Fiber to the X

This section of the chapter looks at different PON configurations. These different configurations can be grouped into what is known as *Fiber to the X*, or *FTTX*. FTTX can be used to describe any optical fiber network that replaces all or part of a copper network.

FTTX is different from a traditional fiber-optic network that would be used for a local area network (LAN) application. One key difference is the number of optical fibers required for each user. In most FTTX applications, only one optical fiber is used. That single optical fiber passes data in both directions. This is very different from a LAN application, where the transmit optical fiber sends data in one direction and the receive optical fiber sends data in the other direction. In a LAN application, both optical fibers can have data passing through them at the same time.

We've discussed transceivers in several chapters in this book. A transceiver is typically a device that has two receptacles, like the 1.25Gbps, 850nm transceiver shown in Figure 13.4. One receptacle mates with the transmit optical fiber and the other mates with the receive optical fiber. This allows the transceiver to be transmitting and receiving simultaneously. This is known as *full-duplex* operation.

FIGURE 13.4
850nm 1.25Gbps transceiver with an LC transmit and a receive receptacle

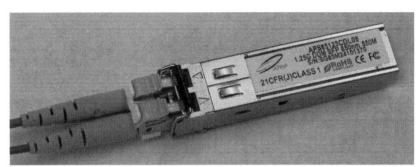

Photo courtesy of Electronic Manufacturers' Agents, Inc.

FTTX systems typically achieve full-duplex operation over a single optical fiber using multiple wavelengths. The downstream laser is always a different wavelength than the upstream laser. The downstream laser is typically the longer wavelength, such as 1490nm or 1550nm (or both), and the upstream laser is typically 1310nm. The use of different wavelengths allows each transceiver to be transmitting and receiving simultaneously just as a full-duplex system with two optical fibers.

FTTX is possible with optical fiber distances up to 20km because optical fiber is capable of transmitting information with a very low loss. The typical loss for an FTTX optical fiber at 1550nm is 0.25dB/km and 0.35dB/km at 1310nm.

Fiber to the Home

A fiber-to-the-home (FTTH) PON uses optical fiber from the central office to the home; there are no active electronics helping with the transmission of data in between the two locations. The central office is a communications switching facility. It houses a large number of complex switches that establish temporary connections between subscriber lines that terminate at the central office.

At the home, a converter box changes the optical signal from the optical fiber into electrical signals. The converter box interfaces with existing home cabling such as coaxial cabling for cable TV, twisted-pair cabling for telephone, and Category 5e or 6 cabling for Internet connectivity.

Fiber to the Building

A fiber-to-the-building (FTTB) PON is very similar to an FTTH PON. It uses optical fiber from the central office to the building and there are no electronics helping with transmission in between. The optical signal from the optical fiber is converted into electrical signals in a converter box at the building. The converter box interfaces with existing cabling such as coaxial cabling for cable TV, twisted-pair cabling for telephone, and Category 5e or 6 cabling for Internet connectivity.

Fiber to the Curb

In a fiber-to-the-curb (FTTC) PON, optical fiber runs from the central office and stops at the curb. The "curb" may be right in front of the house or some distance down the block. The converter box is located where the optical fiber stops, and it changes the optical signal from the optical fiber into electrical signals. These electrical signals are typically brought into the home through some existing copper cabling. The electrical signals may need to be processed by another converter box inside the house to interface with existing cabling such as coaxial cabling for cable TV, twisted-pair cabling for telephone, and Category 5e or 6 cabling for Internet connectivity.

Fiber to the Node

Fiber to the node (FTTN) is sometimes referred to as fiber to the neighborhood. A FTTN PON has only optical fiber from the central office to the node. The node is typically a telecommunications cabinet that serves a neighborhood or section of a neighborhood. The optical signal from the optical fiber is converted into electrical signals inside the telecommunications cabinet. These electrical signals are distributed throughout the neighborhood through existing copper cables to the houses.

Outside Plant Components

This section will discuss the major outside plant components for an FTTX PON. The outside plant components make up the PON infrastructure and are all designed for installation exterior to buildings. The cables connect different access points in the PON. Everything is initiated from the central office or central switching point.

Cables

Several different types of cables are employed in an FTTX PON, including feeder, distribution, and drop cables.

FEEDER CABLES

Feeder cables run from the central switching point to the local convergence point. These cables typically contain multiple ribbons of 12 single-mode optical fibers each. A common *feeder cable* will contain 18 ribbons for a total of 216 single-mode optical fibers.

DISTRIBUTION CABLES

Distribution cables run from the local convergence point to the network access point. Like the feeder cables, they typically contain multiple ribbons of 12 single-mode optical fibers. However, distribution cables do not contain as many optical fibers as feeder cables. A *distribution cable* can have as few as 12 optical fibers or as many as 144. A typical distribution cable has 72 optical fibers.

DROP CABLES

A *drop cable* is a single optical fiber cable that is terminated at the factory, typically with SC connectors on both ends. The cable is environmentally sealed and the connectors are sealed when they are mated. Drop cables, like the one shown in Figure 13.5, are typically available in 15' increments in lengths from 90' to 180' and in 50' increments in lengths from 200' up.

FIGURE 13.5
Single optical fiber
drop cable with SC
connector

Photo courtesy of Corning Cable Systems

Drop cables run from the network access point to the residence or building. They are designed to minimize installation cost and to provide years of trouble-free service.

Local Convergence Point

The *local convergence point (LCP)* is the access point where the feeder cables are broken out into multiple distribution cables. It is typically located in a field-rated cabinet like the 432–optical fiber cabinet shown in Figure 13.6 or the 864–optical fiber cabinet shown in Figure 13.7. A local convergence point services a neighborhood or business park.

Depending on the architecture of the PON, the local convergence point may or may not be the place where the optical signals are split. The optical signals may be split at the network access point with a splitter like the one shown in Figure 13.8. This splitter distributes the optical signal to 32 individual optical fibers. Each optical fiber is terminated with a connector for easy installation and configuration.

FIGURE 13.6
432–optical fiber
field rated local con-
vergence cabinet

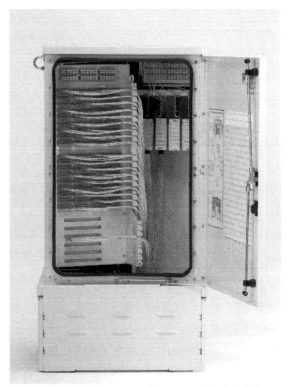

Photo courtesy of Corning Cable Systems

FIGURE 13.7
864–optical fiber
field rated local con-
vergence cabinet

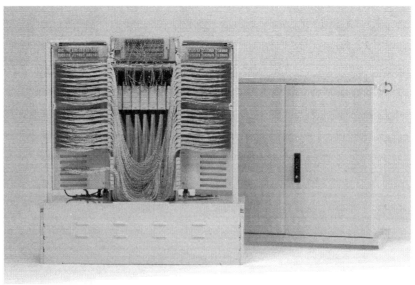

Photo courtesy of Corning Cable Systems

FIGURE 13.8
32-optical fiber
splitter

Photo courtesy of Corning Cable Systems

Network Access Point

The network access point (NAP) is located close to the homes or buildings it services. This is the point where a distribution cable is broken out into multiple drop cables. The NAP is a terminal that serves as a connection point for drop cables. It may be installed in an aerial installation, in a pedestal, or in a hand hole.

Depending on the architecture of the PON, the NAP may or may not house the optical splitter. Figure 13.9 is a photograph of a system terminal. A system terminal serves as the NAP for some FTTX installations. Figure 13.10 is a photograph of an aerial system terminal serving as a NAP.

FIGURE 13.9
System terminal
that serves as a net-
work access point

Photo courtesy of KITCO Fiber Optics

FIGURE 13.10
Aerial system ter-
minal serving as
a network access
point

Photo courtesy of Corning Cable Systems

Network Interface Device

The drop cable runs from the NAP to the network interface device (NID). The NID is typically mounted to the outside of the house or building. It is an all-plastic enclosure designed to house the electronics that support the network. The SC connector on the end of the drop cable mates with the SC connector in the NID.

The passive optical network ends at the NID. The electronics in the NID will interface with existing cabling for television, telephone, and Internet connectivity. Figure 13.11 is a photograph of an NID with the cover opened. You can see the copper and fiber-optic cabling entering and exiting the NID from the bottom.

FIGURE 13.11
Network interface
device with the
cabling entering
and exiting from
the bottom

Shrouded bolt

Twisted
pair wiring

Coaxial cable

Fiber-optic
drop cable

Photo courtesy of KITCO Fiber Optics

The NID is typically secured so the customer does not have access. As you can see in Figure 13.11, even with the cover open, a special tool is required to loosen the shrouded bolt that allows access to the interconnection for the drop cable. Only qualified service personnel should access any area inside the NID. Be aware that accessing any area of the NID may expose you to dangerous electrical or optical power levels.

In Figure 13.12 the interconnection for the drop cable inside the NID is exposed. The drop cable comes up from the bottom and secures to an adapter that is physically attached to the NID. The bidirectional transceiver is located behind the plastic cover. The single optical fiber pigtail from the transceiver routes through the opening near the adapter and mates with the SC connector at the end of the drop cable.

FIGURE 13.12
NID with the electronics section exposed

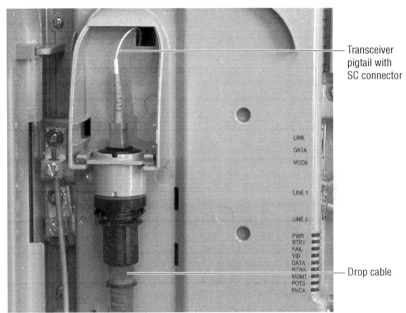

Transceiver pigtail with SC connector

Drop cable

Photo courtesy of KITCO Fiber Optics

PON Standards and Active Equipment

Throughout this book, many different standards organizations and standards have been introduced, covering virtually every aspect of a fiber-optic link. This section focuses on two standards organizations responsible for defining the characteristics of PONs capable of data rates from 100Mbps to 10Gbps. It also provides an overview of the active equipment required make these networks possible.

Globally there are many different *Internet service providers* (ISPs) or telecommunication companies providing consumers with access to the Internet. These companies may use copper, fiber optics, wireless, or a combination to deliver Internet access to the consumer. Because there are so many different providers, so many different existing infrastructures, and so many variations in geography globally, it is impossible to examine all the different PON configurations. Therefore, only one variation of a standardized 1Gbps PON will be explored.

PON Standards

The International Telecommunications Union (ITU) and the Institute of Electrical and Electronics Engineers (IEEE) have both published standards that define the characteristics of PONs capable of data rates up to 10Gbps. The ITU-T G.984 series of documents defines gigabit-capable PONs (GPONs) and the ITU-T G.987 series of documents defines 10-Gigabit-capable PONs (XG-PONs).

IEEE 802.3 specifies Ethernet LAN operation for selected speeds of operation from 1Mbps to 10Gbps. This document is broken into multiple sections. The specifications for Ethernet type are defined in separate clauses. For example, a single-fiber, bidirectional 100Mbps Ethernet is described in clause 58, 1Gbps for PON applications is described in clause 60, and 10Gbps for PON applications is defined in clause 75.

Unlike many of the standards that have been discussed in this book, most of the IEEE 802.3, ITU-T G.984, and ITU-T G.987 series of documents are available at no cost. You can access and download many of these documents at each organization's website:

www.ieee.org

www.itu.int

PON Active Equipment

As we learned earlier in the chapter, the passive optical network ends at the NID. The electronics required to interface with existing cabling for television, telephone, and Internet connectivity are located in the NID. The NID also contains the fiber-optic transceiver like the one shown in Figure 13.13. This bidirectional transceiver has a pigtail that is terminated with an APC SC connector. APC connectors are typically used for PON applications because they minimize back reflections better than PC type connectors do.

FIGURE 13.13
PON Gbps single-fiber transceiver terminated with an APC SC connector

Photo courtesy of Electronic Manufacturers' Agents, Inc.

PON Fiber-Optic Transceiver Optical Characteristics

Numerous fiber-optic transceivers have been developed for PON applications, and it is not possible to examine each one in this book. This section examines the optical characteristics of a single-fiber 1Gbps PON as defined in IEEE 802.3 clause 60 for link lengths up to 10 or 20km. In

802.3, they are referred to as 1000BASE-PX10 for link lengths up to 10km and as 1000BASE-PX20 for lengths up to 20km.

Wavelengths

Both of these PONs use two wavelengths to communicate over one single-mode optical fiber. The upstream wavelength is 1310nm, and the downstream wavelength is 1490nm. The upstream wavelength is transmitted from the NID, and the downstream wavelength is received by the NID.

The detailed label internal to the NID shown in Figure 13.14 describes how to interpret the LED indicators. Notice that the absence of the green NTWK LED indicates that the downstream 1490nm link has not been established. These indications are very helpful to the technician during the initial installation and during troubleshooting. These indicators typically display the results of built-in tests and are often referred to as BITs.

FIGURE 13.14
NID label that
describes how to
interpret the LED
indicators

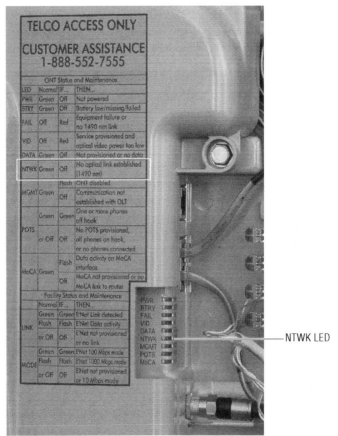

Photo courtesy of KITCO Fiber Optics

NTWK LED

Transmitter Optical Specifications

In Chapter 10, "Fiber-Optic Light Sources and Transmitters," optical characteristics are described in detail from the perspective of the manufacturer's datasheet. However, in this section optical specifications are described from the perspective of the IEEE 802.3 standard. This means that some of the parameter descriptions will vary from those used in Chapter 10.

Table 13.1 lists the 1000BASE-PX10 wavelength range and the maximum and minimum average launch power for the downstream and upstream transmitters. The 1000BASE-PX10-D column describes the downstream transmitter, and the 1000BASE-PX10-U column describes the upstream transmitter. Notice that the minimum and maximum average launch powers for the upstream transmitter are 2dB greater than the downstream transmitter.

TABLE 13.1:   1000BASE-PX10 transmit characteristics

PARAMETER DESCRIPTION	1000BASE-PX10-D	1000BASE-PX10-U	UNIT
Wavelength	1480 to 1500	1260 to 1360	nm
Average launch power (max)	+2	+4	dBm
Average launch power (min)	−3	−1	dBm

IEEE std. 802.3-2012 Section Five

Table 13.2 lists the 1000BASE-PX20 wavelength range and the maximum and minimum average launch power for the downstream and upstream transmitters. The 1000BASE-PX20-D column describes the downstream transmitter, and the 1000BASE-PX20-U column describes the upstream transmitter. Notice that the minimum and maximum average launch powers for the upstream transmitter are 3dB less than the downstream transmitter.

TABLE 13.2:   1000BASE-PX20 transmit characteristics

PARAMETER DESCRIPTION	1000BASE-PX20-D	1000BASE-PX20-U	UNIT
Wavelength	1480 to 1500	1260 to 1360	nm
Average launch power (max)	+7	+4	dBm
Average launch power (min)	+2	−1	dBm

IEEE std. 802.3-2012 Section Five

Receiver Optical Specifications

In Chapter 11, "Fiber-Optic Detectors and Receivers," optical characteristics are described in detail from the perspective of the manufacturer's datasheet. However, in this section optical specifications are described from the perspective of the IEEE 802.3 standard. This means that some of the parameter descriptions will vary from those used in Chapter 11.

Table 13.3 lists the 1000BASE-PX10 wavelength range, the maximum average receiver power, and the maximum receiver sensitivity for the downstream and upstream receivers. The 1000BASE-PX10-D column describes the downstream receiver, and the 1000BASE-PX10-U column describes the upstream receiver. Notice that the maximum receiver sensitivity is identical for each receiver. In addition, the maximum average receive power for the downstream receiver is identical to the minimum average launch power for the downstream transmitter. The same is true for the upstream receiver and transmitter.

TABLE 13.3:   1000BASE-PX10 receive characteristics

PARAMETER DESCRIPTION	1000BASE-PX10-D	1000BASE-PX10-U	UNIT
Wavelength	1260 to 1360	1480 to 1500	nm
Average receive power (max)	−1	−3	dBm
Receiver sensitivity (max)	−24	−24	dBm

IEEE std. 802.3-2012 Section Five

Table 13.4 lists the 1000BASE-PX20 wavelength range, the maximum average receiver power, and the maximum receiver sensitivity for the downstream and upstream receivers. The 1000BASE-PX20-D column describes the downstream receiver, and the 1000BASE-PX20-U column describes the upstream receiver. Notice that downstream receiver has 3dB more sensitivity than the upstream receiver does. Unlike the 10km receivers, the maximum average receive power for the downstream receiver is not identical to the minimum average launch power for the downstream transmitter. The same is true for the upstream receiver and transmitter.

TABLE 13.4:   1000BASE-PX20 receive characteristics

PARAMETER DESCRIPTION	1000BASE-PX20-D	1000BASE-PX20-U	UNIT
Wavelength	1260 to 1360	1480 to 1500	nm
Average receive power (max)	−6	−3	dBm
Receiver sensitivity (max)	−27	−24	dBm

IEEE std. 802.3-2012 Section Five

Chapter 15, "Fiber-Optic System Design Considerations," examines how the fiber-optic transceiver optical characteristics defined in IEEE 802.3 are used in the design of a network.

Radio Frequency (RF) Over Fiber

In Chapter 15, optical fiber is compared to Category 5e and coaxial copper cable in seven different performance areas. These performance areas include bandwidth, attenuation, electromagnetic immunity, size, weight, safety, and security. While optical fiber holds advantages in each of these performance areas, only bandwidth and attenuation apply to this discussion of RF optical networks.

All transmission media lose signal strength over distance. The loss of signal strength or attenuation is typically measured in decibels. Optical fiber systems measure attenuation using optical power. Copper cable systems typically use voltage drop across a defined load at various transmission frequencies to measure attenuation. The key difference here is not that optical fiber uses power and copper uses voltage. The key difference is that attenuation in copper cables is measured at different transmission frequencies. This is not the case with optical fiber, where attenuation is measured with a continuous wave light source.

The attenuation in a copper cable increases as the transmission frequency increases. Table 13.5 lists the attenuation performance of a popular coaxial cable, RG6, at different frequencies for a length of 100m. The data clearly shows the effects that frequency has on attenuation.

TABLE 13.5:   RG6 cable insertion loss*

FREQUENCY (MHz)	RG6 (dB)
1.0	1.0
5.0	1.8
10.0	2.3
20.0	3.3
50.0	5.0
100.0	6.2
200.0	9.2
300.0	11.0
400.0	12.5
500.0	14.0
1000.0	19.4
3000.0	35.1

For a length of 100m (328')

Optical fiber does not suffer the frequency-related attenuation found in coaxial cables. As you learned in Chapter 5, "Optical Fiber Characteristics," ANSI/TIA-568-C.3 does not define the bandwidth for a single-mode optical fiber, and the maximum attenuation for an outside plant OS1 or OS2 single-mode optical fiber is 0.5dB/km. This means that a 3000.0MHz transmission through an OS1 or OS2 optical fiber would be attenuated no greater than 0.05dB over a length of 100m. When compared to the 35.1dB loss over the same distance of RG6 coaxial cable, optical fiber looks like a great choice for RF applications. There are many different applications for RF over fiber. However, this chapter examines only two: wireless telecommunications and analog video.

Fiber to the Antenna

Wireless telecommunication, or the use of mobile, or cell, phones, has become a necessity of life. It represents one of the largest revenue growth services for telecommunications service providers. It also represents many challenges for service providers as 3G and 4G solutions are deployed.

Mobile phones connect to the telephone network through cell sites. Cell sites have antennas on a cell tower, like the one shown in Figure 13.15, that transmit and receive voice or data to and from the mobile user. At a typical cell site, the antennas connect to the RF transceivers through coaxial cables like the ones shown in Figure 13.16. However, at a fiber-to-the-antenna cell site the coaxial cables are replaced with fiber-optic cables similar to the drop cable shown in Figure 13.5.

FIGURE 13.15
Antennas on a cell tower

FIGURE 13.16
Typical coaxial
cable

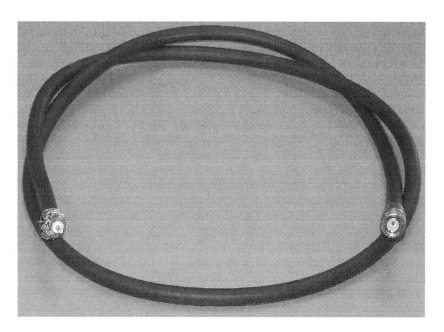

Replacing coaxial cable with fiber-optic cable requires a transmitter and receiver to change the electrical energy from the antenna into optical energy (transmitter) and the optical energy at the base of the cell tower into electrical energy (receiver), as shown in Figures 13.17 and 13.18.

FIGURE 13.17
Transmitter con-
verts the electrical
RF signal into light
energy

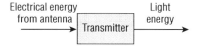

FIGURE 13.18
Receiver converts
the light energy
into an electrical RF
signal

The transmitter that converts the RF signal from the antenna does not use direct modulation as most digital transmitters do. With direct modulation the light source is turned on and off to correspond with the digital signal being input into the transmitter. However, the signal from the antenna is not a digital signal; it is an analog signal.

To transmit the analog signal from the antenna through the optical fiber, a technique known as indirect modulation is used. With indirect modulation, the laser is continuously transmitting at the same wavelength and optical power level. A modulator external to the transmitter varies the laser optical power output level to correspond with the variations in the RF signal from the antenna, as shown in Figure 13.19. This type of modulation, referred to as amplitude modulation (AM), is discussed in detail in Chapter 2, "Principles of Fiber-Optic Transmission."

FIGURE 13.19
Direct and indirect
modulation

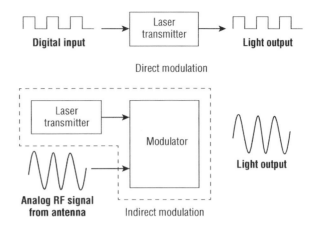

The receiver converts the AM optical signal back to an electrical signal that has the same RF characteristics as the signal from the antenna. However, this signal does not have the loss that the copper coaxial cable would have added.

This process is reversed when the cell site transmits an RF signal. The RF energy to be transmitted is converted to optical energy by the transmitter at the base of the cell tower. The optical signal is transmitted up the tower through a fiber-optic cable to the receiver. The receiver converts the optical signal to an electrical signal, which is amplified. The antenna broadcasts the amplified signal to the mobile user.

Analog Video over Fiber

The big difference between fiber to the antenna and analog video over fiber is the length of the optical fiber. Just like fiber to the antenna, a transmitter is required to turn the video signal into an AM optical signal and a receiver is required to convert it back to an electrical signal. However, the difference in the length of the optical fiber can be significant. The length of a cell tower cable could typically be described in meters whereas the length of cable required for a cable television installation could typically be described in kilometers.

Video over fiber may be employed in FTTX without adding an additional optical fiber. The IEEE 802.3 PON described earlier in this chapter could share the same optical fiber with an analog video transmitter and receiver as long as the video is transmitted at a different wavelength and a multiplexer and demultiplexer is employed to combine and separate the different wavelengths at the transmitter and receiver. This approach employs the same principles described in the "Wavelength Division Multiplexing" section of Chapter 12.

A popular video wavelength is 1550nm. There is 60nm of separation between the 1550nm video wavelength and the 1490nm downstream IEEE 802.3 PON transmission. This much separation can be supported with a wideband multiplexer and demultiplexer. The multiplexer would combine the wavelengths at the central office and the demultiplexer would separate the wavelengths inside the NID. An additional receiver installed in the NID would convert the analog optical signal into the required television RF signal.

The Bottom Line

Identify the different PON configurations. FTTX can be used to describe any optical fiber network that replaces all or part of a copper network.

Master It In this PON, the optical signals are converted right in front of the house or some distance down the block. What type of PON is this?

Identify the cables used in a PON. Several different types of cables are employed in an FTTX PON.

Master It This cable type runs from the central switching point to the local convergence point. What type of cable is it?

Identify the different access points in a PON. Cables connect the different access points in the PON.

Master It What do you call the point where a distribution cable is broken out into multiple drop cables?

Chapter 14

Cable Installation and Hardware

Under the right conditions, an optical fiber will carry light great distances almost instantly. If the cable is not installed properly, though, the light it is supposed to carry may not even travel from one part of an office building to another.

Proper installation depends on a thorough knowledge of the strengths and limits of a fiber-optic cable, as well as the methods for protecting the cable from damage both during the installation and over its lifetime. The cable must endure the pulling force that is necessary for it to be put in place; environmental conditions that threaten to freeze, soak, or otherwise damage it; and the daily stresses that result from its location or position.

The rules governing fiber-optic cable installation are designed to minimize the short-term and long-term stresses on the cable as well as ensure that the installation conforms to codes governing fire safety. Some of these requirements were discussed in Chapter 7, "Fiber-Optic Cables," with regard to the structure of the cable itself. The rules governing cable construction can only be effective, though, if the cable is installed properly.

This chapter describes the requirements for a successful fiber-optic cable installation. We discuss the conditions affecting fiber-optic cable and the methods for installing the cable so that the effects of those conditions are minimized. We describe regulations for electrical safety in cable installations as well as methods for routing cables in different situations. The chapter also examines the ways in which cables are enclosed and terminated, including proper labeling methods.

In this chapter, you will learn to:

♦ Determine the minimum cable bend radius

♦ Determine the maximum cable tensile rating

♦ Determine the fill ratio for a multiple-cable conduit installation

Installation Specifications

Many fiber-optic cables are created for specific types of service and must be installed in accordance with their manufacturer's specifications. Though some specifications are concerned with the particular duty or job a cable will perform, others apply to traits shared by all cables. Two of these specifications are *bend radius* and *tensile strength* or *pull strength*. Both specifications apply to conditions faced by fiber-optic cable while it is being installed and, once it has been installed, to its normal working conditions.

Two standards that apply to cable installation are ANSI/TIA-568-C.0 and ANSI/TIA-568-C.3. Section 5 of ANSI/TIA-568-C.0 addresses copper and optical fiber cabling installation requirements. The installation requirements for optical fiber cabling are described in Section 5.4; minimum bend radius and maximum pulling tension requirements are defined in Section 5.4.1.

Section 4 of ANSI/TIA-568-C.3 defines performance standards for optical fiber cables recognized in premises cabling standards. The physical requirements for optical fiber cables are defined in Section 4.3. Four cable types are covered in this section and in Section 5.4.1 of ANSI/TIA-568-C.0: *inside plant, indoor-outdoor, outside plant,* and *drop cable.*

Inside plant cables   Designed for installation in the building interior.

Indoor-outdoor cables   Designed for installation in the building interior or exterior to the building. These cables are not designed for long-haul outdoor applications. They are typically used to provide connectivity between two buildings. Unlike an outside plant cable, these cables resist the spread of fire as defined in Article 770 of the National Electrical Code (NEC).

Outside plant cables   Designed for outdoor installations. These cables typically do not resist the spread of fire as defined in Article 770 of the NEC. Because of this, the length of outside plant cable within a building cannot exceed 50' (15.2m) from the point of entrance into the building. To comply with Article 770 of the NEC, these cables must also be terminated in an enclosure. This is discussed in detail in the section "Fire Resistance and Grounding" later in this chapter.

Drop cables   Designed to link a drop terminal to a premise terminal. A terminal is a device that is capable of sending and/or receiving information over a communications channel that in this case uses an optical fiber. The service provider typically owns the drop terminal.

Bend Radius

As you have seen in previous chapters, optical fiber depends on the maintenance of total internal reflection to carry an optical signal. Macrobends, or very small radius bends in the optical fiber, can change the light's angle of incidence enough to cause some or a majority of the light to pass into the cladding, severely attenuating the signal or cutting it off completely.

Manufacturers often specify a minimum installation and operational bend radius for their optical fiber cables. The minimum installation bend radius is the short-term bend radius, and the minimum operational bend radius is the long-term bend radius. The minimum short-term bend radius is actually the larger of the two because of the tensile stresses that may be placed on the cable during installation.

Following the manufacturer's guidelines reduces the risk of damage to the cable and optical fiber during and after installation. It also reduces the risk of macrobends. Keep in mind that bends approaching the minimum bend radius might cause some attenuation. However, how much attenuation occurs depends on multiple factors, including cable construction, optical fiber type, and the wavelength of the light source.

Another, more fundamental reason for the minimum bend radius is that the optical fibers, though flexible, are not indestructible. Bending beyond the minimum radius may cause the fiber to break inside the cable, which would require repair at the very least and replacement of the entire cable as a worst-case scenario. In addition, overbending may cause the optical fiber to fail days, months, or years after the installation. It is impossible to predict when an overbending failure may occur. Environmental factors such as temperature, shock, and vibration stress an optical fiber and typically the greater the stress the sooner the optical fiber will fail.

Both ANSI/TIA-568-C.0 and ANSI/TIA-568-C.3 define the bend radius requirements for inside plant, indoor-outdoor, outside plant, and drop cables.

Inside plant cables are broken into three groups:

♦ Cables with four or fewer optical fibers intended for Cabling Subsystem 1

♦ Cables with four or fewer fibers intended to be pulled through *pathways* during installation

♦ All other inside plant cables

Inside plant cables intended for Cabling Subsystem 1 run from an *equipment outlet* to a *distributor* in the *hierarchical star topology*. This distributor may be an optional connection facility, intermediate connection facility, or central connection facility. These cables must support a bend radius of 25mm (1") when not subject to a tensile load. In other words, 25mm is the minimum bend radius for this cable type.

Inside plant cables with four or fewer fibers intended to be pulled through pathways during installation must support a bend radius of 50mm (2") while under a pull load of 220N (50 pound-force [lbf]). For all other inside plant cables, the bend radius is based on the cable diameter and the loading conditions. Under no tensile load or pull load, these cables must support a bend radius of 10 times the cable's outside diameter. At the maximum tensile load rating or pull load defined by the manufacturer, these cables must support a bend radius of 20 times the cable's outside diameter.

Indoor-outdoor cables, outside plant cables, and drop cables under no tensile load or pull load must support a bend radius of 10 times the cable's outside diameter. At the maximum tensile load rating or pull load defined by the manufacturer, these cables must support a bend radius of 20 times the cable's outside diameter.

NOTE   While ANSI/TIA-568-C.0 and ANSI/TIA-568-C.3 provide bend radius requirements, not all cables comply with these requirements. You should always refer to the manufacturer's datasheet for the bend radius and pull load performance for each cable being installed.

Tensile Rating

Recall that there are different types of strength members in fiber-optic cables. The job of the strength member is to ensure that no tensile stress is placed on the optical fiber during and after installation. The strength member does have physical limitations. Tensile forces that exceed the physical limitations the cable was designed to handle can damage the cable and possibly break the optical fiber. Excessive tensile loading may also create macrobends, causing the optical fiber to attenuate a signal.

There are two types of tensile loads:

♦ A *static load* is a tensile load on a cable that does not change. A static load is often referred to as the *operational load*.

♦ A *dynamic load* is a changing tensile load. Dynamic load is referred to as the *installation load*.

ANSI/TIA-568-C.3 Section 4.3 defines the minimum pull strength requirements for inside plant, indoor-outdoor, outside plant, and drop cables. *Pull load*, *pull strength*, and *tensile load* are all the same type of load on a cable, and the terms may be interchanged.

Inside plant cables As you learned in the section "Bend Radius," earlier in the chapter, inside plant cables are broken into three groups. The minimum pull load requirements only apply to cables with four or fewer fibers intended to be pulled through pathways during installation. The minimum pull load for these cables is 220N (50lbf).

Indoor-outdoor cables Indoor-outdoor cables with 12 or fewer fibers must have a minimum pull strength or tensile load rating of 1,335N (300lbf). Cables with more than 12 fibers must have a minimum tensile load rating or pull strength of 2,670N (600lbf).

Outside plant cables The minimum tensile load rating or pull strength for an outside plant cable is not dependent on the fiber count. All outside plant cables must have a minimum pull strength of 2,670N (600lbf).

Drop cables The minimum pull strength for a drop cable depends on how the cable is installed. A drop cable that is directly buried, placed by trenching, or blown into a duct must have a minimum pull strength 440N (100lbf). If the cable is installed by pulling, the minimum pull strength is 1,335N (300lbf).

When installing any cable type, always refer to the manufacturer's datasheet for physical requirement specifications. Some cables may or may not meet the minimum physical requirements defined in ANSI/TIA-568-C.0 and C.3. Other cables may exceed these requirements. The maximum tensile load and minimum bend radius specifications for all the optical fiber cable types defined in ANSI/TIA-568-C.0 and C.3 are summarized in Table 14.1.

TABLE 14.1: ANSI/TIA-568-C.0 and C.3 maximum tensile load and minimum bend radius summary

CABLE TYPE	MAXIMUM LOAD: SHORT TERM	MINIMUM BEND RADIUS: SHORT TERM	MINIMUM BEND RADIUS: LONG TERM
Inside plant cable with four or fewer optical fibers intended for Cabling Subsystem 1	50lbf 220N	2" 50mm	1" 25mm
Inside plant cable with four or fewer fibers intended to be pulled through pathways during installation	50lbf 220N	2" 50mm	1" 25mm
All other inside plant cables	Per manufacturer	20 times the cable's outside diameter	10 times the cable's outside diameter
Indoor-outdoor cables with 12 or fewer fibers	300lbf 1335N	20 times the cable's outside diameter	10 times the cable's outside diameter
Indoor-outdoor cables with than 12 fibers	600lbf 2670N	20 times the cable's outside diameter	10 times the cable's outside diameter

TABLE 14.1:   ANSI/TIA-568-C.0 and C.3 maximum tensile load and minimum bend radius summary *(continued)*

CABLE TYPE	MAXIMUM LOAD: SHORT TERM	MINIMUM BEND RADIUS: SHORT TERM	MINIMUM BEND RADIUS: LONG TERM
Outside plant cables	600lbf 2670N	20 times the cable's outside diameter	10 times the cable's outside diameter
Drop cable installed by pulling	300lbf 1335N	20 times the cable's outside diameter	10 times the cable's outside diameter
Drop cable that is directly buried, placed by trenching, or blown into a duct	100lbf 440N	20 times the cable's outside diameter	10 times the cable's outside diameter

Let's examine the physical requirements specification for two fiber-optic cables. Table 14.2 lists the tensile load and bend radius specifications from a manufacturer's datasheet for two fiber-optic cables. This table defines the minimum bend radius and the maximum tensile load or pull load for each cable. These values are defined for short- and long-term operation.

The cables described in Table 14.2 are inside plant cables. The single-fiber cable is shown in Figure 14.1.

FIGURE 14.1
Single fiber cable
with ST connectors

TABLE 14.2:   Inside plant cable physical requirements specification

CABLE TYPE	DIAMETER	WEIGHT	MINIMUM BEND RADIUS: SHORT TERM	MINIMUM BEND RADIUS: LONG TERM	MAXIMUM LOAD: SHORT TERM	MAXIMUM LOAD: LONG TERM
Single-fiber cable	0.114", 2.9mm	6lb/kft, 9kg/km	1.8",4.5cm	1.2",3.0cm	50lbf.,220N	15lbf.,66N
12-fiber cable	0.26", 6.6mm	27lb/kft, 40kg/km	3.9",9.9cm	2.6",6.6cm	300lbf.,1,334N	90lbf.,396N

The 12-fiber cable is shown in Figure 14.2. Both cables use a tight-buffered fiber surrounded by aramid yarn strength members, with a flexible jacket.

FIGURE 14.2
12-fiber cable

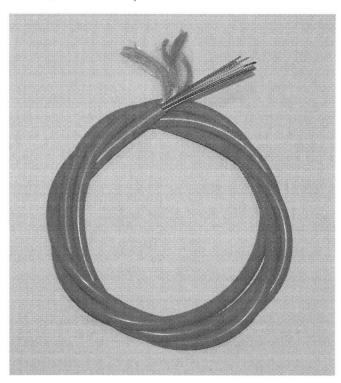

Note that the weight of the each cable over distance is included in Table 14.2. The reason for including the weight is that it can add to the tensile load if the fiber is to be suspended or hung vertically for long distances. In the single-fiber cable, for example, the long-term load limit is 15 pounds, while its weight is 6 pounds per thousand feet. If the cable must support itself for a length of 500', it already has 3 pounds of tensile loading on it.

NOTE   When you're working with cable diameter and bend radius data, it is easy to mistake bend diameter for bend radius. Remember bend diameter is twice the bend radius. If a cable has a diameter of 1″ and the bend radius is 10 times the diameter, the bend radius is 10″ and the bend diameter is 20″. Confusing the two may lead to overbending the cable.

Note also that the manufacturers' specified bend radius for installation of the 12-fiber cable is about 15 times the cable diameter, exceeding the ANSI/TIA-568-C.0 and C.3 minimum requirements. The long-term bend radius is about 10 times the cable diameter, meeting the ANSI/TIA-568-C.0 and C.3 minimum requirements. The installation tensile load is about three times as much as the long-term tensile load. Remember that the figures used in this example are for a cable that is installed inside a building. Cables that will be installed outside of a building will typically have a larger minimum bend radius.

Installation Hardware

Fiber-optic cables and the optical fibers within them have a number of specialized requirements for their installation and termination. From cable-pulling tools to cable-protecting enclosures, installation hardware has been specially designed and built to meet the needs of fiber-optic cables in almost any environment and situation.

Let's take a look at some of the hardware commonly used in fiber-optic installation.

Pulling Eye

Sooner or later, you will need to run a cable through a wall, conduit, or other inaccessible space. An indispensable tool for this job is the pulling eye, shown in Figure 14.3. This device is specially designed to attach to the cable's strength member at one end and a pulling line at the other. The pulling line is fed through the space to be occupied by the cable. The line is then used to pull the eye through the space with the cable attached.

FIGURE 14.3
The pulling eye is used to pull cable through the conduit.

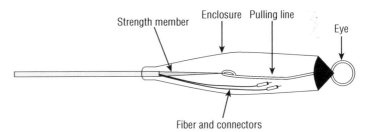

The pulling eye also uses a sheath that encloses the fiber ends to protect them from damage while the cable is being pulled.

Pull Box

Optical fiber is typically small enough, light enough, and flexible enough to be relatively easy to pull through conduits. However, friction is always a concern. Friction on the cable as it is pulled through the conduit will increase the tensile loading. Increases in friction as the cable is pulled through a turn or several turns can cause the tensile load to exceed its maximum.

To make the cable easier to pull and to ease the tensile load on it, pull boxes are installed at intervals in the conduit. Typically, pull boxes are installed after long straight runs and every time a set of turns totals 180° or more. The purpose of a pull box is to create an intermediate opening for pulling the cable to reduce the length that is being pulled through the conduit and to reduce the number of turns through which the cable must be pulled at any one time.

To use a pull box, pull the fiber-optic cable through the box and out the large opening. Be very conscious of how much the cable is being bent while you are doing this. Also be careful not to place a bend in the cable that has a radius less than the minimum bend radius defined by the manufacturer. Once the cable has been pulled as far as it will go, feed it into the other side of the box and down the conduit, again paying close attention to the bend radius. This process minimizes the stress placed on the cable.

There are two types of pull boxes, as shown in Figure 14.4: straight and corner.

FIGURE 14.4
Straight and corner pull boxes ease tensile loading on fiber during installation.

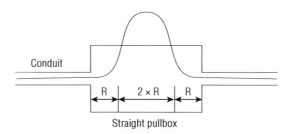

Straight pullbox

R = Manufacturer's specified minimum bend radius

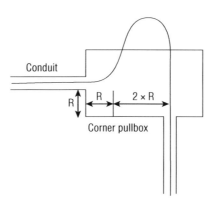

Corner pullbox

Straight pull box   Straight pull boxes are installed in-line in the conduit and should have an opening at least four times the minimum bend radius of the cable being pulled. This length prevents the cable from exceeding its minimum bend radius as the last of it is pulled through the box.

Corner pull box   Corner pull boxes are installed at angles in the conduit and typically require a length of three times the cable's bending radius and a depth that equals or exceeds the bending radius. This requirement prevents the cable from dragging against a sharp turn when it is pulled through.

Splice Enclosures

Any time you have a splice in an optical fiber, whether it is a mechanical or a fusion splice, you must protect it from exposure and strain. *Splice enclosures* take many forms, depending on their location and specific application. Some have been adapted from electrical splice enclosures used in the telecommunications industry for aerial and underground cable, whereas others are designed specifically for optical fibers and are used for indoor installations.

Splice enclosures can typically be placed into two categories:

Radial splice enclosure   With a radial splice enclosure, cables enter and exit the enclosure through the same side, as shown in Figure 14.5.

Axial splice enclosure   With an axial splice enclosure, cables enter and exit the enclosure at opposite ends, as shown in Figure 14.6. Both types can be used for an aerial or direct burial installation. However, the radial is typically used with a pedestal enclosure.

The *pedestal enclosure* shown in Figure 14.7 is used when the fiber-optic cable has been buried underground. The pedestal enclosure is placed over the cables as they enter or exit the ground.

FIGURE 14.5
Radial splice
enclosure

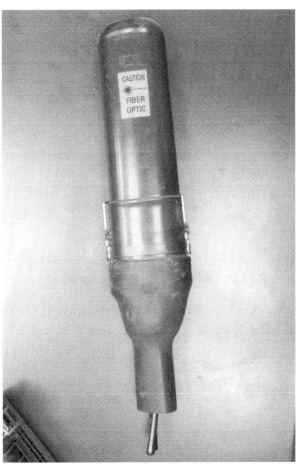

Photo courtesy of KITCO Fiber Optics

FIGURE 14.6
Axial splice
enclosure

Photo courtesy of KITCO Fiber Optics

FIGURE 14.7
Fiber-optic pedestal

Photo courtesy of KITCO Fiber Optics

Splice enclosures designed for outdoor applications like those shown in Figures 14.5 and 14.6 are environmentally sealed. However, indoor splice enclosures may or may not be environmentally sealed. The rack-mounted splice enclosure shown in Figure 14.8 is for indoor applications and is not environmentally sealed.

FIGURE 14.8
Rack-mounted
splice enclosure

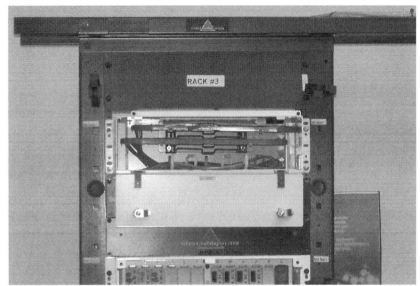

Photo courtesy of KITCO Fiber Optics

Typically, splice enclosures will incorporate the following features:

♦ A strain-relief system that ensures the strength member will carry all of the tensile loading.

♦ Clips incorporated into a panel or tray, or bonded to the enclosure. The clips hold the actual splices in an orderly fashion.

♦ Space for looping the extra optical fiber required to perform the splice outside of the enclosure.

The splice enclosure shown in Figure 14.9 has three mechanical splices and three fusion splices. The clips that hold the splices are bonded to the enclosure. The mechanical splices are held in the top three clips and the fusion splices in the bottom three. Notice that this splice enclosure allows sufficient space for the optical fiber to be bent without exceeding its minimum bend radius.

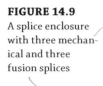

FIGURE 14.9
A splice enclosure
with three mechan-
ical and three
fusion splices

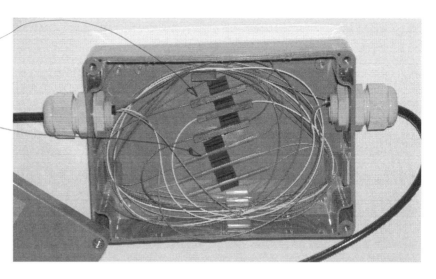

Patch Panels

A *patch panel* is an interconnection point for fiber-optic cables. They allow signals to be routed from one cable to another with a patch cord or jumper. Patch panels are available in many different shapes and sizes. Patch panels are often mounted in a rack, as shown in Figure 14.10.

FIGURE 14.10
A rack-mounted
patch panel

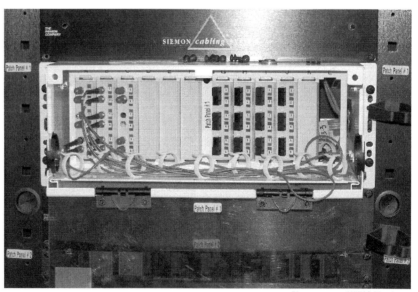

Photo courtesy of KITCO Fiber Optics

However, patch panels may be contained within an enclosure, as shown in Figure 14.11.

The patch panel contained within the wall-mounted enclosure shown in Figure 14.11 allows the interconnection of the four fiber-optic cables entering the box from the top. Signals from one cable can be routed to another simply by making a connection between the two optical fibers with a jumper. The jumpers to the right of the patch panel provide an optical path from one fiber-optic cable to another.

FIGURE 14.11
An enclosure-mounted patch panel

Photo courtesy of KITCO Fiber Optics

Patch panels use a bulkhead-mounted *mating sleeve* or adapter to make the interconnection. These mating sleeves can feature identical connector receptacles or different receptacles. Figure 14.12 shows an ST-to-ST mating sleeve with connectors on each end. Figure 14.13 shows an SC-SC mating sleeve with connectors on each end, and Figure 14.14 shows an LC-LC mating sleeve with connectors on each end.

FIGURE 14.12
ST-to-ST mating sleeve with connectors

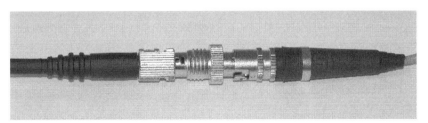

FIGURE 14.13
SC-to-SC mating sleeve with connectors

FIGURE 14.14
LC-to-LC mating sleeve with connectors

Figure 14.15 shows a mating sleeve that mates an ST connector with an SC connector. Mating sleeves with different connector receptacles are often referred to as *hybrid adapters*.

FIGURE 14.15
ST-to-SC mating sleeve or hybrid adapter with connectors

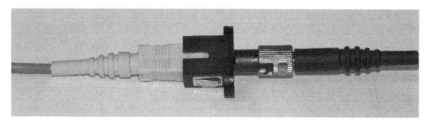

Like connectors, receptacles have dust caps to prevent contamination. Figure 14.16 shows an ST-to-SC hybrid adapter with the dust caps in place.

FIGURE 14.16
ST-to-SC hybrid adapter with dust caps installed

3. **Installation Methods**

Optical fiber has already reached into most of the places that once only knew copper cable. As technology, regulations, and pricing permit, fiber will ultimately replace copper for most signal-carrying applications, even into the home.

Many fiber installations resemble those used for copper wiring and have built on the lessons learned from it. There are some installation requirements and methods unique to fiber, however, that are required to protect the cable and ensure the highest-quality transmission.

Let's look at a typical application for optical fiber that uses a variety of installation methods. In our example, shown in Figure 14.17, a manufacturing plant is using fiber to carry instrumentation and control signals between the production building and another building several hundred meters away (called the control building in the diagram). The simplex cables carrying the signals must be collected in a central area and routed into a multi-fiber cable that runs through trays and ductwork to the outside. The cable then runs from the building and is strung across several poles until it reaches a road. There, the cable runs underground until it enters the next building and is distributed to data collection and control systems.

FIGURE 14.17
A sample fiber
installation
scenario

Let's look at installation for each of these situations.

Tray and Duct

Tray and duct installation is used inside structures and is similar to installation methods used for electrical wiring. Because many optical fiber cables are nonconductive, some of the requirements and restrictions for copper cables do not apply to fiber.

When the optical fiber cable rests in trays or horizontal ductwork, as shown in Figure 14.18, the weight of the cable is usually not a factor as long as the runs remain on the same vertical level. If optical fiber cabling is run vertically, however, the cable will have to support itself or be secured using either cable clamps or hangers. Be sure to follow the cable manufacturer's specifications for vertical cable rise.

FIGURE 14.18
Tray and duct
installation

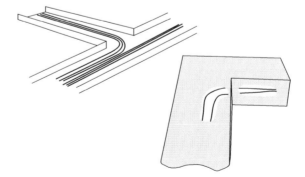

There are two methods for clamping cables for vertical runs:

♦ One type of clamp, which can also be used to secure cables horizontally, secures the cable directly to the surface by placing pressure against it. These should be installed carefully to prevent crushing the cable with excessive force.

♦ The second type of vertical clamp is a tight wire mesh that wraps around the cable and is secured to a hanger. This method has the advantage of reducing pressure against the cable itself while still supporting it. It also allows the cable to be removed more easily, since the mesh wrap is slipped over the hanger, and not permanently attached.

When installing cable in a tray, be aware of whether the tray will also be occupied by other cables, especially electrical cables. Aside from safety considerations surrounding the use of conductive or composite optical fiber cables, the weight of the much larger copper cables can cause macrobends if they are piled on top of the fiber. If there is a risk of this happening, choose an optical fiber cable that will protect optical fiber inside from the crushing weight of other cables.

If you are laying the cable into the tray, rather than pulling it, you can use the manufacturer's minimum long term bend radius specification. If you will be pulling it, however, be sure to leave some extra radius to account for the tensile loading.

Whether you are installing in a tray or in a duct, do not allow the fiber to contact any sharp edges or exceed the minimum bend radius. Ideally, you should be working with components that do not have these hazards built in; however, if these hazards exist you should keep cable tension as low as possible during installation.

Conduit

Conduit installation uses dedicated conduits for the cable runs, which may be installed by feeding pulling lines through the conduit with a fish tape, attaching the cable's strength member to a pulling eye, and then pulling the cable through the conduit.

Conduit may be run inside structures or underground, and in many cases conduits may already be in place for other applications, such as power or telephone lines. When you are installing cable in conduits within a structure, be sure that you have allowed enough room in the conduit for the cable or cables you are installing, as shown in Figure 14.19. If possible, account for future expansion by installing a cable with extra optical fibers.

FIGURE 14.19
The conduit must leave room for fiber to be pulled.

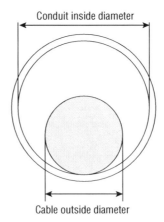

Conduit inside diameter

Cable outside diameter

When preparing for a conduit installation, you must be aware of the *fill ratio* for the conduit. If you are installing cable in existing conduit already occupied by one or more cables containing electrical conductors, you will need to determine the fill ratio of the conduit prior to selecting your cable. The NEC specifies the following maximum fill ratios by cross-sectional area for conduit:

1 cable: 53 percent

2 cables: 31 percent

3 or more cables: 40 percent

To determine the fill ratio, use the following formula:

Fill Ratio = $(OD_2 \text{cable}_1 + OD_2 \text{cable}_2 \dots)$ / $ID_2 \text{conduit}$

where OD is the outside diameter of the cable and ID is the inside diameter of the conduit.

After you have determined the fill ratio of the existing cables in the conduit, select your cable. You need to ensure that the cable you select does not cause the NEC fill ratios to be exceeded. Fortunately, optical fiber is available in many different cable types. If the existing fill ratio will only permit a fiber-optic cable with a very small cross-sectional area, use a ribbon cable.

NOTE   When you're installing fiber-optic cabling in conduit, the NEC fill ratios apply only when they are placed in conduit with electrical power cables or when the fiber-optic cable contains metallic members. A metallic member might be a wire, metal strength, or metal shield.

Let's determine the conduit fill ratio with four existing cables in place. The inside diameter of the conduit is 1.5″. The cable to be installed has an outside diameter of 0.35″. The outside diameters of the existing cables are 0.45″, 0.37″, 0.25″, and 0.55″. To determine the fill ratio of the conduit before the fiber-optic cable is installed, use the following formula:

Fill Ratio = $(OD_2 \text{cable}_1 + OD_2 \text{cable}_2 + OD_2 \text{cable}_3 + OD_2 \text{cable}_4)$ / $ID_2 \text{conduit}$

Fill Ratio = $(0.45^2 + 0.37^2 + 0.25^2 + 0.55^2)$ / 1.5^2

Fill Ratio = $(0.2025 + 0.1369 + 0.0625 + 0.3025)$ / 2.25

Fill Ratio = 0.7044 / 2.25

Fill Ratio = 0.313 or 31.3%

The next step is to determine the conduit fill ratio with the addition of the fiber-optic cable. To determine that fill ratio, use the following formula:

Fill Ratio = $(OD_2 \text{cable}_1 + OD_2 \text{cable}_2 + OD_2 \text{cable}_3 + OD_2 \text{cable}_4 + OD_2 \text{cable}_5)$ / $ID_2 \text{conduit}$

Fill Ratio = $(0.45^2 + 0.37^2 + 0.25^2 + 0.55^2 + 0.35^2)$ / 1.5^2

Fill Ratio = $(0.2025 + 0.1369 + 0.0625 + 0.3025 + 0.1225)$ / 2.25

Fill Ratio = 0.8269 / 2.25

Fill Ratio = 0.3675 or 36.75%

The calculated fill ratio is less than 40 percent, so the conduit is large enough as defined by the NEC.

If a large conduit is already in place and contains other cables or if it is likely that other cables will be run through the conduit in the future, you may want to consider installing an *innerduct*, a conduit sized for the optical fiber cable that will protect it from other activities in the larger conduit.

Direct Burial

To run an outdoor fiber-optic cable out of sight, we can install it by *direct burial*. As the name implies, this method can be as simple as placing a suitable cable directly in the ground. Direct burial methods also include placing a cable within a protective pipe or conduit and burying it.

Direct burial has some advantages. Cables that are buried underground are not visible and do not obstruct the scenery like cables that span telephone poles or buildings. Burying a cable keeps it out of the wind, rain, snow, and ice. The wind from a tornado or hurricane will not damage a buried cable. Unlike aerial cables, a buried cable will not break from the excessive weight of ice from an ice storm.

However, direct burial of fiber-optic cables also has some disadvantages. Fiber-optic cables buried underground are difficult to locate since they do not emit any electromagnetic energy the way a copper cable such as a power line does. Animals that burrow can damage cables buried underground. A direct burial cable typically features an armor layer that protects it from burrowing animals and damage that may be caused when rock is used to cover the cable as is placed in the ground.

To place a direct burial cable in the ground, a backhoe or shovel can be used to dig a trench in which the cable is laid and then covered up. This approach is typically used for short runs. For longer runs that require more efficient methods, a cable-laying plow is available. This device is designed to open a trench, lay the cable, and cover it up again while on the move. It is a more complicated machine, but it is useful when longer distances must be covered.

When using direct burial methods, be sure to dig the trench deep enough to be below the frost line. In some areas, this can be as much as 30″ deep.

Remember that you must contact the local utilities before you begin digging or trenching operations. The location of existing underground cables must be known before you dig.

Aerial

As our cable leaves the production building, it's going to be spliced to a cable with a messenger wire, which will be strung along a series of poles in an aerial installation. These poles already carry power lines, but they will not affect the data on the optical fiber itself because it is an insulator.

If a messenger wire is not incorporated into the cable assembly used, the cable will have to be lashed to messenger wire. A messenger wire is a steel or aluminum wire running between the poles to support the fiber-optic cable. Either way, cables in aerial installations must be able to withstand loading from high winds, ice, birds and climbing animals, and even windblown debris such as branches.

Blown Fiber

Blown fiber is an alternative to installing cables. In new construction or renovation, it is a good idea to consider blown fiber as an alternative to traditional methods.

The three steps to installing blown fiber are as follows:

1. Install the special tubing or conduit.

2. Blow the optical fiber through the tubes from location to location.

3. Terminate the optical fiber.

The optical fiber used in a blown fiber installation is the same optical fiber found in fiber-optic cables. However, the strand or strands of optical fiber will be coated with a protective layer that creates friction as dry gas passes over it. The friction generated by the dry gas as it flows through the tubing pulls the optical fiber or optical fiber bundle through the tubing. During this process no tensile stress is placed on the optical fiber.

The tubing is about 5mm in diameter. Typically, a jacket is applied around two or more tubes, forming what looks like a cable. This cable-like bundle of tubes is installed without the optical fiber. The optical fiber is blown through these tubes after the installation. Figure 14.20 shows two different bundles of tubes and a single tube that is spliced in the center.

FIGURE 14.20
Blown fiber tubing
assemblies

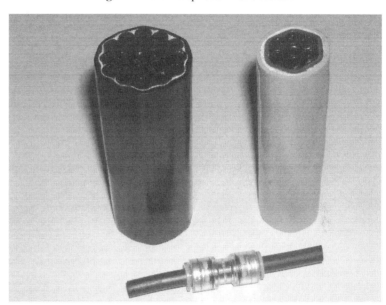

Blown optical fiber is used in buildings, between buildings, and on board ships. The U.S. Navy uses blown optical fiber on aircraft carriers. The life of a ship can exceed 50 years. Removing and installing cables on board a ship is expensive, and there is always the risk that this process may cause damage to other cables or equipment.

The tubing installed for blown optical fiber never needs to be removed because old optical fiber can be pulled out and new fiber blown in. The process of pulling out the old or damaged fiber and blowing in new fiber takes very little time in comparison to the time it would take to remove and reinstall a new cable. Figure 14.21 is a photograph of shipboard enclosure where blown optical tubes are interconnected with a splice.

FIGURE 14.21
Interconnected
blown fiber tubes

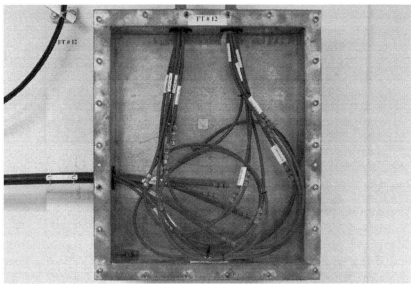

Photo courtesy of KITCO Fiber Optics

Figure 14.22 is a photograph of a shipboard enclosure where the optical fibers in the tubes are terminated.

FIGURE 14.22
Terminated blown
optical fibers

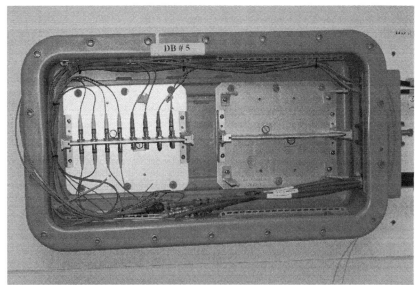

Photo courtesy of KITCO Fiber Optics

LEAVING SLACK FOR REPAIRS

No matter where you install cable, you'll want to leave some slack for the inevitable repairs that will have to be performed. How much slack should you leave? That depends on the situation. Look at the location; the number of turns that the fiber takes as it passes through conduits, trays, or ductwork; and the likelihood that the fiber will be disturbed after you have installed it. Consider the following cautionary tale.

During the construction of a strip mall in Virginia, cable installers ran cable underground in an area where construction was ongoing. After the cable was laid, backhoe operators accidentally pulled up the conduit carrying the cable seven times, requiring the cable to be spliced and re-laid in the ground. After the sixth event, all of the slack in the cable had been used up in repairs and the cable had to be replaced.

For the record, the location of the cable was clearly marked.

Cable Slack

How much slack an installation should have and where the slack should be stored depends on the installation. However, there are some basic guidelines for slack.

When a cable is brought into a building from outside and the cable is not a listed indoor-outdoor type cable, the slack should be outside the building. Typically, 10 meters of slack is sufficient for most installations.

When a cable is brought into a building from outside and the cable is a listed indoor-outdoor type cable, the slack should be inside the building. Typically, 10 meters of slack is sufficient for most installations.

Cables that span the inside of a building between cross-connects should have slack on each end of the cable. As with the other installations, 10 meters of slack is typically sufficient.

All cable slack should be stored in the form of a loop. For an indoor installation, it should be either hung from the wall or placed under a subfloor. For an outdoor installation, it should be placed in the closest suitable outdoor access point.

Fire Resistance and Grounding

As you learned in Chapter 7, Article 770 of the NEC places requirements on fiber-optic cables and their installation within buildings. One requirement is that the cable must be able to resist the spread of fire. This requirement applies to all fiber-optic cables installed in a building. Other requirements address the metallic members in some cable types.

This section will address cable selection to meet fire resistance requirements defined in Article 770 of the NEC. It will also address the installation of fiber-optic cables that contain metallic members.

NOTE This chapter provides an overview of Article 770 and is not a substitute for the NEC. The National Fire Protection Association (NFPA) updates the NEC every three years. The person performing the installation is responsible for compliance with the current version of the NEC. In addition, the installer needs to be aware of local, state, and federal codes that apply to the installation.

Fire Resistance

Article 770 of the NEC states that optical fiber cables installed within buildings shall be listed as being resistant to the spread of fire. These cables are also required to be marked in accordance with Table 14.3.

TABLE 14.3: NEC cable markings, types, locations, and permitted substitutions

MARKING	TYPE	LOCATION	PERMITTED SUBSTITUTIONS
OFNP	Nonconductive optical-fiber plenum cable	Ducts, plenums, other air spaces	None
OFCP	Conductive optical-fiber plenum cable	Ducts, plenums, other air spaces	OFNP
OFNR	Nonconductive optical-fiber riser cable	Risers, vertical runs	OFNP
OFCR	Conductive optical-fiber riser cable	Risers, vertical runs	OFNP, OFCP, OFNR
OFNG	Nonconductive optical-fiber general-purpose cable	General-purpose use except for risers and plenums	OFNP, OFNR
OFCG	Conductive optical-fiber general-purpose cable	General-purpose use except for risers and plenums	OFNP, OFCP, OFNR, OFCR, OFNG, OFN
OFN	Nonconductive optical-fiber general-purpose cable	General-purpose use except for risers, plenums, and spaces used for environmental air	OFNP, OFNR
OFC	Conductive optical-fiber general-purpose cable	General-purpose use except for risers, plenums, and spaces used for environmental air	OFNP, OFCP, OFNR, OFCR, OFNG, OFN

There are two exceptions to this:

♦ Unlisted conductive and nonconductive outside plant optical fiber cables may be installed in building spaces, other than risers, ducts used for environmental air, plenums used for environmental air, and other spaces used for environmental air, as long as the length of the cable within the building, measured from its point of entrance, does not exceed 15 meters (50′) and is terminated in an enclosure.

♦ A nonconductive outside plant optical fiber cable is not required to be listed or marked when the cable enters the building from the outside and is run in any of the following raceways:

  ♦ Intermediate metal conduit (IMC)

  ♦ Rigid metal conduit (RMC)

  ♦ Rigid polyvinyl chloride conduit (PVC)

  ♦ Electrical metallic tubing (EMT)

Plenum cable   Plenum cables, whether conductive or nonconductive, are suitable for use in ducts, plenums, and other space used for environmental air. These cables will have fire resistance and low smoke-producing characteristics.

Riser cable   Riser cables, whether conductive or nonconductive, are suitable for a vertical run in a shaft or from floor to floor. These cables will have fire resistance characteristics capable of preventing the carrying of a fire from floor to floor.

General-purpose cable   General-purpose cables, whether conductive or nonconductive, are resistant to the spread of fire. However, these cables are not suitable for plenum or riser applications.

Grounding

Even though optical fiber cable itself does not carry electrical power, there may be some circumstances in which you will have to contend with electricity. If a cable contains conductive components such as armor or metal strength members and the cable is likely to come in contact with electrical circuits, there is a chance that, in an accident, the metal could become a path for an electrical current, potentially leading to fire or personal injury.

In the event that a conductive component in a cable comes in contact with an electrical current, that current is going to seek the path of least resistance. If you happen to touch the cable or anything to which the cable is connected, that path could be through you. Depending on the voltage and current involved, you could face severe injury or even death.

For this reason, Article 770 places a grounding or isolation requirement on any fiber-optic cable entering a building that contains electrically conductive materials and that may be exposed to or come in contact with electrical light or power conductors. The metallic member of the cable must be grounded as close to the point of entrance as practicable. An alternative to grounding is placing an insulating joint or equivalent device as close to the point of entrance as practicable. The insulating joint interrupts the flow of current through the conductive materials in the cable.

Cable Types

Article 770 of the NEC groups fiber-optic cables into three types:

♦ Nonconductive cables contain no electrically conductive materials.

♦ Conductive cables have conductive components such as metallic strength members, vapor barriers, or armor. These conductive components were not designed into the cable to conduct current and are thus referred to as *non-current-carrying conductive members*.

♦ Composite cables contain optical fibers and electrical conductors designed to carry current. These cables may also have metallic members that were not designed to carry current such as metallic strength members, vapor barriers, or armor.

Hardware Management

Eventually, fiber-optic cables end up in some kind of cabinet, panel, rack, or enclosure. Whether it's a patch panel, rack, splice enclosure, or wall outlet, the preparation of these items is crucial to the performance of the entire network. In these locations, the fiber or cable is terminated and connected in some way to another fiber, a connector, or a piece of hardware, and this process involves a risk of mistakes and mismanagement.

Poor hardware management can lead to confusion, excess strain on cables and fibers, and inefficient troubleshooting. In addition, a poorly organized hardware space reflects badly on the installer.

In this section, we will look at some guidelines for good hardware management.

Cleanliness

A clean working environment is essential to fiber work. Whether you are splicing fibers, terminating cables, or mounting cables in cabinets, it's important to keep dust, trash, water, and other contaminants away from your work area.

If possible, block off the space in which you are working to prevent exposure to contaminants. Use air filters to draw dust and grit out of the air, and try to limit the amount of non-essential work traffic in the area where you are working. Crews have been known to park their work trucks over manholes to keep dirt from the street from falling inside while they work. In other areas, it may be enough to use good housekeeping practices to ensure that the work area won't be contaminated.

Organization

Once a cable enters the telecommunications room, it is likely to be joined by anywhere from one to dozens of other cables. If these cables are not organized from the very beginning, they can quickly become a rat's nest of tangled, interwoven lines.

Many manufacturers offer cable management devices. These devices will help you organize cables for a neater, more user-friendly layout. Often the shortest distance between two points is not the best way to route a cable. Cables should only be routed vertically and horizontally. A cable should never be routed diagonally.

Secure cables in place when required. This helps relieve strain on the cable and keeps it from assuming new and interesting configurations after the cabinet is closed. Figure 14.23 shows cables routed to the top of a 19" rack. These cables are secured to minimize stress and maintain a configuration.

Organization and neatness are also essential for efficient troubleshooting. It's much easier to trace connections and links if the cables are organized and neatly bundled than if they are spread all over the cabinet and tangled together.

FIGURE 14.23
Secured cables
routed to a 19″ rack

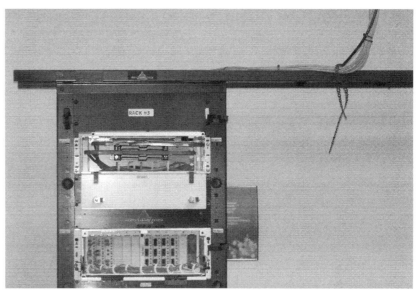

Photo courtesy of KITCO Fiber Optics

Clamps and Cable Ties

Sometimes the best-planned installation can be destroyed by a simple cable tie or clamp. Earlier in the chapter, we discussed the bend radius of a fiber-optic cable. We learned that exceeding the bend radius can cause attenuation or damage the cable or optical fiber. A cable tie or clamp, if improperly applied, can place a sharp bend in a cable. This sharp bend can produce a bend radius small enough to cause attenuation or damage.

When you are installing a cable tie or clamp, be very conscious of the pressure that is being placed on the cable. Look closely at the jacket of the cable to see if you are compressing or crushing it. You do not have to be very strong to damage a cable with a clamp or cable tie. Clamps and cable ties should be installed so they do not place unnecessary pressure on the cable jacket.

It is not always easy to tell that you have placed too much pressure on the jacket with a clamp or cable tie. One way to avoid problems is to test a piece of cable with a light source on one end of the cable and a power meter on the other. Apply the clamp or cable tie to the center of the cable under test and watch the power meter to see if there are any changes in power. If the power meter shows a reduction in power as you apply the clamp or cable tie, the pressure is too great. Testing with a light source and power meter is covered in detail in Chapter 16, "Test Equipment and Link/Cable Testing."

Another way to test, without a light source and power meter, is to apply the clamp or cable tie to the desired pressure. Let it sit for several minutes, and then remove it. Carefully look at the cable after the clamp or cable tie is removed to see if an imprint was left in the jacket. If no imprint was left in the jacket, the pressure was not too great. If an imprint was left, the pressure should be reduced.

6. Labeling Requirements and Documentation

Even the simplest fiber-optic network can become confusing if it is improperly or insufficiently labeled to show where cables originate, how they are connected, and where they go.

It may be tempting to come up with your own system for labeling, but unless you can predict all of the changes that will take place as the system grows, you may soon find yourself with a mess on your hands. In addition, if you don't clearly document your labeling system, it will be difficult if not impossible for others to decipher it.

ANSI/TIA-606-B is the administration standard for telecommunications infrastructures. This document contains detailed information on conventions for labeling optical fiber connections within buildings, datacenters, and between buildings that share a telecommunication network. The standard specifies the methods used to determine how optical fiber channels are identified using a numbering/lettering system. It is also backward compatible with ANSI/TIA-606-A Addendum 1, and it is compatible with the following identifiers in ISO/IEC/TR 14763-2-1:

- ◆ Premises identifiers
- ◆ Space identifiers
- ◆ Closure identifiers
- ◆ Closure port and closure termination point identifiers
- ◆ Cabling identifiers
- ◆ Patch cord and jumper identifiers
- ◆ Pathway system identifiers
- ◆ Cabinet and frame bonding conductor identifiers

Identifiers are used to label telecommunications infrastructure components such as cables, racks, telecommunications rooms, equipment rooms, pathways, and outlets. Each identifier should be a unique set of numbers, letters, or an alphanumeric combination. Identifiers should be tracked in an administrative database. This database may be a simple spreadsheet depending on the number of users.

ANSI/TIA-606-B requires that a label be attached to each end of the cable assembly. These labels need to be clearly visible and durable. The labels should be designed to last as long as the cable assembly and must be printed with a mechanical device designed for this purpose. Handwritten labels are not acceptable.

Most patch panels for copper cross connects have provisions for inserting labels between the wiring blocks, these labels do not need to be color coded as defined in ANSI/TIA-606-A. The use of colored labels is recommended in ANSI/TIA-606-B, but not required. When colored labels are used, the color of the label identifies the origin of the cables. Labels may also contain additional information about the connection such as the cabinet number and port number. Table 14.4 lists the termination type color-coding scheme.

Fiber-optic patch panels like the one shown in Figure 14.24 do not use color-coded labels because there is no provision for inserting a label due to the limited amount of space. The label would be secured to a nearby location such as the enclosure door.

TABLE 14.4:   ANSI/TIA-606-A/B patch panel color-coding scheme

TERMINATION TYPE	LABEL COLOR	TYPICAL APPLICATION TYPE
Demarcation point	Orange	Central office or entrance facility terminations
Network connection	Green	User side of central office or entrance facility connection, equipment room
Common equipment	Purple	Connections to a mainframe, LAN, or multiplexer
Key system	Red	Connections to key telephone systems
First-level backbone	White	Main cross-connect to intermediate cross-connect terminations
Second-level backbone	Gray	Intermediate cross-connect to telecommunications room cable terminations
Campus backbone	Brown	Inter-building backbone terminations
Horizontal	Blue	Horizontal cable terminations
Miscellaneous	Yellow	Alarms, security, or energy management

ANSI/TIA-606B was published in June 2012, and prior to that publication date, labels were generated in accordance with ANSI/TIA-606-A with and without Addendum 1. Addendum 1 added identifiers for datacenter racks and cabinets and was incorporated into ANSI/TIA-606-B. Depending on the size of the datacenter, X and Y coordinates that relate to floor tiles or rows may be used to generate a unique alphanumeric identifier for each location. However, a smaller datacenter may use the number of rows and racks or cabinets to develop the identifier for each location.

All telecommunication spaces require a unique FS.XY identifier. The first character, F, is numeric and identifies the floor of the building. The second character, S, is alphabetical and defines the space. The third character, X, is the X coordinate and is an alphabetical character. The last character, Y, is the Y coordinate and is a numerical character.

Keep in mind that the FS.XY identifier may be longer than four characters with a decimal point in the center. The first character that identifies the floor may require three numbers. The second character that defines the space is typically one letter. The X coordinate is typically two letters, and the Y coordinate is typically two numbers. For example, 15B.AJ08 would be on the 15th floor in room B. The cabinet or rack would be in grid location AJ08.

Additional information is required to locate the patch panel and the specific port within the datacenter. This information follows the FS.XY identifier and is separated by a hyphen and a colon. For example, 1C.AB02-20:03 would be on the first floor in room A. The cabinet or rack would be in grid location AB02. The patch panel would be 20 rack units from the bottom of the rack or cabinet and the specific port would be 03.

FIGURE 14.24
SC-SC patch panel
inside an enclosure

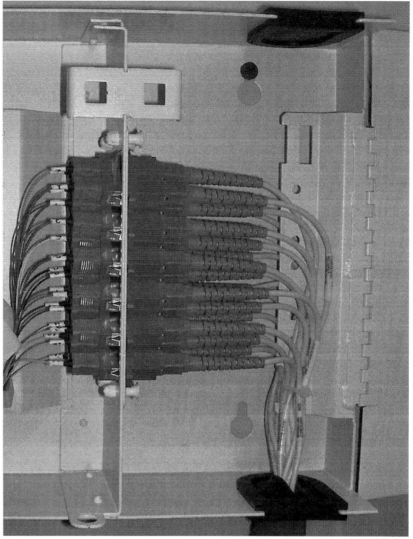

Photo courtesy of MicroCare

An installer or technician should also be familiar with identifiers prior to Addendum 1. As an example of labeling according to ANSI/TIA-606-A without Addendum 1, imagine working with a wall outlet that has a port labeled:

1A-B002

This particular number, known as a horizontal link identifier, immediately communicates the following information:

1: The fiber originates in a telecommunications room (TR) on the first floor of the building.

A: The TR is designated as A.

B: The fiber originates in patch panel B in TR A.

002: The fiber originates in port 002 of the patch panel.

Using this labeling system, we can very quickly trace the fiber to a specific connection point, even in a very large datacenter or building. Horizontal link identifiers are also described in ANSI/TIA-606B.

While the previous examples are basic, they show that by using a consistent labeling system, it is possible to communicate fiber routings accurately. A copy of ANSI/TIA-606-B may be obtained from the TIA website: `www.tiaonline.org`.

Documentation

Properly labeling cables, racks, telecommunications rooms, equipment rooms, pathways, and outlets is not enough; you must take the time to document this information and any supporting information in a database. Depending on the complexity, this database may be a simple spreadsheet or custom software package. Many companies define documentation requirements in their quality manual. Good documentation can speed up troubleshooting, system repair, and modification.

Polarity

Polarity is required to ensure the proper operation of bidirectional fiber optic communication systems that use separate transmit and receive optical fibers. Polarity is discussed in ANSI/TIA-568-C.0 and C.3; however, neither document provides a definition. Absent a standardized definition, this text defines polarity in a fiber-optic network as the positioning of connectors and adapters to ensure that there is an end-to-end transmission path between the transmitter and the receiver of a channel. A channel as defined in ANSI/TIA-568-C.0 is an end-to-end transmission path between two points at which application-specific equipment is connected.

ANSI/TIA-568-C.3 defines two polarity schemes for duplex patch cords, A-to-A and A-to-B. The most commonly deployed scheme is A-to-B; therefore this section focuses on that scheme. An A-to-B duplex patch cord should be oriented so that position A connects to position B on one fiber and position B connects to position A on the other fiber, as shown in Figure 14.25.

FIGURE 14.25
A-to-B duplex
patch cord

ANSI/TIA-568-C.0 Annex B outlines methods that are used to maintain polarity in a fiber-optic cabling system. The A-to-B polarity scheme would implement a method known as consecutive fiber positioning, and the A-to-A polarity scheme would implement the reverse-pair positioning method. As you learned earlier, A-to-B is the most commonly deployed polarity scheme, so this section will focus on consecutive fiber positioning. Information about the reverse-pair positioning and array polarity systems covered in Annex B may be obtained from the TIA website: www.tiaonline.org.

When employing either polarity scheme, connector positions at the ends of a duplex patch cord must be identified as position A and position B. This is typically accomplished with raised lettering on the latch that holds the two connectors at each end of the patch cord together. A raised letter latch holding two SC connectors together is shown in Figure 14.26. Because of the physical size of an SC connector, the raised lettering is easy to ready. However, this is not the case with small form factor connectors, such as the latched LC pair shown in Figure 14.27.

When employing the A-to-B polarity scheme, the position A connector is always plugged into the receive port on the transceiver and the position B connector is always plugged into the transmit port, as shown in Figure 14.28. Be aware that a manufacturer may use text or a symbol to mark the transmit and receive ports. The transmit port may be marked with the text "TX" and the receive port with the text "RX." When symbols are used, they typically look like an arrow-head, such as those shown in Figure 14.28. The arrow points outward on the transmit port and inward on the receive port.

FIGURE 14.26
A raised letter latch holding two SC connectors together

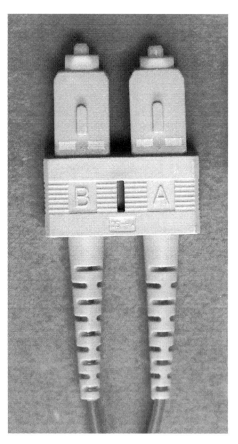

FIGURE 14.27
A raised letter latch holding two LC connectors together

FIGURE 14.28
Arrows on a transceiver identify the transmit and receive ports.

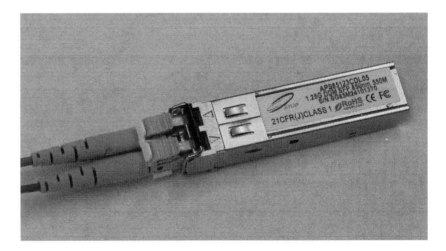

To understand how to employ consecutive-fiber positioning on a pair of channels, a cabling configuration drawing similar to the one shown in Figure 14.29 must be developed. Working from the switch receive port to the workstation transmit port, the position B connector always mates with the position A connector. It is the opposite for the switch transmit port; the position A connector always mates with the position B connector.

FIGURE 14.29
Cabling configuration for a basic fiber-optic network

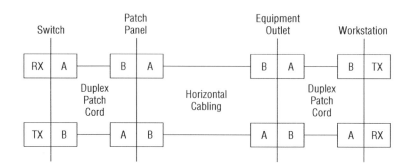

Working from the workstation receive port back to the switch transmit port, the position B connector always mates with the position A connector. It is the opposite for the workstation transmit port; the position A connector always mates with the position B connector. Regardless of which end of the channel you start from, the position A connector will always mate with the position B connector, and vice versa.

The Bottom Line

Determine the minimum cable bend radius.   Remember that cable manufacturers tend to define the minimum bend radius when the cable is stressed and unstressed. These numbers are typically in compliance with ANSI/TIA-568-C.0 and C.3.

Master It   Refer to the following table and determine the minimum installation bend radius and the minimum operational bend radius for the 24-fiber cable.

CABLE TYPE	DIAMETER	WEIGHT	MINIMUM BEND RADIUS: SHORT TERM	MINIMUM BEND RADIUS: LONG TERM	MAXIMUM LOAD: SHORT TERM	MAXIMUM LOAD: LONG TERM
12-fiber cable	0.26", 6.6mm	27lb/kft, 40kg/km	3.9", 9.9cm	2.6", 6.6cm	300lb., 1334N	90lb., 396N
24-fiber cable	0.375", 9.5mm	38.2lb/kft, 56.6kg/km	5.5", 14cm	3.7", 9.4cm	424lb., 1887N	127lb., 560N

Determine the maximum cable tensile rating.   Remember that cable manufacturers typically define a dynamic and static pull load.

Master It   Refer to the following table and determine the maximum dynamic tensile load and the maximum static tensile load for the 12-fiber cable.

CABLE TYPE	DIAMETER	WEIGHT	MINIMUM BEND RADIUS: SHORT TERM	MINIMUM BEND RADIUS: LONG TERM	MAXIMUM LOAD: SHORT TERM	MAXIMUM LOAD: LONG TERM
12-fiber cable	0.26", 6.6mm	27lb/kft, 40kg/km	3.9", 9.9cm	2.6", 6.6cm	300lb., 1334N	90lb., 396N
18-fiber cable	0.325", 8.3mm	33.8lb/kft, 50kg/km	4.9", 12.4cm	3.25", 8.3cm	375lb., 1668N	112lb., 495N

Determine the fill ratio for a multiple-cable conduit installation. When preparing for a multiple cable conduit installation, you must be aware of the fill ratio for the conduit. If you are installing cable in existing conduit already occupied by one or more cables that contain electrical conductors, you will need to determine the fill ratio of the conduit prior to selecting your cable.

Master It You need to install one cable into a conduit that has three existing electrical cables. The inside diameter of the conduit is 1". The cable to be installed has an outside diameter of 0.25". The outside diameters of the existing cables are 0.28", 0.32", and 0.38". Determine the fill ratio of the conduit before and after the fiber-optic cable is installed and verify that this installation will not overfill the conduit as defined by the NEC.

Chapter 15

Fiber-Optic System Design Considerations

Up to this point, you have learned about transmitters, receivers, couplers, attenuators, connectors, splices, and fiber-optic cable. These pieces are some of the building blocks of a fiber-optic system. A basic fiber-optic system contains a transmitter, receiver, fiber-optic cable, and connectors.

There are many ways to approach fiber-optic system design, and there are many different fiber-optic systems. Fortunately, industry standards simplify fiber-optic system design. This chapter focuses on the basic design considerations for a fiber-optic system, compares optical fiber to copper, explains how to break down a fiber-optic link to analyze performance, and shows how to prepare a power budget.

In this chapter, you will learn to:

♦ Calculate the bandwidth for a length of optical fiber

♦ Calculate the attenuation for a length of optical fiber

♦ Calculate the power budget and headroom for a multimode fiber-optic link

♦ Calculate the power budget and headroom for a single-mode fiber-optic link

The Advantages of Optical Fiber over Copper

Before you begin to learn about fiber-optic system design, we will examine seven cabling performance areas and compare the performance of optical fiber to twisted pair and coaxial copper cable. Performance in this comparison will be evaluated in the areas of bandwidth, attenuation, *electromagnetic immunity*, size, weight, safety, and security. The fiber-optic system will operate at an 850nm wavelength with a VCSEL source.

Performance data for the laser-optimized optical fiber will be taken from the ANSI/TIA-568-C.3 and ISO/IEC 11801 standards. It will be compared to both Category 6A twisted pair cable and RG6 coaxial cable. Performance data for the Category 6A cable will be taken from the ANSI/TIA-568-C.2 Commercial Building Telecommunication Cabling Standard Part 2: Balanced Twisted-Pair Cabling Components, and the performance data for the RG6 coaxial cable is derived by averaging the values found in several manufacturers' datasheets.

All the comparisons in this chapter are based on latest standards and datasheets that were available as of this writing.

Bandwidth

Bandwidth is a popular buzzword these days. We are bombarded with commercials advertising high-speed downloads from a *cable modem*, satellite, or fiber-to-the-home (FTTH). The competition to sell us bandwidth is fierce. As you learned in Chapter 1, "History of Fiber Optics and Broadband Access," while the basic broadband speed was defined with a 3Mbps download speed, more than 94 percent of the homes in America exceed 10Mbps. More than 75 percent have download speeds greater than 50Mbps, 47 percent have download speeds greater than 100Mbps, and more than 3 percent enjoy download speeds greater than one billion bits (a gigabit) per second (Gbps).

In earlier chapters, we looked at the physical properties of the optical fiber and fiber-optic light source that limit bandwidth. You learned that single-mode systems with laser transmitters offer the greatest bandwidth over distance and that multimode systems with LED transmitters offer the least bandwidth and are limited in transmission distance. You also learned that the bandwidth of the optical fiber is inversely proportional to its length. In other words, as the length of the optical fiber increases, the bandwidth of the optical fiber decreases.

So how does length affect the bandwidth of a copper cable? Well, when it comes to cable length, copper suffers just as optical fiber does. When the length of a copper cable is increased, the bandwidth for that cable decreases. So if both copper and optical fiber lose bandwidth over distance, why is optical fiber superior to copper? To explain that, we will look at the minimum bandwidth requirements defined in ANSI/TIA-568-C.3 and ISO/IEC 11801 for multimode optical fiber, ANSI/TIA-568-C.2 for unshielded twisted-pair (UTP) Category 6A cable, and the datasheets for RG6 coaxial cable.

Remember that the values defined in ANSI/TIA-568-C.2, ANSI/TIA-568-C.3, and ISO/IEC 11801 are the minimum values that a manufacturer needs to achieve. There are many manufacturers offering optical fiber and copper cables that exceed these minimum requirements. However, for this comparison, only values defined in these two standards will be used.

One of the problems in doing this comparison is cable length. ANSI/TIA-568-C.2 defines Category 6A performance at a maximum physical length of 100m at frequencies up to 500MHz. ANSI/TIA-568-C.3 and ISO/IEC 11801 does not define a maximum bandwidth at a maximum optical fiber length. As you learned in Chapter 5, "Optical Fiber Characteristics," ANSI/TIA-568-C.3 and ISO/IEC 11801 define multimode optical fiber bandwidth two ways depending on the type of light source. In this comparison the 850nm VCSEL is used with 50/125μm laser-optimized optical fiber (OM4, Type A1a.3, TIA 492AAAD). The minimum effective modal bandwidth-length product in MHz · km is 4,700 for this combination, as shown in Table 15.1.

NOTE ANSI/TIA-568-C.1 states that the maximum horizontal cable length should be limited to 90m regardless of whether copper or fiber is used. The 90m distance allows for an additional 5m at each end for equipment cables or patch cords. The total combined length cannot exceed 100m.

To do this comparison, we will need to calculate the bandwidth limitations for 100m of OM4 laser-optimized multimode optical fiber, as shown here:

Bandwidth-length product = 4,700 MHz · km

Optical fiber length = 0.1km

Bandwidth in MHz = 4,700 ÷ 0.1

Bandwidth = 47,000MHz

TABLE 15.1:   Characteristics of ANSI/TIA-568-C.3 and ISO/IEC 11801-recognized optical fibers

OPTICAL FIBER CABLE TYPE	WAVELENGTH (NM)	MAXIMUM ATTENUATION (DB/KM)	MINIMUM OVERFILLED MODAL BANDWIDTH-LENGTH PRODUCT (MHZ · KM)	MINIMUM EFFECTIVE MODAL BANDWIDTH-LENGTH PRODUCT (MHZ · KM)
62.5/125µm Multimode OM1 Type A1b TIA 492AAAA	850 1300	3.5 1.5	200 500	Not Required Not Required
50/125µm Multimode OM2 TIA 492AAAB Type A1a.1	850 1300	3.5 1.5	500 500	Not Required Not Required
850nm laser-optimized 50/125µm Multimode OM3 Type A1a.2 TIA 492AAAC	850 1300	3.5 1.5	1500 500	2000 Not Required
850nm laser-optimized 50/125µm Multimode OM4 Type A1a.3 TIA 492AAAD	850 1300	3.5 1.5	3500 500	4700 Not Required

TABLE 15.1: Characteristics of ANSI/TIA-568-C.3 and ISO/IEC 11801-recognized optical fibers *(continued)*

OPTICAL FIBER CABLE TYPE	WAVELENGTH (NM)	MAXIMUM ATTENUATION (DB/KM)	MINIMUM OVERFILLED MODAL BANDWIDTH-LENGTH PRODUCT (MHZ · KM)	MINIMUM EFFECTIVE MODAL BANDWIDTH-LENGTH PRODUCT (MHZ · KM)
Single-mode	1310	0.5	N/A	N/A
Indoor-outdoor	1550	0.5	N/A	N/A
OS1				
Type B1.1				
TIA 492CAAA				
OS2				
Type B1.3				
TIA 492CAAB				
Single-mode	1310	1.0	N/A	N/A
Inside plant	1550	1.0	N/A	N/A
OS1				
Type B1.1				
TIA 492CAAA				
OS2				
Type B1.3				
TIA 492CAAB				
Single-mode	1310	0.5	N/A	N/A
Outside plant	1550	0.5	N/A	N/A
OS1				
Type B1.1				
TIA 492CAAA				
OS2				
Type B1.3				
TIA 492CAAB				

The 100m OM4 laser-optimized multimode optical fiber has a bandwidth 94 times greater than the Category 6A cable. This clearly shows that optical fiber offers incredible bandwidth advantages over Category 6A cable.

NOTE   How do we get from Gbps to MHz? Remember that in Chapter 2, "Principles of Fiber-Optic Transmission," you learned about bits and symbols and defined baud rate as the number of symbols per second. Baud as defined in IEEE 802.3 is a unit of signaling speed, expressed as the number of times per second the signal can change the electrical state of the transmission line or other medium. Depending on the type of encoding, a signal event may represent a single bit, more, or less than one bit. For most digital fiber-optic applications, the baud rate equals the bit rate. In other words, a 1Gbps transmission could change states up to one billion times per second or transmit up to one billion symbols per second.

As you learned in Chapter 2, one complete cycle of a square wave contains two states or two symbols, a high symbol and a low symbol. Therefore, a 500MHz square wave transmits one billion symbols per second, or 1Gbps.

Now that we have seen how the bandwidth of OM4 laser-optimized multimode optical fiber with an 850nm VCSEL source greatly exceeds that of Category 6A cable, let's do a comparison with RG6 coaxial cable. This cable has been chosen because it is widely used in homes and buildings for video distribution. Remember the performance data for the RG6 coaxial cable is taken from the average of several manufacturers' datasheets. ANSI/TIA-568-C does not define performance parameters for RG6 coaxial cable.

An RG6 coaxial cable with a transmission frequency of 4GHz has roughly the same 100m transmission distance characteristics as a Category 6A cable at 500MHz. We know from the previous comparison that a 100m length of 50/125μm multimode optical fiber will support transmission frequencies up to 47GHz over that same distance with an 850nm light source. In this comparison, the laser-optimized multimode optical fiber offers a bandwidth advantage 11.75 times greater than the RG6 coaxial cable.

These comparisons have demonstrated the bandwidth advantages of optical fiber over copper cable. The comparisons were done at very short distances because at this point we have not addressed how attenuation in a copper cable changes with the transmission frequency, whereas in an optical fiber, attenuation is constant regardless of the transmission frequency.

Attenuation

All transmission mediums lose signal strength over distance. As you know, this loss of signal strength is called attenuation and is typically measured in decibels. Optical fiber systems measure attenuation using optical power. Copper cable systems typically use voltage drop across a defined load at various transmission frequencies to measure attenuation. The key difference here is not that optical fiber uses power and copper uses voltage. The key difference is that attenuation in copper cables is measured at different transmission frequencies. This is not the case with optical fiber, where attenuation is measured with a continuous wave light source.

The attenuation in a copper cable increases as the transmission frequency increases. Table 15.2 shows the maximum worst pair insertion loss for a 100m horizontal Category 5e and Category 6A cable as defined in ANSI/TIA-568-C.2. This table clearly shows the effects that transmission frequency has on a copper cable.

TABLE 15.2: Horizontal cable insertion loss, worst pair*

FREQUENCY (MHZ)	CATEGORY 5E (DB)	CATEGORY 6A (DB)
1.0	2.0	2.1
4.0	4.1	3.8
8.0	5.8	5.3
10.0	96.5	5.9
16.0	8.2	7.5
20.0	9.3	8.4
25.0	10.4	9.4
31.25	11.7	10.5
62.5	17.0	15.0
100.0	22.0	19.1
200.0		27.6
300.0		34.3
400.0		40.1
500.0		45.3

For a length of 100m (328')

The maximum allowable attenuation in an optical fiber is defined in ANSI/TIA-568-C.3 and ISO/IEC 11801. Table 15.3 shows the attenuation portion of the optical fiber cable transmission performance parameters. This table defines attenuation for both multimode and single-mode optical fibers. You will notice that there is no column for transmission frequency in this table. That is because optical fiber does not attenuate as the transmission frequency increases like copper cable does.

TABLE 15.3: ANSI/TIA-568-C.3 and ISO/IEC 11801 optical fiber cable attenuation performance parameters

OPTICAL FIBER CABLE TYPE	WAVELENGTH (NM)	MAXIMUM ATTENUATION (DB/KM)
50/125µm multimode	850	3.5
	1300	1.5

TABLE 15.3: ANSI/TIA-568-C.3 and ISO/IEC 11801 optical fiber cable attenuation performance parameters *(continued)*

OPTICAL FIBER CABLE TYPE	WAVELENGTH (NM)	MAXIMUM ATTENUATION (DB/KM)
62.5/125µm multimode	850	3.5
	1300	1.5
Single-mode inside plant cable	1310	1.0
	1550	1.0
Single-mode outside plant cable	1310	0.5
	1550	0.5

Now that you know how optical fiber and copper cable attenuate, let's do a comparison. The first comparison will put optical fiber up against Category 6A cable. The distance will be 100m and the transmission frequency will be 500MHz.

Table 15.2 lists 45.3dB for the worst-case attenuation for the Category 6A cable at 500MHz. This means the Category 6A cable loses 99.997 percent of its energy over a distance of 100m at that transmission frequency. As you learned in Chapter 5, the attenuation of an optical fiber is different for each wavelength. Table 15.3 lists the maximum attenuation in dB/km. We can solve for the maximum attenuation at both wavelengths as shown here:

Attenuation coefficient at 850nm = 3.5dB/km

Optical fiber length = 0.1km

Attenuation for the length of optical fiber = $0.1 \times 3.5 = 0.35$

Maximum attenuation at 850nm for 100m of optical fiber is 0.35dB

Attenuation coefficient at 1300nm = 1.5dB/km

Optical fiber length = 0.1km

Attenuation for the length of optical fiber = $0.1 \times 1.5 = 0.15$

Maximum attenuation at 1300nm for 100m of optical fiber is 0.15dB

The equations show that multimode optical fiber loses 7.7 percent of the light energy from the transmitter over the 100m distance at 850nm. At 1300nm, the loss is only 3.4 percent. However, the Category 6A cable loses 99.997 percent of the transmitted signal over that distance. This means that the Category 6A cable loses roughly 130 times more energy than the optical fiber 850nm and roughly 300 times more energy at 1300nm.

This comparison clearly shows the attenuation advantages of optical fiber over Category 6A cable.

Now let's compare the same optical fiber at the same wavelengths to an RG6 coaxial cable. For this comparison, the RG6 attenuation characteristics are taken from the average of several different manufacturers' published datasheets, as shown in Table 15.4.

TABLE 15.4: RG6 cable insertion loss*

FREQUENCY (MHZ)	RG6 (DB)
1.0	0.8
5.0	1.7
100	6.4
200	9.5
500	14.0
1,000	21.3
3,000	38.8
4,500	49.0

For a length of 100m (328')

As you look at Table 15.4, you can see that RG6 coaxial cable easily outperforms Category 5e and 6A cable. However, it does not even begin to approach the multimode optical fiber operating at 850nm or 1300nm. The RG6 coaxial cable has 14dB of attenuation at a transmission frequency of 500MHz over a distance of 100m. Whereas the Category 6A cable lost 99.997 percent of its signal strength, in the previous comparison the RG6 coaxial cable lost only 96 percent of its energy over the same distance.

These comparisons should make it clear that optical fiber has an enormous attenuation advantage over copper cable. The comparisons were done only with multimode optical fiber. Outside-plant, single-mode optical fiber greatly outperforms multimode optical fiber. Single-mode optical fiber links are capable of transmission distances greater than 80km without re-amplification. Based on the previous comparisons, Category 6A cable would require re-amplification every 100m at a transmission frequency of 500MHz. The RG6 coaxial cable would require re-amplification every 325m at a transmission frequency of 500MHz.

An 80km RG6 coaxial cable link transmitting a frequency of 500 MHz would require roughly 246 amplifiers to re-amplify the signal. However, the single-mode optical fiber link would not require re-amplification. Optical fiber links are unsurpassed in transmission distance.

Electromagnetic Immunity

Electromagnetic interference (EMI) is electromagnetic energy, sometimes referred to as *noise*, which causes undesirable responses, degradation, or complete system failure. Systems using copper cable are vulnerable to the effects of EMI because a changing electromagnetic field will induce current flow in a copper conductor. Optical fiber is a dielectric or an insulator, and current does not flow through insulators. What this means is that EMI has no effect on the operation of an optical fiber.

Let's take a look at some examples of EMI-induced problems in a copper system. A Category 6A cable has four pairs of twisted conductors. The conductors are twisted to keep the impedance uniform along the length of the cable and to decrease the effects of EMI by canceling out opposing fields. Two of the conductor pairs may be used to transmit and two of the conductor pairs may be used to receive.

For this example, only one pair is transmitting and one pair is receiving. The other two pairs in the cable are not used. Think of each pair of conductors as an antenna. The transmitting pair is the broadcasting antenna and the receiving pair is picking up that transmission just like your car radio antenna.

The data on the transmitting pair is broadcast and picked up by the receiving pair. This is called *crosstalk*. If enough current is induced into the receiving pair, the operation of the system can be affected. This is one reason why it is so critical to maintain the twists in a Category 6A cable.

Would we have this problem with an optical fiber? Does crosstalk exist in optical fiber? Regardless of the number of transmitting and receiving optical fibers in a cable assembly, crosstalk does not occur. To have crosstalk in an optical fiber cable assembly, light would have to leave one optical fiber and enter one of the other optical fibers in the cable assembly. Because of total internal reflection, under normal operating conditions light never leaves the optical fiber. Therefore, crosstalk does not exist in optical fiber cable assemblies.

Now let's take a look at another EMI scenario. Copper cables are being routed through a manufacturing plant. This manufacturing plant houses large-scale electromechanical equipment that generates a considerable amount of EMI, which creates an EMI-rich environment.

Routing cables through an EMI-rich environment can be difficult. Placing the cables too close to the EMI-generating source can induce unwanted electrical signals strong enough to cause systems to function poorly or stop operating altogether. Copper cables used in EMI-rich environments typically require electrical shielding to help reduce the unwanted electrical signals. In addition to the electrical shielding, the installer must be aware of the EMI-generating sources and ensure that the copper cables are routed as far as possible from these sources.

Routing copper cables through an EMI-rich environment can be challenging, time-consuming, and expensive. However, optical fiber cables can be routed through an EMI-rich environment with no impact on system performance. The fiber-optic installer is free to route the optical fiber as efficiently as possible. Optical fiber is very attractive to every industry because it is immune to EMI.

Size and Weight

You must take into account the size of any cable when preparing for an installation. Often fiber-optic cables will be run through existing conduits or raceways that are partially or almost completely filled with copper cable. This is another area where small fiber-optic cable has an advantage over copper cable.

Let's do a comparison and try to determine the reduced-size advantage that fiber-optic cable has over copper cable. As you learned in Chapter 4, "Optical Fiber Construction and Theory," a coated optical fiber is typically 250µm in diameter. You learned that ribbon fiber-optic cables sandwich up to 12-coated optical fibers between two layers of Mylar tape. Eighteen of these ribbons stacked on top of each other form a rectangle roughly 4.5mm by 3mm. This rectangle can be placed inside a buffer and surrounded by a strength member and jacket to form a cable. The overall diameter of this cable would be 16.5mm (0.65"), only slightly larger than a bundle of four RG6 coaxial cables or a bundle of four Category 6A cables.

So how large would a copper cable have to be to offer the same performance as the 216 optical fiber ribbon cable? That would depend on transmission distance and the optical fiber data rate. Because we have already discussed Category 6A performance, let's place a bundle of Category 6A cables up against the 216 optical fiber ribbon cable operating at a modest 2.5Gbps data rate over a distance of just 100m.

A Category 6A cable contains four conductor pairs and each pair is capable of a 500MHz transmission over 100m. As you learned earlier in this chapter, a 500MHz transmission carries 1 billion symbols per second. If each symbol is a bit, the 500MHz Category 6A cable is capable of a 1Gbps transmission rate. When the performance of each pair is combined, a single Category 6A cable is capable of a 4Gbps transmission rate over a distance of 100m.

Now let's see how many Category 6A cables will be required to provide the same performance as the 216 optical fiber ribbon cable. The 216 optical fiber ribbon cable has a combined data transmission rate of 540Gbps (2.5Gbps × 216). When we divide 540Gbps by 4Gbps, we see that 135 Category 6A cables are required to equal the performance of this modest fiber-optic system.

When 135 Category 6A cables are bundled together, they are roughly 1.7" in diameter. As noted earlier in this chapter, the 216 optical fiber ribbon cable is approximately the size of four Category 6A cables bundled together. The Category 6A bundle thus has a volume roughly 33.75 times greater than the 216 optical fiber ribbon cable. In other words, Category 6A bundles need 33.75 times more space in the conduit than the 216 optical fiber ribbon cable.

The comparison we just performed is very conservative. The distance we used was kept very short and the transmission rate for the optical fiber was kept low. We can get even a better appreciation for the cable size reduction fiber-optic cable offers if we increase the transmission distance and the data rate.

In this comparison, let's increase the transmission distance to 1,000m and the data transmission rate to 10Gbps. The bandwidth of a copper cable decreases as distance increases, just as with fiber-optic cables. Because we have increased the transmission distance by a factor of 10, it's fair to say that the Category 6A cable bandwidth will decrease by a factor of 10 over 1000m.

With a reduction in bandwidth by a factor of 10, we will need ten times more Category 6A cables to equal the old 2.5Gbps performance. In other words, we need 1,350 Category 6A cables bundled together. In this comparison, however, the bandwidth has been increased from 2.5Gbps to 10Gbps. This means we have to quadruple the number of Category 6A cables to meet the bandwidth requirement. We now need 5,400 Category 6A cables bundled together. Imagine how many cables we would need if the transmission distance increased to 80,000m. We would need a whopping 432,000 Category 6A cables bundled together.

These comparisons vividly illustrate the size advantage that fiber-optic cables have over copper cables. The advantage becomes even more apparent as distances increase. The enormous capacity of such a small cable is exactly what is needed to install high-bandwidth systems in buildings where the conduits and raceways are almost fully populated with copper cables.

Now that we have calculated the size advantages of optical fiber over Category 6A cable, let's look at the weight advantages. It is pretty easy to see that thousands or tens of thousands of Category 6A cables bundled together will outweigh a ribbon fiber-optic cable 0.65" in diameter. It's difficult to state exactly how much less a fiber-optic cable would weigh than a copper cable performing the same job—there are just too many variables in transmission distance and data rate. However, it's not difficult to imagine the weight savings that fiber-optic cables offer over copper cables. These weight savings are being employed in commercial aircraft, military aircraft, and the automotive industries, just to mention a few.

Security

We know that optical fiber is a dielectric and because of that it is immune to EMI. So why is an optical fiber secure and virtually impossible to tap? Because of total internal reflection, optical fiber does not radiate.

In Chapter 5, you learned about macrobends. Excessive bending on an optical fiber will cause some of the light energy to escape the core and cladding. This light might penetrate the coating, buffer, strength member, and jacket. This energy is detectable by means of a *fiber identifier*.

Fiber identifiers detect light traveling through an optical fiber by inserting a macrobend. Photodiodes are placed against the jacket or buffer of the fiber-optic cable on opposite sides of the macrobend. The photodiodes detect the light that escapes from the fiber-optic cable. The light energy detected by the photodiodes is analyzed by the electronics in the fiber identifier. The fiber identifier can typically determine the presence and direction of travel of the light.

If the fiber identifier can insert a simple macrobend and detect the presence and direction of light, why is fiber secure? Detecting the presence of light and determining the source of the light does not require much optical energy. However, as you learned earlier in this chapter, a fiber-optic receiver typically has a relatively small window of operation. In other words, the fiber-optic receiver typically needs at least 10 percent of the energy from the transmitter to accurately decode the signal on the optical fiber. Inserting a macrobend in a fiber-optic cable and directing 10 percent of the light energy into a receiver is virtually impossible. A macrobend this severe would also be very easy to detect with an *optical time-domain reflectometer* (OTDR). There is no transmission medium more secure than optical fiber.

NOTE   The fiber identifier and OTDR are covered in detail in Chapter 16, "Test Equipment and Link/Cable Testing."

Safety

Electrical safety is always a concern when working with copper cables. Electrical current flowing through copper cable poses shock, spark, and fire hazards. Optical fiber is a dielectric that cannot carry electrical current; therefore, it presents no shock, spark, or fire hazard.

Because optical fiber is a dielectric, it also provides electrical isolation between electrical equipment. Electrical isolation eliminates ground loops, eliminates the potential shock hazard when two pieces of equipment at different potentials are connected together, and eliminates the shock hazard when one piece of equipment is connected to another with a ground fault.

Ground loops are typically not a safety problem; they are usually an equipment operational problem. They create unwanted noise that can interfere with equipment operation. A common example of a ground loop is the hum or buzz you hear when an electric guitar is plugged into an amplifier with a defective copper cable or electrical connection. Connecting two pieces of equipment together with optical fiber removes any path for current flow, which eliminates the ground loop.

Using copper cable to connect two pieces of equipment that are at different electrical potentials poses a shock hazard. It's not uncommon for two grounded pieces of electrical equipment separated by distance to be at different electrical potentials. Connecting these same two pieces of equipment together with optical fiber poses no electrical shock hazard.

If two pieces of electrical equipment are connected together with copper cable and one develops a ground fault, there is now a potential shock hazard at both pieces of equipment. Everyone is likely to experience, or hear of someone experiencing, a ground fault at least once in their life. A common example is when you touch an appliance such as an electric range or washing machine and experience a substantial electrical shock. If the piece of equipment that shocked you was connected to another piece of equipment with a copper cable, there is a possibility that someone touching the other piece of equipment would also be shocked. If the two pieces of equipment were connected with optical fiber, the shock hazard would exist only at the faulty piece of equipment.

Nonconductive fiber-optic cables offer some other advantages, too. They do not attract lightning any more than any other dielectric. They can be run through areas where faulty copper cables could pose a fire or explosion hazard. The only safety requirement that Article 770 of the NEC places on nonconductive fiber-optic cables addresses the type of jacket material. When electrical safety, spark, or explosion hazards are a concern, there is no better solution than optical fiber.

NOTE Article 770 of the NEC is discussed in detail in Chapter 7, "Fiber-Optic Cables."

Basic Fiber-Optic System Design Considerations

Before beginning a basic fiber-optic system design, you must answer three questions:

♦ How much data needs to be moved today?

♦ How much data will need to be moved in the future?

♦ What is the transmission distance?

Throughout this book, you have learned how the physical properties of light and of the optical fiber determine bandwidth and transmission distance. Now let's take some of these lessons learned and apply them in the design of a basic fiber-optic system.

The release of a laser-optimized multimode optical fiber standard and the improvements in VCSEL bandwidth are changing the traditional approaches to designing a fiber-optic system. In the past, high data rates such 1Gbps or 10Gbps were typically accomplished using single-mode optical fiber. However, the VCSEL, when used with laser optimized multimode optical fiber, is capable of these types of data rates over several hundred meters.

Many local area network applications do not require transmission distances greater than several hundred meters. This means the entire network could be designed using laser-optimized multimode optical fiber. As you have learned in earlier chapters, multimode optical fiber has a much larger core than single-mode optical fiber. The larger multimode core is more forgiving than the smaller single-mode core at interconnections. A small misalignment at a single-mode interconnection could prevent the system from operating, whereas that same amount of misalignment would have little or no effect on the multimode interconnection.

We mentioned earlier that before you begin a design, the current data rate, future data rate, and transmission distance must be known. Let's look at three rules of thumb that will help you design a link to meet your current data rate, future data rate, and transmission distance requirements.

Rule 1 For current and future data rates up to 1Gbps with a transmission distance no greater than 860m, choose laser-optimized multimode optical fiber.

Rule 2 For current and future data rates greater than 1Gbps up to 10Gbps and transmission distances no greater than 300m, choose laser-optimized multimode optical fiber.

Rule 3 For all other applications, choose a laser transmitter and single-mode optical fiber.

You may remember from Chapter 10, "Fiber-Optic Light Sources and Transmitters," that LED transmitters are typically not designed for data rates exceeding 155Mbps, whereas laser transmitters that support data rates from 155Mbps through 10Gbps are readily available.

The three rules of thumb just discussed are not all encompassing. They are intended to be a guideline and are based on IEEE 802.3 Ethernet standards and Fibre Channel standards for a single laser and single optical fiber. There are applications where these rules of thumb do not apply; however, those applications are not basic systems and may require more than one laser or optical fiber.

Design to a Standard

We have learned about the three rules of thumb that will help you design a link to meet your current data rate, future data rate, and transmission distance requirements. Now it is time to select a local area network (LAN) or storage network (SAN) that meets your needs.

This section provides physical layer design information based on the IEEE 802.3 Standard for Ethernet and Fibre Channel standards. It outlines the maximum supportable distance and maximum allowable channel insertion loss based on the optical fiber type and the data rate. IEEE 802.3 defines channel insertion loss for fiber-optic links as the static loss of light through a link between a transmitter and receiver. It includes the loss of the fiber, connectors, and splices and, for passive optical network links, optional power splitter/combiner.

The maximum channel insertion loss should not be confused with the link power budget, which is covered later in this chapter. The maximum channel insertion loss and the link power budget are similar in value for data rates of 155Mbps or less that use an LED light source. However, when the data rates increase and the transmitter contains a laser light source, power penalties not attributed to link attenuation must be accounted for. These power penalties include modal noise, relative intensity noise (RIN), intersymbol interference (ISI), mode partition noise, extinction ratio, and eye-opening penalties.

How power penalties are calculated goes beyond the scope of this book. Fortunately, the IEEE 802.3 and Fibre Channel standards account for power penalties and any other unallocated margin in the link power budget when channel insertion loss is calculated. Therefore, when we design a fiber-optic link to these standards, we only need to ensure that maximum optical fiber length and maximum channel attenuation are not exceeded.

When designing a link to IEEE 802.3 or Fibre Channel standards, keep in mind that the maximum optical fiber length may also be described as the channel length because a channel as defined in ANSI/TIA-568-C.0 is an end-to-end transmission path between two points at which application specific equipment is connected. Regardless of the type of equipment used, the channel length is the total amount of optical fiber that is required to connect the transmitter at one end to the receiver at the other end, and vice versa. Channel length includes all the cable plant optical fiber in addition to all the jumpers or patch cords required to complete the end-to-end transmission path.

As mentioned earlier, channel insertion loss for a fiber-optic link includes the loss of the fiber, connectors, and splices, and, for passive optical network links, optional power splitter/combiner. Later in this chapter, you will learn how to calculate a power budget using industry standards.

The power budget calculation will include the total link loss for the optical fiber, interconnections, and splices at different wavelengths. The calculated total link loss is also the calculated channel insertion loss.

Table 7 in ANSI/TIA-568-C.0 lists the channel attenuation and maximum supportable distances for many of the popular Ethernet and Fibre Channel networks. The information contained in this table was used to complete this section of the book.

DESIGNING TO IEEE 802.3

IEEE 802.3 is an international standard for local and metropolitan area networks (LANs and MANs). This very widely used standard was first published in 1985 and specified a data rate of just 10Mbps. As of this writing, the latest edition was IEEE Std 802.3 2012. This document is broken into six sections and can be obtained at no cost from the IEEE website:

    http://standards.ieee.org/about/get/802/802.3.html

In this section, two 1Gbps LANs and two 10Gbps LANs are described. For each data rate there is a short wavelength 850nm LAN that uses only multimode optical fiber and a long wavelength 1310nm LAN that can use multimode or single-mode optical fiber. The use of single-mode optical fiber allows for greater transmission distances.

If you select a long wavelength LAN and choose to use multimode optical fiber, a single-mode fiber offset-launch mode-conditioning patch cord must be used at each transmitter to connect the transmitter to the cable plant. The mode-conditioning patch cord must have single-mode optical fiber on one end and multimode optical fiber on the other end. The multimode optical fiber must be the same type as the optical fiber used for the cable plant.

The mode-conditioning patch cord must be connected as shown in Figure 15.1. In addition, it must meet specifications described in Table 15.5.

FIGURE 15.1
Mode-conditioning patch cord installation between the equipment and the cable plant

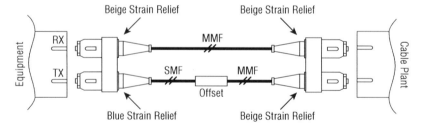

TABLE 15.5:   Single-mode fiber offset-launch mode conditioner specifications

DESCRIPTION	62.5/125 µM	50/125 µM	UNIT
Maximum insertion loss	0.5	0.5	dB
Coupled power ratio (CPR)	28 < CPR < 40	12< CPR < 20	dB
Optical center offset between single-mode and multimode optical fibers	17 < Offset < 23	10 < Offset < 16	µm
Maximum angular offset	1	1	degree

Source: IEEE 802.3, 2012

1000BASE-SX

This short wavelength 1Gbps LAN features a VCSEL laser transmitter that operates a nominal wavelength of 850nm. It supports both 62.5/125µm and 50/125µm multimode optical fiber. The maximum allowable channel attenuation and the maximum supportable distance for each optical fiber type are described in Table 15.6.

TABLE 15.6:   1000BASE-SX

PARAMETER	62.5/125 µM TIA 492AAAA OM1	50/125 µM TIA 492AAAB OM2	50/125 µM LASER-OPTIMIZED TIA 492AAAC OM3
Maximum channel attenuation (dB)	2.6	3.6	4.5
Maximum supportable distance (meters)	275	550	800

Source: ANSI/TIA-568-C.0

1000BASE-LX

This long wavelength 1Gbps LAN features a Fabry-Pérot laser transmitter that operates a nominal wavelength of 1310nm. It supports both 62.5/125µm and 50/125µm multimode optical fiber, and single-mode optical fiber. The maximum allowable channel attenuation and the maximum supportable distance for each optical fiber type are described in Table 15.7. Remember a mode-conditioning patch cord must be connected as shown in Figure 15.1 when single-mode optical fiber is used.

TABLE 15.7:   1000BASE-LX

PARAMETER	62.5/125 µM TIA 492AAAA OM1	50/125 µM TIA 492AAAB OM2	50/125 µM LASER-OPTIMIZED TIA 492AAAC OM3	TIA 492CAAA OS1 TIA 492CAAB OS2
Maximum channel attenuation (dB)	2.3	2.3	2.3	4.5
Maximum supportable distance (meters)	550	550	550	5,000

Source: ANSI/TIA-568-C.0

10GBASE-S

This short wavelength 10Gbps LAN features a VCSEL laser transmitter that operates a nominal wavelength of 850nm. It supports both 62.5/125μm and 50/125μm multimode optical fiber. The maximum allowable channel attenuation and the maximum supportable distance for each optical fiber type are described in Table 15.8.

TABLE 15.8: 10GBASE-S

PARAMETER	62.5/125 μM TIA 492AAAA OM1	50/125 μM TIA 492AAAB OM2	50/125 μM LASER-OPTIMIZED TIA 492AAAC OM3
Maximum channel attenuation (dB)	2.4	2.3	2.6
Maximum supportable distance (meters)	33	82	300

Source: ANSI/TIA-568-C.0

10GBASE-L

This long wavelength 10Gbps LAN features a distributed feedback (DFB) laser transmitter that operates a nominal wavelength of 1310nm. It only supports single-mode optical fiber; therefore a mode-condition patch cord is not required. The maximum allowable channel attenuation and the maximum supportable distance are described in Table 15.9.

TABLE 15.9: 10GBASE-L

PARAMETER	TIA 492CAAA OS1 TIA 492CAAB OS2
Maximum channel attenuation (dB)	6.2
Maximum supportable distance (meters)	10,000

Source: ANSI/TIA-568-C.0

FIBRE CHANNEL

While IEEE 802.3 is a widely used standard for LANs and MANs, Fibre Channel is the foundation for over 90 percent of all SAN installations globally. However, unlike the IEEE 802.3 standard, Fibre Channel standards cannot be obtained free of charge. Additional information on Fibre Channel can be found at the Fibre Channel Industry Association website:

www.fibrechannel.org

In this section, two 1062Mbaud SANs and two 10512Mbaud SANs are described. For each data rate there is a short wavelength 850nm SAN that uses only multimode optical fiber and a long wavelength 1310nm SAN that uses single-mode optical fiber. The use of single-mode optical fiber allows for greater transmission distances. Keep in mind that unlike Ethernet, Fibre Channel does not offer any long wavelength SANs that operate with multimode or single-mode optical fiber.

100-MX-SN-I

This short wavelength 1062Mbaud SAN features a VCSEL laser transmitter that operates a nominal wavelength of 850nm. It supports both 62.5/125µm and 50/125µm multimode optical fiber. The maximum allowable channel attenuation and the maximum supportable distance for each optical fiber type are described in Table 15.10.

TABLE 15.10:   100-MX-SN-I

PARAMETER	62.5/125 µM TIA 492AAAA OM1	50/125 µM TIA 492AAAB OM2	50/125 µM LASER-OPTIMIZED TIA 492AAAC OM3
Maximum channel attenuation (dB)	3.0	3.9	4.6
Maximum supportable distance (meters)	300	500	860

Source: ANSI/TIA-568-C.0

100-SM-LC-L

This long wavelength 1062Mbaud SAN features a Fabry-Pérot laser transmitter that operates a nominal wavelength of 1310nm. It only supports single-mode optical fiber; therefore a mode-condition patch cord is not required. The maximum allowable channel attenuation and the maximum supportable distance are described in Table 15.11.

TABLE 15.11:   100-SM-LC-L

PARAMETER	TIA 492CAAA OS1 TIA 492CAAB OS2
Maximum channel attenuation (dB)	7.5
Maximum supportable distance (meters)	10,000

Source: ANSI/TIA-568-C.0

1200-MX-SN-I

This short wavelength 10215Mbaud SAN features a VCSEL laser transmitter that operates a nominal wavelength of 850nm. It supports both 62.5/125µm and 50/125µm multimode optical fiber. The maximum allowable channel attenuation and the maximum supportable distance for each optical fiber type are described in Table 15.12.

TABLE 15.12: 1200-MX-SN-I

PARAMETER	62.5/125 µM TIA 492AAAA OM1	50/125 µM TIA 492AAAB OM2	50/125 µM LASER-OPTIMIZED TIA 492AAAC OM3
Maximum channel attenuation (dB)	2.4	2.2	2.6
Maximum supportable distance (meters)	33	82	300

Source: ANSI/TIA-568-C.0

1200-SM-LL-L

This long wavelength 10215Mbaud SAN features a DFB laser transmitter that operates a nominal wavelength of 1310nm. It supports only single-mode optical fiber. The maximum allowable channel attenuation and the maximum supportable distance are described in Table 15.13.

TABLE 15.13: 1200-SM-LL-L

PARAMETER	TIA 492CAAA OS1 TIA 492CAAB OS2
Maximum channel attenuation (dB)	6.0
Maximum supportable distance (meters)	10,000

Source: ANSI/TIA-568-C.0

Link Performance Analysis

Now that we have discussed fiber-optic link design considerations and the advantages of optical fiber over copper cable, let's look at how to analyze the performance of a fiber-optic link. This section focuses on link performance analysis using the ANSI/TIA-568-C.3 and ISO/IEC 11801.

In this chapter, ANSI/TIA-568-C.3 and ISO/IEC 11801 are used to define cable transmission performance. ANSI/TIA-568-C.3 is used to define optical fiber components' performance for premises applications. TIA-758-B is used to define splice performance for outside plant applications.

When analyzing link performance using ANSI/TIA-568-C.3, ISO/IEC 11801, and TIA-758-B, you are performing a worst-case analysis. In other words, your link should perform no worse than the performance levels defined in the standard. Typically, a link will outperform the standards or greatly exceed the standards. The standards set the minimum requirements for cable transmission performance, splice performance, and connector performance.

There are typically three different methodologies that can be used to develop a loss budget or power budget:

Worst-case method   The worst-case method that will be discussed in this chapter is commonly viewed as the most conservative engineering approach. When using this method, you use worst-case maximum value loss for each passive component. The minimum transmitter output power and minimum receiver sensitivity are also used.

Statistical method   The statistical method is less conservative than the worst-case method. Instead of using worst-case values for each component, the statistical method distributes the properties for each component in some manner around the mean value for that component type. The goal of this method is to more accurately quantify the predicted performance of the fiber-optic link.

Numerical method   The numerical method is the most calculation-intensive method and typically requires the use of a computer program to accomplish. This method looks at the environmental conditions of each component and the performance is calculated based on these conditions.

Cable Transmission Performance

As you saw earlier in this chapter, ANSI/TIA-568-C.3 and ISO/IEC 11801 address the performance of 50/125µm multimode optical fiber, 62.5/125µm multimode optical fiber, and single-mode inside and outside plant optical fiber. Maximum attenuation and minimum bandwidth-length product is defined for each fiber type by wavelength, as shown in Table 15.14.

TABLE 15.14:   Characteristics of ANSI/TIA-568-C.3 and ISO/IEC 11801-recognized optical fibers

OPTICAL FIBER CABLE TYPE	WAVELENGTH (NM)	MAXIMUM ATTENUATION (DB/KM)	MINIMUM OVERFILLED MODAL BANDWIDTH-LENGTH PRODUCT (MHZ · KM)	MINIMUM EFFECTIVE MODAL BANDWIDTH-LENGTH PRODUCT (MHZ · KM)
62.5/125µm	850	3.5	200	Not required
Multimode	1300	1.5	500	Not required
OM1				
Type A1b				
TIA 492AAAA				

TABLE 15.14:   Characteristics of ANSI/TIA-568-C.3 and ISO/IEC 11801-recognized optical fibers *(continued)*

OPTICAL FIBER CABLE TYPE	WAVELENGTH (NM)	MAXIMUM ATTENUATION (DB/KM)	MINIMUM OVERFILLED MODAL BANDWIDTH-LENGTH PRODUCT (MHZ · KM)	MINIMUM EFFECTIVE MODAL BANDWIDTH-LENGTH PRODUCT (MHZ · KM)
50/125μm	850	3.5	500	Not required
Multimode	1300	1.5	500	Not required
OM2				
TIA 492AAAB				
Type A1a.1				
850nm laser-optimized	850	3.5	1500	2000
50/125μm	1300	1.5	500	Not required
Multimode				
OM3				
Type A1a.2				
TIA 492AAAC				
850nm laser-optimized	850	3.5	3500	4700
50/125μm	1300	1.5	500	Not required
Multimode				
OM4				
Type A1a.3				
TIA 492AAAD				
Single-mode	1310	0.5	N/A	N/A
Indoor-outdoor	1550	0.5	N/A	N/A
OS1				
Type B1.1				
TIA 492CAAA				
OS2				
Type B1.3				
TIA 492CAAB				

TABLE 15.14: Characteristics of ANSI/TIA-568-C.3 and ISO/IEC 11801-recognized optical fibers *(continued)*

OPTICAL FIBER CABLE TYPE	WAVELENGTH (NM)	MAXIMUM ATTENUATION (DB/KM)	MINIMUM OVERFILLED MODAL BANDWIDTH-LENGTH PRODUCT (MHZ · KM)	MINIMUM EFFECTIVE MODAL BANDWIDTH-LENGTH PRODUCT (MHZ · KM)
Single-mode	1310	1.0	N/A	N/A
Inside plant	1550	1.0	N/A	N/A
OS1				
Type B1.1				
TIA 492CAAA				
OS2				
Type B1.3				
TIA 492CAAB				
Single-mode	1310	0.5	N/A	N/A
Outside plant	1550	0.5	N/A	N/A
OS1				
Type B1.1				
TIA 492CAAA				
OS2				
Type B1.3				
TIA 492CAAB				

Splice and Connector Performance

ANSI/TIA-568-C.3 addresses the performance of fusion or mechanical optical fiber splices for premises applications also known as *inside plant* applications. The standard states that a fusion or mechanical splice shall not exceed a maximum optical insertion loss of 0.3dB when measured in accordance with ANSI/EIA/TIA-455-34-A, Method A (factory testing), or ANSI/TIA-455-78-B (field-testing).

This section also defines the minimum *return loss* for mechanical or fusion optical fiber splices. Multimode mechanical or fusion splices shall have a minimum return loss of 20dB while single-mode mechanical or fusion splices shall have a minimum return loss of 26dB when measured in accordance with ANSI/TIA/EIA-455-107-A. The minimum single-mode return loss for broadband analog video CATV applications is 55dB when measured in accordance with ANSI/TIA/EIA-455-107-A.

ANSI/TIA-568-C.3, Annex A describes the optical fiber connector performance requirements. It states that all multimode connectors, adapters, and cable assemblies shall meet the requirements at both 850nm and 1300nm ± 30nm wavelengths. All single-mode connectors, adapters, and cable assemblies shall meet the requirement at both 1310nm and 1550nm ± 30nm wavelengths.

Annex A also states the maximum insertion loss for a mated connector pair. The maximum insertion loss of a mated multimode or single-mode connector pair is 0.75dB. Multimode-mated pairs shall be tested in accordance with FOTP-171 methods A1 or D1, or FOTP-34 method A2. Single-mode mated pairs shall be tested in accordance with FOTP-171 methods A3 or D3, or FOTP-34 method B.

Outside plant splice performance is addressed in TIA-758-B. TIA-758-B states that splice insertion loss shall not exceed 0.1dB mean (0.3dB maximum) when measured with OTDR testing.

USING A POWER BUDGET TO TROUBLESHOOT A FIBER-OPTIC SYSTEM

This chapter explains how to calculate a power budget for a multimode fiber-optic link and a single-mode fiber-optic link. A power budget tells the fiber-optic installer or technician the maximum allowable loss for every component in the fiber-optic link. The sum of these losses equals the maximum acceptable loss for the link per the ANSI/TIA-568-C.3, ISO/IEC 11801, and TIA-758-B standards.

The maximum acceptable loss for the link is the worst-case scenario. Every component in the fiber-optic link should outperform the ANSI/TIA-568-C.3 standard for premises applications and TIA-758-B for outside plant applications. Links with a total loss just under the maximum acceptable loss may not function properly. This is especially true with short links.

To illustrate, we'll tell a story about a short fiber-optic link on a submarine. Of course, all fiber-optic links on a submarine are short. This link, when tested with the light source and power meter, had a total loss under the maximum allowable value. However, when the link was connected to the receiver and transmitter, the fiber-optic system did not function properly.

Because the link was short, the loss for the optical fiber was negligible. The calculated loss came from the multiple interconnections. The maximum allowable loss for the interconnections was 3.0dB. The measured loss for the link was 2.1dB. This is 0.9dB below the maximum allowable. The typical performance for a link with this number of interconnections would have been less than 1.0dB.

One of the interconnections in the link had a loss much greater than the maximum allowable. This bad interconnection was reflecting roughly 20 percent of the light energy from the transmitter back at the transmitter. This back-reflected energy was reflected toward the receiver at the transmitter interconnection. The receiver detected the back-reflected energy from the transmitter interconnection, causing the system to malfunction.

Just because the total loss for a fiber-optic link is below the maximum allowable doesn't mean your system will perform without problems. If your system is having problems, start by evaluating each component in the fiber-optic link.

Power Budget

Now that the link performance requirements have been identified, let's put them to use in a power budget. A power budget as defined in IEEE Standard 802.3 is the minimum optical power available to overcome the sum of attenuation plus power penalties of the optical path between the transmitter and receiver. These losses are calculated as the difference between the minimum transmitter launch power and the minimum receive power.

MULTIMODE LINK ANALYSIS

Table 15.15 lists some of the typical optical characteristics for a 1300nm LED transmitter that could be used for a 100BASE-FX LAN application. In this table, we see that manufacturers typically list three values for optical output power. There is a typical value and there are the maximum and minimum values. The minimum value represents the least amount of power that the transmitter should ever output. This is the output power level we will use in calculating our power budget.

TABLE 15.15:   LED transmitter optical characteristics

PARAMETER		SYMBOL	MIN.	TYP.	MAX.	UNIT
Optical output power	BOL	P_0	−19	−16.8	−14	dBm avg.
62.5/125µm, NA = 0.275 fiber	EOL		−20			
Optical output power	BOL	P_0	−22.5	−20.3	−14	dBm avg.
50/125µm, NA = 0.20 fiber	EOL		−23.5			
Optical extinction ratio				0.001–50	0.03–35	%dB
Optical output power at logic "0" state		P_0 ("0")			−45	dBm avg.
Center wavelength		λ_c	1270	1300	1380	nm
Spectral width—FWHM		$\Delta\lambda$		137	170	nm
Optical rise time		t_r	0.6	1.0	3.0	ns
Optical fall time		t_f	0.6	2.1	3.0	ns
Duty cycle distortion contributed by the transmitter		DCD		0.02	0.6	ns_{P-P}
Duty cycle jitter contributed by the transmitter		DDJ		0.02	0.6	ns_{P-P}

NOTE   Some of the parameters in this table are outside the scope of this book. You do not have to understand all aspects of the datasheet, just the information you need.

Table 15.16 lists some of the typical optical characteristics for a 1300nm receiver that could be used for a 100BASE-FX LAN application. In this table, the manufacturer lists the maximum and minimum optical input power, and under each of those, there is a typical value and a minimum or maximum value. The receiver will perform best when the input power is between the minimum value for maximum optical input power and maximum value for minimum optical input power.

TABLE 15.16:   LED receiver optical characteristics

PARAMETER	SYMBOL	MIN.	TYP.	MAX.	UNIT
Optical input power minimum at window edge	$P_{IN\,Min.}$ (W)		−33.5	−31	dBm avg.
Optical input power maximum	$P_{IN\,Max.}$	−14	−11.8		dBm avg.
Operating wavelength	λ_c	1270		1380	nm

NOTE   Remember from Chapter 10 that the manufacturer-stated optical output power of a typical laser transmitter is measured after 1m of optical fiber. Therefore, we do not have to account for the connector loss at the transmitter. Remember from Chapter 11, "Fiber-Optic Detectors and Receivers," that the manufacturer typically states minimum optical input power for the receiver at the window edge. This means we do not have to account for the connector loss at the receiver.

To determine the power budget for the transmitter defined in Table 15.15 and the receiver defined in Table 15.16, we need to choose an optical fiber. In this example we will use a 62.5/125µm multimode optical fiber. We will also use the EOL value for transmitter output power to account for the aging of the light source.

The minimum EOL value for the optical output power is −20dBm for a 62.5/125µm multimode optical fiber. The maximum value for the minimum optical input power is −31dBm. The difference is 11dB. This is the window of operation where the receiver will provide the best performance. This is also our power budget.

To better understand a power budget, let's take a look a basic fiber-optic link, as shown in Figure 15.2, and calculate the maximum loss allowable. After we have calculated the maximum allowable loss, we can compare maximum loss to the power budget and determine the minimum power to the receiver. If the minimum power to the receiver falls within the window of operation, the link should support low bit error rate data transmission.

The link shown in Figure 15.2 has a 490m span of 62.5/125µm multimode optical fiber. Each end of the optical fiber is connectorized and plugged into a patch panel. On the other side of each patch panel, there is a 5m patch cord. One patch cord connects to the transmitter and the other to the receiver. The total link length is 500m.

FIGURE 15.2

Basic fiber-optic link

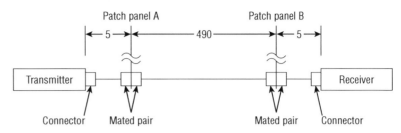

Distance in meters (not to scale)

You saw earlier in Table 15.14 that ANSI/TIA-568-C.3 and ISO/IEC 11801 define the maximum attenuation for a 62.5/125μm optical fiber at both 850nm and 1300nm. Because the transmitter selected for this example has a 1300nm output, we will only evaluate link performance at 1300nm. The maximum allowable attenuation per ANSI/TIA-568-C.3 and ISO/IEC 11801 is 1.5dB per km of 62.5/125μm optical fiber. The total link length was 500m with a maximum allowable attenuation of 0.75dB, as shown here:

Attenuation coefficient at 1300nm = 1.5dB/km

Optical fiber length = 0.5 km

Attenuation for the length of optical fiber = $0.5 \times 1.5 = 0.75$

Maximum attenuation for 500m of optical fiber is 0.75dB

The link shown in Figure 15.2 has six connectors. One connector is plugged into the transmitter, one is plugged into the receiver, and the remaining four are mated together in pairs at each patch panel. When evaluating connector loss, only the mated pairs are accounted for. The loss at the transmitter and receiver is ignored.

Because the transmitter and receiver manufacturers have accounted for connector loss, we only have to account for mated pair loss. In the link shown in Figure 15.2, there are two mated pairs, one at each patch panel. As you learned earlier in this chapter in the "Splice and Connector Performance" section, ANSI/TIA-568-C.3 states that the maximum allowable insertion loss for a mated connector pair is 0.75dB. Because there are two mated pairs on our link, the maximum interconnection loss for our link is 1.5dB.

At this point, we have identified the maximum allowable loss for the optical fiber and the interconnections. The maximum allowable loss for the link is the sum of these two losses. The maximum allowed loss for the link in Figure 15.2 is 2.25dB.

Now that all the performance parameters of the link have been identified, let's determine the minimum power that should be available to the receiver. The minimum power available to the receiver is equal to the minimum optical output power of the LED transmitter minus the loss of the link. The minimum optical output power for the LED transmitter is –20dBm, the link loss is 2.25dB, and the minimum power available at the receiver should be –22.25dBm. The power budget for the link in Figure 15.2 was 11dB. The maximum loss allowable for this link is 2.25dB. The loss for this link is small and provides 8.75dB of headroom for the life of the LED transmitter.

The link we are going to examine in Figure 15.3 is a 50/125μm multimode optical fiber link with a splice. The transmitter for this link is an 850nm VCSEL that could be used for a 1000BASE-SX LAN described in Table 15.6 or a 100-MX-SN-I 1062Mbaud SAN described in

Table 15.10. The minimum EOL optical output power for the 850nm VCSEL transmitter is −8dBm. The minimum optical input power for the receiver is −17dBm.

Looking at Figure 15.3, we see two patch panels with 800m of 50/125μm OM3 multimode optical fiber between them. There is a mechanical splice 400m away from patch panel A. Each end of the optical fiber is connectorized and plugged into the patch panel. Patch cords 5m in length each are used to connect the VCSEL transmitter to one patch panel and the receiver to the other patch panel.

FIGURE 15.3

Multimode fiber-optic link with splice

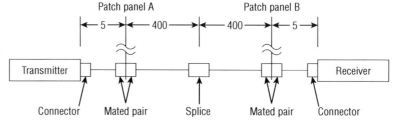

Distance in meters (not to scale)

Table 15.17 is a blank power budget calculation table. This table can be used to calculate the power budget for a link, sum all the losses in the link, and calculate headroom.

TABLE 15.17:   Blank multimode power budget calculation table

STEP NO.	DESCRIPTION	VALUE
1	Minimum EOL optical output power.	dBm
2	Minimum optical input power.	dBm
3	Subtract line 2 from line 1 to calculate the power budget.	dB
4	km of optical fiber.	
5	Number of interconnections.	
6	Number of splices.	
7	Multiply line 4×3.5.	
8	Multiply line 4×1.5.	
9	Multiply line 5×0.75.	
10	Multiply line 6×0.3.	
11	Add lines 7, 9, and 10 for total link loss at 850nm.	dB
12	Add lines 8, 9, and 10 for total link loss at 1300nm.	dB

TABLE 15.17:   Blank multimode power budget calculation table *(continued)*

STEP NO.	DESCRIPTION	VALUE
13	Subtract line 11 from line 3 for headroom at 850nm.	dB
14	Subtract line 12 from line 3 for headroom at 1300nm.	dB

Table 15.18 has all the values for the link filled in. The power budget for the link is 9dB. The loss for the link is approximately 4.6dB and the headroom for the link is approximately 4.4dB. This link does not meet the requirements established for a 1000BASE-SX LAN in Table 15.6 because it is 10m too long and it exceeds the maximum channel attenuation by 0.1dB. However, it does meet the requirements established for a 100-MX-SN-I 1062Mbaud SAN, as described in Table 15.10.

TABLE 15.18:   Completed multimode calculation table for an 850nm VCSEL transmitter

STEP NO.	DESCRIPTION	VALUE
1	Minimum EOL optical output power.	–8dBm
2	Minimum optical input power.	–17dBm
3	Subtract line 2 from line 1 to calculate the power budget.	9.0dB
4	km of optical fiber.	0.810
5	Number of interconnections.	2.0
6	Number of splices.	1.0
7	Multiply line 4×3.5.	2.8
8	Multiply line 4×1.5.	
9	Multiply line 5×0.75.	1.5
10	Multiply line 6×0.3.	0.3
11	Add lines 7, 9, and 10 for total link loss at 850nm.	4.6dB
12	Add lines 8, 9, and 10 for total link loss at 1300nm.	
13	Subtract line 11 from line 3 for headroom at 850nm.	4.4dB
14	Subtract line 12 from line 3 for headroom at 1300nm.	

Up to this point, we have only looked at calculating the power budget for a link using a table. However, the power budget can be plotted graphically. The graphical representation of the losses in the link very much resembles the trace of an OTDR. The OTDR is covered in detail in Chapters 16 and 17.

Figure 15.4 is the graphical representation of the link we analyzed in Figure 15.3. When analyzing a link graphically, power gain or loss is noted on the vertical scale and distance is noted on the horizontal scale. Power gain and loss may be drawn to scale; however, there is not usually sufficient room on your paper to draw distance to scale.

FIGURE 15.4

Graphical representation of a multimode fiber-optic link

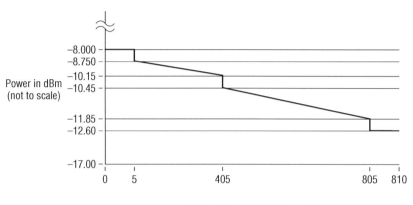

The first step in setting up the graphical representation of the link is to assign values for the vertical axis. As shown in Table 15.18, the maximum value assigned is the minimum EOL optical output power, –8dBm. The minimum value assigned is the minimum optical input power for the receiver, –17dBm.

The next step is to assign values to the vertical axis that represent the before and after power levels for the interconnections. Working from the transmitter to the receiver, the first interconnection is a mated connector pair at patch panel A. There is a 5m patch cord between the transmitter and patch panel A. At a wavelength of 850nm, the maximum loss for the 5m patch cord is only 0.0175dB. This very small value represents only a fraction of a percent of the loss for this link and can be ignored in the graphical representation. Typically, you can ignore all patch cord loss when generating a graphical representation of a link. This simplifies the drawing and reduces the time required to generate the drawing.

The amount of light energy entering the first mated pair is –8dBm. The maximum loss allowed by ANSI/TIA-568-C.3 for a mated pair is 0.75dB. Using this value, the power exiting the mated pair should be no less than –8.75dBm. In a linear fashion working from top to bottom, the next value recorded on the vertical axis is –8.75dBm.

The next value on the vertical axis represents the light energy entering the mechanical splice. There is 400m of optical fiber from the first mated pair to the mechanical splice. The loss for the 400m of optical fiber is 1.4dB. The light energy entering the mechanical splice is –10.15dBm and is recorded on the vertical axis. The maximum loss allowed by ANSI/TIA-568-C.3 for a mechanical splice is 0.3dB. Using this value, the power exiting the mechanical splice should be no less than –10.45dBm, which is recorded on the vertical axis.

Between the mechanical splice and patch panel B is 400m of optical fiber. The loss of the 400m of optical fiber is 1.4dB. The light energy entering the mated pair at patch panel B is –11.85dBm, which should be recorded on the vertical axis. As we already know, ANSI/TIA-568-C.3 allows a maximum loss of only 0.75dB for a mated pair. The minimum amount of light energy exiting the mated pair should be –12.6dBm. If we ignore the loss for the 5m patch cord between patch panel B and the receiver, the power available to the receiver is –12.6dBm. This is the last value recorded on the vertical axis.

The next step in this process is to record the horizontal axis values. Working from left to right starting at the transmitter, the first value recorded on the horizontal axis is zero. It is from this point that we will record the successive segments of optical fiber.

The first segment of optical fiber is the 5m patch cord. With an overall length of 810m, it's not practical to draw the 5m patch cord to scale unless you have a very large sheet of paper. Looking at Figure 15.4, you can see that the first segment on the horizontal axis is long enough to clarify that the 5m patch cord is a separate segment of optical fiber. The second value recorded on the horizontal axis is 5m.

The second segment is a 400m span of optical fiber from patch panel A to the mechanical splice. The next value recorded on the horizontal axis is 405m.

The third segment of optical fiber is a 400m span from the mechanical splice to patch panel B. The next value recorded on the horizontal axis is 805m.

The last segment of optical fiber is the 5m patch cord that connects the receiver to patch panel B. The last value recorded on the horizontal axis is 810m.

With all the values for the vertical and horizontal axes recorded, the last step is to draw a series of lines that graphically represent the recorded values. (You may want to use a straight edge for this.) The first line segment will be a horizontal line from –8.0dBm on the vertical axis to the 5m point on the horizontal axis. This line represents the patch cord that connects the transmitter to patch panel A. The line is horizontal because the patch cord loss is insignificant. At the end of this, a vertical line will be drawn that runs from –8.0dBm to –8.75dBm; it represents the loss for the mated pair at patch panel A.

The third line will extend from the 5m point on the horizontal axis to the 405m point. It will be a sloping line from left to right that goes from –8.75dBm on the vertical axis to –10.15dBm. This line represents the loss for the 400m section of optical fiber from patch panel A to the mechanical splice. At the end of this line, a vertical line is drawn from –10.15dBm to –10.45dBm to represent the loss for the mechanical splice.

The fifth line will extend horizontally from the 405m point to the 805m point. This line will slope from –10.45dBm on the vertical axis to –11.85dBm; it represents the loss for the 400m section of optical fiber from the mechanical splice to patch panel B. At the end of this line, a vertical line will be drawn from –11.85dBm to –12.6dBm, representing the loss for the mated pair at patch panel B.

The last line drawn is a horizontal line from the 805m point on the horizontal axis to the 810m point. This line represents the patch cord from patch panel B to the receiver. There is no slope to this line because the loss for the patch cord is insignificant. If everything has been drawn correctly, your drawing should look like Figure 15.4.

SINGLE-MODE LINK ANALYSIS

Figure 15.5 is a basic single-mode optical fiber link with 4510m of optical fiber spliced in three locations between two patch panels located in separate buildings. The transmitter is connected to patch panel A with a 5m patch cord and the receiver is connected to patch panel B with a 5m patch cord. There is an inside plant splice 50m from patch panel A, another 25m from patch panel B, and an outside plant splice 2000m from patch panel A.

FIGURE 15.5
Single-mode fiber-optic link

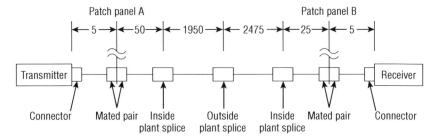

Distance in meters (not to scale)

Before we can analyze the link and do a power budget, we need to look at how manufacturers typically describe the performance characteristics for laser transmitters and receivers. Table 15.19 lists some of the typical optical power characteristics for a 1310nm laser transmitter that could be used for a 1000BASE-LX LAN described in Table 15.7 or a 100-SM-LC-L 1062Mbaud SAN described in Table 15.11. In this table, we see that manufacturers usually list three values for optical output power. There is a typical value, and there are maximum and minimum values. The minimum value represents the least amount of power the transmitter should ever output. This is the output power level we will use in calculating our power budget.

TABLE 15.19:   Laser transmitter optical characteristics

PARAMETER	SYMBOL	MIN.	TYP.	MAX.	UNIT
Optical output Power 9μm SMF	P_{OUT}	–10	–6	–3	dBm
Center wavelength	λ_c	1260		1360	nm
Spectral width—rms	D_c		1.8	4	nm rms
Optical rise time	t_r		30	70	ps
Optical fall time	t_f		150	225	ps
Extinction ratio	E_R	8.2	12		dB
Optical output eye	Compliant with eye mask Telecordia GR-253-CORE				
Back-reflection sensitivity				–8.5	dB
Jitter generation	pk to pk			70	mUI
	RMS			7	mUI

Table 15.20 lists some of the typical optical power characteristics for a 1310nm receiver that could be used for a 1000BASE-LX LAN described in Table 15.7 or a 100-SM-LC-L 1062Mbaud SAN described in Table 15.11. In this table, the manufacturers list the maximum and minimum optical input power, and under each of those, there is a typical and a minimum or maximum value. The receiver will perform best when the input power is between the minimum value for maximum optical input power and maximum value for minimum optical input power.

TABLE 15.20: Laser receiver optical characteristics

PARAMETER	SYMBOL	MIN.	TYP.	MAX.	UNIT
Receiver sensitivity	P_{IN} MIN		−23	−19	dBm avg.
Receiver overload	P_{IN} MAX	−3	+1		dBm avg.
Input operating wavelength	λ	1260		1550	nm
Signal detect—asserted	P_A		−27.3	−19.5	dBm avg.
Signal detect—deasserted	P_D	−35	−28.7		dBm avg.
Signal detect—hysteresis	P_H	0.5	1.4	4	dB
Reflectance			−35	−27	dB

With all the transmitter and receiver values defined, we can calculate the power budget using the power budget calculation table (Table 15.21). The minimum value for the maximum optical input power is −10dBm. The maximum value for the minimum optical input power is −19dBm. The difference is 9dB; however, the manufacturer of the transmitter did not state an EOL value. Since the EOL value was not provided, we should assume the transmitter output power will decay 1.5dB. This value is defined as the industry convention (see Chapter 10). With transmitter output power decay added in, the power budget is 7.5dB for this transmitter-receiver combination.

TABLE 15.21: Blank single-mode power budget calculation table

STEP NO.	DESCRIPTION	VALUE
1	Minimum EOL optical output power.	dBm
2	Minimum optical input power.	dBm
3	Subtract line 2 from line 1 to calculate the power budget.	dB
4	km of inside plant optical fiber.	
5	km of outside plant optical fiber.	

TABLE 15.21: Blank single-mode power budget calculation table *(continued)*

STEP NO.	DESCRIPTION	VALUE
6	Number of interconnections.	
7	Number of inside plant splices.	
8	Number of outside plant splices.	
9	Multiply line 4 × 1.0.	
10	Multiply line 5 × 0.5.	
11	Multiply line 6 × 0.75.	
12	Multiply line 7 × 0.3.	
13	Multiply line 8 × 0.1.	
14	Add lines 9, 10, 11, 12, and 13 for total link loss.	dB
15	Subtract line 14 from line 3 for headroom.	dB

Now that the power budget has been calculated, the power budget calculation table can be completed. The link shown in Figure 15.5 contains two mated pairs, two inside plant splices, one outside plant splice, and 4510m of optical fiber. The maximum allowable loss for the two mated pairs is 1.5dB. The maximum allowable loss for the two inside plant splices is 0.6dB, and for the outside plant splice it is 0.1dB.

The 4510m of optical fiber is broken up into inside plant fiber and outside plant fiber. The 4425m of optical fiber between the two inside plant splices will be evaluated as outside plant and the remaining 85m will be evaluated as inside plant. The maximum allowable loss for the outside plant optical fiber is rounded down to 2.2dB. The maximum allowable loss of the inside plant optical fiber is 0.085, which is rounded up to 0.1dB. The maximum allowable loss for this link is rounded up to 4.5dB. Our power budget was 7.5dB; that leaves us 3.0dB of headroom. Table 15.22 is the completed power budget calculation table with the rounded values.

This link meets the requirements for a 1000BASE-LX LAN described in Table 15.7 or a 100-SM-LC-L 1062Mbaud SAN described in Table 15.11.

TABLE 15.22: Completed single-mode power budget calculation table

STEP NO.	DESCRIPTION	VALUE
1	Minimum EOL optical output power.	–11.5dBm
2	Minimum optical input power.	–19.0dBm
3	Subtract line 2 from line 1 to calculate the power budget.	7.5dB

TABLE 15.22: Completed single-mode power budget calculation table *(continued)*

STEP NO.	DESCRIPTION	VALUE
4	km of inside plant optical fiber.	0.085
5	km of outside plant optical fiber.	4.425
6	Number of interconnections.	2
7	Number of inside plant splices.	2
8	Number of outside plant splices.	1
9	Multiply line 4×1.0.	0.1
10	Multiply line 5×0.5.	2.2
11	Multiply line 6×0.75.	1.5
12	Multiply line 7×0.3.	0.6
13	Multiply line 8×0.1.	0.1
14	Add lines 9, 10, 11, 12, and 13 for total link loss.	4.5dB
15	Subtract line 14 from line 3 for headroom.	3.0dB

With the power budget calculation table completed, you should be able to quickly sketch out the graphical representation of the link. Figure 15.6 is the completed graphical representation for the link we just discussed without rounding.

FIGURE 15.6

Graphical representation of a single-mode fiber-optic link

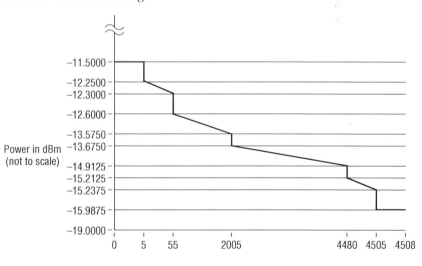

Power in dBm (not to scale)

Distance in meters (not to scale)

The Bottom Line

Calculate the bandwidth for a length of optical fiber.   The minimum effective bandwidth-length product in MHz · km as defined in TIA-568-C.3 and ISO/IEC 11801.

Master It   Refer to the following table and determine the minimum effective modal bandwidth for 300 meters of OM4 optical fiber at a wavelength of 850nm.

Optical fiber cable type	Wavelength (nm)	Maximum attenuation (dB/km)	Minimum overfilled modal bandwidth-length product (MHz · km)	Minimum effective modal bandwidth-length product (MHz · km)
62.5/125μm Multimode OM1 Type A1b TIA 492AAAA	850 1300	3.5 1.5	200 500	Not required Not required
50/125μm Multimode OM2 TIA 492AAAB Type A1a.1	850 1300	3.5 1.5	500 500	Not required Not required
850nm laser-optimized 50/125μm Multimode OM3 Type A1a.2 TIA 492AAAC	850 1300	3.5 1.5	1500 500	2000 Not required
850nm laser-optimized 50/125μm Multimode OM4 Type A1a.3 TIA 492AAAD	850 1300	3.5 1.5	3500 500	4700 Not required
Single-mode Indoor-outdoor OS1 Type B1.1 TIA 492CAAA OS2 Type B1.3 TIA 492CAAB	1310 1550	0.5 0.5	N/A N/A	N/A N/A

Optical fiber cable type	Wavelength (nm)	Maximum attenuation (dB/km)	Minimum overfilled modal bandwidth-length product (MHz · km)	Minimum effective modal bandwidth-length product (MHz · km)
Single-mode Inside plant OS1 Type B1.1 TIA 492CAAA OS2 Type B1.3 TIA 492CAAB	1310 1550	1.0 1.0	N/A N/A	N/A N/A
Single-mode Outside plant OS1 Type B1.1 TIA 492CAAA OS2 Type B1.3 TIA 492CAAB	1310 1550	0.5 0.5	N/A N/A	N/A N/A

Calculate the attenuation for a length of optical fiber.   The maximum allowable attenuation in an optical fiber is defined in TIA-568-C.3 and ISO/IEC 11801.

Master It   Refer to the following table and determine the maximum allowable attenuation for 5.6km of outside plant single-mode optical fiber.

Optical fiber cable type	Wavelength (nm)	Maximum attenuation (dB/km)
50/125μm multimode	850 1300	3.5 1.5
62.5/125μm multimode	850 1300	3.5 1.5
Single-mode inside plant cable	1310 1550	1.0 1.0
Single-mode outside plant cable	1310 1550	0.5 0.5

Calculate the power budget and headroom for a multimode fiber-optic link.   Before we can analyze the link and do a power budget, we need to know the optical performance characteristics for transmitter and receiver.

Master It Refer to the following graphic and two tables to calculate the power budget and headroom for the multimode fiber-optic link with 50/125μm multimode optical fiber.

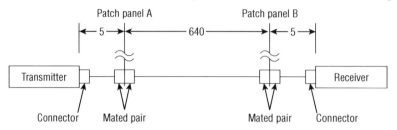

Distance in meters (not to scale)

Parameter	Symbol		Min.	Typ.	Max.	Unit
Optical output power	BOL	P_0	−19	−16.8	−14	dBm avg.
62.5/125μm, NA = 0.275 fiber	EOL		−20			
Optical output power	BOL	P_0	−22.5	−20.3	−14	dBm avg.
50/125μm, NA = 0.20 fiber	EOL		−23.5			
Optical extinction ratio				0.001–50	0.03–35	%dB
Optical output power at logic "0" state	P_0 ("0")				−45	dBm avg.
Center wavelength	λ_c		1270	1300	1380	nm
Spectral width—FWHM	$\Delta\lambda$			137	170	nm
Optical rise time	t_r		0.6	1.0	3.0	ns
Optical fall time	t_f		0.6	2.1	3.0	ns
Duty cycle distortion contributed by the transmitter	DCD			0.02	0.6	ns_{p-p}

Parameter	Symbol	Min.	Typ.	Max.	Unit
Duty cycle jitter contributed by the transmitter	DDJ		0.02	0.6	ns_{p-p}
Optical input power minimum at window edge	$P_{IN\ Min.}$ (W)		−33.5	−31	dBm avg.
Optical input power maximum	$P_{IN\ Max.}$	−14	−11.8		dBm avg.
Operating wavelength	λ_c	1270		1380	nm

Calculate the power budget and headroom for a single-mode fiber-optic link. Before we can analyze the link and do a power budget, we need to know the optical performance characteristics for transmitter and receiver.

> **Master It** Refer to the following graphic and two tables to calculate the power budget and headroom for the single-mode fiber-optic link.

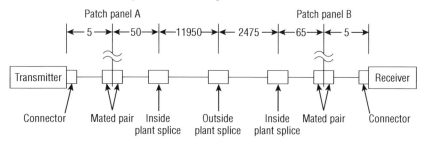

Distance in meters (not to scale)

Parameter	Symbol	Min.	Typ.	Max.	Unit
Optical output Power 9µm SMF, EOL	P_{OUT}	−5	−2	0	dBm
Center wavelength	λ_c	1260		1360	nm
Spectral width—rms	$\Delta\lambda$		1.8	4	nm rms
Optical rise time	t_r		30	70	ps
Optical fall time	t_f		150	225	ps
Extinction ratio	E_R	8.2	12		dB

Parameter	Symbol	Min.	Typ.	Max.	Unit
Optical output eye	Compliant with eye mask Telecordia GR-253-CORE				
Back-reflection Sensitivity				–8.5	dB
Jitter generation	pk to pk			70	mUI
	RMS			7	mUI
Receiver sensitivity	PIN_{MIN}		–23	–19	dBm avg.
Receiver overload	$P_{IN} MAX$	0	+1		dBm avg.
Input operating wavelength	λ	1260		1550	nm
Signal detect—asserted	P_A		–27.3	–19.5	dBm avg.
Signal detect—deasserted	P_D	–35	–28.7		dBm avg.
Signal detect—hysteresis	P_H	0.5	1.4	4	dB
Reflectance			–35	–27	dB

Chapter 16

Test Equipment and Link/Cable Testing

This chapter describes the basic operation and application of essential fiber-optic test equipment. The basic theory, operation, and application of each piece of test equipment will be explained. Many of the test methods described in this chapter are based on current industry standards. How to test to these standards is described in detail.

In this chapter, you will learn to:

- ◆ Perform optical loss measurement testing of a cable plant
- ◆ Determine the distance to a break in the optical fiber with an OTDR
- ◆ Measure the loss of a cable segment with an OTDR
- ◆ Measure the loss of an interconnection with an OTDR
- ◆ Measure the loss of a fusion splice or macrobend with an OTDR

Calibration Requirements

The routine calibration of test equipment is essential to achieving accurate test results. Calibration is always required for test equipment that makes absolute measurements, such as an optical power meter. However, not all test equipment requires calibration. For example, calibration is not required for a continuity tester; however, it is necessary for a stabilized light source.

Optical power meters should be calibrated to the three principle wavelength regions: 850nm, 1300nm, and 1550nm. The calibration should be traceable to the National Institute of Standards and Technology (NIST) calibration standard or some other standard. A calibration sticker similar to the one shown in Figure 16.1 should be clearly visible and not expired. The calibration is expired when the posted due date has passed.

FIGURE 16.1
Calibration sticker
on a piece of test
equipment

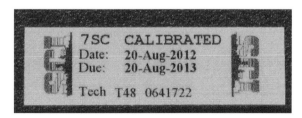

7 SC   CALIBRATED
Date: 20-Aug-2012
Due: 20-Aug-2013

Tech T48   0641722

NIST in general does not require or recommend any set calibration intervals. Many companies adopt their own measurement assurance programs and define calibration intervals based on various factors such as:

♦ Equipment manufacturer recommendations

♦ Contractual requirements

♦ Regulatory requirements

♦ Environmental factors

♦ Inherent stability of the device

To access NIST publications, go to www.nist.gov.

Continuity Tester

The continuity tester is an essential piece of test equipment for every fiber-optic toolkit. It is also one of the least expensive tools in your toolkit. This low-cost tool will allow you to quickly verify the continuity of an optical fiber.

Many of the continuity testers available today are just modified flashlights. Some flashlights, such as the one pictured in Figure 16.2, have been modified to use an LED instead of an incandescent lamp. Others may just use a brighter incandescent lamp, like the tester pictured in Figure 16.3.

FIGURE 16.2
LED continuity
tester

FIGURE 16.3
Incandescent continuity tester

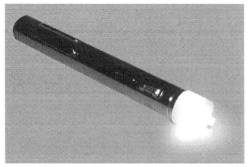

The job of the continuity tester is to project light from the LED or incandescent lamp into the core of the optical fiber. This is usually accomplished by attaching a receptacle to the lamp end of the flashlight. The receptacle is designed to center and hold the connector ferrule directly above the LED or incandescent lamp. When the connector ferrule is inserted into the receptacle, the endface is typically just above the light source. This approach eliminates the need for a lens to direct light into the core of the optical fiber. However, it directs only a fraction of the light emitted by the lamp or LED into the core of the optical fiber.

Because there is no lens used to focus light energy into the optical fiber, the continuity tester works best with multimode optical fiber. The measured optical output power of an LED continuity tester is typically less than –36dBm when measured at the end of one meter of 62.5/125μm optical fiber. The optical output power is reduced another 3–4dB when used with 50/125μm optical fiber and 16–20dB when used with single-mode optical fiber. The continuity tester works best with multimode optical fiber; however, it can be used with single-mode optical fiber. For the best results with single-mode optical fiber, dim the lights in the test area if possible.

The low output power of the continuity tester combined with the high attenuation of visible light by an optical fiber limit the length of optical fiber that can be tested. A multimode optical fiber will attenuate a 650nm light source roughly 7dB per kilometer. This high attenuation limits the use of the continuity tester to multimode optical fibers typically no greater than 2km in length.

LED continuity testers have a couple of advantages over incandescent lamp testers. They typically feature a red (635–650nm) LED that is easy to see. They require far less power from the batteries than an incandescent lamp. An LED continuity tester may provide 10 or more times longer battery life compared to an incandescent lamp. The LED lamp will never need replacing, unlike an incandescent lamp, which may last only 10 hours. LED lamps are also shock resistant.

Whether you are using an LED or an incandescent lamp continuity tester, operation and testing are identical. The continuity tester can test only for breaks in the optical fiber. It does not have sufficient power to identify the location of a break.

The first step when using the continuity tester is to clean and visually inspect the endface of the connector before inserting it into the continuity tester. You need to visually inspect the connector to verify that there is no endface damage. A shattered endface will significantly reduce the light coupled into the core of the optical fiber under test.

After the connector has been cleaned and inspected, you need to verify that the continuity tester is operating properly. Turn the continuity tester on and verify that it is emitting light. This could save you from embarrassment. You don't want to tell the customer that there is a break in their fiber-optic cable when the real problem is dead batteries in your continuity tester.

Depending on where the other end of the fiber-optic cable to be tested is located, you may need an assistant to help you. With the continuity tester turned on, insert the ferrule of the connector under test into the receptacle. If light is being emitted from the other end of the optical fiber, there is good continuity. This means only that there are no breaks in the optical fiber. It does not mean that there are no macrobends or high-loss interconnections in the fiber-optic cable or link under test.

The continuity tester is often used to verify that there are no breaks in a reel of fiber-optic cable before it is installed. There are a few ways you could approach testing the reel. One way would be to install a connector on either end of the cable. The other end of the cable should have the jacket and strength member stripped back so that the buffer is exposed. You should remove a small amount of buffer to expose the optical fiber under test. This will allow you to clearly see the light from the continuity tester, ensuring accurate results. To test the optical fiber in the reel, turn on the continuity tester and insert the ferrule of the connector under test into the receptacle. If light is being emitted from the other end of the optical fiber, there is good continuity.

Another approach is to use a pigtail with a mechanical splice. The pigtail would have a connector on one end that will mate with the continuity tester receptacle. The other end should have the jacket, strength member, and buffer stripped back so that 10 to 15mm of optical fiber is exposed, as shown in Figure 16.4. The optical fiber should have a perpendicular cleave.

FIGURE 16.4
Pigtail prepared for temporary mechanical splice

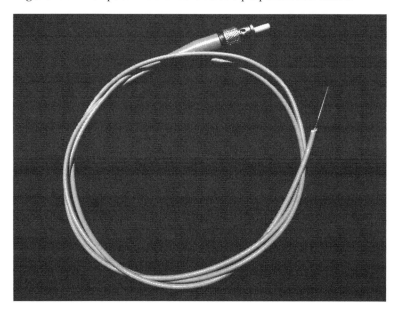

One end of the cable under test should be prepared just like the nonconnectorized end of the pigtail. The other end should have the jacket and strength member stripped back so that the buffer is exposed. A small amount of buffer should be removed to expose the optical fiber under test.

To test for continuity, insert the pigtail connector into the continuity tester receptacle. Turn on the continuity tester and verify that light is being emitted from the exposed optical fiber on the opposite end of the pigtail. Insert the exposed optical fiber from the pigtail into one side of a mechanical splice. Insert the optical fiber from the cleaved end of the cable into the other side of the mechanical splice and check for continuity. Typically, one mechanical splice can be reused for many optical fibers depending on the type of splice.

Another method that works well with multimode optical fiber is simply projecting light into the optical fiber with a continuity tester. As with the other techniques discussed, both ends of the cable should have the jacket and strength member stripped back so that the buffer is exposed. A small amount of buffer should be removed to expose the optical fiber under test and one end of the optical fiber should have a perpendicular cleave.

After the fiber-optic end is cleaved, simply insert the buffered optical fiber into the continuity tester receptacle as shown in Figure 16.5 to test the continuity of the entire length of fiber. Use caution when performing this step to ensure that the optical fiber end is not broken off.

FIGURE 16.5
Testing the continuity of a single fiber in a 12-fiber breakout cable

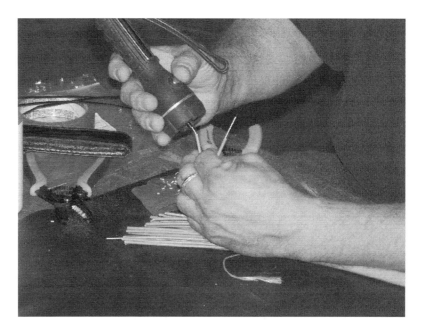

While performing the steps outlined previously, always work over a nonreflective black surface like the one shown in Figure 16.5. The nonreflective black mat makes it easier to keep track of optical fiber ends. Always use a labeled container designated only for fiber-optic waste to dispose of bare optical fibers. Refer to Chapter 6, "Safety," for additional safety information.

Visual Fault Locator

Like the continuity tester, the *visual fault locator (VFL)* is an essential tool for every fiber-optic toolkit. Unlike the continuity tester, it is not one of the least expensive tools in your toolkit. The VFL will allow you to quickly identify breaks or macrobends in the optical fiber and identify a poor fusion splice in multimode or single-mode optical fiber.

The big difference between the continuity tester and the VFL is the light source and optical output power of the light source. The VFL typically uses a red (635–650nm) laser light source. The optical output power of the laser is typically 1mW or less. Because of the high optical output power, you should never view the output of the VFL directly.

The VFL is available in different shapes and sizes. Some may look like a pen like the one shown in Figure 16.6, others may be built into an optical time-domain reflectometer (OTDR) or other piece of test equipment like the one shown in Figure 16.7, and some may look like a small test equipment box like the one shown in Figure 16.8.

FIGURE 16.6
VFL in the shape of a pen

Photo courtesy of KITCO Fiber Optics

FIGURE 16.7
VFL built into
a piece of test
equipment

FIGURE 16.8
VFL with LC
receptacle

The VFL fills the core of the optical fiber with light from the laser. The light from the laser escapes the optical fiber at a break or macrobend. The light escaping from the optical fiber will typically illuminate the buffer surrounding the optical fiber. Macrobends are not always visible through the jacket but are typically visible through the buffer. Breaks may be visible through the jacket of the fiber-optic cable depending on jacket color, thickness, number of optical fibers in the cable, and amount of strength member.

Unlike the continuity tester, the VFL is not limited to testing multimode optical fibers 2km or less in length. The VFL can be used to verify continuity of multimode or single-mode optical fiber longer than 2km. Due to attenuation of the 635–650nm laser light source by the optical fiber, macrobends may not be detectable beyond 1km in multimode optical fiber and 500m in single-mode optical fiber. The same holds true for finding breaks in the optical fiber through the jacket of the fiber-optic cable.

The first step before using the VFL for testing an optical fiber is to measure the output power of the VFL. This can be done with a multimode test jumper and an optical power meter. Connect the output of the VFL to the test jumper and the other end of the test jumper to the power meter input. Set the power meter at the shortest wavelength possible. This is typically 850nm. The measured optical output power of the VFL should be no less than –7dBm with a 62.5/125µm optical fiber. If the measured optical output power is less than –7dBm, replace the battery and retest. Most VFLs will have a measured optical output power greater than –3dBm with a fresh battery. Optical output power below –7dBm will limit the effectiveness of the VFL to identify breaks and macrobends.

The second step before using the VFL is to clean and visually inspect the endface of the connector on the optical fiber to be tested before inserting it into the VFL. You need to visually inspect the connector to verify that there is no endface damage. A shattered endface will not allow light to be properly coupled into the optical fiber under test. A shattered endface may also damage the VFL. Some VFLs use a short pigtail from the laser to the receptacle. The other end of the pigtail has a connector that will mate with the receptacle; this is referred to as a contact VFL. A shattered endface can damage the optical fiber in the connector on the end of the pigtail.

NOTE   There are two types of VFLs: contact and non-contact. With a contact VFL, the optical fiber under test will make contact with the VFL. However, with a non-contact VFL the optical fiber under test will not touch the VFL.

To identify a macrobend in the optical fiber, visually inspect the length of the cable under test. If you locate a bend in the cable that is glowing, this is a macrobend. Figure 16.9 shows simplex cordage with a macrobend.

FIGURE 16.9
The light from a VFL identifying a macrobend in simplex cordage

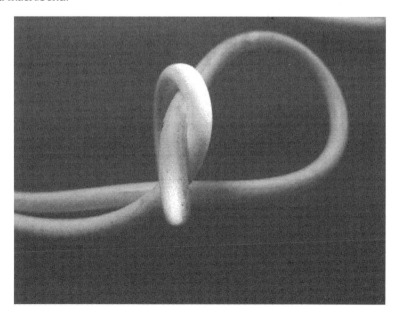

To identify a break in the optical fiber, visually inspect the length of the cable under test. If you locate a small spot on the cable that is emitting red light, this is a break. Figure 16.10 shows multimode cordage with a break.

FIGURE 16.10
The light from a VFL identifying a break in simplex cordage

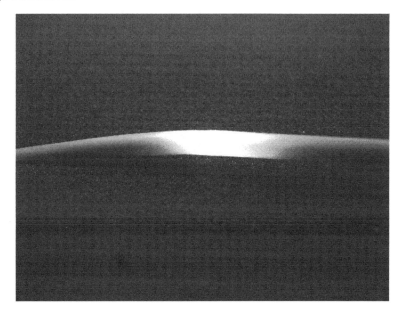

The VFL can be used to test the continuity of an optical fiber in the same manner described in the "Continuity Tester" section earlier in this chapter. The VFL, however, will couple 1,000 times (30dB) or more light energy into the optical fiber than the LED continuity tester. You should never view the output of the VFL or the endface of a connector being tested by the VFL directly.

Using the VFL to test for continuity can produce false results. A single break in an optical fiber does not stop all the light from the VFL cold. Some of the light will typically reach the end of the optical fiber. In most cases, so much light reaches the end of the optical fiber that the person viewing the light might believe there is good continuity when in fact the optical fiber is broken.

A break in the optical fiber will typically attenuate the light from the VFL by 20–30dB. As you learned earlier in the chapter, the VFL will couple 30dB or more light energy into the optical fiber than the LED continuity tester. The maximum loss from a break in an optical fiber typically never exceeds 30dB. Therefore, the light energy from the VFL after passing through a single break with 30dB of loss will be equal to or greater than the light energy from an LED continuity tester. While the VFL can be used to test continuity, the continuity tester is typically the best tool for that job.

The VFL is often used in conjunction with an OTDR to help identify the actual location of the fault. Chapter 17, "Troubleshooting and Restoration," describes how to use the VFL to troubleshoot macrobends or breaks in an optical fiber.

NOTE   Good light and bad light: When working with other installers or technicians, you may hear them refer to good light or bad light while using a VFL for continuity testing. The term *good light* implies there is good continuity and *bad light* implies there is not good continuity. The use of the good light/bad light technique will produce false results simply because the VFL may couple 1,000 times (30dB) or more light energy into the optical fiber than the LED continuity tester. The light energy from the VFL will pass through a single break and typically look brighter than the light energy from an unbroken optical fiber being tested with an LED continuity tester. This is why a VFL is not called a continuity tester and a continuity tester is not called a VFL.

Fiber Identifier

The fiber identifier acts as the fiber-optic installer or technician's infrared eyes. By placing a slight macrobend in an optical fiber or fiber-optic cable, it can detect infrared light traveling through the optical fiber and determine the direction of light travel. Some fiber identifiers can also detect test pulses from an infrared (800–1700nm) light source or measure the power in the optical fiber.

The fiber identifier typically contains two photodiodes that are used to detect the infrared light. The photodiodes are mounted so that they will be on opposite ends of the macrobend of the optical fiber or fiber-optic cable being tested; this is often referred to as the optical head. Figure 16.11 shows the location of the photodiode assemblies in the fiber identifier. The photodiode assemblies look like two small glass lenses.

FIGURE 16.11
Photodiode assemblies in the fiber identifier

The fiber identifier can typically be used with coated optical fiber, tight-buffered optical fiber, a single optical fiber cable, or a ribbon cable. Each of these must be placed in the center of the photodiodes during testing. Selecting the correct attachment for the optical fiber or optical fiber cable type under test typically does this. Figure 16.12 shows three attachments. Some fiber identifiers have proprietary optical head designs that do require attachments to be interchanged when working with different diameter buffers or cables.

FIGURE 16.12
Fiber identifier
optical fiber and
fiber-optic cable
attachments

To test an optical fiber or fiber-optic cable, select the correct attachment if required and install it on the fiber identifier. Figure 16.13 shows the fiber identifier ready to test a 900μm tight-buffered optical fiber. Center the tight-buffered optical fiber over the photodiode assemblies and insert a macrobend into the tight-buffered optical fiber by pushing the slide into the tight-buffered optical fiber and compressing it, as shown in Figure 16.14.

FIGURE 16.13
Fiber identifier
ready to test a
900μm tight-
buffered optical
fiber

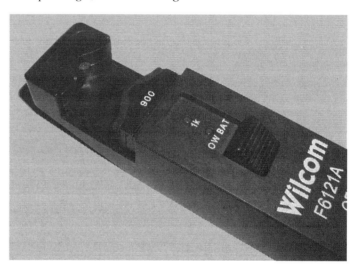

If there is sufficient infrared light energy traveling through the optical fiber, one of the directional arrows on the fiber identifier should illuminate. The directional arrow is pointing in the direction the light is traveling. If a test pulse is being transmitted through the optical fiber, the directional arrow and test pulse indicator should illuminate. If the fiber identifier contains a power meter, the optical power level should be displayed.

The fiber identifier works best with coated optical fiber or tight-buffered optical fiber. Like the VFL, the fiber identifier may not always work with single optical-fiber cables. How well it works with a cable depends on the amount of strength member within the cable, jacket thickness, jacket color, and amount of light energy available from the optical fiber.

The fiber identifier can be used by itself or in conjunction with an OTDR. It can be used to identify traffic in a working fiber-optic link or can be used to help identify the location of a fault. Fault location techniques are discussed in detail in Chapter 17.

FIGURE 16.14
Fiber identifier
compressing a
tight-buffered opti-
cal fiber

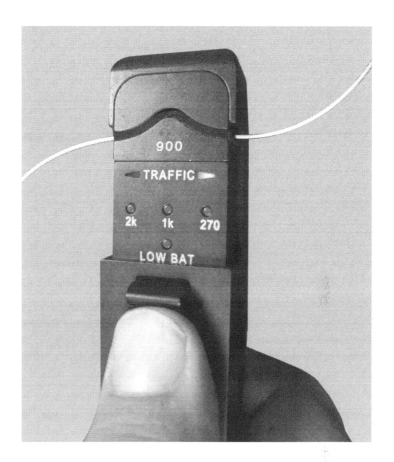

Inline Optical Power Monitoring

Inline power monitoring is not a new concept in telecommunications. It has been deployed for decades in radio frequency (RF) applications measuring the forward and reverse power in coaxial copper cables. However, it is relatively new to fiber-optic applications. As of this writing, AIR6552/1, Inline Optical Power Monitoring, was under development within SAE International's AS-3 Fiber Optics and Applied Photonics Committee. This standard and its slash sheets are covered in detail in the "Emerging Testing Standards" section at the end of this chapter.

NOTE   *Slash sheets* are typically a family of documents where the main document such as AIR6552 will refer to an overall/general aspect of the document and then the slash sheet document(s) will refer to specific items within that overall document.

There are many applications for inline optical power monitoring, such as measuring forward optical power, reverse optical power, controlling optical power in an amplifier, anticipating failures, and detecting fluctuations, to name a few.

Inline optical power monitoring is different from using a fiber identifier, which places a macrobend in the optical fiber to detect infrared light traveling through the optical fiber. It requires an inline power tap like the one shown in Figure 16.15 to be placed inline with the optical fiber. The inline power tap is described in detail in Chapter 12, "Passive Components and Multiplexers."

FIGURE 16.15

Inline optical fiber power tap

Photo courtesy of EigenLight Corporation

An inline power tap like the one shown in Figure 16.15 is a directional sensor that can only measure the light traveling in one direction. This power tap can be placed inline to measure forward power or reverse power. Two taps placed in series could be configured to measure forward and reverse power simultaneously. An example of this is shown in the block diagram in Figure 16.16, where two taps are used to measure the forward power into an interconnection and the reverse power or back reflection from the interconnection. While other test equipment such as an OTDR can be used to evaluate the light energy, being reflected back from an interconnection, only the power tap allows this to be performed in real time using the light energy from the transmitted signal.

FIGURE 16.16

Inline optical fiber power taps in series measuring forward and reverse power

Inline Optical Power Monitor

Inline optical power monitors like the one shown in Figure 16.17 are optically passive devices that provide a continuous readout of optical power either absolute or relative traveling through the optical fiber in one direction. They are typically available with pigtails like the unit shown in Figure 16.17 or receptacles that accept connectors like the unit shown in Figure 16.18. Unlike the fiber identifier, which can be used with single-mode or multimode optical fiber, the inline optical power monitor is optimized for a specific optical fiber type.

This book cannot cover all the different ways an inline optical power monitor can be used. Some typical uses include measuring the output power of the transmitter or the input power of the receiver. That same meter could be reversed so light travels in the opposite direction, allowing it to measure the power being reflected back toward the transmitter.

Multiple inline optical power monitors can be configured to support many types of measurements. Two common applications include measuring the gain of an optical amplifier, as shown in the block diagram in Figure 16.19, and measuring the power delivered to each receiver in a single-fiber bidirectional application, as shown in the block diagram in Figure 16.20.

FIGURE 16.17
Inline optical
power monitor with
pigtail

Photo courtesy of EigenLight Corporation

FIGURE 16.18
Inline optical
power monitor with
receptacles

Photo courtesy of EigenLight Corporation

FIGURE 16.19
Block diagram of an
inline optical power
monitor measuring
amplifier gain

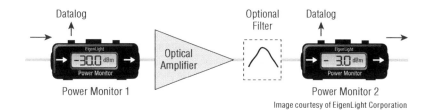

Image courtesy of EigenLight Corporation

FIGURE 16.20
Block diagram of
inline optical power
monitors measur-
ing receiver power

Image courtesy of EigenLight Corporation

Inline Network Sensors

Inline networks sensors are similar to inline optical power monitors; however, they allow real-time access to the output of the transimpedance amplifier and typically have adjustable gain. The transimpedance amplifier in the inline network sensor functions very much as the transimpedance amplifier discussed in Chapter 11, "Fiber-Optic Detectors and Receivers." It converts the light energy received by the photodiode in the power tap into electrical current.

The electrical current from the transimpedance amplifier could be measured by a number of devices, such as a voltmeter, oscilloscope, or analog-to-digital converter, to name a few. The gain of each transimpedance amplifier is typically adjustable to compensate for higher power levels when measuring forward power or lower power levels when measuring reverse power. Multiple sensors may be incorporated into a single configurable rack mount enclosure, as they are in the unit shown in Figure 16.21.

FIGURE 16.21
1U rack mount enclosure with three independently adjustable inline network sensors

Photo courtesy of Engineering Development Laboratory, Inc.

The unit shown in Figure 16.21 features three power taps, each terminated at the front panel with FC receptacles. The output of each transimpedance amplifier can be measured across the banana receptacles on the front panel. The gain of each transimpedance amplifier is adjustable, so this unit can be configured to measure a wide range of forward or reverse power levels.

Optical Return Loss Test Set

Optical return loss (ORL) testing is performed with an optical return loss test set. ORL testing measures the amount of optical light energy that is reflected back to the transmit end of the fiber-optic cable. The energy being reflected back, or back reflections, comes from mechanical interconnections, passive devices, fiber ends, and Rayleigh scattering caused by impurities within the optical fiber.

Besides reducing the amount of light transmitted, back reflections can cause various laser light source problems. They can cause the laser's output wavelength to vary. They can also cause fluctuations in the laser's optical output power and possibly permanently damage the laser light source.

The impact that back reflections have on the laser light source can cause problems in analog and digital systems. In digital systems, they can increase the BER. In analog systems, they reduce the *signal-to-noise ratio (SNR)*.

The ORL test set measures return loss using an *optical continuous wave reflectometer (OCWR)*. A light source within the ORL test set continuously transmits light through a directional coupler, as shown in Figure 16.22. Light energy returned from the optical fiber is directed to the photodiode of a power meter. The light energy measured by the power meter is the return loss.

FIGURE 16.22

Directional coupler
in an ORL test set

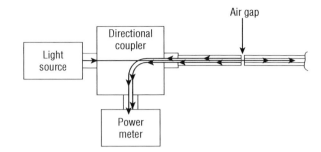

ORL measurements of a fiber-optic link should be taken with all patch cords and equipment cords in place. All system equipment should be turned off. The receive connector should remain plugged into the equipment receiver. The transmit connector should be unplugged from the equipment transmitter and plugged into the ORL test set after the test set has been calibrated, as shown in Figure 16.23. The test set should be calibrated as described in the manufacturer's operation manual. Before performing ORL testing, clean and inspect all connectors. Allow sufficient time for the ORL test set to warm up and stabilize.

FIGURE 16.23

ORL test set connected to fiber-optic link and system equipment

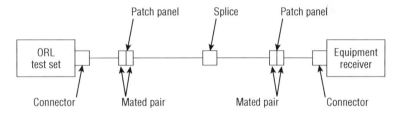

ORL test sets are available for both multimode and single-mode optical fiber. The multimode ORL test uses an LED light source with a typical output power of –20dBm. The return loss measurement range is usually between 10 and 45dB. Accuracy is normally within 0.5dB.

The single-mode ORL test set uses a laser source with a typical output power of –10dBm. The return loss measurement range is usually between 0 and 50dB or 0 and 60dB. Accuracy in the 0 to 50dB range is normally 0.5dB with a decrease in the 0 to 60dB range of 1dB.

Stabilized Light Source and Optical Power Meter

The stabilized light source and optical power meter work hand in hand. They are typically referred to as an *optical loss test set (OLTS)*. Many different OLTSs are available. Some support only multimode testing, some single-mode, and a few can be used for both. The multimode OLTS is typically the lowest priced, followed by the single-mode and then the combination multimode and single-mode.

Since the fourth edition of this text, encircled flux launch conditions for all multimode *insertion loss* measurements has been added to IEC 61280-4-1 Edition 2 and TIA-526-14-B. Encircled flux is not required for insertion loss testing performed in accordance with TIA-526-14-A. The addition of the encircled flux requirement does not render existing test equipment obsolete, and the publication of TIA-526-14-B does not mean that TIA-526-14-A will not be called out in an installation or testing contract.

As of this writing, no known multimode stabilized light sources meet the encircled flux requirement without the addition of a modal controller on the transmit jumper. The encircled flux requirement, modal controllers, and mode filters are covered in detail later in this chapter. This section explains the basic operation of the multimode and single-mode OLTS.

Multimode OLTS

The multimode OLTS is available with either a VCSEL or an LED stabilized light source. This section discusses the differences between applications of these test sets.

VERTICAL-CAVITY SURFACE-EMITTING LASER (VCSEL) OLTS

As discussed in earlier chapters, ANSI/TIA-568-C.3 defines optical fiber transmission performance parameters. This standard includes laser-optimized optical fiber, which was not included in ANSI/TIA/EIA-568-B.3. It would seem that the best source to test laser-optimized optical fiber would be the VCSEL, and several manufacturers do offer VCSEL light sources. However, multimode optical fiber test standards such as TIA-TSB-140 still specify an LED light source. The VCSEL light source should not be used for testing to TIA-526-14 revision A or B.

LED OLTS

The multimode LED OLTS consists of an LED-based stabilized light source and an optical power meter. The typical light source consists of an 850nm LED and a 1300nm LED, like the one shown in Figure 16.24. The optical power meter is typically selectable for 850nm, 1300nm, and 1550nm, like the one shown in Figure 16.25.

FIGURE 16.24
Multimode 850nm and 1300nm light source

Two types of LED light sources are available: filtered and unfiltered. A filtered light source does not overfill the optical fiber. An unfiltered light source does overfill the optical fiber. Multimode insertion loss measurements being done in accordance with TIA-526-14A should be performed with a light source that meets the launch requirements of ANSI/TIA-455-78-B. This launch requirement can be achieved within the test equipment, which is the case with the filtered light source or with an external mandrel wrap on the transmit test jumper. Most LED light sources in use today are unfiltered and require the mandrel wrap. The mandrel wrap is discussed in detail later in this chapter.

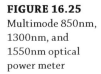

FIGURE 16.25
Multimode 850nm,
1300nm, and
1550nm optical
power meter

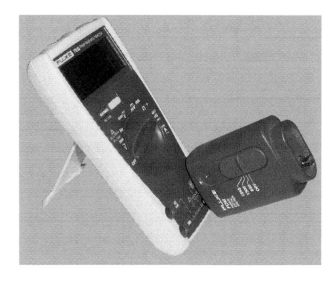

NOTE   An overfilled light source has a Category 1 Coupled Power Ratio (CPR) value as defined in TIA-526-14A, Annex A Tables A.1 and A.2. Table A.1 defines the CPR value range for an 850nm light source, and Table A.2 defines the CPR value range for a 1300nm light source. A mandrel wrap must be used on the transmit test jumper of a Category 1 CPR light source.

Multimode insertion loss measurements being performed in accordance with TIA-526-14-B require a modal controller on the transmit jumper. The modal controller, which replaces the mandrel, is discussed in detail later in this chapter.

The optical output power of an unfiltered LED light source is typically –20dBm for both the 850nm and 1300nm LEDs when measured with a 62.5/125µm optical fiber. The optical output power increases roughly 4dB when used with 100/140µm optical fiber and decreases roughly 3dB when used with a 50/125µm optical fiber.

NOTE   Remember that ANSI/TIA-568-C.3 recognizes only 50/125µm or 62.5/125µm multimode optical fiber.

Most multimode light sources are designed to operate between the temperatures of 0° to +40° C or 0° to +50° C. Within these temperature ranges, they should provide a stable optical output power that typically varies less than ±0.2dB over an eight-hour period. To achieve this performance, the light source should be allowed to warm up before beginning testing. The manufacturer specifies warm-up times.

The multimode power meter typically uses a single photodiode with a switch to select the proper wavelength. The selector switch compensates for the responsivity of the photodiode at the various wavelengths. The output of the optical power meter is in decibels (dB) referenced to one milliwatt (mW), where 0dBm is equal to 1mW of optical power.

Many power meters will support both multimode and single-mode testing. Acceptable optical fiber types may range from 9/125µm through 100/140µm. The optical input power range is typically +3dBm to –50dBm.

As with the light sources, most optical power meters are designed to operate between the temperatures of 0° and +40° C or 0° and +50° C. When operated within this temperature range, absolute accuracy is typically within ±0.25dB, relative accuracy within ±0.15dB, and repeatability within ±0.04dB. The values stated here are typical. Actual values can be obtained from the manufacturer's data sheet.

IS MY POWER METER BROKEN?

After years of watching students test various fiber-optic links with an OLTS, I have seen many confused looks when the reading on the power meters drops to a very low value and never stabilizes. As discussed in this chapter, the manufacturer defines the optical input power range for the power meter. This value is typically between +3dBm and –50dBm.

The output power of an LED light source is typically –20dBm with a 62.5/125µm optical fiber and –23dBm with a 50/125µm optical fiber. A break in an optical fiber will cause significant attenuation, typically between 20 and 30dB. If a light source coupled –23dBm into the fiber-optic link under test and the break added 30dB of attenuation, the maximum light energy after the break would equal –53dBm, which is below the power range for the power meter. The actual power would be less because of the losses from the optical fiber and interconnections.

When the light energy into the power meter falls below the lowest value defined by the manufacturer, the power meter typically appears to be hunting. In other words, the numbers slowly increase and decrease; they never seem to stop changing. When this happens to you, remember that the power meter is not broken; however, the optical fiber is probably broken.

Single-Mode OLTS

The single-mode OLTS consists of a laser-based stabilized light source and an optical power meter. The typical light source consists of a 1310nm laser and a 1550nm laser. Some models have a separate output port for each laser, and others combine both wavelength lasers into a single output port. Another popular laser combination is 1550nm and 1625nm.

The optical output power of the laser is typically around –5dBm into a 9/125µm single-mode optical fiber. Some single-mode light sources have the ability to attenuate the output of the laser. Most single-mode light sources are designed to operate between the temperatures of 0° to +40° C or 0° to +50° C. Within these temperature ranges, they should provide a stable optical output power that typically varies no more than ±0.1dB over an eight-hour period. To achieve this performance, the light source should be allowed to warm up before beginning testing. The manufacturer specifies warm-up times.

The single-mode optical power meter is typically designed to handle a range of wavelengths. It may measure common multimode optical fiber wavelengths and single-mode optical fiber wavelengths. A typical power meter will be selectable for 850nm, 1300nm, 1310nm, 1480nm, 1550nm, and 1625nm.

Single-mode optical power meters are typically designed to operate between the temperatures of 0° to +40° C or 0° to +50° C. When operated within this temperature range, they provide the best accuracy. Absolute accuracy is typically within ±0.25dB when measured at +25° C with an optical input power of –10dBm, relative accuracy within ±0.15dB, and repeatability within ±0.04dB. The values stated here are typical. Actual values can be obtained from the manufacturer's data sheet.

WHY IS THE LOSS FOR THIS CABLE ASSEMBLY DIFFERENT EACH TIME I MEASURE IT?

One thing I have always enjoyed is the friendly competition between the students during lab. As the students get to know each other and openly communicate, they compete for the lowest-loss cable assembly. Sometimes there is a substantial gap between the loss measurements and the winner is clearly defined. Other times the gap is not measurable with the equipment in the lab and the results vary slightly each time a cable assembly is measured.

In this chapter, the typical values for absolute accuracy, relative accuracy, and repeatability for a power meter were defined. The stability of the light source was also defined. Each of these parameters has a range, and absolute accuracy is typically the largest. What these parameter ranges mean is that insertion loss measurements will vary. There is no such thing as a perfect piece of test equipment.

Patch Cord

ANSI/TIA-568-C.3, section 6 sets the performance specifications for optical fiber patch cords recognized in premises cabling standards. The optical fiber *patch cord* is used at cross-connections to connect optical fiber links. They are also used as equipment or work area cords to connect telecommunications equipment to horizontal or backbone cabling.

A patch cord is a length of optical fiber cable with connectors on both ends. It uses the same connector type and optical fiber type as the optical fiber cabling that it is connected to. The patch cord must meet the cable transmission performance requirements and physical cable specifications of sections 4.2 and 4.3.1 of ANSI/TIA-568-C.3. The patch cord must also meet the connector and adapter requirements of section 5.2 of ANSI/TIA-568-C.3.

Optical fiber patch cords used for either cross-connection or interconnection to equipment shall have a termination configuration defined in section 6.4 of ANSI/TIA-568-C.3. Section 6.4 describes configurations for simplex, duplex, and array patch cords. A simplex patch cord is a single-fiber cable with simplex connection terminations. A duplex patch cord is a two-fiber cable with duplex connectors. An array patch cord is a multifiber cable with array connectors on each end. Array connectors feature a single ferrule that contains multiple optical fibers arranged in a row or rows and columns. This connector is covered in detail in Chapter 9, "Connectors."

Test Jumper

The terms *patch cord* and *jumper* are often interchanged. A test jumper can be a single-fiber cable or a multifiber cable. This section focuses on multimode and single-mode test jumpers as described in ANSI/TIA-568-C.0, Annex E.

In the fiber-optic industry, the *test jumper* has several names. The test jumper connected to the light source is typically called the *transmit jumper*. The test jumper connected to the optical power meter is typically called the *receive jumper*. The U.S. Navy calls the test jumper a *measurement quality jumper (MQJ)*. You may also see the test jumper referred to as a *reference jumper*. Regardless of the name, the test jumper is a critical part of your optical power measurement equipment.

Test jumpers must have a core diameter and numerical aperture nominally equal to the optical fiber being tested. You cannot test a 50/125μm link with a 62.5/125μm test jumper, or vice versa. Per ANSI/TIA-568-C.0, jumpers shall be no less than 1m and no greater than 5m in length. The termination of the test jumper shall be compatible with the cable being tested.

Test jumpers should be cleaned and inspected prior to making measurements. The endface of each connector should be evaluated under a microscope. There should be no scratches, notches, or chips in the endface of the optical fiber. You should also clean and inspect each connector of the cable under test. Mating a dirty or damaged connector with a test jumper connector can destroy the test jumper connector.

Test jumpers should be tested for insertion loss prior to performing any of the ANSI/TIA-526-14 revision A or B methods. How they are tested depends on which standard or standards are called out by the customer. It may be one of the methods described in ANSI/TIA-526-14 revision A or B or a modified version of one of those methods. The maximum acceptable loss may be defined by the customer; when it is not, 0.4dB is typically a good value. This value is not defined by the ANSI/TIA-568-C.0 standard but serves as a good rule of thumb. Many test jumpers will have a measured loss of less than 0.2dB.

When verifying the quality of a test jumper in accordance with ANSI/TIA-568-C.0, clean and inspect the test jumpers. Verify that your test jumpers have the same optical fiber type as the cable plant you are going to test. While you are cleaning and inspecting the test jumpers, turn on the light source and optical power meter. Set both of them to the test wavelength. Allow sufficient time for the light source and optical power meter to warm up and stabilize. Warm-up information can be found in the manufacturer's data sheet.

After sufficient time has passed for the light source and optical power meter to stabilize, you need to measure the loss of the test jumpers. Connect test jumper 1 as shown in Figure 16.31. When testing multimode optical fiber, test jumper 1 must have a modal controller, and when testing single-mode optical fiber a single 30mm loop is required. Record the optical power displayed on the optical power meter. Disconnect test jumper 1 from the optical power meter; then connect test jumper 2 as shown in Figure 16.32 and record the optical power displayed on the optical power meter. The loss for test jumper 2 in that direction is the difference between the recorded values.

Remove test jumper 2 and interchange the ends. Take the end that was mating with test jumper 1 and connect it to the power meter. Mate the other end with test jumper 1. Record the optical power displayed on the optical power meter. The loss for test jumper 2 in that direction is the difference between the recorded values. If either value exceeds a specified limit, test jumper 2 cannot be used for testing the cable plant.

Test jumpers should be treated with care, especially multiple fiber measurement quality jumpers (MQJs) like the one shown in Figure 16.26, which can be very expensive. The measured loss of the test jumper will increase gradually as it is used. However, you can typically continue to use the test jumper as long as the loss does not exceed 0.4dB and the endface remains free from defects. This value is not defined by the ANSI/TIA-568-C.0 standard but serves as a good rule of thumb.

FIGURE 16.26
Multiple-fiber MQJ

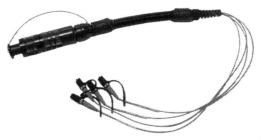

Photo courtesy of KITCO Fiber Optics

Launch Conditions, Mode Filters, and Encircled Flux

Since the fourth edition of this book, *encircled flux* launch conditions for all multimode insertion loss measurements has been added to IEC 61280-4-1 Edition 2, ANSI/TIA-568-C.0, and TIA-526-14-B. The addition of this requirement arose from the need to improve the accuracy of insertion loss measurements. As the data rates of multimode networks have increased, the maximum allowable channel attenuation has decreased. A channel is the end-to-end transmission path between two pieces of equipment such as a transmitter and receiver.

The maximum allowable channel attenuation for a 100BASE-FX, Ethernet with OM3 optical fiber is 6dB while the maximum allowable attenuation for a 1000BASE-LX Ethernet with the same optical fiber is 2.3dB. Both Ethernets operate at 1300nm, but there is a 3.7dB difference in the maximum allowable channel attenuation between 1000BASE or 1Gbps Ethernet and the 100BASE or 100Mbps Ethernet. When the data rate increases to 10Gbps, the maximum allowable channel attenuation for some Ethernet applications drops to 1.9dB.

Variations in the launch conditions of multimode light sources are well known. Prior to the introduction of encircled flux, the *mandrel wrap* or *mode filter* was required for light sources that did not meet the launch requirements of ANSI/TIA-455-78-B. It was required because the attenuation in a short link of multimode optical fiber may be higher than calculated when making insertion loss measurements due to the power lost in the high-order modes. The high-order modes are caused by an LED light source that overfills the optical fiber.

Even when light sources met the launch requirements of ANSI/TIA-455-78-B, it was still difficult to get repeatable insertion loss measurements from one piece of test equipment to another. The variations obtained with different pieces of test equipment combined with the reduced channel attenuation led to the development of encircled flux requirement.

The encircled flux requirement is based on defined lower and upper boundaries of encircled flux values at different distances from the center of the optical fiber core. It corresponds to the ratio between the transmitted power at a given distance from the center of the core and the total power injected into the optical fiber. It is intended to define the light source's near-field output profile and ensure repeatable test results from one piece of test equipment to another.

Just because the encircled flux requirement has been introduced, that does not mean you must throw away your mandrel wraps. The customer typically defines how a link or channel will be tested, and they may call out testing to TIA-526-14-A instead of TIA-526-14-B. In addition, low-bandwidth systems with a large loss budget will not require encircled flux unless specified by the customer. It is for these reasons that both the encircled flux modal controller and mandrel wrap are discussed in this book.

When you are measuring the insertion loss of a multimode link with an encircled flux requirement, you must use a transmit jumper with a modal controller that meets the requirements of IEC 61280-4-1 Edition 2. Unlike the mandrel wrap, you cannot pick one up for a few dollars and wrap a test jumper around it. A modal controller typically has pigtails, and as with the mandrel wrap, you need a different one for each optical fiber core diameter. However, they are far more expensive than a mandrel wrap. Like the mode filter, the modal controller is a passive optical device; no power is required for the modal controller to operate.

When you're measuring the insertion loss of a multimode link that does not have an encircled flux requirement and using an LED source that overfills the optical fiber, a mandrel wrap must be used on the transmit test jumper. The transmit test jumper should have five nonoverlapping turns around a smooth mandrel. The diameter of the mandrel depends on the cable diameter and optical fiber core diameter, as shown in Table 16.1. Figure 16.27 shows a transmit test jumper with a mandrel wrap attached to an LED light source.

FIGURE 16.27
Transmit test jumper with mandrel wrap

TABLE 16.1: Mandrel diameters for multimode optical fiber

FIBER CORE SIZE (μm)	MANDREL DIAMETER FOR 900μm BUFFERED OPTICAL FIBER (MM)	MANDREL DIAMETER FOR 2MM JACKETED CABLE	MANDREL DIAMETER FOR 2.4MM JACKETED CABLE	MANDREL DIAMETER FOR 3MM JACKETED CABLE
50	25	23	23	22
62.5	20	18	18	17

An unfiltered LED light source overfills an optical fiber by launching high- and low-order modes into the core and cladding. The high-order modes experience more attenuation than the low-order modes at interconnections, splices, and bends. This causes higher-than-expected loss in short multimode links. Figure 16.28 shows high-order modes from the LED source entering the core and the cladding of the optical fiber.

The mandrel wrap takes advantage of the high loss of high-order modes caused by excessive bending of the optical fiber. The mandrel wrap removes the high-order modes by inserting a series of macrobends in the transmit test jumper. Figure 16.29 shows the high-order modes in the core of the optical fiber refracting into the cladding because of a 90-degree macrobend.

The five small-radius nonoverlapping loops around the mandrel wrap shown in Figure 16.27 significantly attenuate the high-order modes with minimal attenuation of the low-order modes. The mandrel wrap is actually a low-order or low-pass mode filter. It allows the low-order modes to pass with very little attenuation while greatly attenuating the high-order modes. The mandrel wrap should be on the transmit test jumper at every stage of testing when you're making insertion loss measurements.

FIGURE 16.28
LED source overfilling a multimode optical fiber

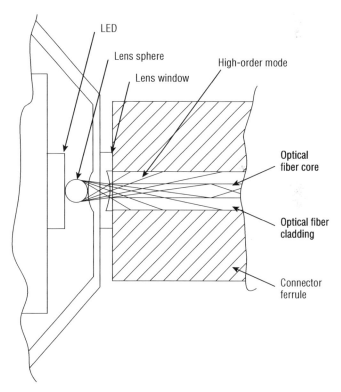

FIGURE 16.29
Macrobend
attenuating loosely
coupled modes in a
multimode optical
fiber

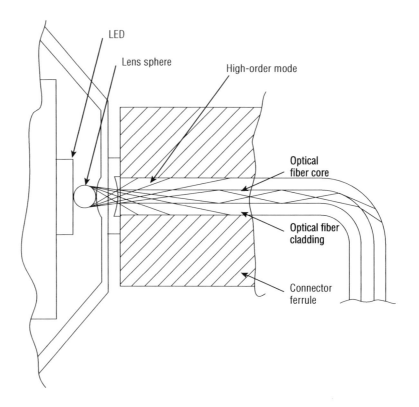

ANSI/TIA-526-14 Optical Loss Measurement Methods

ANSI/TIA-526-14 revisions A and B provide three methods for testing the *cable plant*. The difference between the three methods is how the reference power measurement is taken. Each method uses a different number of test jumpers and is described in detail in this section.

Test methods may be described using either the letter or the number of jumpers. One standard may call out a one test jumper method whereas another may call out method B. A good installer or technician is fluent in both. Therefore, in this section both approaches will be interchanged.

This section of the text does not show a mandrel wrap or modal conditioner on the transmit jumper in the illustrations. This section focuses on how to perform each test method. The test methods are performed exactly the same regardless of whether a mode filter or modal controller is required.

The three test methods described in this section each measure the optical loss of the cable plant at different points, as shown in Figure 16.30. The two-test jumper method, or method A, measures the loss of the cable plant plus one connection loss. The one-test jumper method, or method B, measures the loss of the cable plant plus two connection losses. The three-test jumper method, or method C, measures the loss of the cable plant only. Only the one-test jumper method can be used when testing to ANSI/TIA-568-C.0.

FIGURE 16.30
Cable plant mea-
sured values for
methods A, B, and C

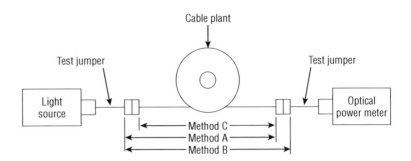

The first step before performing any of the optical power loss measurements is to clean and inspect the test jumpers and connectors at the ends of the cable plant to be tested. Verify that your test jumpers have the same optical fiber type as the cable plant you are going to test.

While you are cleaning and inspecting the test jumpers and cable plant connectors, turn on the light source and optical power meter. Set both of them to the test wavelength. Allow sufficient time for the light source and optical power meter to warm up and stabilize. Warm-up information can be found in the manufacturer's data sheet.

After sufficient time has passed for the light source and optical power meter to stabilize, you need to measure the loss of the test jumpers. Connect test jumper 1 as shown in Figure 16.31. Record the optical power displayed on the optical power meter. This number is the reference power measurement. This number is typically around –20dBm with a 62.5/125μm multimode optical fiber and –23dBm with a 50/125μm multimode optical fiber. These numbers can vary from OLTS to OLTS.

FIGURE 16.31
Test jumper 1 opti-
cal reference power
measurement

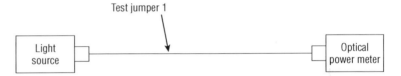

Connect test jumper 2 as shown in Figure 16.32. Record the optical power displayed on the optical power meter. The difference between the reference power measurement and this value is the connection loss for test jumpers 1 and 2. The difference between this value and the reference power measurement should be less than or equal to 0.4dB. If this number is greater than 0.4dB, clean the connectors again and retest. If the loss still exceeds 0.4dB, replace test jumper 2. Repeat this test for test jumper 3 if method C is being used.

FIGURE 16.32
Test jumper 2 opti-
cal reference power
measurement

Method A, Two-Test Jumper Reference

Method A uses two jumpers to establish the optical power reference. Connect test jumpers 1 and 2 as shown in Figure 16.32. Record the optical power displayed by the optical power meter. This number is the reference power measurement. This number is typically around –20dBm with a 62.5/125μm multimode optical fiber and –23dBm with a 50/125μm multimode optical fiber. These numbers can vary from OLTS to OLTS.

After recording the reference power measurement, separate the test jumpers at their point of interconnection, as shown in Figure 16.33. Do this without disturbing their attachment to the light source and optical power meter. Attach the test jumpers to the cable plant as shown in Figure 16.34.

FIGURE 16.33
Test jumpers 1 and 2 separated

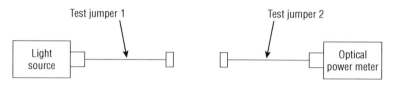

FIGURE 16.34
Test jumpers 1 and 2 connected to the cable plant

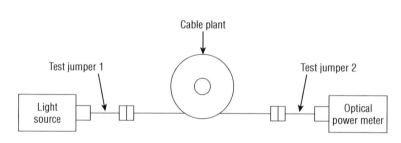

With the test jumpers attached, record the optical power displayed on the optical power meter. This is your test power measurement. The loss for the cable plant is the difference between the reference power measurement and the test power measurement.

To obtain the optical power loss for the cable plant, subtract the test power measurement from the reference power measurement. If the reference power measurement was –20.4dBm and the test power measurement was –21.6dBm, the optical power loss for the cable plant would be 1.2dB.

NOTE Remember the mandrel wrap is always installed on the test jumper connected to the light source when an unfiltered light source is used.

Method B, One-Test Jumper Reference

Method B uses one jumper to establish the optical power reference. However, two test jumpers are required to perform the test. This method is required by ANSI/TIA-568-C.0. Link attenuation using this method is equal to cable attenuation plus connector insertion loss plus splice insertion loss. The required wavelengths and directions are discussed in the section "Link Segment and Cabling Subsystem Performance Measurements" later in this chapter.

Connect test jumper 1 as shown earlier in Figure 16.31. Record the optical power displayed by the optical power meter. This number is the reference power measurement. This number is typically around –20dBm with a 62.5/125µm multimode optical fiber and –23.5dBm with a 50/125µm multimode optical fiber. These numbers can vary from OLTS to OLTS.

After recording the reference power measurement, disconnect the test jumper from the optical power meter. Do not disturb the jumper's attachment to the light source. Attach test jumper 2 to the optical power meter as shown in Figure 16.33. Attach the test jumpers to the cable plant as shown in Figure 16.34.

With the test jumpers attached, record the optical power displayed on the optical power meter. This is your test power measurement. The loss for the cable plant is the difference between the reference power measurement and the test power measurement.

To obtain the optical power loss for the cable plant, subtract the test power measurement from the reference power measurement. If the reference power measurement was –20.7dBm and the test power measurement was –23.6dBm, the optical power loss for the cable plant would be 2.9dB.

Method C, Three-Test Jumper Reference

Method C uses three jumpers to establish the optical power reference. Connect test jumpers 1, 2, and 3 as shown in Figure 16.35. Record the optical power displayed by the optical power meter. This number is the reference power measurement. This number is typically around –20dBm with a 62.5/125µm multimode optical fiber and –23.5dBm with a 50/125µm multimode optical fiber. These numbers can vary from OLTS to OLTS.

FIGURE 16.35
Method C reference power measurement

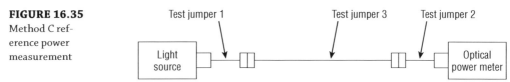

After recording the reference power measurement, separate the test jumpers at their point of interconnection, as shown in Figure 16.36. Do this without disturbing their attachment to the light source and optical power meter. Attach the test jumpers to the cable plant as shown earlier in Figure 16.34.

FIGURE 16.36
Removal of test jumper 3 from method C reference power measurement

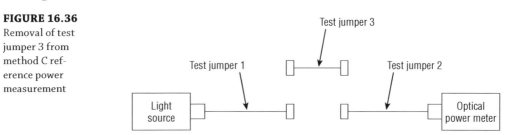

NOTE   Test jumper 3 is not used during testing. It is used only to establish reference power.

With the test jumpers attached, record the optical power displayed on the optical power meter. This is your test power measurement. The loss for the cable plant is the difference between the reference power measurement and the test power measurement.

To obtain the optical power loss for the cable plant, subtract the test power measurement from the reference power measurement. If the reference power measurement was –20.2dBm and the test power measurement was –21.6dBm, the optical power loss for the cable plant would be 1.4dB.

Patch Cord Optical Power Loss Measurement

Multimode patch cord optical loss power measurement is performed using the steps described in ANSI/TIA-526-14, method A. The patch cord is substituted for the cable plant. Because patch cords are typically no longer than 5m, the loss for the optical fiber is negligible and testing can be performed at 850nm or 1300nm. The loss measured in this test is the loss for the patch cord connector pair. ANSI/TIA-568-C.3 states that the maximum loss for a connector pair is 0.75dB.

After setting up the test equipment as described in ANSI/TIA-526-14, method A, clean and inspect the connectors at the ends of the patch cords to be tested. Verify that your test jumpers have the same optical fiber type and connectors as the patch cords you are going to test. The transmit jumper should have a mandrel wrap or modal conditioner depending on the revision ANSI/TIA-526-14 being used for testing. Ensure that there are no sharp bends in the test jumpers or patch cords during testing.

Because both patch cord connectors are easily accessible, optical power loss should be measured in both directions. The loss for the patch cord is the average of the two measurements. If the loss for the patch cord exceeds 0.75dB in either direction, the patch cord needs to be repaired or replaced.

Connector Insertion Loss Measurement

Connector insertion loss measurement isolates the loss of a single connector on a cable assembly. It may be referred to as connector loss. Many times cable assemblies are shipped from the manufacturer with the insertion loss for each connector listed on the packaging. The package shown in Figure 16.37 contains a duplex multimode patch cord. In the upper-left corner of the package, a label lists the insertion loss measurements for each connector.

Connect test jumper 1 as shown earlier in Figure 16.31. Record the optical power displayed by the optical power meter. This number is the reference power measurement. This number is typically around –20dBm with a 62.5/125µm multimode optical fiber and –23.5dBm with a 50/125µm multimode optical fiber. These numbers can vary from OLTS to OLTS.

After recording the reference power measurement, disconnect the test jumper from the optical power meter. Do not disturb the jumper's attachment to the light source. Attach the cable to be tested to the optical power meter as shown in Figure 16.38 and record the optical power displayed on the optical power meter. This is your test power measurement. The loss for the connector shown in Figure 16.38 is the difference between the reference power measurement and the test power measurement. The transmit jumper should have a mandrel wrap or modal conditioner depending on the revision of ANSI/TIA-526-14 being used for testing.

To obtain the connector insertion loss, subtract the test power measurement from the reference power measurement. If the reference power measurement was –20.7dBm and the test power measurement was –20.9dBm, the connector insertion loss would be 0.2dB.

FIGURE 16.37
Duplex multimode
fiber patch cable
with connector
insertion loss mea-
surements on the
packaging

FIGURE 16.38
Test jumper and
cable under test for
connector insertion
loss measurement

Link Segment and Cabling Subsystem Performance Measurements

Link segment performance measurements are described in ANSI/TIA-568-C.0. Multimode link attenuation should be measured using the reference methods specified in TIA-526-14-B and single-mode link attenuation should be measured using the reference methods specified in TIA-526-7. The one-jumper reference method is preferred for both multimode and single-mode; however, other methods described in TIA-526-14-B or TIA-526-7 may be applied. Ensure that the applied test method is included in testing documentation.

When measuring single-mode link attenuation in accordance with TIA-526-7 Method A.1 follow Method B, the one-test jumper reference procedure described earlier in the chapter. To comply with ANSI/TIA-568-C.0, a single turn 30mm in diameter loop must be applied to the transmit test jumper to ensure single-mode operation as described in ANSI/TIA-455-78B.

ANSI/TIA-568-C.0 outlines cabling subsystem performance measurements. There are three cabling subsystems. Cabling Subsystem 1 describes the cabling from the equipment outlet to a connection facility, an intermediate connection facility, or a central connection facility. Cabling Subsystem 2 describes the cabling from a connection facility to either an intermediate connection facility or a central connection facility when an intermediate connection facility is not used. Cabling Subsystem 3 describes the cabling from the intermediate connection facility to the central connection facility.

Cabling Subsystem 1 Testing Requirements   These link segments are only required to be tested in one direction at one wavelength. Multimode optical fiber link segments can be tested at either 850nm or 1300nm. Single-mode optical fiber link segments can be tested at either 1310nm or 1550nm.

Cabling Subsystem 2 and 3 Testing Requirements   These link segments are required to be tested in at least one direction at both operational wavelengths. Multimode optical fiber link segments must be tested at 850nm and 1300nm. Single-mode optical link segments shall be tested at 1310nm and 1550nm.

NOTE   A link as defined in ANSI/TIA-568-C.0 is a transmission path between two points not including equipment and cords. A link segment does not include any active or passive devices other than the optical fiber cabling, connectors, and splices if required.

Tier 1 Testing

ANSI/TIA-568-C.0 Annex E provides guidelines for field testing length, loss, and polarity of optical fiber cabling. Section E.2.2 describes Tier 1 testing. Tier 1 testing requires the following:

♦ Link attenuation measurement

♦ Fiber length verification

♦ Polarity verification

Link Attenuation Measurement   Link attenuation measurement must be performed with an OLTS. When testing multimode optical fiber, attenuation measurements should be taken in accordance with TIA-526-14, Method B, which was covered earlier in this chapter. When testing to TIA-526-14-A, ensure the light source meets the launch requirements of ANSI/TIA-455-78B. This can be achieved either within the stabilized light source by the manufacturer or with the edition of a mandrel wrap as described earlier in this chapter. When testing to TIA-526-14-B, ensure the transmit jumper has a modal controller.

When testing single-mode optical fiber, attenuation measurements should be taken in accordance with TIA-526-7-A, Method A.1. A single 30mm diameter loop is often applied to the source test jumper to ensure single-mode operation.

Fiber Length Verification   Fiber length verification may be obtained by using the sequential markings on the cable jacket or sheath as described in Chapter 7, "Fiber-Optic Cables." It may also be obtained using an OLTS that has length measurement capability.

Polarity Verification   Polarity verification is performed to ensure the optical path from the transmit port of one device is routed to the receive port of another device. The polarity may be verified using an OLTS, VFL, or continuity tester.

Documentation of OLTS Testing

As you test the cable plant with the OLTS, you need to properly document the results of the test. The documentation should include at a minimum the following:

◆   The date of the testing

◆   Test personnel

◆   Field test instrument manufacturer

◆   Field test instrument serial number

◆   Field test instrument calibration date

◆   Test jumper type

◆   Test jumper length

◆   The number or identification of fiber or cable tested

◆   The test procedure, method used, wavelengths, directions

◆   Link loss results

Tier 2 Testing

Tier 2 testing as defined in ANSI/TIA-568-C.0 Annex E is optional. It can be used to supplement Tier 1 testing with the addition of an OTDR trace of the link. However, it is not a replacement for the Tier 1 link attenuation measurement performed with an OLTS. Tier 2 testing with the OTDR can characterize the elements of a link including:

♦   Fiber segment length

♦   Attenuation uniformity and attenuation rate

♦   Interconnection location and insertion loss

♦   Splice location and insertion loss

♦   Marcobends

Tier 2 testing documentation is described at the end of the next section.

Optical Time-Domain Reflectometer

So far in this chapter, you have learned about tools and test equipment that can be used to test a fiber-optic link or cable. Of all the tools and test equipment discussed, none provides more information about the fiber-optic cable or link than the *optical time-domain reflectometer (OTDR)*. The OTDR can be used to evaluate the loss and reflectance of interconnections and splices. It will measure the attenuation rate of an optical fiber and locate faults.

In general there are two types of OTDRs. One is often referred to as a long-haul OTDR and the other as a high-resolution OTDR. A long-haul OTDR is designed for use with cable lengths typically associated with telecommunication systems. They may range from 50m to many kilometers. A high-resolution OTDR is designed for use with short cable segment lengths such as those found on an aircraft or ship and may also be used in longer optical fiber length telecommunication applications. The high-resolution OTDR can also be used to evaluate optical sensors.

The use of high-resolution optical reflectometry to characterize and map the optical link is described in AIR6552/2, which as of this writing was under development within SAE International's AS-3 Fiber Optics and Applied Photonics Committee. This standard and its slash sheets are covered in detail in "Emerging Testing Standards" later in this chapter.

This section of the chapter focuses on basic OTDR theory, setup, and testing that applies to both long-haul and high-resolution OTDRs. The OTDR is a complex piece of test equipment with many variables that must be programmed correctly before testing can be performed. Proper OTDR setup and cable preparation will ensure accurate test results. Always consult the manufacturer's documentation before using any OTDR.

OTDR Theory

The OTDR is nothing more than a device that launches a pulse or pulses of light into one end of an optical fiber and records the amount of light energy that is reflected back. Unlike all the test equipment discussed up to this point, the OTDR provides a graphical representation of what is happening in the fiber-optic link or cable under test. With the OTDR, the fiber-optic link or cable is no longer a black box. The OTDR shows how light passes through every segment of the fiber-optic link.

Light reflecting back in an optical fiber is the result of a reflection or backscatter. Reflections are when the light traveling through the optical fiber encounters changes in the refractive index. These reflections are called Fresnel reflections. Backscatter, or Rayleigh scattering, results from evenly distributed compositional and density variations in the optical fiber. Photons are scattered along the length of the optical fiber. The photons that travel back toward the OTDR, as shown in Figure 16.39, are considered backscatter.

FIGURE 16.39
Backscattered
photons

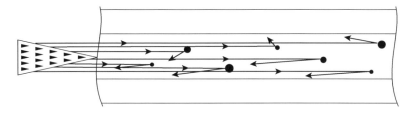

OTDRs come in many different shapes and sizes, as shown in Figure 16.40. The newer models may be almost pocket-sized whereas others require a shoulder strap. The small OTDR is lighter and easier to carry; however, the small screen permits only a limited viewing area. Some OTDRs may not have a screen and may be controlled with a PC via the USB port.

FIGURE 16.40
Large and small
OTDRs

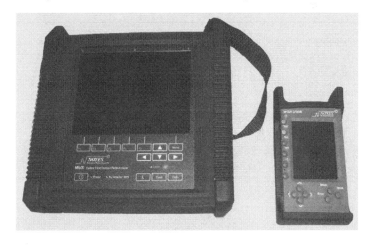

The OTDR is typically a battery-powered device. It is battery powered because many places where the OTDR is used have no electrical power available. It is a good idea to bring a charged spare battery with you when you are performing testing.

Many OTDRs can be configured to test both multimode and single-mode optical fibers. A typical OTDR may hold up to three light source modules, like the OTDR shown in Figure 16.41. Modules can be added or removed as testing requirements change. Some OTDRs even contain a VFL like the one shown in Figure 16.42.

FIGURE 16.41
Light source module locations on an OTDR

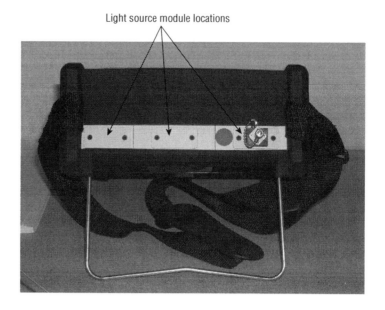

Light source module locations

FIGURE 16.42
The VFL is built into this OTDR.

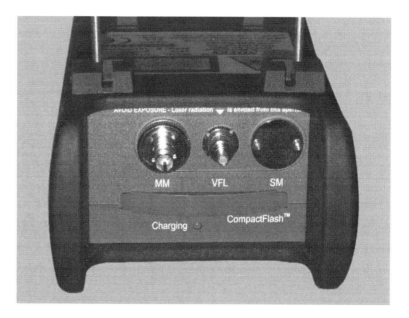

A generic OTDR can be broken up into eight basic components: the *directional coupler, laser, timing circuit, single-board computer, digital signal processor (DSP), analog-to-digital converter, sample-and-hold circuit,* and *avalanche photodiode.* Figure 16.43 is a block diagram of the OTDR showing how light is launched from the laser through the directional coupler into the optical fiber. The directional coupler channels light returned by the optical fiber to the avalanche photodiode.

FIGURE 16.43
OTDR block
diagram

FIGURE 16.43
OTDR block
diagram

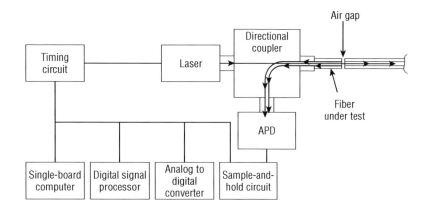

The avalanche photodiode converts the light energy into electrical energy. The electrical energy is sampled at a very high rate by the sample-and-hold circuit. The sample-and-hold circuit maintains the instantaneous voltage level of each sample long enough for the analog-to-digital converter to convert the electrical value to a numerical value. The numerical value from the analog-to-digital converter is processed by the DSP and the result is sent to the single-board computer to be stored in memory and displayed on the screen. This entire process is typically repeated many times during a single test of an optical fiber and coordinated by the timing circuit.

For the OTDR to produce accurate results, the refractive index of the optical fiber under test must be known. The refractive index of an optical fiber is different for each wavelength tested. The operator must enter the correct refractive index into the OTDR for each wavelength. The correct refractive index for a fiber-optic cable can be obtained from the manufacturer.

The OTDR samples light energy from the optical fiber at precise intervals. Each sample taken by the OTDR represents the round-trip time for light energy in the optical fiber. Let's assume that the OTDR is taking 500 million samples per second, or one sample every two nanoseconds. If the refractive index for the optical fiber under test were equal to 1.5, every five samples would represent the distance of 1m, as shown in Figure 16.44.

FIGURE 16.44
OTDR sampling at
2ns rate

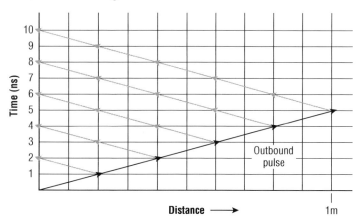

The following formula can be used to find distance based on time and refractive index. In this formula, the speed of light is rounded up to 3×10^8 m/s:

Distance = ((time in ns)/2) × (speed of light in free space)/refractive index

For our example:

Distance = (10 ns/2) × (3×10^8)/1.5

Distance = 1m

OTDR Display

The OTDR displays time or distance on the horizontal axis and amplitude on the vertical axis, as shown in Figure 16.45. The horizontal axis can typically be programmed to display distance in feet, meters, miles, or kilometers. The vertical axis is not programmable. The vertical axis displays relative power in decibels.

FIGURE 16.45
OTDR display

The OTDR creates a trace like the one shown in Figure 16.46. The trace shows event loss, event reflectance, and optical fiber attenuation rate. The OTDR can horizontally and vertically zoom in on any section of the trace. This permits a more detailed inspection of the optical fiber or event.

The light pulses from the OTDR produce a blind spot or dead zone, like the one shown in Figure 16.47. The dead zone is an area where the avalanche photodiode has been saturated by the reflectance of a mechanical interconnection. The size of the dead zone depends on the length of each light pulse and the amount of light reflected back toward the avalanche photodiode.

FIGURE 16.46
Event-filled OTDR
trace

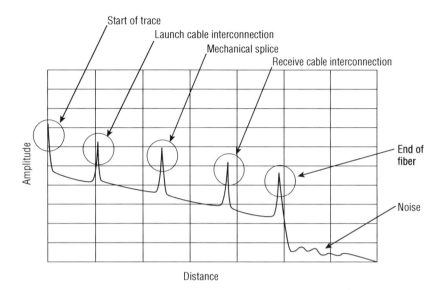

FIGURE 16.47
OTDR trace dead
zone

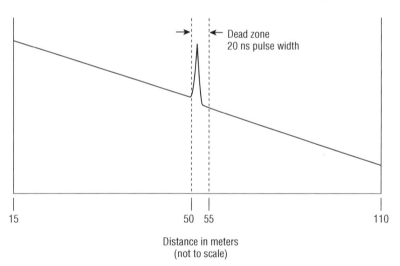

An interconnection that has very little reflectance will not saturate the avalanche photodiode. This type of interconnection will produce the smallest dead zone. Dividing the pulse width in nanoseconds by 10 can approximate the length of the light pulse as seen by the OTDR in meters. A pulse width of 20ns will yield an ideal dead zone of 2m whereas a pulse width of 2ns will yield an ideal dead zone of 20cm. Pulse widths less than 10ns are typically available only on high-resolution OTDRs like the one shown in Figure 16.48. The trace shown on the OTDR display in Figure 16.48 contains multiple reflective events over a distance of approximately 17m.

The ratio of the dead zone to the pulse width depends on the amount of light energy reflected back toward the OTDR. Poor interconnections or interconnections with an air gap typically saturate the avalanche photodiode and produce a greater-than-ideal dead zone, as much as 10 times greater than ideal.

FIGURE 16.48
High-resolution
OTDR capable of
2ns pulse widths
displaying multiple
reflective events
over a distance of
approximately 17m

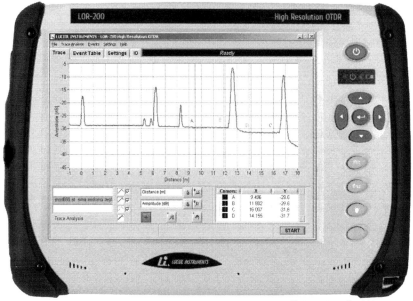

OTDR Setup

Before testing, the OTDR needs to be set up correctly to provide the most accurate readings. When setting up the OTDR, you must select the correct fiber type, wavelength or wavelengths, range and resolution, pulse width, averages, refractive index, thresholds, and backscatter coefficient. This process takes only a couple of minutes and ensures the most accurate results possible.

There are many different OTDRs on the market, and it is impossible to describe the setup for each. This section focuses on general setup parameters for both long-haul and high-resolution OTDRs. Some OTDRs may have additional parameters, and some may not include all of the parameters discussed.

FIBER TYPE

Each light source or light source module in a long-haul or high-resolution OTDR is designed for one or several specific optical-fiber types. A multimode module should not be used to test a single-mode optical fiber, and vice versa. Before heading for the test site, ensure that your OTDR has the correct module for the optical fiber to be tested.

WAVELENGTH

The wavelength that a long-haul or high-resolution OTDR can test with depends on the light source module or modules in your OTDR. Some light source modules contain a single laser whereas others contain two different wavelength lasers. A light source with two lasers allows testing of the optical fiber at two wavelengths without disconnecting the cable under test. This simplifies testing and reduces testing time.

RANGE AND RESOLUTION

The distance range of an unzoomed trace displayed on a long-haul or high-resolution OTDR and the distance between data points is determined by range and resolution. As a general rule of thumb, the OTDR range should be set to 1.5 times the length of the fiber-optic link. If the range is set too short, the results will be unpredictable and the entire link may not be displayed. If the range is set too long, the trace will fill only a small portion of the display area.

Increasing the range automatically increases the distance between the data points. When the distance between the data points is increased, resolution is reduced. When setting range, choose the first range value that exceeds your fiber-optic link length. Selecting a 2km range on a long-haul OTDR for a 1.3km link will yield more accurate results than selecting a 20km range.

PULSE WIDTH

The pulse width determines the size of the dead zone and the maximum length optical fiber that can be tested. A short pulse width produces a small dead zone. However, a short pulse width reduces the length of optical fiber that can be tested.

The pulse width should be selected so that the trace never disappears into the noise floor. If the pulse width is set properly, the trace will stay smooth until the end of the fiber-optic link. If the pulse is set too wide, events may be lost in the dead zone.

AVERAGES

When setting the averages parameter on an OTDR, select the number of averages that produce a smooth trace in the least amount of time. If too few averages are taken, the trace will appear noisy because the noise floor is too high. If too many samples are taken, the trace will be smooth; however, valuable testing time will be wasted.

REFRACTIVE INDEX

As mentioned earlier in the chapter, the manufacturer should specify the refractive index or the group index of refraction for an optical fiber. The refractive index of similar optical fibers does not vary much from manufacturer to manufacturer. Most values are typically within 1 percent of each other. If the value for the optical fiber you are testing is not known, use the values shown in Table 16.2. Entering a low refractive index will produce measurements that are too long, and entering a high refractive index will produce measurements that are too short.

TABLE 16.2:   Refractive index default values

WAVELENGTH	REFRACTIVE INDEX
850nm	1.4960
1300nm	1.4870
1310nm	1.4675
1550nm	1.4681

THRESHOLDS

Thresholds can typically be set for end of fiber, event loss, and reflectance. Many OTDRs have a default value preset for each of these. Depending on the testing you are performing, the default values may be too sensitive or not sensitive enough.

Most OTDRs will generate event tables automatically based on the threshold settings. To capture a majority of the events in the event table, set the thresholds on the sensitive side. A good starting place would be to set the values as shown in Table 16.3. If these values are too sensitive, you can always increase them.

TABLE 16.3:   Threshold default values

EVENT	DEFAULT VALUE
End of fiber	6.0dB
Event loss	0.05dB
Reflectance	−65dB

BACKSCATTER COEFFICIENT

The OTDR uses the backscatter coefficient to calculate reflectance. As with the refractive index, the optical fiber manufacturer specifies backscatter coefficient. Backscatter coefficient does vary from manufacturer to manufacturer. However, the variation is typically not that great, and the default values shown in Table 16.4 can be used with good results when the manufacturer's specified value is not known.

TABLE 16.4:   Backscatter coefficient default values

WAVELENGTH	BACKSCATTER COEFFICIENT
850nm	68.00
1300nm	76.00
1310nm	80.00
1550nm	83.00

Cable Plant Test Setup

Now that the OTDR has been configured with the correct light source modules and the setup parameters have been entered, the OTDR can be connected to the fiber-optic link. There are many different ways that the OTDR can be used to test a single optical-fiber or fiber-optic link. In this chapter, the test setup will include a *launch cable* and *receive cable*. Launch and receive cables may also be referred to as *pulse suppression cables, launch and receive fibers, dead zone cables, test fiber box*, or *access jumpers*.

Launch and receive cables allow the OTDR to measure the insertion loss at the near and far ends of the fiber-optic link. Like test jumpers, launch and receive cables should be constructed of optical fiber similar to the optical fiber under test. Launch and receive cables are available from several manufacturers and come in all shapes and sizes. Some launch and receive cables are designed for rugged environments, like the cable shown in Figure 16.49. Others are designed to be compact, almost pocket-sized, like the ring-type shown in Figure 16.50.

FIGURE 16.49
Rugged launch/
receive cable

Photo courtesy of OptiConcepts

FIGURE 16.50
Compact ring-type
launch/receive
cable

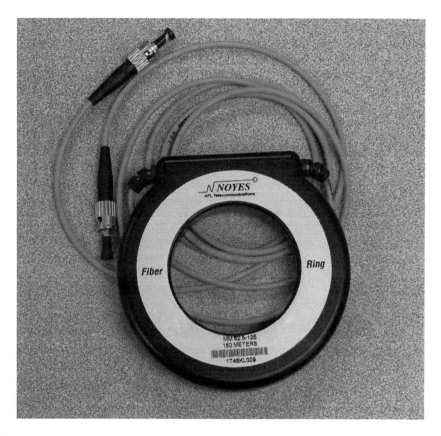

To achieve the compact size, manufacturers coil up coated optical fiber and add a buffer, strength member, and jacket at each end to protect the optical fiber, as shown in Figure 16.51. The open ring-type launch/receive cable shown in Figure 16.51 contains 150m of multimode optical fiber.

You should select the length of the launch and receive cables based on the manufacturer's recommendation or the pulse width you are using to test the optical fiber. Launch and receive cables are typically the same length. A good rule of thumb is to have at least 1m of launch or receive cable for each nanosecond of pulse width, with the minimum length cable being 100m. A 2km link being tested with a 20ns pulse width would require at least a 100m launch and receive cable. A 10km link being tested with a 200ns pulse would require at least a 200m launch and receive cable.

When you are testing generic telecommunications cabling in accordance with ANSI/TIA-568-C.0 Annex E, a launch cable length of 100 meters for multimode optical fiber or 300 meters for single-mode optical fiber is typically acceptable in the absence of manufacturer recommendations.

After the launch and receive cables have been selected, they can be attached to the fiber-optic link, as shown in Figure 16.52. Remember to clean and inspect the connector endfaces on the cable under test and on the launch and receive cables. Dirty or damaged endfaces on the launch and receive cables can damage the endface of the connectors that you are going to test. A dirty or damaged endface on a launch or receive cable typically results in a high-loss interconnection.

FIGURE 16.51
Open ring-type
launch/receive
cable

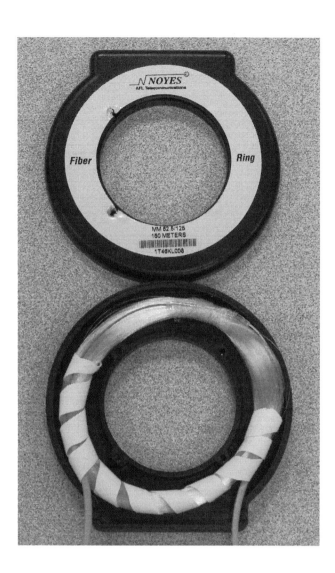

FIGURE 16.52
OTDR connected
to fiber-optic link
under test with
launch and receive
cables

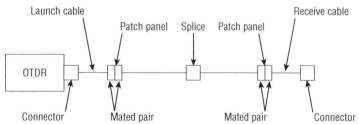

IF YOUR PULSE SUPPRESSION CABLES DON'T MATCH, YOU MAY SEE A GHOST

One day I was watching some students test a reel of multifiber cable with the OTDR. The Army had recently used this cable in the desert. Each fiber in the cable had been terminated on both ends.

After cleaning and visually inspecting the connectors, the pulse suppression cables were connected to both ends of the cable. The OTDR was connected to the launch cable and each optical fiber was tested. Looking at the OTDR trace, the students concluded that each optical fiber had multiple breaks because the OTDR trace had multiple back reflections every 100 meters.

I asked the students to test the continuity of each optical fiber with the continuity tester. Each optical fiber had good continuity. I then asked the students the core diameter of the optical fiber they were testing and the pulse suppression cables.

It turned out that the repetitive pulses or ghosts on the OTDR trace were caused by the core diameter mismatch between the pulse suppression cables and the optical fiber under testing. The students were testing a 50/125μm optical fiber with 62.5/125μm pulse suppression cables.

Testing and Trace Analysis

Testing the cable plant or a fiber-optic link with the OTDR should be done in both directions. Multimode optical fiber should be tested at both 850nm and 1300nm. Single-mode optical fiber should be tested at both 1310nm and 1550nm. Event loss and optical fiber attenuation is typically the average of the bidirectional values.

There are many standards related to testing both multimode and single-mode optical fiber with an OTDR. This section describes several techniques that will allow you to measure the distances and losses in a typical cable plant or fiber-optic link. We will focus on short links—the type you would typically find in a commercial building.

All measurement techniques discussed in this chapter use the 2-point method. All OTDRs can measure loss using the 2-point method, but not all OTDRs can automatically perform 2-point subtraction. This section assumes that your OTDR can perform 2-point subtraction. If your OTDR can't, you will have to perform the calculations manually.

Some OTDRs are capable of *least-squares averaging (LSA)*. LSA can be used to measure attenuation. This section does not address LSA measurements. If your OTDR can perform LSA measurements, consult the operator's manual for how to perform them.

WARNING   When testing with the OTDR, always observe the manufacturer's precautions. Never directly view the end of an optical fiber being tested with an OTDR. Viewing the end of an optical fiber being tested with an OTDR directly or with a microscope may cause eye damage.

BASELINE TRACE

The first step in trace analysis is to generate the baseline trace. You should not press the test button on the OTDR to generate a baseline trace until you ensure that:

♦ All connectors have been cleaned and inspected and are undamaged.

♦ Launch and receive cables have optical fiber similar to the optical fiber under test.

♦ Launch and receive cables are the correct length.

♦ Launch and receive cables are properly connected to each end of the fiber-optic link under test.

♦ The correct fiber type, wavelength, range and resolution, pulse width, averages, refractive index, and backscatter coefficient have been entered into the OTDR.

Figure 16.53 is a drawing of a baseline trace as presented on the OTDR screen. This baseline trace contains a launch cable, horizontal segment, and receive cable. The launch and receive cables are 100m in length. The horizontal segment is 85m in length.

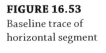

FIGURE 16.53

Baseline trace of horizontal segment

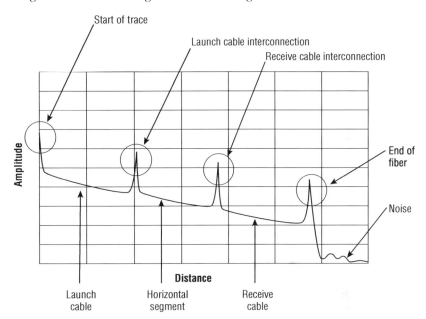

Looking at the trace from left to right, you will notice there is a large back reflection at the input to the launch cable. Because a 20ns pulse width was selected, the trace is smooth within 10m. The smooth trace slopes gradually to the back reflection caused by the connector pair where the launch cable and horizontal segment are connected.

The trace becomes smooth again 10m after the interconnection back reflection. The trace remains smooth up to the back reflection caused by the connector pair where the receive cable and horizontal segment interconnect. The trace again becomes smooth 10m after the interconnection back reflection until a large back reflection is generated by the end of the receive cable. The receive cable back reflection is followed by a large reduction in amplitude, and then the trace disappears into the noise floor.

MEASURING THE ATTENUATION OF A PARTIAL LENGTH OF OPTICAL FIBER

When using the 2-point method, the first thing you should do is measure the attenuation for a partial segment of the cable under test. This should be done for each wavelength that you are testing the optical fiber with. It has to be done only one time for each cable type being tested. The data should be recorded—this information will be needed to accurately measure interconnection loss.

After taking the baseline trace, position the two cursors on a smooth section of the optical fiber. The longer the section, the more accurate your results will be, because noise will have less impact on the overall measurement. The trace in Figure 16.54 has the A and B cursors 50m apart on the smooth section of the horizontal segment of the trace. The loss for this 50m segment at a wavelength of 850nm is approximately 0.14dB.

FIGURE 16.54
Measuring the attenuation of a cable segment using the 2-point method

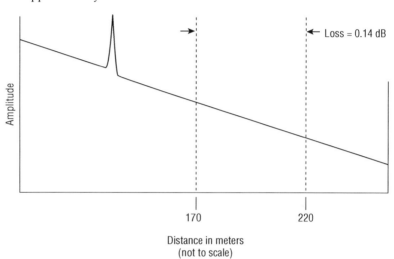

NOTE Because of ripple on the smooth trace, you will notice that the A-B loss changes slightly every time a cursor moves. Ripple caused by noise will cause variations from measurement to measurement. This variation may be as much as 0.05dB.

MEASURING THE DISTANCE TO THE END OF THE OPTICAL FIBER

A break in an optical fiber in a fiber-optic cable looks like the end of the optical fiber on the OTDR display because the OTDR cannot see beyond the break. Regardless of the number of optical fiber breaks in a cable, the OTDR can see only the one closest to it. Because a break looks like the end of the optical fiber on the OTDR, the same method is employed to measure the distance to a break in an optical fiber or measure the distance to the end of an optical fiber.

Light from the OTDR exiting the optical fiber under test typically produces a strong back reflection. The strong back reflection is caused by the Fresnel reflection when the light from the optical fiber hits the air. However, a break in an optical fiber is not always exposed to the air. The break may be exposed to index matching gel from a mechanical splice, gel from a loose buffer, or water.

The same method can be used to find the distance to the end of an optical fiber whether or not a Fresnel reflection happens. Figure 16.55 has two OTDR traces. The top trace is the end of an optical fiber with a Fresnel reflection. The bottom trace is the end of an optical fiber without a Fresnel reflection.

FIGURE 16.55
Measuring the distance to the end of an optical fiber using the 2-point method

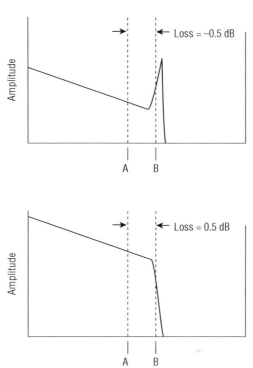

To measure the distance to the end of the optical fiber after zooming in on the back reflection, place the A cursor on a smooth section of the trace just in front of the back reflection or the drop in the trace. Move the B cursor toward the A cursor until it is in the leading edge of the back reflection. Keeping moving the B cursor toward the A cursor until the A-B loss is ±0.5dB. It should be 0.5dB of loss for the nonreflective trace and 0.5dB of gain for the reflective trace. The length for the entire span is the distance for the B cursor.

MEASURING THE LENGTH OF A CABLE SEGMENT

The first step in measuring the length of a cable segment is to horizontally zoom in on the cable segment, as shown in Figure 16.56. Place the cursors in the leading edge of the reflective events for that segment. The cursors should intersect the leading edge of the reflective event at the same vertical height above the smooth part of the trace, as shown in Figure 16.56. The distance between cursors is the length of the cable segment.

FIGURE 16.56
Measuring the
length of a cable
segment

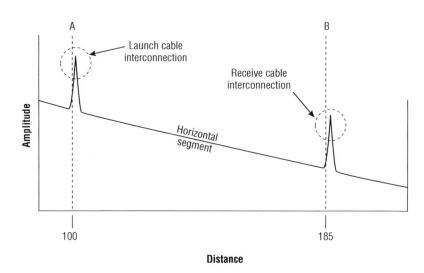

MEASURING INTERCONNECTION LOSS

The first step in measuring interconnection loss is to horizontally zoom in on the interconnection. Position the A cursor just in front of the back reflection; then position the B cursor on a smooth area on the trace after the interconnection back reflection. The loss for the interconnection and the optical fiber between the cursors is displayed on the OTDR screen. The loss for the interconnection shown in Figure 16.57 is 0.4dB. The distance between the A and B cursors is 50m.

FIGURE 16.57
Measuring inter-
connection loss
with the OTDR

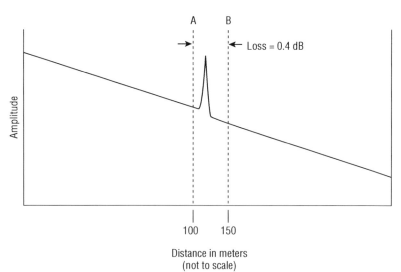

NOTE The B cursor should always be in a smooth section of the trace. When a long pulse width is used, the dead zone may be very large. The distance between the A and B cursors may need to be several hundred meters. In that case, you measure the attenuation of a partial length of optical fiber no less than several hundred meters.

To find the loss for only the interconnection, the loss for the optical fiber between cursors A and B needs to be subtracted from the A-B loss displayed by the OTDR. This loss was previously at 0.14dB for 50m. To find the actual interconnection loss, subtract 0.14dB from the 0.4dB A-B loss. The loss for only the interconnection is approximately 0.26dB.

MEASURING THE LOSS OF A FUSION SPLICE OR MACROBEND

A fusion splice or macrobend does not produce a back reflection. A macrobend will always appear as a loss in the form of a dip in the smooth trace. A fusion splice may appear as a dip or a bump in the trace. The fusion splice will appear as a dip in the trace when tested in both directions when the optical fibers fusion-spliced together have the same backscatter coefficient.

When optical fibers with different backscatter coefficients are fusion-spliced together, the splice may appear as a loss in one direction and a gain in the form of a bump in the trace when tested in the other direction. This is referred to as a *gainer*. To find the loss for the fusion splice, the splice must be tested in both directions and the results averaged together. The losses or loss and gain should be added together and the sum divided by 2. This is the loss for the fusion splice.

To find the loss of a fusion splice or macrobend, horizontally zoom in on the event. The loss from a fusion splice or macrobend is typically very small and will require vertical zoom in addition to horizontal zoom. Place the A cursor on the smooth part of the trace before the dip in the trace. Place the B cursor on the smooth part of the trace after the dip, as shown in Figure 16.58. The loss for this event is 0.25dB. The loss for the event includes the loss for the fusion splice or macrobend plus the 50m of optical fiber between the cursors. Subtract the loss for the 50m of optical fiber that was previously measured at 0.14dB from the event loss. The loss for this fusion splice or macrobend is 0.11dB.

FIGURE 16.58
Measuring the loss of a fusion splice or macrobend

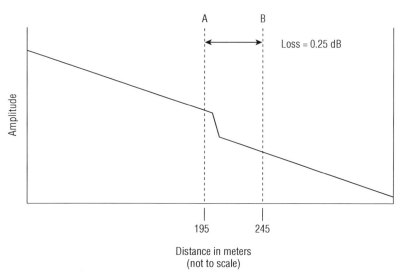

To find the gain of a fusion splice, horizontally zoom in on the event. The gain from a fusion splice is typically very small and will require vertical zoom in addition to horizontal zoom. Place the A cursor on the smooth part of the trace before the bump in the trace. Place the B cursor on the smooth part of the trace after the bump, as shown in Figure 16.59. The gain for this event is 0.15dB. The gain for the event includes the gain for the fusion splice plus the 50m of optical fiber between the cursors. Add the value for the loss for the 50m of optical fiber that was previously measured at 0.14dB to the event gain. The gain for this fusion splice is 0.29dB.

FIGURE 16.59

Measuring the gain of a fusion splice

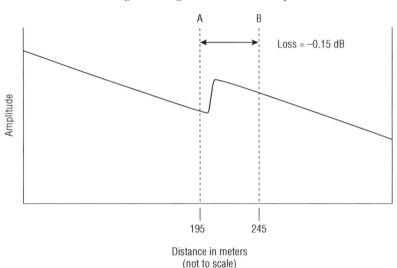

MEASURING THE LOSS OF A CABLE SEGMENT AND INTERCONNECTIONS

To find the loss for a cable segment plus the interconnections, the exact length of the segment must be known. This can be done by using the method described to measure the length of a cable segment. The length of the cable segment in this example, as shown in Figure 16.60, is 85m.

The first step is to horizontally zoom in on the cable segment. After zooming in on the cable segment, place the A cursor on a smooth section of the trace just in front of the leftmost cable segment interconnection back reflection. Place the B cursor on the smooth part of the trace after the rightmost cable segment interconnection back reflection. Looking at the A-B distance on the OTDR, move the B cursor to the right until the distance equals the length of the cable segment plus 50m, as shown in Figure 16.60 earlier. The A-B loss is 1.5dB.

To find the loss for the cable segment plus the interconnections, subtract the loss for the 50m of optical fiber from the 1.5dB loss for the cable segment and interconnections. The previously measured loss for 50m of optical fiber at 850nm is 0.14dB. The loss for the cable segment and the interconnections is 1.36dB.

FIGURE 16.60
Measuring the loss of a cable segment and interconnections

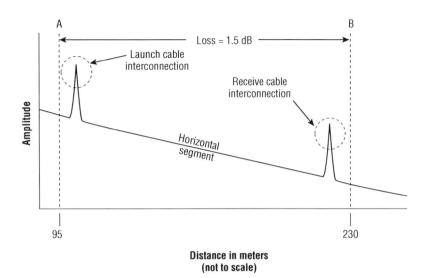

Documentation of OTDR Testing

As you test the cable plant with the OTDR, you need to properly document the results of the test. The documentation should include at a minimum the following:

♦ The date of the testing

♦ Test personnel

♦ The number or identification of fiber or cable tested

♦ The test procedure number

♦ OTDR make, model, serial number, and calibration dates

♦ Pulse width

♦ Range and resolution

♦ Number of averages

♦ Tested wavelength(s)

♦ Type and length of launch cable

Emerging Testing Standards

In April 2013, SAE International's AS-3 Fiber Optics and Applied Photonics Committee began work on a new optical performance monitoring system standard AIR6552 that outlines the sensors and basic architecture required to detect, localize, and isolate impairments as well as assist in failure prediction of the physical layer of a fiber-optic network. This fiber-optic network end-to-end data link evaluation system standard is also known as NEEDLES. Because NEEDLES focuses on the physical layer, it can be applied to any fiber-optic network, including:

♦ Ethernet

♦ ATM

♦ RF over fiber

♦ Passive Optical Networks

   ♦ FTTH

   ♦ FTTB

   ♦ FTTC

   ♦ FTTN

Since NEEDLES is a multisensor architecture, it has a main document AIR6552 and several slash sheets. Each slash sheets addresses a particular type of sensor. As of this writing, the main document and three slash sheets are under development.

AIR6552/1, which was discussed earlier in this chapter, addresses inline optical power monitoring. Inline optical power monitoring can be used to control optical power, anticipate failures, and detect fluctuations. This document establishes methods to obtain, store, and access data about the health of a fiber-optic network using commercially available inline optical power monitoring sensors.

AIR6552/2, which was discussed earlier in this chapter, addresses high-resolution optical reflectometry. High-resolution optical reflectometry can characterize and map the optical link. This document establishes methods to obtain, store, and access data about the health of a fiber-optic network using a commercially available high-resolution reflectometer.

AIR6552/3, which was discussed in Chapter 11, addresses transceiver health monitoring. Transceiver health monitoring can provide real-time information about the physical, electrical, and optical characteristics of the transceiver. This document establishes methods to obtain, store, and access data about the health of a fiber-optic network using commercially available sensors located in or near the transceiver.

The Bottom Line

Perform optical loss measurement testing of a cable plant. ANSI/TIA-526-14 revisions A and B provide three methods for testing the cable plant. The difference between the three methods is how the reference power measurement is taken. Each method uses a different number of test jumpers. Method A uses two test jumpers to obtain the reference power measurement, method B uses one test jumper, and method C uses three test jumpers.

Master It Before testing the cable plant using method B, you obtained your reference power measurement of –23.5dBm. Your test power measurement was –27.6dBm. What was the loss for the cable plant?

Determine the distance to a break in the optical fiber with an OTDR. A break in an optical fiber in a fiber-optic cable is not the end of the optical fiber. However, a break in an optical fiber looks like the end of the optical fiber on the OTDR display. The same method is employed to measure the distance to a break in an optical fiber and to measure the distance to the end of an optical fiber.

Master It You are troubleshooting fiber-optic cables with the OTDR. One optical fiber appears to be broken. Refer to the following graphic and determine the distance to the break in the optical fiber.

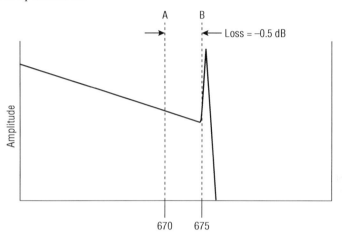

Measure the loss of a cable segment with an OTDR. To find the loss for a cable segment plus the interconnections, the exact length of the segment must be known.

Master It You are testing a new fiber-optic cable installation with the OTDR and the customer has requested the loss information for a cable segment including the interconnections. Refer to the following graphic and determine the cable segment. The A-B distance equals the length of the cable segment plus 50m.

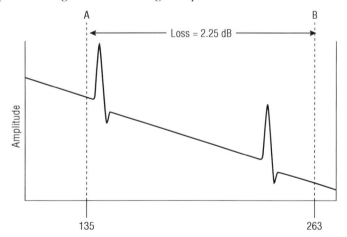

Measure the loss of an interconnection with an OTDR. When measuring the loss for an interconnection with the OTDR, you must remember to subtract the loss for the optical fiber.

Master It You are testing a fiber-optic cable installation with the OTDR and the customer has requested the loss information for each interconnection. Refer to the following graphic and determine the cable segment.

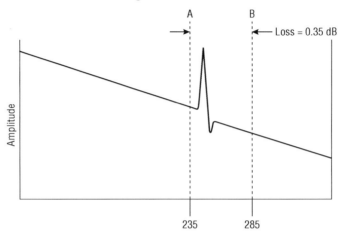

Measure the loss of a fusion splice or macrobend with an OTDR. Fusion splices or macrobends do not produce a back reflection. A macrobend will always appear as a loss in the form of a dip in the smooth trace. A fusion splice may appear as a dip or a bump in the trace.

Master It You are testing a fiber-optic cable installation with the OTDR and see a dip in the trace. Since there are no splices in this cable, you determine that this event must be a macrobend. Refer to the following graphic and determine the loss for the macrobend.

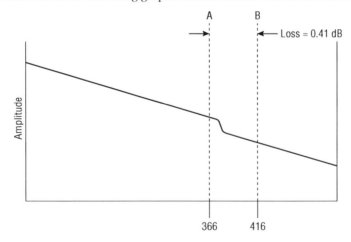

Chapter 17

Troubleshooting and Restoration

So far you have learned about fiber-optic theory, cables, connectors, splices, transmitters, receivers, and many different passive components. We have shown you how to predict the performance of a link and test that link to industry standards using various pieces of test equipment. However, we have not discussed what to do when the link doesn't work.

This chapter takes the knowledge and skills you have learned up to this point and applies them to troubleshooting a fiber-optic link or cable. You will learn basic techniques that will allow you to quickly analyze and determine the fault. You will learn fault location techniques with devices as simple as a flashlight and as complicated as an OTDR.

In this chapter, you will learn to:

- ◆ Identify a high-loss interconnection with an OTDR
- ◆ Identify a high-loss mechanical splice with an OTDR
- ◆ Identify a high-loss fusion splice or macrobend with an OTDR

Optical Fiber Type Mismatch

Optical fiber type mismatch can cause attenuation and or back reflection problems at interconnections. With the increased use of pre-polished connectors and changes in cordage colors over the years, it is not uncommon to encounter an optical fiber type mismatch. This section describes techniques for determining the cable optical fiber type and the pre-polished connector optical fiber type.

Cable Optical Fiber Type Mismatch

As you learned in Chapter 7, "Fiber-Optic Cables," in addition to the NEC cable marking optical fiber cables typically have a number of other markings and codes. Markings that appear on the jacket of the cable help identify what is inside the cable and where it originated. In addition, TIA-598-C provides color-coding schemes for premises jackets and optical fibers within a fiber-optic cable.

Because this section focuses only on identifying the optical fiber type, the color-coding scheme for individual fibers bundled in a cable will not be discussed.

DETERMINING THE CABLE OPTICAL FIBER TYPE

Working in the field you will encounter cabling that has been in place for many years. This cabling may not comply with the current industry standards, and that can make determining the optical fiber type within a cable challenging. Understanding how the 598-color code has changed over the years will be helpful in correctly identifying the optical fiber type.

Revision A of the 598-color code is formally known as TIA/EIA-598-A. As of this writing, differences exist between revision A and the current revision (C). The primary differences lie in the color-coding of premises cable jackets. Table 17.1 shows the revision A color-coding used on premises cable jackets to indicate the type of fiber they contain. Table 17.2 shows the revision C color-coding.

NOTE Simplex and duplex premises cables that contain only one type of optical fiber are typically referred to as *cordage*.

TABLE 17.1: TIA/EIA-598-A premises cable jacket colors

FIBER TYPE AND CLASS	DIAMETER	JACKET COLOR
Multimode 1a	50/125µm	Orange
Multimode 1a	62.5/125µm	Slate
Multimode 1a	80/125µm	Blue
Multimode 1a	100/140µm	Green
Single-mode IVa	All	Yellow
Single-mode IVb	All	Red

TABLE 17.2: TIA-598-C premises cable jacket colors

FIBER TYPE	JACKET COLOR FOR NONMILITARY APPLICATIONS	JACKET COLOR FOR MILITARY APPLICATIONS
Multimode (50/125µm)	Orange	Orange
Multimode (50/125µm) Laser-optimized	Aqua	
Multimode (62.5/125µm)	Orange	Slate
Multimode (100/140µm)	Orange	Green
Single-mode (NZDS)	Yellow	Yellow
Polarized maintaining Single-mode	Blue	

When you compare Table 17.1 to Table 17.2, you can see a few differences. One key difference is that revision A has only one color code column, whereas Table 17.2 has columns for military and nonmilitary applications. Revision A also has an additional multimode core size, 80/125µm. Both revision A and C have a single-mode optical fiber type that is not described the other revision.

The single-mode IVa optical fiber type in revision A is a non-zero-dispersion-shifted optical fiber. The IVb single-mode optical fiber is a dispersion-shifted optical fiber. The polarization maintaining single-mode optical fiber described in revision C is not found in revision A.

The only jacket colors that are unique within revisions A and C are yellow, aqua, red, and slate. Regardless of how old the cabling is, a yellow jacket identifies a non-zero-dispersion-shifted single-mode optical fiber, an aqua jacket identifies a laser-optimized multimode optical fiber, a red jacket identifies a dispersion-shifted single-mode optical fiber, and a slate jacket identifies 62.5/125µm multimode optical fiber.

Jacket colors orange, blue, and aqua are the problem colors. An orange jacket can represent a 50/125µm or 62.5/125µm multimode optical fiber. When working with orange jackets, always read the text printed on the jacket to confirm the core diameter. Blue can represent an 85/125µm multimode optical fiber or a polarization maintaining single-mode optical fiber, and aqua may be mistaken for blue. When working with blue jackets, always read the text printed on the jacket to determine if it is 80/125µm multimode, 50/125µm multimode, or single-mode optical fiber.

When working with premises cables that are not considered cordage, the jacket color is not a concern regardless of when the cable was manufactured and which 598 standard it complied with. The text printed on the jacket should identify the optical fiber classification (multimode or single-mode) and the optical fiber sizes for multimode optical fiber only. For newer multimode optical fiber, the jacket text should also identify whether the multimode optical fiber is laser optimized.

Remember, you can always contact the cable manufacturer using the phone number printed on the cable to determine the optical and geometrical performance of the optical fiber.

Connector Optical Fiber Type Mismatch

Whether the cable is terminated with a pre-polished connector or an oven-cured connector, mistakes that indicate the wrong optical fiber type happen. This section describes techniques to determine whether the correct color strain relief was installed when the cable was terminated. In the event that a pre-polished connector was used to terminate the cable, these techniques can be used to determine whether the optical fiber in the connector is the same as the optical fiber in the cable.

CONNECTOR COLOR CODE

Multimode optical fiber should be terminated with the correct multimode connector, and single-mode optical fiber should be terminated with the correct single-mode connector. One method to determine the type of multimode or single-mode connector is to use the color code defined in ANSI/TIA-568-C.3. The color code in Table 17.3 should be used unless color coding is used for another purpose. This table lists the strain relief and adapter housing color for different optical fiber types.

TABLE 17.3: Multimode and single-mode connector and adapter identification

OPTICAL FIBER TYPE OR CONNECTOR TYPE	STRAIN RELIEF AND ADAPTER HOUSING COLOR
850nm laser-optimized 50/125µm optical fiber	Aqua
50/125µm optical fiber	Black
62.5/125µm optical fiber	Beige
Single-mode optical fiber	Blue
Angled contact ferrule single-mode connectors	Green

Connector color codes work only if there were no errors during assembly. Many installers will not discard a strain relief with a failed connector. Over time, they may acquire a collection of strain reliefs and accidently install the wrong color strain relief while terminating the cable. When you suspect a problem, do not rely solely on the color of the strain relief for optical fiber identification. You will need to examine the endface with an inspection microscope.

Because of the dramatic differences in core size, it is easy to tell the difference between a multimode and single-mode optical fiber with an inspection microscope. However, telling the difference between 62.5/125µm optical fiber and 50/125µm optical fiber with an inspection microscope is a little more challenging, and even an experienced installer or technician may guess incorrectly without something to reference.

One technique is to have 62.5/125µm fiber and 50/125µm optical fiber simplex patch cords in your toolbox. When trying to determine the core size of the optical fiber in question, compare it to the two different patch cords in your toolbox. Using this approach, you should be able to determine the optical fiber core diameter.

Another technique is to carry a flexible tape measure. First, measure the diameter of the cladding on the video inspection microscope display and record that value in your notebook. Then measure the diameter of the core and record that value in your notebook. If the core diameter is approximately 50 percent of the cladding diameter, it is a 62.5/125µm multimode optical fiber. If the core diameter is approximately 40 percent of the cladding diameter, it is a 50/125µm multimode optical fiber.

Inspection and Evaluation

Good inspection and evaluation skills are essential for anyone attempting to troubleshoot a fiber-optic link or cable. Often the cause of a problem is basic and can be discovered with a thorough inspection. For many troubleshooting scenarios, expensive test equipment is not necessary.

This section focuses on the visual inspection of the connector ferrule, endface, receptacle, and adapter (which is also referred to as an alignment sleeve or mating sleeve).

Connector Inspection

The weakest link in any fiber-optic installation is likely to be the connector. When you look at a fiber-optic connector, you don't see anything exciting. Fiber-optic connectors look simple. They have few moving parts and do not require electricity. How hard can it be to properly install a fiber-optic connector?

As you learned in Chapter 9, "Connectors," the fiber-optic connector provides a way to connect an optical fiber to a transmitter, receiver, or other fiber-optic device. The optical fiber is physically much smaller than the connector and virtually invisible to the naked eye. The naked eye cannot tell the difference between a perfectly polished fiber-optic connector and a fiber-optic connector in which the optical fiber is shattered.

Fiber-optic connectors can have poor performance for the many reasons that are described in detail in Chapter 9. However, in our experience poor surface finishes and dirt are the primary causes. We cannot stress enough how important it is to make every effort possible to ensure that the connector is kept clean.

Dirt particles on the connector endface are not the only cause of poor connector performance. Dirt anywhere on the connector ferrule or inside the connector receptacle or adapter can also cause poor performance. Occasionally epoxy that runs onto the side of the connector ferrule during oven curing goes unnoticed, preventing the connector ferrule from aligning properly in the connector receptacle or mating sleeve. This can result in a high-loss interconnection.

It is a good idea to have an eye loupe like the one shown in Figure 17.1 in your tool bag. The eye loupe will allow you to spot small dirt or epoxy particles that may be on the connector ferrule. Excess epoxy needs to be removed from the connector ferrule before the connector is inserted into a receptacle or adapter. The endface of the connector should be cleaned following the procedures in Chapter 9. After the connector has been properly cleaned, it should be evaluated with an inspection microscope.

FIGURE 17.1
Eye loupe

Photo courtesy of KITCO Fiber Optics

NOTE   Before evaluating the endface of a connector with an inspection microscope, ensure the microscope alignment sleeve is not contaminated by routinely cleaning it with a cleaning stick or swab as described in the "Receptacle and Adapter Inspection and Cleaning" section later in this chapter.

Connector Endface Evaluation

Various standards have been developed to assess the endface of polished fiber-optic connectors. This section provides an overview of two standards, ANSI/TIA-455-57-B and IEC 61300-3-35. In addition, microscope magnification levels, the effects of different abrasives, endface finish, and geometry are covered.

The visual inspection of the endface of a fiber-optic connector can be properly performed only with an inspection microscope. As you recall from Chapter 9, many different inspection microscopes are available. Typically, a multimode connector can be evaluated with a 100X microscope, while a single-mode connector requires a minimum of 200X magnification. A 400X microscope works even better for both multimode and single-mode. Figure 17.2 shows two handheld microscopes. The smaller microscope is a 100X and the larger is a 400X.

FIGURE 17.2
400X and 100X
microscopes

ANSI/TIA-455-57-B

ANSI/TIA-455-57-B was published in 1994 and reaffirmed in 2005. When a standard is reaffirmed, it is not changed or updated. This standard provides a guideline for examination of an optical fiber endface. Figure 17.3 contains possible core and cladding fiber endface results. Each endface in Figure 17.3 is labeled with a number and a letter. The number and letter combinations are listed in Table 17.4. This table is used to identify always acceptable, usually acceptable, and often acceptable endface results.

TIP   Remember that a dirt particle invisible to the naked eye can damage the optical fiber in a fiber-optic connector beyond repair.

FIGURE 17.3
ANSI/TIA-455-57-B core and cladding endface results

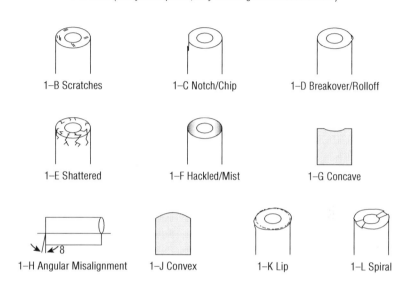

1–A Good (always acceptable; may have slight indent/score mark)

1–B Scratches   1–C Notch/Chip   1–D Breakover/Rolloff

1–E Shattered   1–F Hackled/Mist   1–G Concave

1–H Angular Misalignment   1–J Convex   1–K Lip   1–L Spiral

Typical fiber endface results with core and cladding

TABLE 17.4:   ANSI/TIA-455-57-B core and cladding endface results

	APPEARANCE ACCEPTABILITY		
Figure	**Always acceptable**	**Usually acceptable[1,2]**	**Often acceptable[1]**
34.3	1-A	1-B, 1-C, 1-D, 1-G, 1-H	1-E, 1-F, 1-J, 1-K, 1-L

1. *Tighter or looser limits may be specified by the fiber-optic test procedure (FOTP).*

2. *Unless defect extends into the core.*

Looking at Table 17.4, you will see that one endface result— endface finish 1-A—is always acceptable. Note that 1-A is drawn to have no imperfections, or to be "cosmetically perfect." However, a slight indent or score mark is acceptable.

IEC 61300-3-35

As you learned in Chapter 9, IEC 61300-3-35 is a standard that provides methods for quantitatively evaluating the endface quality of a polished fiber optic connector. This standard breaks the endface into four zones—core, cladding, adhesive or epoxy ring, and contact zone, as shown in Figure 17.4. The diameters of the zones are defined in Table 17.5.

FIGURE 17.4

Approximation of the four endface zones

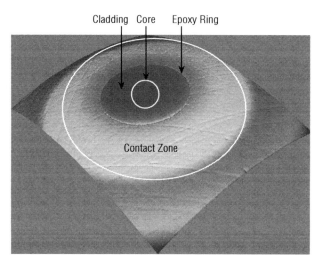

Cladding Core Epoxy Ring

Contact Zone

TABLE 17.5:   IEC 61300-3-35 endface measurement regions

ZONE	SINGLE-MODE DIAMETER	MULTIMODE DIAMETER
Core	0 to 25μm	0 to 65μm
Cladding	25 to 120μm	65 to 120μm
Epoxy Ring	120 to 130μm	120 to 130μm
Contact Zone	130 to 250μm	130 to 250μm

This standard uses an inspection procedure flowchart and various pass/fail criteria for different optical fiber and endface types. Only a portion of these will be examined in this text. Imperfections are broken into two groups: scratches and defects. A scratch is a permanent linear surface feature whereas a defect may be anything that is not a scratch. Regardless of the optical fiber type or endface, scratches and defects are permitted in the 10μm wide epoxy ring zone.

No defects are permitted in the core zone of APC single-mode or PC single-mode connectors with a return loss of 45dB. PC single-mode connectors with a return loss of 26dB are permitted to have two defects 3μm in size. PC single-mode connectors with a return loss of 45dB may not have any scratches in the core zone either.

As of this writing, this standard was under revision. It is unknown whether an additional zone will be included in the next revision.

ENDFACE EVALUATION TECHNIQUES

When you are evaluating the endface of a connector with a microscope, the magnification level of the scope has everything to do with the detection of imperfections. Earlier, we mentioned that most multimode fiber-optic connectors can be properly evaluated with a 100X inspection microscope. Viewing a multimode endface with this magnification level will reveal only gross defects—small scratches from the polishing abrasive are typically undetectable.

The same connector viewed with a 400X inspection microscope may show many small scratches that were caused by the polishing abrasive. The endface that looked cosmetically perfect or defect-free with 100X magnification might look really ugly with 400X magnification. Imagine how that endface would look with 600X or 800X magnification.

Typically, a multimode fiber-optic connector endface does not have to be cosmetically perfect to provide a low-loss interconnection. In my classes, I (Bill) demonstrate this to the students. I take a patch cord with cosmetically perfect connectors on each end and measure the loss. Then I take another patch cord with connector endfaces that have been polished with only a 1μm abrasive and measure the loss. The students are surprised when the cosmetically perfect patch cord has more loss than the less-than-cosmetically-perfect patch cord.

For many applications, a multimode connector endface does not need to be polished with an abrasive finer than 1μm. A 1μm abrasive will leave visible scratches in the endface when viewed with a 400X microscope. However, this same endface may appear scratch-free when viewed with a 100X microscope.

To evaluate the endface of a multimode connector, you need to divide the endface into three parts: the core, the inner cladding, and the outer cladding. As shown in Figure 17.5, the outer cladding is the area from the center of the cladding to the epoxy ring. The inner cladding is the area from the center of the cladding to the core.

FIGURE 17.5
Sections of a multi-mode connector endface

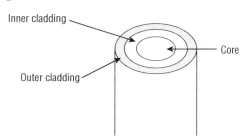

Field-terminated multimode connectors are typically polished with an aluminum oxide abrasive. As you learned in Chapter 9, aluminum oxide is harder than the optical fiber but softer than the ceramic ferrule. Because the aluminum oxide does not polish down the ceramic ferrule during the polishing process, chipping around the epoxy ring extends into the outer cladding. Sometimes this chipping may extend in excess of 10μm from the epoxy ring. This is typical, which means that this is acceptable.

TIP A good rule of thumb when evaluating a multimode connector is to accept all endface finishes that have no defects in the core and inner cladding. The connector endface should be defect-free when viewed with a 100X microscope. Minor scratches are acceptable when viewed with a 200X or a 400X microscope.

Figure 17.6 shows a multimode endface polished with a 1μm aluminum oxide abrasive viewed with a 400X video microscope. Note the minor chipping along the epoxy ring and the minor scratches on the inner cladding and core. This is an acceptable multimode endface. This connector endface viewed with a 100X microscope would look cosmetically perfect.

FIGURE 17.6
Multimode connec-
tor endface polished
with a 1μm alumi-
num oxide abrasive
viewed with a 400X
microscope

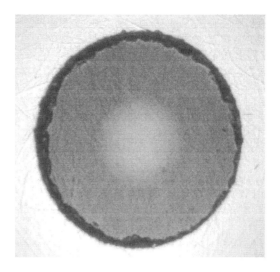

Single-mode connector evaluation requires 400X magnification. The endface of a single-mode connector should look cosmetically perfect, like the endface shown in Figure 17.7. Remember from Chapter 9 that a single-mode connector endface is typically finished with a diamond abrasive. A diamond abrasive polishes down the ferrule and the glass.

FIGURE 17.7
Single-mode con-
nector endface
with no defects
viewed with a 400X
microscope

Photo courtesy of MicroCare

A single-mode endface polished with diamond abrasive should have a thin epoxy ring in comparison to an endface polished with aluminum oxide. There should be no pitting or chipping of the optical fiber along the epoxy ring; however, the epoxy ring may appear to have pitting or chipping. The inner cladding, outer cladding, and core should be defect-free when viewed with a 400X microscope.

NOTE A multimode connector endface that has been polished with a diamond abrasive should not have any pitting or chipping on the optical fiber along the epoxy ring.

In some applications, single-mode terminations are performed in the field with an aluminum oxide abrasive. These terminations may have endface finishes similar to a field-terminated multimode endface. The endface should be evaluated as you would evaluate a multimode endface.

Up to this point, we have focused only on the cosmetics of the endface and we have not covered its geometry. When a connector is viewed with a microscope, you see only a two-dimensional view—you can't see depth or height. As you learned in Chapter 9, virtually all connector ferrules have a radius. This means that a properly polished optical fiber should have the same radius as the ferrule.

Neither ANSI/TIA-455-57-B nor IEC 61300-3-35 provides guidance when it comes to measuring or evaluating ferrule or endface radius. Ferrule or endface radius cannot be measured with a microscope; it can be measured only with an interferometer. An endface may be defect-free during a visual inspection and yet it may be overpolished. Overpolishing is typically impossible to detect with a visual inspection. However, it is easy to detect with an interferometer.

Interferometers generate a 3D map of the endface and quantitatively define three critical parameters: radius of curvature, apex offset, and fiber undercut or protrusion. These critical parameters are required by Telcordia GR-326 to evaluate connector endface geometry for single-mode connectors and jumper assemblies. The overpolished endface shown in Figure 17.8 would appear defect-free during a visual inspection. The visual inspection would not reveal how overpolishing created a significant undercut.

FIGURE 17.8

3D endface image of an overpolished endface with significant undercut

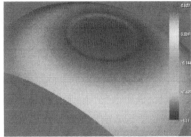

Photo courtesy of PROMET International Inc.

In the early stages of troubleshooting, a visual inspection of the connector endface is all that is necessary. If the connector endface meets the requirements established by ANSI/TIA-455-57-B or IEC 61300-3-35, you can go to the next step: checking optical fiber continuity. You don't need to be concerned about the connector endface radius, unless testing with the OTDR reveals a high interconnection loss or high interconnection back reflection.

High interconnection loss or back reflection from a mated connector pair with acceptable endface finishes is usually caused by a change in the refractive index. This may be the result of an air gap or contamination. An air gap has several causes:

♦ Dirt that is not allowing the ferrule to properly seat in the receptacle

♦ The connector not being inserted properly into the receptacle

♦ One or both connectors having a concave or flat endface finish as opposed to a convex one

Contamination is often caused from handling. If you touch the connector endface, oil from your skin will be deposited on the endface. The oil on the endface of the connector does not disappear when the connector is mated with another connector. It creates a thin film between the two optical fibers. This thin oil film, as shown in Figure 17.9, has a different refractive index compared to the optical fiber; the change in refractive index causes a Fresnel reflection.

FIGURE 17.9
Single-mode con-
nector endface with
skin oil viewed with
a 200X microscope

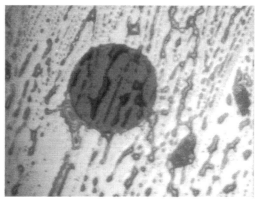

Photo courtesy of MicroCare

Receptacle and Adapter Inspection and Cleaning

As discussed in Chapter 10, "Fiber-Optic Light Sources and Transmitters," and Chapter 11, "Fiber-Optic Detectors and Receivers," many fiber-optic transceivers feature receptacles that mate with the connectors. Sometimes the problem is as simple as a connector not being fully inserted into the receptacle and latched. The transceiver shown in Figure 17.10 has ST connector receptacles. If you look closely at the photograph, you will see that the connector on the right has not been fully inserted and latched. However, the connector on the left has.

FIGURE 17.10
The ST connector
on the right is not
fully mated with
the receptacle.

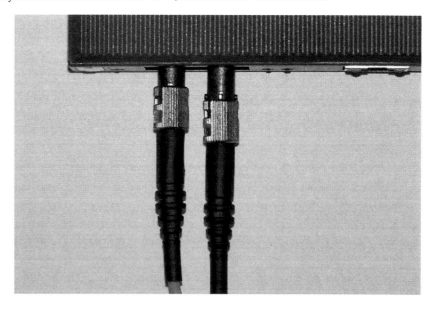

Detecting when a small form factor connector such as an LC is not fully inserted into the receptacle and latched can be challenging. LC connectors are typically used with small-form factor pluggable (SFP) transceivers like the one shown in Figure 17.11. The SFP transceiver has a small handle that is used to remove the transceiver from the equipment. This handle hides part of the latching mechanism, as shown Figure 17.12, making it difficult to visually determine that the receive LC connector is not latched. Figure 17.13 shows the unlatched receive LC connector with the handle removed from view.

FIGURE 17.11
SFP transceiver
with LC connectors

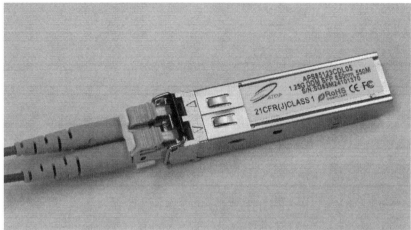

Photo courtesy of Electronic Manufacturers' Agents, Inc.

FIGURE 17.12
SFP transceiver
handle hiding the
unlatched receive
LC connector

FIGURE 17.13
SFP transceiver
with unlatched
receive LC connec-
tor and the handle
removed from view

NOTE The terms *alignment sleeve*, *mating sleeve*, and *adapter* are interchangeable. ANSI/TIA-568-C defines an adapter as a mechanical device designed to align and join two optical fiber connectors (plugs) to form an optical connection. However, sometimes the term *adapter* is used when referring to a device that mates two different connector types such as an SC and ST. Every adapter contains an alignment sleeve that aligns the connector ferrules or plugs.

When the connector fails to properly mate with the transmitter or the receiver, problems may occur. The same is true at the adapter when two connectors do not mate properly. Figure 17.14 shows two SC connectors in an adapter. The connector on the right has not been fully inserted into the adapter. This results in an air gap that produces a Fresnel reflection.

FIGURE 17.14
SC-to-SC mating
with connector
on right not fully
mated

Receptacles and adapters get dirty. If a dirty connector is plugged into a transceiver receptacle, the dirt from the endface of the connector will transfer to the contact area inside the receptacle. This could cause a problem and affect the operation of the transmitter or receiver. It is difficult to examine the mating surface inside a receptacle. If the endface of the connector you just removed was dirty, the receptacle should be cleaned.

There are different cleaning products available to clean the contact area inside the receptacle. These same products can also be used to clean the endface of a connector in the alignment sleeve of an adapter, the alignment sleeve, and the other areas of the adapter. In this section, we will examine how to clean these hard-to-reach areas.

CLEANING AN ENDFACE OR CONTACT AREA WITH A CLEANING STICK OR SWAB

The connector cleaning stick or swab is a simple and inexpensive tool that can be used to clean the endface of a connector in an alignment sleeve or the contact area in a receptacle. Multiple manufacturers offer a variety of cleaning sticks or swabs. The cleaning swabs shown in Figure 17.15 were engineered to clean MT ferrule endfaces and alignment sleeves. The cleaning sticks shown in Figure 17.16 were engineered to clean 2.5mm, 2mm, 1.6mm, and 1.25mm endfaces and alignment sleeves.

FIGURE 17.15
Cleaning swabs engineered to clean MT ferrule endfaces and alignment sleeves

Photo courtesy of ITW Chemtronics

FIGURE 17.16
Cleaning sticks
engineered to
clean 2.5mm,
2mm, 1.6mm, and
1.25mm endfaces
and alignment
sleeves

Photo courtesy of MicroCare

When using a cleaning stick or swab, you should select one that has a cleaning tip diameter similar to the connector ferrule the receptacle or alignment sleeve was designed to accept. Ferrule diameters are listed in Chapter 9.

When working with cleaning sticks or swabs, be careful not to touch the cleaning tip with your fingers. Without touching the tip, lightly moisten the tip of the cleaning stick or swab with a cleaning fluid engineered for fiber-optic applications like the fluids shown in Figure 17.17 or Figure 17.18. Unlike isopropyl alcohol, these cleaning fluids will not leave a residue.

FIGURE 17.17
Fiber-optic splice
and connector
cleaner

Photo courtesy of MicroCare

FIGURE 17.18
Fiber-optic cleaner

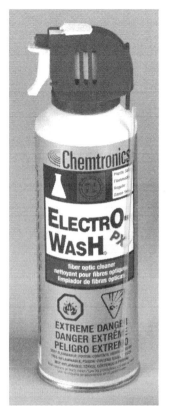

Photo courtesy of ITW Chemtronics

To clean the contact area in a receptacle, insert the moistened tip into the receptacle or mating sleeve until it makes contact with the mating surface, as shown in Figure 17.19. Rotate the cleaning stick or swab in a clockwise direction 10 revolutions, applying varying pressure to create a gentle pumping action. Remove and dispose of the stick or swab; never use a cleaning stick or swab twice.

To clean the endface of a connector in an alignment sleeve, insert the moistened tip into the alignment sleeve as shown in Figure 17.20 until it makes contact with the endface. Rotate the cleaning stick or swab in a clockwise direction 10 revolutions, applying varying pressure to create a gentle pumping action. Remove and dispose of the stick or swab; never use a cleaning stick or swab twice.

To clean the alignment sleeve in an unpopulated adapter, insert the moistened tip into one side of the alignment sleeve, as shown in Figure 17.20, and rotate the cleaning stick or swab in a clockwise direction 10 revolutions. Remove and dispose of the stick or swab. Insert the moistened tip of a new stick or swab into the other side of the alignment sleeve. Rotate the cleaning stick or swab in a clockwise direction 10 revolutions. Remove and dispose of the stick or swab.

FIGURE 17.19
Using a cleaning swab to clean the contact area in a connector receptacle

FIGURE 17.20
Using a cleaning stick to clean the endface of an SC connector in an alignment sleeve

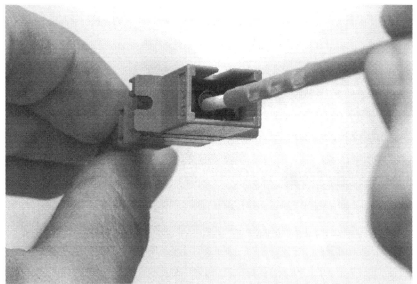

Photo courtesy of MicroCare

The same technique can be used to clean the alignment sleeve in an inspection microscope. To clean the alignment sleeve in an inspection microscope, insert the moistened tip into the alignment sleeve as shown in Figure 17.21 and rotate the cleaning stick or swab in a clockwise direction 10 revolutions. Remove and dispose of the stick or swab.

FIGURE 17.21
Using a cleaning swab to clean an inspection microscope alignment sleeve

To clean the other areas of the adapter, insert the moistened tip into one side of the adapter, as shown in Figure 17.22. While rotating the cleaning stick or swab in a clockwise direction, move it around, ensuring that contact is made with 100 percent of the surface area. Remove and dispose of the stick or swab.

FIGURE 17.22
Using a cleaning
swab to clean an
adapter

Photo courtesy of ITW Chemtronics

CLEANING AN ENDFACE WITH A TAPE TYPE CLEANER

A tape type cleaner is also a simple and inexpensive cleaning tool that can reach the contact area in a receptacle or the endface to a connector in a mating sleeve. Unlike the cleaning stick, the tape type cleaner is used without a cleaning fluid. Several manufacturers offer these cleaners; some cleaners are refillable and some are disposable.

The tape type cleaner uses a small cleaning cloth wrapped around a spool stored in the cleaner. The cleaning cloth is exposed at the tip of the cleaner, as shown in Figure 17.23. To clean a connector endface or contact area inside a receptacle, insert the tip of the cleaner as shown in Figure 17.24. When using this cleaner, gently push the handle toward the contact surface. This will cause the cleaning cloth to move across the endface of the connector. Stop pushing when you hear a click. This type of cleaner is typically good for 500 uses before it needs to be replaced.

FIGURE 17.23
Cleaning cloth
exposed at the tip
of the cleaner

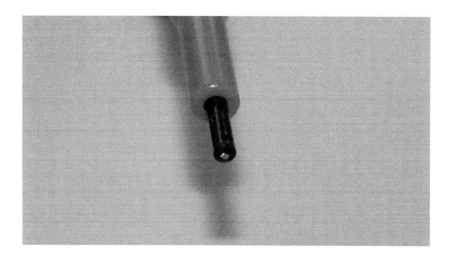

FIGURE 17.24
Cleaner tip inserted
in ST-to-SC adapter

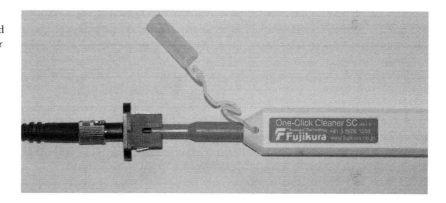

Continuity Tester Fault Location Techniques

As you learned in Chapter 16, "Test Equipment and Link/Cable Testing," the continuity tester is really no more than a flashlight. There are many different continuity testers on the market. Some use red LED light sources; others use incandescent lights. If you don't have a continuity tester, you can just use a flashlight. Figure 17.25 shows a rugged incandescent continuity tester. As you can see, this is a modified flashlight.

NOTE   When trying to locate faults, take good notes that you can reference during the troubleshooting process. Do not rely solely on your memory; a good technician always carries a notebook.

FIGURE 17.25
Rugged and focus-
able incandescent
continuity tester

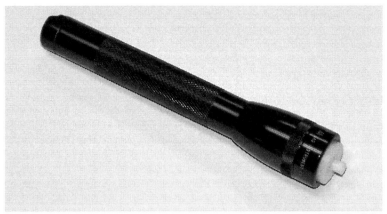

Photo courtesy of KITCO Fiber Optics

The key to using a flashlight is alignment. Remember, only light that enters an optical fiber through the cone of acceptance will propagate the length of the optical fiber. You may have to practice aligning the flashlight on a known good patch cord before testing a longer fiber-optic cable. It is much easier to use a flashlight to test multimode optical fiber than single-mode optical fiber.

The only drawback to using a flashlight, besides manual alignment, is the fact that you have to hold the connector and flashlight together. This could be a real challenge when you are working by yourself and the other end of the fiber-optic cable is in another building. If that is the case, you may have to ask your boss to break down and spend the $30 on a continuity tester.

As you learned in Chapter 16, the optical output power of an LED or incandescent continuity tester is relatively low when compared to an LED or laser transmitter. Most of the LED continuity testers I have examined output somewhere between –30dBm and –40dBm when measured at the end of one meter of 62.5/125μm optical fiber. The output of a focusable incandescent continuity tester when measured at the end of one meter of 62.5/125μm optical fiber can exceed –30dBm.

The rugged continuity tester shown in Figure 17.25 uses a mirror to focus the light. If focused properly, this continuity tester will provide an output greater than –25dBm when measured at the end of 1m of 62.5/125μm optical fiber. Regardless of the light source, all continuity testers should be made eye safe so that you can directly view the connector endface emitting the light energy.

So what can you do with a continuity tester? Basically, you can check the continuity of an optical fiber—that is, check to see if an optical fiber allows light to pass from one end to the other. A continuity tester tests only for breaks in the optical fiber. Macrobends or microbends cannot be identified with a continuity tester.

Let's work through an example and take a step-by-step approach to troubleshooting a fiber-optic link with a continuity tester. As you already know and as has been stated in this book many times, the first step in troubleshooting a fiber-optic link is to clean the connector endface. If you have forgotten how to clean a connector endface, refer to Chapter 9.

With a clean connector in hand, the next step is to examine the connector endface with an inspection microscope. As we discussed earlier in this chapter, a 100X microscope is typically adequate for multimode applications. Single-mode applications require a 400X microscope.

If your examination of the endface of each connector on opposite ends of the fiber-optic cable shows no defects, you can go to the next step. However, if your examination reveals a defect, there is no sense in testing continuity until the connector is repaired or replaced.

At this point, the connectors at each end of the fiber-optic cable are clean and you are ready to test cable continuity. Remember, before attaching the connector to the continuity tester verify that the continuity tester works. To verify continuity tester operation, turn it on and see if it emits light. This seems like beginner-level stuff, but how many times have you heard the story about the technician who shows up to fix a computer only to find that it's simply not plugged in?

CAN A FLASHLIGHT KEEP YOUR COMPANY FROM GOING OUT OF BUSINESS?

Over the years, many students have shared stories about their experiences in the field. The story that really applies to this chapter is about a small company that made its livelihood doing aerial copper cable installations.

One day, this company took a job doing an aerial installation of a multifiber cable. Because this company had never installed a fiber-optic cable, they handled it like a copper cable. After several days of hard work, the cable was in place and the installation company called a fiber-optic contractor to complete the installation.

The first thing the fiber-optic contractor did was test the cable with the OTDR. The first optical fiber tested was broken several hundred meters from the OTDR. The next optical fiber tested was also broken. Testing from the opposite end of the cable revealed at least two breaks in each optical fiber tested.

This small company had to install a new fiber-optic cable and absorb the cost of the cable and the labor. Needless to say, the company almost went out of business because of that. Eventually they determined that the fiber-optic cable had minor damage from what could have been a forklift.

The company never wanted to make this mistake again, so they sent their sharpest employee to our fiber-optic installer course. You can imagine the look on this person's face when we demonstrated how to use a continuity tester to look for a break in an optical fiber.

The moral of the story? Always test your fiber-optic cable before installation. If your company doesn't have an OTDR, use a continuity tester. If you don't have a continuity tester, use a flashlight.

Now that you know the continuity tester works, plug in or attach a connector at one end of the fiber-optic cable. With the continuity tester on, observe the connector at the other end of the fiber-optic cable. If the connector at the other end of the fiber-optic cable is not emitting light, the optical fiber in the fiber-optic cable is broken. If the connector at the other end of the fiber-optic cable is emitting light, you have continuity and know that the optical fiber in not broken.

Figure 17.26 shows a portion of a patch panel with two ST type adapter. Each adapter is populated on the back side. In other words, a fiber-optic cable is connectorized and plugged into the back side of each adapter. On the other end of each fiber-optic cable, a continuity tester is attached and turned on. Adapter A shows good continuity. No light is emitted from adapter B, indicating a broken optical fiber.

FIGURE 17.26
Continuity test
of two fiber-optic
cables

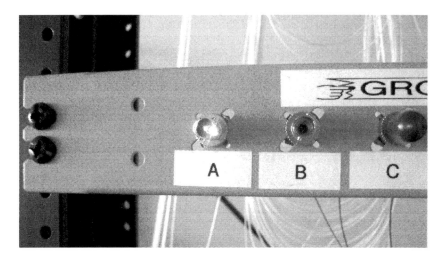

So far you have learned how to test the continuity of the optical fiber in a fiber-optic cable. Now let's look at the limitations of the continuity tester. Adapter A is emitting light, so you know that there is no break in the optical fiber. Unfortunately, that is all you can tell about the optical fiber in this cable.

Our eyes respond to light in a *logarithmic* manner, which means that we can't detect small changes in optical power. A change of 0.5dB or 1.0dB, for example, is not detectable. This means that we can't look at the light being emitted and tell if there is a macrobend or microbend in the optical fiber. The continuity tester is just that, a continuity tester. It can only tell you that the optical fiber is not broken.

Another drawback to the continuity tester is the fact that it uses visible light. You may remember from Chapter 5, "Optical Fiber Characteristics," that visible light is a poor choice for fiber-optic transmission because of attenuation. The high attenuation of visible light in an optical fiber limits the use of the continuity tester to cables no longer than 2km.

Remember that the continuity tester is basically a flashlight with no lens system to direct the light energy into the core of the optical fiber. Back in Chapter 10, we explained that a 62.5μm core accepts more light energy than a 50μm core when plugged into the same transmitter. This means that the continuity tester will not direct much light energy into the core of a single-mode optical fiber.

The continuity tester can be used with single-mode optical fiber. However, the light emitted from the connector at the opposite end of the cable will be dim in comparison to the multimode optical fiber. This is one application where the performance of the multimode optical fiber literally outshines the performance of the single-mode optical fiber.

When you are using the continuity tester with single-mode optical fiber, you may want to test a short cable to see how well it works before testing a longer cable. Dimming the lights can help you see the light emitted from the connector endface. If you cannot dim the lights, shroud the connector with your hand as shown in Figure 17.27 to block ambient light.

FIGURE 17.27
Shroud the connector with your hands to reduce ambient light during a continuity test.

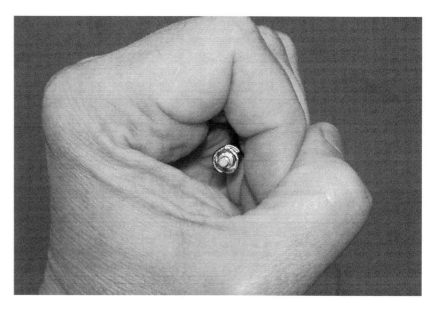

Continuity Tester Polarity Verification Techniques

Polarity problems exist in both copper and fiber-optic networks. The feedback received over the years from many technicians servicing both types of networks is that fiber-optic networks experience polarity problems more often than copper networks. Polarity problems in a fiber optic network are not the result of a dirty connector, broken optical fiber, or a macrobend; they are the result of improper labeling and or human error.

A polarity problem exists when a transmitter is not connected to the correct receiver, or vice versa. Verifying polarity in even the simplest fiber-optic network can become confusing if it is improperly or insufficiently labeled to show where cables originate, how they are connected, and where they go. Two standards that address labeling are ANSI/TIA-568-C and ANSI/TIA-606-B. Both are covered in detail in Chapter 14, "Cable Installation and Hardware."

A continuity tester is the best tool to assist in polarity verification because it emits visible eye–safe light. Depending on the complexity of the network, verifying the polarity of a channel may be time consuming. It is a good idea to bring a notebook and record your observations as you work your way from one end of the channel to the other.

NOTE ANSI/TIA-568-C.0 Annex B is titled "Maintaining Optical Fiber Polarity." This annex does not provide a definition for optical fiber polarity, nor does ANSI/TIA-568-C.O Section 3, "Definition of Terms, Acronyms, and Abbreviations, and Units of Measure."

Polarity in fiber-optic networks may sound out of place since in physics polarity typically implies the condition of electrical poles or magnetic poles. Without a definition for polarity from ANSI/TIA-568-C.0, the following definition may be used when describing polarity:

Polarity: In a fiber-optic network, it is the positioning of connectors and adapters to ensure that there is an end-to-end transmission path between the transmitter and the receiver of a channel.

As you learned in Chapter 14, ANSI/TIA-568-C defines two polarity schemes for duplex patch cords: A-to-A and A-to-B. The most commonly deployed scheme is A-to-B; therefore this section focuses on that scheme. Because there are many different network configurations, we will only describe polarity verification techniques for the configuration shown in Figure 17.28.

FIGURE 17.28

Cabling configuration for a basic fiber-optic network

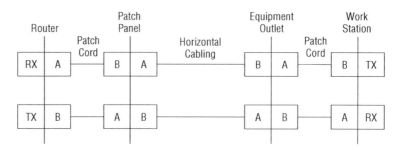

NOTE ANSI/TIA-568-C defines a channel as an end-to-end transmission path between two points at which application-specific equipment is connected.

Each end of a duplex patch cord should identify position A and position B. This is typically accomplished with raised lettering on the latch that holds together the two connectors at each end of the patch cord. A raised letter latch holding together two SC connectors is shown in Figure 17.29. Because of the physical size on an SC connector, the raised lettering is easy to read. However, this is not the case with small form factor connectors such as the latched LC pair shown in Figure 17.30.

FIGURE 17.29

A raised letter latch holding two SC connectors together

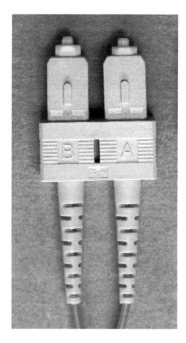

FIGURE 17.30
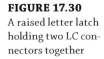

A raised letter latch holding two LC connectors together

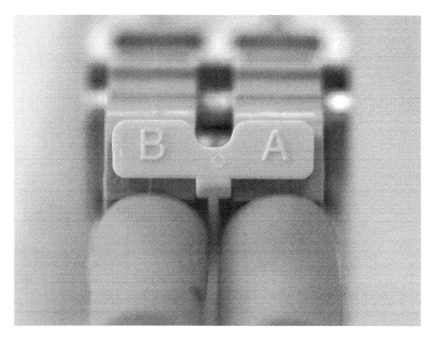

The first step in verifying the polarity of the channels shown in Figure 17.28 is to de-energize the equipment at both ends of the network. Since latched duplex patch cords are used, unplug both connectors from both ends of the patch cords at each end of the network. Using the continuity tester, verify that connectors on each end of a patch cord are oriented so position A goes to position B. You can do so by inserting the ferrule of the A position connector into the continuity tester as shown in Figure 17.31.

With the connector inserted, energize the continuity tester and check to see if light is exiting the optical fiber in position B at the opposite end of the patch cord. If light is exiting the optical fiber in position B, the polarity is correct. If light is exiting the optical fiber in position A, the polarity is not correct. With the continuity tester still attached and energized, unlatch both connectors, swap locations, and relatch. Verify light is exiting from the optical fiber in position B. Repeat this test for the other patch cord and correct as necessary.

With both patch cords properly configured, the next step is to verify the polarity of the horizontal cabling. To minimize access to other horizontal cabling, you should work from the equipment outlet to the patch panel. Remove the cover of the equipment outlet and plug both connectors at one end of the patch cord into the receptacles on the equipment outlet. Do not disturb the horizontal cabling connections.

Insert the ferrule of the A position connector at the end of the patch cord into the continuity tester as shown earlier in Figure 17.31. With the connector inserted, energize the continuity tester and check to see if light is exiting the horizontal cabling optical fiber in position B at the patch panel. If light is exiting the optical fiber in position B, the polarity is correct. If light is exiting the optical fiber in position A, the polarity is not correct. With the continuity tester still attached and energized, unlatch both horizontal cabling connectors at the back of the equipment outlet, swap locations, and relatch. Light should be exiting from the optical fiber in position B. Reinstall the equipment outlet cover.

FIGURE 17.31
The "A" position
connector ferrule
inserted into the
continuity tester

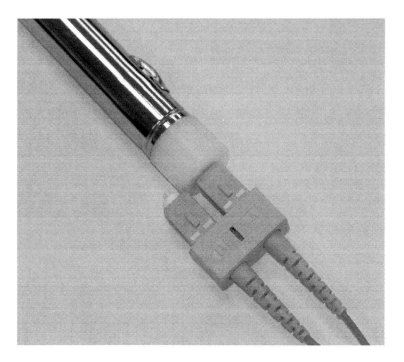

With the both patch cords and the horizontal cabling configured correctly, reinstall the patch cords starting at the router end of the network. Plug the patch cord into the patch panel and then into the router; ensure all interconnections are fully mated. At the workstation end of the network, plug the patch cord into the equipment outlet and then into the workstation; ensure all interconnections are fully mated.

Visual Fault Locator

The continuity tester is a valuable piece of test equipment. However, due to its low output power, it is limited to testing the continuity of the optical fiber. The visual fault locator (VFL) is similar to the continuity tester; however, it uses a visible laser source instead of an LED or incandescent lamp. The powerful laser in the VFL allows you to visibly identify breaks or macrobends in the optical fiber.

VFLs come in all shapes and sizes. Some are the size and shape of a large pen, like the one shown in Figure 17.32, whereas others may have the size and shape of a digital multimeter, like the one shown in Figure 17.33. Regardless of the design, all of them perform the same function. VFLs are also incorporated into some OTDRs. Incorporating the VFL into the OTDR reduces the number of pieces of test equipment that you need to carry to the job site.

The laser light sources used in most VFLs typically output 1mW of optical energy somewhere in the 650nm range. The output of the VFL is roughly 1,000 to 10,000 times greater than the output of an LED continuity tester. This means that you should never look directly at the endface of a connector emitting light from the VFL. The VFL is an eye hazard, and proper safety precautions need to be taken when operating the VFL. You can refer to Chapter 23, "Safety," to find out detailed information on light source classification and safety.

FIGURE 17.32
Pen-type VFL

Photo courtesy of KITCO Fiber Optics

FIGURE 17.33
Multimeter-
style VFL

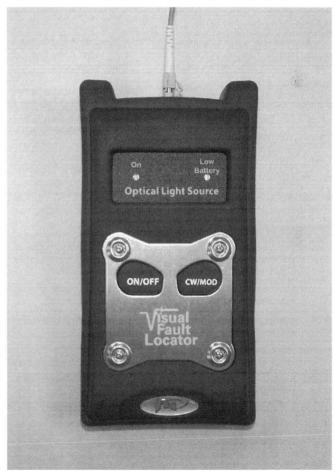

Photo courtesy of UrsaNav, Inc.

Now let's take a look at how to employ the VFL to locate faults in a fiber-optic cable. As with the continuity tester, the first thing you will need to do is clean the connector endface and inspect it with a microscope. If the endface finish is acceptable, the VFL can be connected to a connector at one end of the fiber-optic cable. The connector at the other end of the fiber-optic cable should not be viewed directly during this testing.

The VFL is designed to fill the core of an optical fiber with visible light. Depending on the cable type, the VFL may illuminate faults in an optical fiber through the buffer, strength member, and even jacket. However, the VFL performs best with tight-buffered optical fiber.

Figure 17.34 shows a broken tight-buffered optical fiber. You can see in the photograph how the VFL illuminates the break. The light energy from the VFL penetrates the buffer, allowing the person performing the test to quickly identify exactly where the fault is located.

FIGURE 17.34
Broken tight-buffered optical fiber

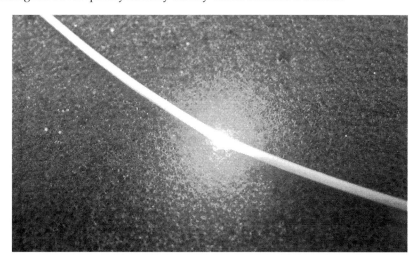

The VFL and the OTDR work hand in hand when it comes to locating breaks in an optical fiber. The OTDR can provide the operator with the distance to the break. The VFL allows the operator to see the break in the optical fiber.

Fiber-optic cables are not the only place where the optical fiber may break. The optical fiber may break inside the connector or connector ferrule. Unless the optical fiber is broken at the endface of the connector, it is not visible with a microscope.

Often, students connect cables that look great when viewed with the microscope but fail continuity testing. When this happens, the hardest part is determining which connector contains the break in the optical fiber. Without a VFL in the classroom, students would have to cut the cable in half and use the continuity tester to identify the bad connection.

The VFL will often identify the bad termination or connector. Figure 17.35 shows an ST connector with a broken optical fiber in the ferrule of the connector. Looking at the photograph, you can see the VFL illuminating the break in the optical fiber. The output of the VFL is so powerful that it penetrates the ceramic ferrule. Figure 17.36 shows an ST connector without a break in the optical fiber.

FIGURE 17.35
Broken optical fiber in ST connector ferrule being tested with VFL

FIGURE 17.36
Good ST connector being tested with VFL

The VFL can also be used to locate a macrobend in an optical fiber. However, macrobends do not allow nearly as much light to penetrate the buffer and jacket as does a break in the optical fiber. Locating a macrobend with the VFL may require darkening the room. Figure 17.37 is a photograph of a severe macrobend in a tight-buffered optical fiber. You can see from the photograph how the VFL illuminates the macrobend.

FIGURE 17.37
Macrobend in a tight-buffered optical fiber

Macrobends and high-loss fusion splices appear the same on an OTDR trace. The VFL allows the identification of a high-loss fusion splice. Figure 17.38 shows three fusion splices and three mechanical splices inside a splice enclosure. The splice illuminated by the VFL is a high-loss fusion splice. Believe it or not, the loss from this splice was not great enough to impact system performance. The loss for this splice was only 0.6dB.

FIGURE 17.38
High-loss fusion splice being tested with VFL

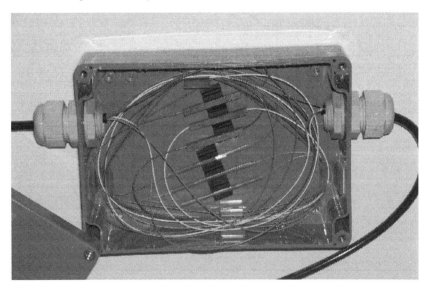

Fiber Identifier

The fiber identifier is a piece of test equipment that allows you to see through the jacket, strength member, and buffer of the fiber-optic cable. As you learned in Chapter 16, the fiber identifier is designed to place a macrobend in the fiber-optic cable under test.

Photodiodes in the fiber identifier detect light penetrating through the fiber-optic cable. The electronics in the fiber identifier measure the detected light energy and display the direction of light travel through the optical fiber.

The fiber identifier is used very much like the VFL when it comes to troubleshooting. One key difference is that the fiber identifier replaces your eyes. Another difference is that fiber-optic cable under test typically does not have to be disconnected from an active circuit—it can remain plugged into the transmitter and receiver. The infrared light traveling through the optical fiber during normal operation is often enough to perform most tests. However, sometimes an additional infrared light source is required to adequately troubleshoot.

Up to this point, we have only discussed troubleshooting with test equipment that emits visible light. You may have noticed that we keep mentioning that the fiber identifier works with infrared light. We have not mentioned using visible light with the fiber identifier. Visible light can be used with the fiber identifier; however, most fiber identifiers perform best with longer wavelength (800–1700nm) light sources.

Figure 17.39 shows a fiber identifier clamped around a tight-buffered optical fiber. This tight-buffered optical fiber is connected between a 1310nm laser transmitter and receiver. You can see in the photograph that the arrow is pointing to the direction the light is traveling through the optical fiber. This is an example of a fiber identifier being used to detect traffic on an optical fiber.

The fiber identifier can also be used with an external light source. Often the external light source is an OTDR. Many OTDR manufacturers build or program in a pulsed output function. When set for a pulsed output, the OTDR emits a continuous pulse train at a predetermined frequency. The electronics in the fiber identifier can detect preset frequencies and illuminate the corresponding LED. This feature can be very helpful when you are trying to identify an unmarked tight-buffered optical fiber within a bundle of tight-buffered optical fibers. It can also be helpful when you are trying to approximate the location of a break in the optical fiber.

The fiber identifier can be used with the OTDR to narrow down the location of a break in an optical fiber when a VFL is not available or when the light from the VFL is not visible through the jacket of the fiber-optic cable. If the index of refraction is correct, the OTDR should provide an accurate distance to the fault. Remember from Chapter 16 that the OTDR measures the length of optical fiber to the fault, not the length of fiber-optic cable. The cable length may be shorter than the optical fiber length. This is especially true if loose buffer cable exists in the fiber-optic link.

Once you have found the approximate location of the fault with the OTDR, set the OTDR or infrared light source to pulse at a predetermined frequency. Clamp the fiber identifier on the faulted fiber-optic cable several meters before the approximate location of the fault. Check the fiber identifier for the predetermined frequency. If the fiber identifier does not detect the predetermined frequency, move it several meters closer to the OTDR or infrared light source and recheck for the predetermined pulse. If you have selected the correct fiber-optic cable to test and you're confident about the distance to the fault, you should detect the predetermined frequency.

If you still do not detect the frequency, double-check everything and retest. If you still do not detect the predetermined frequency, there may not be enough optical energy for the fiber identifier to function properly.

FIGURE 17.39
Fiber identifier

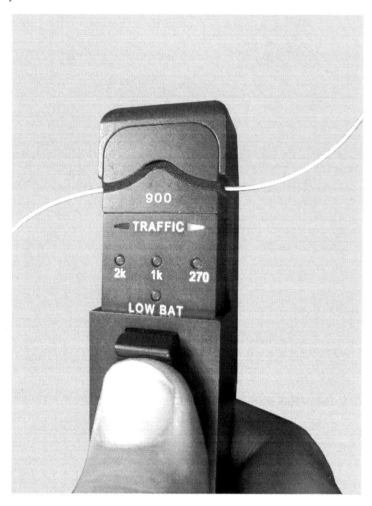

If you are able to detect the predetermined frequency, move the fiber identifier down the fiber-optic cable away from the OTDR or infrared light source in one-meter increments. Continue to do this until the fiber identifier no longer detects the predetermined pulse. You now know within one meter where the break in the optical fiber is located. At this point, you may want to disconnect the OTDR or infrared light source and connect the visible fault locator. The visible fault locator may illuminate the exact location of the fault. If the visible fault locator does not illuminate and conditions permit, darken the area around the fault. This may allow you to see the illuminated fault.

OTDR Fault Location Techniques

This section will show you how to use the OTDR to troubleshoot or locate faults. Fiber-optic links exist in virtually every place imaginable. We do not attempt to address every possible scenario; rather, we discuss how to look for and find typical faults in a fiber-optic link.

NOTE   This section is not designed to teach you how to operate the OTDR or use an OTDR to test a fiber-optic link or cable. OTDR operation and OTDR fiber-optic link and cable testing are covered in Chapter 16.

Many troubleshooting scenarios never require an OTDR. The OTDR may be used exclusively to troubleshoot or it may be brought in after it is determined that there is a fault in a fiber-optic link. Remember that the OTDR is used to locate the fault in a fiber-optic cable or to evaluate connector or splice performance. A faulty fiber-optic cable can typically be identified with a simple continuity test.

Let's say you just finished testing a fiber-optic link with a fiber-optic source and power meter. You know the length of the link and you have calculated the loss for the link, connectors, and splice as outlined in Chapter 15, "Fiber-Optic System Design Considerations." Your calculated loss for the link is 2.5dB @ 1300nm, as shown in Table 17.6. However, your measured loss with the fiber-optic source and power meter is 3.75dB.

Remember from Chapter 15 that there are typically three methodologies that can be used to develop a loss budget or power budget: the worst-case method, the statistical method, and the numerical method. The worst-case method is commonly viewed as the most conservative engineering approach. This method uses the worst-case maximum value loss for each passive component. This is the method we are going to use to calculate the loss for the link under test.

TABLE 17.6:   Completed multimode ANSI/TIA-568-C.3 worst-case loss calculation

#	DESCRIPTION	VALUE
1	Minimum EOL optical output power.	
2	Maximum optical input power.	
3	Subtract line 2 from line 1 to calculate the power budget.	
4	Kilometers of optical fiber.	0.467
5	Number of interconnections.	2.0
6	Number of splices.	1.0
7	Multiply line 4 × 3.5.	1.64
8	Multiply line 4 × 1.5.	0.7
9	Multiply line 5 × 0.75.	1.5
10	Multiply line 6 × 0.3.	0.3

TABLE 17.6:   Completed multimode ANSI/TIA-568-C.3 worst-case loss calculation *(continued)*

#	DESCRIPTION	VALUE
11	Add lines 7, 9, and 10 for total link loss at 850nm.	3.44 dB
12	Add lines 8, 9, and 10 for total link loss at 1300nm.	2.5 dB
13	Subtract line 11 from line 3 for headroom at 850nm.	
14	Subtract line 12 from line 3 for headroom at 1300nm.	

The minimum transmitter output power and minimum receiver sensitivity are not required to calculate the worst-case loss for the link. These values are not used because we are concerned only with the losses from the optical fiber, connectors, and or splices. The worst-case loss should be calculated at both wavelengths and the link should be tested at both wavelengths, as described in Chapter 16.

As mentioned several times in this chapter, the first thing you should do is clean and perform a visual inspection of the connectors. You should also clean and inspect the receptacle that each connector is plugged into. After cleaning, retest with the fiber-optic source and power meter. If the cleaning does not improve link performance to the point where it meets or exceeds the worst-case calculated loss, it's time to start troubleshooting.

You learned earlier in this chapter that a cosmetically perfect connector may not necessarily result in a low-loss connection. A connector with a great endface finish can have a distorted geometry, which could increase the loss in an interconnection. Remember that you can't see height or depth with a two-dimensional microscope. But with the OTDR, you can quickly measure the loss for a mated connector pair and look at the light energy reflected back toward the OTDR.

Chapter 16 showed you how to test a fiber-optic link with the OTDR. You probably remember that launch and receive cables must be attached to both ends of the fiber-optic link under test. Ensure that these cables are long enough for your application, as described in Chapter 16.

With the launch and receive cables attached, test the fiber-optic link as described in Chapter 16. After the link has been tested, the high-loss component or components need to be identified. The measured loss for this link with the fiber-optic source and power meter was 1.25dB greater than the worst-case calculated value. Remember that the calculated value is the worst-case scenario. A well-constructed fiber-optic link should test well under the calculated value.

The next step is to evaluate the OTDR trace. There are many possible causes for the loss. The only thing that testing has revealed so far is that there are no breaks in the optical fiber in the link. If there was a break, the loss for the link when testing with the light source and power meter would typically be in excess of 30dB.

We tend to approach any troubleshooting scenario looking for a single fault. However, at times you will discover more than one fault. This troubleshooting scenario is going to be approached as if there were a single fault.

Based on the light source and power meter testing, this link suffers from higher than acceptable loss. This loss could be caused by an interconnection, splice, or macrobend. It's the OTDR operator's job to be able to identify which of these three possible causes is the problem.

In Chapter 16, you learned how to measure the loss for interconnections and splices using the OTDR. To find the loss for the interconnection or splice, you had to subtract the loss for the length of optical fiber between cursors A and B. In this section, you will learn to look at the OTDR trace and quickly notice an event. This event could be a high-loss interconnection, splice, or macrobend.

When an event occurs that involves significant loss, there is no need to bring the cursors to the event and measure the exact loss. The vertical scale on the OTDR is there to help the operator visually approximate the amount of loss. A good OTDR operator can detect an event in the trace and use the vertical scale to quickly determine if that event exceeds the loss limitations defined in ANSI/TIA-568-C.3.

The operator has to learn to visualize the trace beyond the event. In other words, the operator has to know what the trace would look like if the event had very low loss versus the high loss shown on the OTDR display. Doing this will allow the operator to quickly identify high-loss events without having to move the cursors.

Let's say that a bad interconnection is the problem. A bad interconnection would look like Figure 17.40 on the OTDR. This bad interconnection can be quickly identified by following the sloping line as it enters and exits the interconnection back reflection. Visualize the sloping line traveling through the back reflection and exiting like the dashed line in Figure 17.40. The dashed line represents how a low loss interconnection would appear.

FIGURE 17.40

Bad interconnection

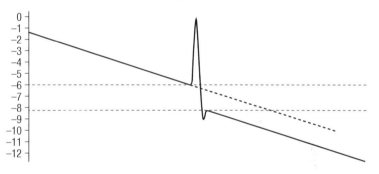

Remember from Chapter 16 that the OTDR display has a vertical and horizontal axis. The vertical axis shows relative power in decibels. Look at the amount of energy entering the interconnection and exiting the interconnection. Use the vertical scale on the left to approximate the loss without having to move the A and B cursors. This interconnection has a loss of approximately 2.25dB, as shown by the horizontal lines extending from the vertical axis in Figure 17.40.

Remember that the maximum allowable loss for an interconnection, according to ANSI/TIA-568-C.3, is 0.75dB. After this interconnection is repaired, the measured loss for the link should improve by no less than 1.5dB (2.25dB – 0.75dB = 1.5dB).

The next possible problem is a bad splice. This link contains one mechanical splice and does not contain a fusion splice. A poor fusion splice looks very similar to a macrobend and will be covered in a moment. A bad mechanical splice, shown in Figure 17.41, looks just like a bad interconnection. You should immediately notice that there is virtually no difference between the bad interconnection trace and this trace. The only difference is the amount of loss. When you look at the energy entering the splice and the energy leaving the splice, you'll see that a loss of 1.75dB is realized. This loss is 1.45dB greater than the maximum amount allowable by ANSI/TIA-568-C.3.

FIGURE 17.41
Bad mechanical
splice

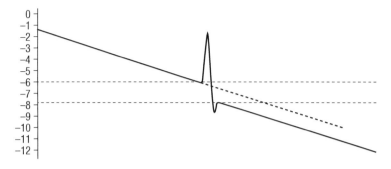

The last possible problem with this fiber-optic link is a macrobend. Although 1.25dB or greater is a significant loss and not a very typical macrobend loss, this amount of loss is possible from a macrobend. Macrobends and bad fusion splices look the same on the OTDR. Neither the fusion splice nor the macrobend produces a back reflection. Remember that back reflections are produced only from mechanical interconnections or breaks in the optical fiber. As with the previous two examples, look at the energy entering the macrobend and the energy exiting the macrobend, as shown in Figure 17.42. The loss for this macrobend is 1.8dB. When this macrobend is repaired, the measured loss for the fiber-optic link should decrease by 1.8dB.

FIGURE 17.42
Macrobend

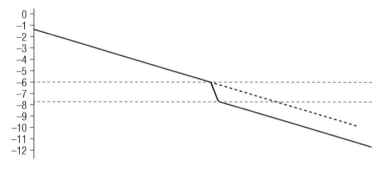

We have not yet addressed using the OTDR to find a break in the optical fiber. A break in an optical fiber can be detected with a continuity tester or visible fault locator. However, the OTDR can indicate the exact location of the break. You find the break in an optical fiber with the OTDR just as you find the length of an optical fiber. Chapter 16 describes in detail how to measure the length of an optical fiber.

Restoration Practices

So far in this chapter, you have learned about some of the tools available to troubleshoot a fiber-optic link. In Chapter 10 and Chapter 11, you learned how to understand the operating specifications of a fiber-optic transmitter and receiver. This section will outline a logical approach to troubleshooting a single fault to restore a fiber-optic link that has stopped working. We will proceed under the assumption that you have access to all the tools described in this chapter.

The first step in any restoration is to ask the customer three questions:

♦ What do they believe the problem is?

♦ When do they believe the problem first occurred?

♦ What was the last thing done to the system?

The best technicians know that many problems are caused by operator error or customer error. For example, the problem may have been caused by the customer changing the connections at the patch panel and making an improper connection. Or the customer might have just rearranged the furniture in their office and damaged a fiber-optic cable in the process.

After speaking with the customer, you should begin to assess the situation and attempt to identify the fault. Fiber-optic cables do not just break, splices do not just go bad, and interconnections do not just fail unless someone has handled them. However, electronics do go bad and electronics do fail. Once you have identified the fiber-optic link that has the problem, the next step is to make sure that all cables are connected properly.

If all the cables are connected properly, the next step is to check whether the electronics are functioning properly. Most fiber-optic transmitters output light energy as soon as they are powered up. In other words, the transmitter will output a series of pulses, typically at a constant frequency, when power is applied. The first step is to look for this light energy at the receiver end of the fiber-optic link and measure it. The measured energy should be within the input specification range for the receiver, which is covered in detail in Chapter 11. If the measured energy is within the specification for the receiver, then the problem is most likely the receiver. If the measured energy is not within the range of the receiver, then the problem may be in the fiber-optic link or the transmitter.

The next step is to measure the optical output power of the transmitter and verify that it is within the specification, as discussed in detail in Chapter 10. If the measured output power of the transmitter is within specification, the problem is somewhere in the fiber-optic link. If the measured output power of the transmitter is not within specification, the problem is with the transmitter.

If the transmitter and receiver are functioning to specification, the next place to look is the fiber-optic link. The first things to test in the fiber-optic link are the patch cords. You should clean and examine the connectors on each patch cord as described earlier in this chapter, and test the patch cords for loss as described in Chapter 16. You should also clean and inspect the connectors on the back side of the patch panel. Faulty patch cords should be replaced and bad connections on the main cable span should be repaired. If any repairs are made, the system should be retested.

After you inspect the patch cords and connectors, if no faults have been located, the next step is to test the entire cable span. Fiber-optic links should be well documented, though that is not always the case. The best scenario would be having that documentation for a comparison to your testing. However, if the customer can't provide documentation, you should still be able to analyze the link and locate faults.

The next step is to connect the OTDR as outlined in Chapter 16 and test the cable plant. If documentation exists, compare the OTDR traces and repair the faulty cable or splice. If documentation does not exist, evaluate the OTDR trace as you saw in Chapter 16 and earlier in this chapter and repair the faulty cable or splice. You may need to employ the VFL or fiber identifier to help you locate the exact physical location of the fault.

The Bottom Line

Identify a high-loss interconnection with an OTDR. A bad interconnection can be quickly identified by following the sloping line as it enters and exits the interconnection back reflection. Visualize the sloping line traveling through the back reflection and exiting.

Master It You are troubleshooting a new fiber-optic cable installation with the OTDR and the trace shown in the following graphic. The back reflection is the result of an interconnection. Using only the relative decibel information on the vertical scale, approximate the loss for the interconnection and determine whether this interconnection meets the loss requirements defined by ANSI/TIA-568-C.3.

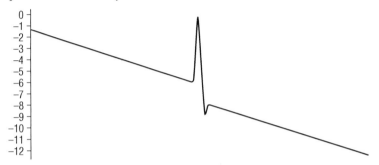

Identify a high-loss mechanical splice with an OTDR. A high-loss mechanical splice can be quickly identified by following the sloping line as it enters and exits the splice back reflection. Visualize the sloping line traveling through the back reflection and exiting.

Master It You are troubleshooting an older fiber-optic cable installation with the OTDR and the trace shown in the following graphic. The back reflection is the result of a mechanical splice. Using only the relative decibel information on the vertical scale, approximate the loss for the splice and determine whether this splice meets the loss requirements defined by ANSI/TIA-568-C.3.

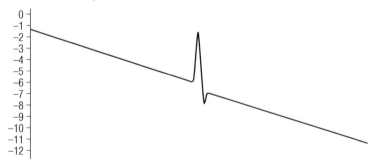

Identify a high-loss fusion splice or macrobend with an OTDR. A high-loss fusion splice or macrobend can be quickly identified by the dip in the trace. These events do not produce a back reflection.

Master It   You are troubleshooting a new fiber-optic cable installation with the OTDR and the trace shown in the following graphic. The dip in the trace is the result of a macrobend. Using only the relative decibel information on the vertical scale, approximate the loss for the macrobend.

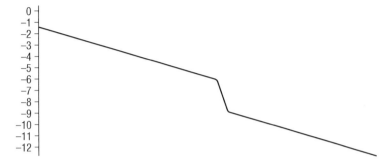

Appendices

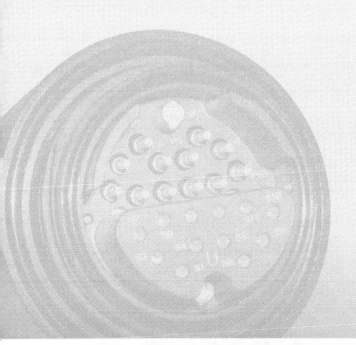

Appendix A

The Bottom Line

Each of "The Bottom Line" sections in the chapters suggests exercises to deepen skills and understanding. Sometimes there is only one possible solution, but often you are encouraged to use your skills and creativity to create something that builds on what you know and lets you explore one of many possibilities.

Chapter 1: History of Fiber Optics and Broadband Access

Recognize the refraction of light.   Refraction is the bending of light as it passes from one material into another.

> **Master It**   You are cleaning your pool with a small net at the end of a pole when you notice a large bug that appears to be 2' below the surface. You place the net where you believe the bug to be and move it through the water. When you lift the net from the pool, the bug is not in the net. Why did the net miss the bug?
>
> **Solution**   Remember refraction is easily observed by placing a stick into a glass of water. When viewed from above, the stick appears to bend because light travels more slowly through the water than through the air. The same is true for the bug. The bug was viewed from above the water and refraction bent the light so the bug appeared to be in a different location than it actually was.

Identify total internal reflection.   In 1840, Colladon and Babinet demonstrated that bright light could be guided through jets of water through the principle of total internal reflection.

> **Master It**   You just removed your fish from a dirty 10-gallon aquarium you are preparing to clean when a friend shows up with a laser pointer. Your friend energizes the laser pointer and directs the light into the side of the tank. The laser light illuminates the small dirty particles in the tank and you and your friend observe the light entering one end of the tank and exiting the other. As you friend aims the laser pointer at an angle toward the surface of the water, the light does not exit; instead it bounces off the surface of the water at an angle. Why did this happen?
>
> **Solution**   Remember that Colladon and Babinet demonstrated that bright light could be guided through jets of water through the principle of total internal reflection. In their demonstration, light from an arc lamp was used to illuminate a container of water. You essentially re-created their experiment. Instead of an arc lamp and a small barrel of water, you had a laser pointer and a 10-gallon aquarium full of dirty water. The light reflected off the surface of the tank because of total internal reflection. You were able to

see this happen, because the water was cloudy. Had you attempted this after the tank was cleaned the laser would be very difficult to view because there would be no particles to scatter the light. Scattering will be covered in detail in Chapter 5, "Optical Fiber Characteristics."

Detect crosstalk between multiple optical fibers.   Crosstalk or optical coupling is the result of light leaking out of one fiber into another fiber.

Master It   You have bundled six flexible clear plastic strands together in an effort to make a fiber-optic scope that will allow you to look into the defroster vent in your car in hope of locating your missing Bluetooth headset. You insert your fiber-optic scope into the defroster vent and are disappointed with the image you see. What is one possible cause for the poor performance of your fiber-optic scope?

Solution   Remember that in 1930, Lamm bundled commercially produced fibers and managed to transmit a rough image through a short stretch of the first fiber-optic cable. The process had several problems, including the fact that the fiber ends were not arranged exactly, and they were not properly cut and polished. Another issue was to prove more daunting. The image quality suffered from the fact that the quartz fibers were bundled against each other. This meant that the individual fibers were no longer surrounded by a medium with a lower index of refraction. Much of the light from the image was lost to crosstalk. Crosstalk or optical coupling is the result of light leaking out of one fiber into another fiber.

Recognize attenuation in an optical fiber.   In 1960 after the invention of the laser, optical fibers were all but ruled out as a transmission medium because of the loss of light or attenuation.

Master It   You are troubleshooting a clog in a drainpipe and because of the size of your flashlight and the location of the drain; you cannot illuminate the drain adequately to see the clog. You decide to modify a flashlight to illuminate the inside of drainpipe in hope of identifying the source of the clog. You purchase a small diameter flexible clear plastic rod approximately 12" in length and secure one end over the bright LEDs in the center of the flashlight. After powering up the flashlight, you are disappointed that the light exiting the optical fiber is much dimmer than you expected; however, you still attempt to identify the clog. Why did the fiber-optic flashlight fail to illuminate the inside of the drainpipe?

Solution   Remember that in 1960 optical fibers were all but ruled out as a transmission medium because of their attenuation. However, in 1966 Charles K. Kao announced that flaws in the manufacturing processes caused signal losses in glass fibers. The plastic rod purchased for the flashlight was not engineered to transmit light and experienced the high attenuation that limited the use of optical fibers prior to 1970 when Corning developed an optical fiber that achieved Kao's target of only 20 decibels per kilometer. Decibels are covered in Chapter 2, "Principles of Fiber-Optic Transmission."

Chapter 2: Principles of Fiber-Optic Transmission

Calculate the decibel value of a gain or loss in power.   The decibel is used to express gain or loss relative to a known value. In fiber optics, the decibel is most commonly used to describe signal loss through the link after the light has left the transmitter.

Master It   The output power of a transmitter is 100μW and the power measured at the input to the receiver is 12.5μW. Calculate the loss in dB.

Solution   To calculate the decibel value of a gain or loss in signal power, use the following equation:

$dB = 10Log_{10} (P_{out} \div P_{in})$

$dB = 10Log_{10} (12.5 \div 100)$

$dB = 10Log_{10} 0.125$

$dB = -9.03$

$Loss = 9.03dB$

Calculate the gain or loss in power from a known decibel value.   If the gain or loss in power is described in dB, the gain or loss can be calculated from the dB value. If the input power is known, the output power can be calculated, and vice versa.

Master It   The loss for a length of optical fiber is 4dB. Calculate the loss in power from the optical fiber and the output power of the optical fiber for an input power of 250μW.

Solution   First solve for the loss, then for the output power. The equation would be written as follows:

$(P_{out} \div P_{in}) = antilog (dB \div 10)$

$(P_{out} \div 250μW) = antilog (\div 4 \div 10)$

$(P_{out} \div 250μW) = antilog \div 0.4$

$(P_{out} \div 250μW) = 0.398$

$P_{out} = 0.398 \times 250μW$

$P_{out} = 99.5μW$

Calculate the gain or loss in power using the dB rules of thumb.   When calculating gain or loss in a system, it is useful to remember the three rules of thumb shown in Table 19.3.

Master It   A fiber-optic link has 7dB of attenuation. The output power of the transmitter is 100μW. Using the dB rules of thumb calculate the power at the input to the receiver.

Solution   To calculate the optical power at the input to the receiver, apply the 7dB rule by dividing 100μW by 5 as shown in the equation below:

$P_{in} = 100μW \div 5$

$P_{in} = 20μW$

The optical power at the input to the receiver is 20μW

Convert dBm to a power measurement. In fiber optics, dBm is referenced to 1mW of power.

Master It A fiber-optic transmitter has an output power of ÷8dBm. Calculate the output power in watts.

Solution To calculate the power in watts, the equation would be written as:

$÷8 = 10\text{Log}_{10} (P ÷ 1\text{mW})$

$÷0.8 = \text{Log}_{10} (P ÷ 1\text{mW})$

$0.158 = P ÷ 1\text{mW}$

$P = 0.158 × 1\text{mW}$

$P = 158.5\mu\text{W}$

Convert a power measurement to dBm. In fiber optics, dBm is referenced to 1mW of power.

Master It A fiber-optic transmitter has an output power of 4mW. Calculate the output power in dBm.

Solution To calculate the power in dBm, the equation would be written as:

$\text{dBm} = 10\text{Log}_{10} (4\text{mW} ÷ 1\text{mW})$

$\text{dBm} = 10\text{Log}_{10} 4$

$\text{dBm} = 10 × 0.6$

$\text{dBm} = 6$

The power value is 6dBm

This answer can be checked by locating 4mW in the first column of Table 19.4.

Chapter 3: Basic Principles of Light

Convert various wavelengths to corresponding frequencies. The relationship between wavelength and frequency can be expressed with the formula $\lambda = v ÷ f$.

Master It If an electromagnetic wave has a frequency of 94.7MHz, what is its wavelength?

Solution By dividing the speed of the electromagnetic wave, 300,000km/s, by the frequency, 94.7MHz, we arrive at the wavelength, 3.17 meters.

Convert various frequencies to corresponding wavelengths. The relationship between frequency and wavelength can be expressed with the formula $f = v ÷ \lambda$.

Master It If an electromagnetic wave has a wavelength of 0.19 meters, what is its frequency?

Solution By dividing the speed of the electromagnetic wave, 300,000km/s, by the wavelength, 0.19 meters, we arrive at the frequency, 1.58GHz.

Calculate the amount of energy in a photon using Planck's constant. The amount of energy in each photon depends on the electromagnetic energy's frequency: the higher the frequency, the more energy in the particle. To express the amount of energy in a photon, we use the equation $E = hf$.

Master It   Calculate the energy of a photon at a frequency of 193.55THz.

Solution   Using Planck's constant, 6.626×10^{-34} joule-seconds, and the formula $E = hf$, you can solve for the energy as shown in this equation:

$(6.626 \times 10^{-34}) \times (1.9355 \times 10^{14}) = 1.282 \times 10^{-19}$ W

Calculate the speed of light through a transparent medium using its refractive index.   The index of refraction is a relative value, and it is based on the speed of light in a vacuum, which has an index of refraction of 1. The index of refraction can be calculated using the equation $n = c/v$.

Master It   If light is passing through a medium with a refractive index of 1.33, what will the velocity of the light through that medium be?

Solution   To find the speed of light in a medium with a known refractive index, we divide the speed of light, 300,000km/s, by the medium's refractive index, in this case, 1.33. The result is a speed of 225,564km/s.

Use Snell's law to calculate the critical angle of incidence.   The incident angle required to produce a refracted angle of 90° is called the critical angle. To find the critical angle of two materials, we can use a modified version of Snell's equation:

$\theta_c = \arcsin(n_2 \div n_1)$

Master It   Calculate the critical angle for two materials where n_2 is 1.45 and n_1 is 1.47.

Solution   The critical angle can be found by dividing n_2 by n_1, then finding the arcsin of the result as shown in the equation below.

$\theta_c = \arcsin(1.45 \div 1.47) = 80.54°$

Calculate the loss in decibels from a Fresnel reflection.   Augustin-Jean Fresnel determined how to calculate the amount of light lost from a Fresnel reflection using the equation $\rho = ((n_1 - n_2) \div (n_1 + n_2))^2$.

Master It   Calculate the loss in decibels from a Fresnel reflection that is the result of light passing from a refractive index of 1.51 into a refractive index of 1.45.

Solution   To calculate the loss in decibels from a Fresnel reflection, we need to use two equations shown here. While the refractive index of air was not given in the problem, it can be obtained from Table 20.1.

$\rho = ((1.51 - 1.45) \div (1.51 + 1.45))^2 = 0.02^2 = 0.0004$

To calculate the loss in decibels, we use the equation

$dB = 10\text{Log}_{10} (1 - 0.0004)$

plugging in the Fresnel reflection value from above,

$dB = 10\text{Log}_{10} 0.9996 = -0.0017$

This gives us a loss of 0.0017dB.

Chapter 4: Optical Fiber Construction and Theory

Select the proper optical fiber coating for a harsh environment. The coating is the true protective layer of the optical fiber; it absorbs the shocks, nicks, scrapes, and even moisture that could damage the cladding. Without the coating, the optical fiber is very fragile. The coating found on an optical fiber is selected for a specific type of performance or environment.

> **Master It** Choose a coating for an optical fiber that must operate in temperatures as high as 175° C.

> **Solution** Coatings found on optical fibers include acrylate, silicone, carbon, and polyimide. Of these four, acrylate has the lowest operating temperature while carbon and polyimide have the highest. Silicone has a much higher operating temperature than acrylate—up to 200° C. While polyimide can operate at temperatures up to 350° C, it is not required for this application. Remember carbon cannot be used by itself and is always combined with another coating.

Identify an industry standard that defines the specific performance characteristics on an optical fiber. There are many standards related to optical fiber. This chapter focused on the standards that define the performance of optical fibers used in the telecommunications industry published by TIA, ITU, ISO, and the IEC.

> **Master It** You have been placed in charge of selecting a contractor to install fiber optic cabling in a new building. Which standard or standards should you review prior to selecting the best contractor?

> **Solution** The ANSI/TIA-568-C standards and the ISO/IEC 11801 standard are applicable to premises cabling components and the new building falls under the definition of premises.

Identify an optical fiber from its refractive index profile. The refractive index profile graphically represents the relationship between the refractive index of the core and that of the cladding. Several common profiles are found on optical fibers used in telecommunications, avionics, aerospace, and space applications.

> **Master It** The refractive index profile of an optical fiber is shown in the following graphic. Determine the type of optical fiber.

> **Solution** The optical fiber in the graphic has a graded index with an optical trench. The graded index indicates multimode and the optical trench indicates bend insensitive; this is a BIMMF.

Calculate the numerical aperture of an optical fiber. To find the numerical aperture of an optical fiber, you need to know the refractive index of the core and the cladding.

Master It   The core of an optical fiber has a refractive index of 1.51 and the cladding has a refractive index of 1.46. Calculate the numerical aperture.

Solution   Using the formula shown here, square the value for the refractive index of the core and then do the same for the cladding. Subtract the cladding value from the core value and take the square root of that value. Your answer should be 0.385.

$$NA= \sqrt{1.51^2 - 1.46^2} = 0.385$$

Calculate the number of modes in an optical fiber.   To find the number of modes in an optical fiber, you need to know the diameter of the core, the wavelength of the light source, and the numerical aperture.

Master It   Calculate the number of possible modes in an optical fiber with the following specifications:

Core diameter = 62.5µm

Core refractive index = 1.48

Cladding refractive index = 1.46

Wavelength of the light source = 1300nm

Solution   First, determine the numerical aperture of the fiber's core using this equation:

$$NA= \sqrt{1.48^2 - 1.46^2} = 0.242$$

Next, use the equation $M = (D \times \pi \times NA/\lambda)^2/2$ to find the number of modes.

$M = (62.5 \times 10^{-6} \times \pi \times 0.242/1300 \times 10^{-9})^2/2 = 667.9$ modes.

Since you cannot have a fraction of a mode, the number of possible modes is 667.

Chapter 5: Optical Fiber Characteristics

Calculate the attenuation in dB for a length of optical fiber.   The attenuation values for a length of optical fiber cable can be calculated using the attenuation coefficient for a specific type of optical fiber cable. This information can be found in the manufacturer's data sheet or in a standard.

Master It   Calculate the maximum attenuation for a 22.5km ITU-T G.657.B.3 optical fiber cable at 1550nm.

Solution   First, locate the attenuation coefficient for an ITU-T G.657.B.3 optical fiber cable at 1550nm in Table 22.10. Then multiply the attenuation coefficient by the length as shown here:

Attenuation for 22.5km of optical fiber = $22.5 \times 0.3 = 6.75$

The maximum attenuation for 22.5km of ITU-T G.657.B.3 optical fiber at 1550nm is 6.75dB.

Calculate the usable bandwidth for a length of optical fiber.   The product of bandwidth and length (MHz · km) expresses the information carrying capacity of a multimode optical fiber. Bandwidth is measured in megahertz (MHz) and the length is measured in kilometers. The MHz · km figure expresses how much bandwidth the fiber can carry per kilometer of its length. The fiber's designation must always be greater than or equal to the product of the bandwidth and the length of the optical fiber.

Master It   Refer to Table 22.4 and calculate the usable bandwidth for 425 meters of 50/125μm OM3 multimode optical fiber using an OFL at a wavelength of 850nm.

Solution   The minimum bandwidth-length product for a 50/125μm OM3 multimode optical fiber at 850nm using an OFL is 1500MHz · km. To find the usable bandwidth for this optical fiber at a length of 425 meters, you must change the equation around and solve as shown here:

Bandwidth-length product = 1500MHz · km

Optical fiber length = 0.425km

Bandwidth in MHz = 1500 ÷ 0.425

Bandwidth = 3529.4MHz

Calculate the total macrobending loss in a single-mode fiber-optic cable.   The macrobending loss is defined for a number of turns of optical fiber around a mandrel at a specified radius and wavelength.

Master It   A single-mode ITU-T G.657.A2 bend-insensitive fiber-optic cable is installed in a home with two small 15mm 90° radius bends. The wavelength of the light source is 1625nm.

Solution   The macrobending loss in Table 22.8 is for optical fiber that has not been placed in a cable. The macrobending loss will vary when the optical fiber is placed in a cable and that information must be obtained from the cable manufacturer. However, to demonstrate how macrobending loss is calculated it is assumed that optical fiber values listed in the standards are the same as the cable loss.

Referencing Table 22.8 we see that the loss for ten 15mm radius turns at 1625nm is 0.1dB. The total macrobending loss (MBL) will be the sum of the two losses. However, we cannot simply add up the dB values from Table 22.8 because each bend is not a full 360° turn; both 15mm radius bends are 90° and the bending loss is not listed for a single turn but a number of turns. First, we must first establish the single turn loss using the below formula:

MBL ÷ number of turns = dB loss per turn

0.1dB ÷ 10 = 0.01dB

Then we have to calculate the loss for the partial bends using the following formula:

(Bend in degrees ÷ 360) × dB loss per turn = MBL

(90 ÷ 360) × 0.01dB = 0.0025dB

Then we calculate the total loss using this formula:

MBL 1 + MBL 2 = Total MBL

0.0025dB + 0.0025dB = 0.005dB

Total macrobending loss = 0.005dB

Calculate the total macrobending loss in a multimode fiber-optic cable.   The macrobending loss is defined for a number of turns of optical fiber around a mandrel at a specified radius and wavelength.

Master It   A generic multimode bend-insensitive fiber-optic cable is installed in a home with three small 7.5mm radius, 120° bends. The wavelength of the light source is 850nm.

Solution   The macrobending loss in Table 22.1 is for a generic optical fiber that has not been placed in a cable. The macrobending loss will vary when the optical fiber is placed in a cable and that information must be obtained from the cable manufacturer. However, to demonstrate how macrobending loss is calculated it is assumed that optical fiber values listed are the same as the cable loss.

Referencing Table 22.1 we see that the loss for two 7.5mm radius turns at 850nm is ≤0.2dB. The total macrobending loss (MBL) will be the sum of the two losses. However, we cannot simply add up the dB values from Table 22.1 because each bend is not a full 360° turn; both 7.5mm radius bends are 120° and the bending loss is not listed for a single turn but a number of turns. First, we must establish the single turn loss using the following formula:

MBL ÷ number of turns = dB loss per turn

≤0.2dB ÷ 2 = ≤0.1dB

Then we have to calculate the loss for the partial bends using this formula:

(Bend in degrees ÷ 360) × dB loss per turn = MBL

(120 ÷ 360) × 0.1dB = 0.033dB

Then we calculate the total loss using this formula:

MBL 1 + MBL 2 = Total MBL

0.033dB + 0.033dB = 0.066dB

Total macrobending loss = 0.066dB

Calculate the acceptance angle for an optical fiber.   The acceptance angle defines the acceptance cone. Light entering the core of the optical fiber at an angle greater than the acceptance angle will not propagate the length of the optical fiber. For light to propagate the length of the optical fiber, it must be at an angle that does not exceed the acceptance angle.

Master It   Determine the acceptance angle of an optical fiber with a core refractive index of 1.48 and a cladding refractive index of 1.45.

Solution   To determine the acceptance angle of an optical fiber with a core refractive index of 1.48 and a cladding refractive index of 1.45, we must determine the NA of the fiber using this formula:

$$\text{NA} = \sqrt{[(n_1)^2 - (n_2)^2]}$$

The acceptance angle is the value of θ.

NA = sinθ

So

θ = arcsinNA = arcsin 0.2965 = 17.25

The acceptance angle is 17.25°.

Determine the latest published revision of a standard using the Internet. It is important to be aware of standards that have been superseded.

Master It Using the Internet determine the latest revision and publication date for ANSI/TIA-568-C.3.

Solution Unless you work with industry standards on a regular basis and keep track of the progress of the many standards working groups, it is difficult to know which standard is the most current. The best way to determine this is to visit the TIA website at `www.tiaonline.org/` and click on the Standards tab. In the Standards tab, click Buy TIA Standards and enter **TIA-568-C.3** in the search box. As of this writing the latest revision was ANSI/TIA-568-C.3-1, published December 2011.

Chapter 6: Safety

Classify a light source based on optical output power and wavelength. The OSHA Technical Manual describes the Optical Fiber Service Group (SG) Designations based on ANSI Z136.2. These SG designations relate to the potential for ocular hazards to occur only when the OFCS is being serviced. OSHA also classifies standard laser systems.

Master It An OFCS does not fall into any of the SG designations defined in the OSHA Technical Manual because during servicing it is possible to be exposed to laser emissions greater than 750mW. What classification would be assigned to this OFCS based on the OSHA Technical Manual?

Solution If the total power for an OFCS is at or above 0.5W, it does not meet the criteria for an optical fiber SG designation and should be treated as a standard laser system. A Class IV laser hazard as defined in the OSHA Technical Manual has a continuous wave output above 500mW, or +27dBm. This OFCS would be Class IV.

Identify the symptoms of exposure to solvents. Solvents can cause organ damage if inhaled or ingested. The molecules that make up most solvents can take the place of oxygen in the bloodstream and find their way to the brain and other organs. As the organs are starved of oxygen, they can become permanently damaged.

Master It Your co-worker spilled a bottle of liquid several minutes ago and now appears impaired. What has your co-worker potentially been exposed to?

Solution Because solvent molecules can displace oxygen molecules in the blood, they can starve the brain of oxygen and give the appearance of intoxication. Your co-worker may have been exposed to a solvent.

Calculate the proper distance the base of a ladder should be from a wall. Remember non-self-supporting ladders, such as extension ladders, require a firm vertical support near the work area and a stable, nonslip surface on which to rest the ladder. When setting up a non-self-supporting ladder, it is important to place it at an angle of 75½° to support your weight and be stable.

Master It A non-self-supporting ladder is 10′ in length. How far should the feet of the ladder be from the wall?

Solution A good rule for finding the proper angle is to divide the working height of the ladder (the length of the ladder from the feet to the top support) by 4, and place the feet of the ladder that distance from the wall. This ladder was 10′ in length so the ladder's feet should be 2.5′ from the wall.

Chapter 7: Fiber-Optic Cables

Determine the cable type from the NEC markings. Article 770 of the NEC states that optical fiber cables installed within buildings shall be listed as being resistant to the spread of fire. These cables are also required to be marked.

Master It You have been asked to install the two cables shown on the next page in a riser space. From the markings on the cables, determine the cable types and determine if each cable can be installed in a riser space.

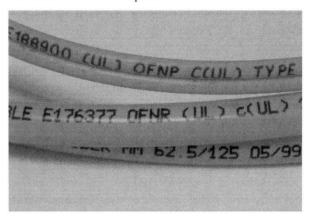

Solution The cable marked OFNR is a nonconductive riser cable and the cable marked OFNP is a nonconductive plenum cable. Per the NEC cable substitution guide in Figure 7.30 both cables can be installed in a riser space. However, only the cable marked OFNP can be installed in the plenum space.

Identify the fiber number from the color code. TIA-598-C defines a color code for optical fibers within a cable assembly.

Master It You have been asked to identify the fourth and seventh tight-buffered optical fibers in the 12-fiber cable shown here. There are no numbers on the tight buffers; however, each buffer is colored as defined in TIA-598-C. What are the colors of the fourth and seventh tight-buffered optical fibers?

Solution   Table 24.2 list the optical fiber color codes as defined in TIA-598-C. The color for fiber number four is brown and the color for fiber number seven is red.

Determine the fiber number from the cable markings.   The color-coding schemes for premises jackets and optical fibers within a fiber-optic cable described in TIA-598-C are not always used.

Master It   You have been asked to identify simplex cable number nine in the 12-fiber cable shown here. Describe the location of that cable.

Solution   Simplex cable number nine is located in the upper-left corner.

Identify the optical fiber type from the color code.   Premises cable jacket colors are defined in TIA-598-C.

Master It   It is your first day on the job and your supervisor has asked you to get some laser-optimized multimode optical fiber from the back of the van. The van contains several different cables, each with a different jacket color. What is the jacket color of the laser-optimized cable?

p 141

7.3

Solution   Table 24.3 list the premises jacket colors as defined in TIA-598-C. The color for the laser-optimized cable is aqua.

Determine the optical fiber type from the cable markings.   Premises cable jacket colors are defined in TIA-598-C. However installed cables may not comply with revision C of this standard.

Master It   You have been asked to identify the optical fiber type in the cable shown here. The jacket color is slate and this cable was not intended for military applications. What type of optical fiber is used in this cable?

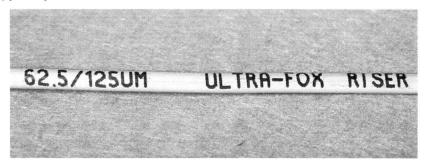

Solution   This cable was manufactured in 1999, 6 years before TIA-598-C was published. It does not comply with this standard. However, the optical fiber type is printed on the jacket. The optical fiber in this simplex cable is 62.5/125μm multimode.

Determine the cable length using sequential markings.   Many markings are found on a cable. Sequential markings typically appear every 2 feet or every 1m.

Master It   You have been asked to determine the length of fiber-optic cable coiled up on the floor. The sequential markings for this cable are meters. The marking on one end of the cable is 0235 and the marking on the other end is 0485. How long is this cable?

Solution   To find the length of the cable, determine the distance between the markings as shown here:

485m – 235m = 250m

Chapter 8: Splicing

Determine if the splice loss is from an intrinsic factor.   Even when fibers are manufactured within specified tolerances, there are still slight variations from one optical fiber to another. These variations can affect the performance of the splice even though the optical fibers are perfectly aligned when mated.

Master It   Two optical fibers are about to be fusion spliced together, as shown here. Determine what type of problem exists with the optical fibers about to be spliced.

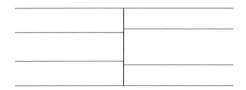

Solution A close inspection of the graphic reveals that the core diameter of the left optical fiber is slightly smaller than the right optical fiber. This core diameter mismatch is the result of an intrinsic factor.

Determine if the splice loss is from an extrinsic factor. Extrinsic factors that affect optical fiber splice performance are factors related to the condition of the splice itself, external to the optical fiber.

Master It Two optical fibers are about to be fusion spliced together, as shown here. Determine what type of problem exists with the optical fibers about to be spliced.

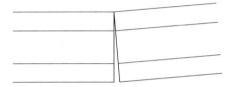

Solution A close inspection of the graphic reveals that there is some angular misalignment. The angular misalignment is the result of an extrinsic factor.

Calculate the potential splice loss from a core diameter mismatch. A core diameter mismatch loss results when the core diameter of the transmitting optical fiber is greater than the core diameter of the receiving optical fiber.

Master It Calculate the worst-case loss for a splice where the transmitting optical fiber has a core diameter of 51μm and the receiving optical fiber has a core diameter of 49.5μm.

Solution We can calculate the worst-case loss percentage for a splice that joins a 50.5μm core and 49.5μm core using the following formula:

$$\text{Loss} = [(d_1)^2 - (d_2)^2] / (d_2)^2$$

where d1 is the diameter of the transmitting core and d2 is the diameter of the receiving core.

$$50.5^2 - 49.5^2 = 100$$

$$100/49.5^2 = 0.0408 \text{ or } 4.08\%$$

Using this formula, the decibel loss can be calculated:

$$dB = 10\text{Log}_{10}(P_{out} / P_{in})$$

Note that the decibel formula is a ratio, where the power output is divided by the power input. We know that the receiving optical fiber will not accept 6.15 percent of the light from the transmitting optical fiber. Subtracting 6.15 percent from 100 percent, as shown here, results in a difference of 93.85 percent, which can be written as 0.9385.

$100\% - 4.08\% = 95.92\%$ or 0.9592

$10\text{Log}_{10} (0.9592 / 1) = -0.181\text{dB}$

Worst-case loss = 0.181dB

Troubleshoot a fusion splice. How well a splice performs depends on many variables. These variables can be broken into two groups: intrinsic factors and extrinsic factors.

Master It Two optical fibers have been fused together as shown here. Determine the type of problem and potential causes.

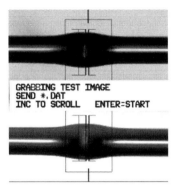

Solution A close inspection of the of the fusion splicer display reveals that the optical fiber ends that have been joined are larger in diameter then optical fiber. This is typically caused by dirt or entrapped air creating a bubble or bubbles. The result is a high-loss fusion splice. To prevent bubbles, verify that you have selected the appropriate splicing program or profile for the optical fiber you are splicing and ensure you are properly cleaning the fusion splicer and cleaver.

Chapter 9: Connectors

Evaluate connector endface polishing. To evaluate the endface of a fiber-optic connector, you need a microscope designed for this application.

Master It You are inspecting the multimode endface shown here with a 200X inspection microscope. Is there anything wrong with this connector endface?

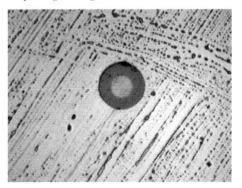

Solution This connector has oil from a fingertip on the endface; however, the endface has a good polish. After cleaning and re-inspection, this connector will be ready to be put to use.

Evaluate connector endface cleaning. A connector that has been cleaned properly will be free of any dirt or contamination.

Master It You are inspecting the connector endfaces shown in these two graphics. Which of these connectors has oil on the endface?

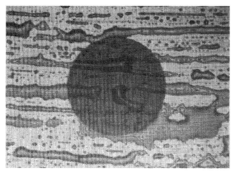

Photo courtesy of MicroCare

Photo courtesy of MicroCare

Solution The connector endface shown in the first graphic has oil on it, and the connector endface shown in the second graphic has debris from a cotton shirt on it. Both of these connectors need to be cleaned and re-inspected.

Evaluate connector endface geometry. The key measurements provided by the interferometer are radius of curvature, apex offset, and fiber undercut or protrusion. These critical parameters are required by Telcordia GR-326 to evaluate connector endface geometry.

Master It You are evaluating a machine polished endface with the interferometer. The radius of curvature measured by the interferometer is 25mm. Does this endface radius of curvature fall within the range specified by Telcordia GR-326?

Solution The minimum and maximum radius of curvature values as defined in Telcordia GR-326 are 7mm and 22mm. Values below or above this range increase the risk of optical fiber damage; 25mm exceeds the maximum acceptable value.

Identify an optical fiber type from the color of the connector strain relief.   Multimode and single-mode connectors and adapters can be identified using the color code in ANSI/ TIA-568-C.3.

Master It   You need to mate two 850nm laser-optimized 50/125µm fiber-optic cables together with a jumper at a patch panel. What color strain reliefs would the correct jumper have?

Solution   850nm laser-optimized 50/125µm optical fiber should have an aqua-colored strain relief.

Chapter 10: Fiber-Optic Light Sources and Transmitters

Determine the minimum optical output power for an LED transmitter.   The minimum optical output power of an LED transmitter should be defined in the manufacturer's data sheet.

Master It   Refer to the following table to determine the BOL and EOL values for the minimum optical output power of a 50/125µm optical fiber.

PARAMETER		SYMBOL	MIN.	TYP.	MAX.	UNIT
Optical Output Power	BOL	P_0	–22	–19.8	–17	dBm avg.
62.5/125µm, NA = 0.275 Fiber	EOL		–23			
Optical Output Power	BOL	P_0	–25.5	–23.3	–17	dBm avg.
50/125µm, NA = 0.20 Fiber	EOL		–26.5			
Optical Extinction Ratio				0.001	0.03	%
				–50	–35	dB
Optical Output Power at Logic "0" State		P_0("0")			–45	dBm avg.
Center Wavelength		λ_c	1270	1308	1380	nm
Spectral Width—FWHM		$\Delta\lambda$		137	170	nm
Optical Rise Time		t_r	0.6	1.0	3.0	ns
Optical Fall Time		t_f	0.6	2.1	3.0	ns
Duty Cycle Distortion Contributed by the Transmitter		DCD		0.02	0.6	ns_{p-p}
Data-Dependent Jitter Contributed by the Transmitter		DDJ		0.02	0.6	ns_{p-p}

Solution To find the minimum optical output power for a specific optical fiber, locate the Optical Output Power row for that optical fiber in the table. In the Min. column, there is a BOL and EOL value. The higher number, –25.5 dBm, is the BOL value and the lower number, –26.5 dBm, is the EOL value.

Determine the maximum optical output power for an LED transmitter. The manufacturer's data sheet should contain the maximum value for the optical output power of an LED transmitter.

Master It Refer to the following table to determine the maximum optical output power of a 62.5/125µm optical fiber.

PARAMETER		SYMBOL	MIN.	TYP.	MAX.	UNIT
Optical Output Power	BOL	P_0	–25	–21.8	–19	dBm avg.
62.5/125µm, NA = 0.275 Fiber	EOL		–25			
Optical Output Power	BOL	P_0	–27.5	–25.3	–19	dBm avg.
50/125µm, NA = 0.20 Fiber	EOL		–28.5			
Optical Extinction Ratio				0.001	0.03	%
				–50	–35	dB
Optical Output Power at Logic "0" State		$P_0("0")$			–45	dBm avg.
Center Wavelength		λ_c	1270	1308	1380	nm
Spectral Width—FWHM		$\Delta\lambda$		137	170	nm
Optical Rise Time		t_r	0.6	1.0	3.0	ns
Optical Fall Time		t_f	0.6	2.1	3.0	ns
Duty Cycle Distortion Contributed by the Transmitter		DCD		0.02	0.6	ns_{p-p}
Data-Dependent Jitter Contributed by the Transmitter		DDJ		0.02	0.6	ns_{p-p}

Solution To find the maximum optical output power for a specific optical fiber, locate the Optical Output Power row for that optical fiber in the table. The number in the Max. column, –19 dBm, is the value for the maximum optical output power.

Determine the minimum optical output power for a laser transmitter. The manufacturer's data sheet should contain the minimum, maximum, and typical value for the optical output power of a laser transmitter.

Master It Refer to the following table to determine the BOL and EOL values for the minimum optical output power.

PARAMETER	SYMBOL	MIN.	TYP.	MAX.	UNIT
Optical Output Power 9µm SMF	P_{OUT}	–6	0	+3	dBm
Center Wavelength	λ_c	1260		1360	nm
Spectral Width—rms	$\Delta\lambda$		1.8	4	nm rms
Optical Rise Time	t_r		30	70	ps
Optical Fall Time	t_f		150	225	ps
Extinction Ratio	E_R	8.2	12		dB
Optical Output Eye	Compliant with eye mask Telecordia GR-253-CORE				
Back-Reflection Sensitivity				–8.5	dB
Jitter Generation	pk to pk			70	mUI
	RMS			7	mUI

Solution   To find the minimum optical output power, locate the Optical Output Power row in the table. In the Min. column, there is a value; since BOL and EOL are not stated, assume this value is the BOL value. To approximate the EOL value, subtract 1.5dB. The BOL value is –6dBm and the EOL value is –7.5dBm.

Determine the maximum optical output power for a laser transmitter.   The manufacturer's data sheet should contain the maximum value for the optical output power of a laser transmitter.

Master It   Refer to the following table to determine the maximum optical output power.

PARAMETER	SYMBOL	MIN.	TYP.	MAX.	UNIT
Optical Output Power 9µm SMF	P_{OUT}	–12	–8	–5	dBm
Center Wavelength	λ_c	1260		1360	nm
Spectral Width—rms	$\Delta\lambda$		1.8	4	nm rms
Optical Rise Time	t_r		30	70	ps
Optical Fall Time	t_f		150	225	ps
Extinction Ratio	E_R	8.2	12		dB

PARAMETER	SYMBOL	MIN.	TYP.	MAX.	UNIT
Optical Output Eye	Compliant with eye mask Telecordia GR-253-CORE				
Back-Reflection Sensitivity				–8.5	dB
Jitter Generation	pk to pk			70	mUI
	RMS			7	mUI

Solution To find the maximum optical output power, locate the Optical Output Power row in the table. The number in the Max. column is the value for the maximum optical output power. There is no BOL or EOL value for the maximum optical output power. The maximum value is –5dBm.

Chapter 11: Fiber-Optic Detectors and Receivers

The dynamic range of the LED receiver is the difference between the minimum value for the maximum optical input power and the maximum value for the minimum optical input.

Master It Refer to the following table and calculate the dynamic range for the LED receiver.

PARAMETER	SYMBOL	MIN.	TYP.	MAX.	UNIT
Optical input power minimum at window edge	$P_{IN\,Min.}$ (W)		–30.5	–29	dBm avg.
Optical input power maximum	$P_{IN\,Max.}$	–17	–13.8		dBm avg.
Operating wavelength	λ	1270		1380	nm

Solution To calculate the dynamic range, subtract the minimum value for the maximum optical input power ($P_{IN\,Max}$) from the maximum value for the minimum optical input power ($P_{IN\,Min}$) as shown in this formula:

$P_{IN\,Max} - P_{IN\,Min}$ = dynamic range

–17dBm – –29dBm = 12dB

Dynamic range = 12dB

Calculate the dynamic range for a laser receiver. The dynamic range of the laser receiver is the difference between the minimum value for the maximum optical input power and the maximum value for the minimum optical input.

Master It   Refer to the following table and calculate the dynamic range for the laser receiver.

PARAMETER	SYMBOL	MIN.	TYP.	MAX.	UNIT
Receiver sensitivity	$P_{IN\,Min}$		−20	−18	dBm avg.
Receiver overload	$P_{IN\,Max}$	−3	+3		dBm avg.
Input operating wavelength	λ	1260		1570	nm
Signal detect—asserted	P_A		−24.3	−16.5	dBm avg.
Signal detect—deasserted	P_D	−32	−25.7		dBm avg.
Signal detect—hysteresis	P_H	0.5	1.4	4	dB
Reflectance			−35	−27	dB

Solution   To calculate the dynamic range, subtract the minimum value for the maximum optical input power ($P_{IN\,Max}$) from the maximum value for the minimum optical input power ($P_{IN\,Min}$) as shown in this formula:

$P_{IN\,Max} - P_{IN\,Min}$ = dynamic range

−3dBm − −18dBm = 15dB

Dynamic range = 15dB

Chapter 12: Passive Components and Multiplexers

Calculate the output power of a real tee coupler port.   Remember that manufacturers may or may not provide the insertion loss combined with the splitting ratio.

Master It   The manufacturer provided the insertion loss combined with the loss from the splitting ratio for a 90:10 tee coupler with an insertion loss in decibels of 0.63/11. Calculate the power available at each output port.

Solution   The manufacturer states that the insertion loss in decibels is 0.63/11. This means that the power at the 90 percent port will be 0.63dB less than the input power to the coupler and the power at the 10 percent port will be 11dB less than the input power to the coupler.

If the input power to the coupler is −10dBm, the power at each port can be calculated by subtracting the insertion loss for each port as defined by the manufacturer from −10dBm as shown here:

−10dBm − 0.63dB = −10.63dBm

−10dBm − 11dB = −21dBm

The output power at the 10 percent port is –21dBm and the output power at the 90 percent port is –10.63dBm. Remember this calculation did not take into account the interconnection losses and uniformity.

Calculate the output power of a real star coupler. Remember that unlike the tee coupler, a manufacturer typically only provides the insertion loss for the coupler and does not take into account the loss for the splitting ratio.

Master It Calculate the output power at each port of a star coupler with one input port and 10 output ports. Take into account the interconnection insertion losses. Assume the insertion loss for the coupler is 1.7dB and each interconnection insertion loss is 0.25dB. The input power to the coupler is –5dBm.

Solution The total loss per port is the sum of the insertion losses for the interconnections (LI) and the insertion loss for the coupler (LC).

$0.25dB \times 2 = 0.5dB$

$0.5dB + 1.7dB = 2.2dB$

Then account for the star coupler insertion loss and the interconnection losses. This is done by subtracting 2.2dB from the input power of –5dBm. The remaining power available to the ports is –7.2dBm, as shown here:

$-5dBm - 2.2dB = -7.2dBm$

Then split the remaining power using the decibel rules of thumb covered in Chapter 2. Since this has 10 output ports, each port will receive 10 percent of the energy because the energy is distributed evenly between the output ports. A loss of 90 percent is a change of 10dB. The output power at each port will equal –7.2dBm minus 10dB. Each output port will have a power output of –17.2dBm, as shown here:

$-7.2dBm - 10dB = -17.2dBm$

Keep in mind that couplers are not perfect and the amount of power actually available will vary depending on the uniformity.

Calculate attenuator values. Remember that if the maximum optical input power for a fiber-optic receiver is exceeded, the receiver will not operate properly.

Master It Calculate the minimum attenuation required to prevent the receiver from being overloaded. Assume the transmitter power is –3dBm, the total losses are 6dB, and the maximum optical input power the receiver can tolerate is –14dBm.

Solution To calculate the minimum attenuation required we need to subtract all the known losses from the output power of the transmitter, as shown here:

Transmitter power (TP) = –3dBm

Receiver maximum optical input power (MP) = –14dBm

Total losses (TL) = 6dB

Minimum attenuation required = MP + TL – TP

$-14dBm + 6dB - -3dBm = -5dB$

At a minimum, a 5dB attenuator is required. However, an attenuator with a larger value could be used as long as it did not over-attenuate the signal. Refer to Chapter 11 to determine the dynamic range of a receiver.

Chapter 13: Passive Optical Networks

Identify the different PON configurations.   FTTX can be used to describe any optical fiber network that replaces all or part of a copper network.

Master It   In this PON, the optical signals are converted right in front of the house or some distance down the block. What type of PON is this?

Solution   In an FTTC PON, optical fiber runs from the central office and stops at the curb. The curb may be right in front of the house or some distance down the block. Where the optical fiber stops is where the optical signal from the optical fiber is converted into electrical signals.

Identify the cables used in a PON.   Several different types of cables are employed in an FTTX PON.

Master It   This cable type runs from the central switching point to the local convergence point. What type of cable is it?

Solution   Feeder cables run from the central switching point to the local convergence point.

Identify the different access points in a PON.   Cables connect the different access points in the PON.

Master It   What do you call the point where a distribution cable is broken out into multiple drop cables?

Solution   The network access point (NAP) is located close to the homes or buildings it services. This is the point where a distribution cable is broken out into multiple drop cables.

Chapter 14: Cable Installation and Hardware

Determine the minimum cable bend radius.   Remember that cable manufacturers tend to define the minimum bend radius when the cable is stressed and unstressed. These numbers are typically in compliance with ANSI/TIA-568-C.0 and C.3.

Master It   Refer to the following table and determine the minimum installation bend radius and the minimum operational bend radius for the 24-fiber cable.

CABLE TYPE	DIAMETER	WEIGHT	MINIMUM BEND RADIUS: SHORT TERM	MINIMUM BEND RADIUS: LONG TERM	MAXIMUM LOAD: SHORT TERM	MAXIMUM LOAD: LONG TERM
12-fiber cable	0.26″, 6.6mm	27lb/kft, 40kg/km	3.9″ 9.9cm	2.6″ 6.6cm	300lb., 1334N	90lb., 396N
24-fiber cable	0.375″, 9.5mm	38.2lb/kft, 56.6kg/km	5.5″ 14cm	3.7″ 9.4cm	424lb., 1887N	127lb., 560N

Solution   The short-term minimum bend radius in the table is the installation bend radius. The minimum installation bend radius for the 24-fiber cable is 5.5″.

The long-term minimum bend radius in the table is the operational bend radius. The minimum operational bend radius for the 24-fiber cable is 3.7″.

Determine the maximum cable tensile rating.   Remember that cable manufacturers typically define a dynamic and static pull load.

Master It   Refer to the following table and determine the maximum dynamic tensile load and the maximum static tensile load for the 12-fiber cable.

CABLE TYPE	DIAMETER	WEIGHT	MINIMUM BEND RADIUS: SHORT TERM	MINIMUM BEND RADIUS: LONG TERM	MAXIMUM LOAD: SHORT TERM	MAXIMUM LOAD: LONG TERM
12-fiber cable	0.26″, 6.6mm	27lb/kft, 40kg/km	3.9″, 9.9cm	2.6″, 6.6cm	300lb., 1334N	90lb., 396N
18-fiber cable	0.325″, 8.3mm	33.8lb/kft, 50kg/km	4.9″, 12.4cm	3.25″, 8.3cm	375lb., 1668N	112lb., 495N

Solution   The short-term maximum load in the table is the dynamic tensile load. The dynamic tensile load for the 12-fiber cable is 300lb.

The long-term maximum load in the table is the static tensile load. The static tensile load for the 12-fiber cable is 90lb.

Determine the fill ratio for a multiple-cable conduit installation.   When preparing for a multiple cable conduit installation, you must be aware of the fill ratio for the conduit. If you are installing cable in existing conduit already occupied by one or more cables that contain electrical conductors, you will need to determine the fill ratio of the conduit prior to selecting your cable.

Master It   You need to install one cable into a conduit that has three existing electrical cables. The inside diameter of the conduit is 1″. The cable to be installed has an outside diameter of 0.25″. The outside diameters of the existing cables are 0.28″, 0.32″, and 0.38″. Determine the fill ratio of the conduit before and after the fiber-optic cable is installed and verify that this installation will not overfill the conduit as defined by the NEC.

Solution   The NEC specifies the following fill ratios by cross-sectional area for conduit:

1 cable: 53 percent

2 cables: 31 percent

3 or more cables: 40 percent

The first step is to determine the conduit fill ratio with the existing cables. To determine tha t fill ratio, use the following formula:

Fill Ratio = $(OD^2 cable_1 + OD^2 cable_2 + OD^2 cable_3)$ / $ID^2 conduit$

Fill Ratio = $(0.28^2 + 0.32^2 + 0.38^2)$ / 1^2

Fill Ratio = $(0.0784 + 0.1024 + 0.1444)$ / 1

Fill Ratio = 0.3252 / 1

Fill Ratio = 0.3252 or 32.52%

The next step is to determine the conduit fill ratio with the addition of the fiber-optic cable. To determine that fill ratio, use the following formula:

Fill Ratio = $(OD^2 cable_1 + OD^2 cable_2 + OD^2 cable3 + OD^2 cable_4)$ / $ID^2 conduit$

Fill Ratio = $(0.28^2 + 0.32^2 + 0.38^2 + 0.25^2)$ / 1^2

Fill Ratio = $(0.0784 + 0.1024 + 0.1444 + 0.0625)$ / 1

Fill Ratio = 0.3877 / 1

Fill Ratio = 0.3877 or 38.77%

The calculated fill ratio is less than 40 percent, so the conduit is large enough as defined by the NEC.

Chapter 15: Fiber-Optic System Design Considerations

Calculate the bandwidth for a length of optical fiber.   The minimum effective bandwidth-length product in MHz · km as defined in TIA-568-C.3 and ISO/IEC 11801.

Master It   Refer to the following table and determine the minimum effective modal bandwidth for 300 meters of OM4 optical fiber at a wavelength of 850nm.

Optical fiber cable type	Wavelength (nm)	Maximum attenuation (dB/km)	Minimum overfilled modal bandwidth-length product (MHz · km)	Minimum effective modal bandwidth-length product (MHz · km)
62.5/125µm Multimode OM1 Type A1b TIA 492AAAA	850 1300	3.5 1.5	200 500	Not required Not required
50/125µm Multimode OM2 TIA 492AAAB Type A1a.1	850 1300	3.5 1.5	500 500	Not required Not required
850nm laser-optimized 50/125µm Multimode OM3 Type A1a.2 TIA 492AAAC	850 1300	3.5 1.5	1500 500	2000 Not required
850nm laser-optimized 50/125µm Multimode OM4 Type A1a.3 TIA 492AAAD	850 1300	3.5 1.5	3500 500	4700 Not required
Single-mode Indoor-outdoor OS1 Type B1.1 TIA 492CAAA OS2 Type B1.3 TIA 492CAAB	1310 1550	0.5 0.5	N/A N/A	N/A N/A

Optical fiber cable type	Wavelength (nm)	Maximum attenuation (dB/km)	Minimum overfilled modal bandwidth-length product (MHz · km)	Minimum effective modal bandwidth-length product (MHz · km)
Single-mode Inside plant OS1 Type B1.1 TIA 492CAAA OS2 Type B1.3 TIA 492CAAB	1310 1550	1.0 1.0	N/A N/A	N/A N/A
Single-mode Outside plant OS1 Type B1.1 TIA 492CAAA OS2 Type B1.3 TIA 492CAAB	1310 1550	0.5 0.5	N/A N/A	N/A N/A

Solution   The steps to calculate the minimum bandwidth for 700m of 850nm laser-optimized 50/125μm multimode optical fiber are:

Bandwidth-length product = 4,700 MHz · km

Optical fiber length = 0.3km

Bandwidth in MHz = 4,700 ÷ 0.3

Bandwidth = 15,666.67MHz

Calculate the attenuation for a length of optical fiber.   The maximum allowable attenuation in an optical fiber is defined in TIA-568-C.3 and ISO/IEC 11801.

Master It   Refer to the following table and determine the maximum allowable attenuation for 5.6km of outside plant single-mode optical fiber.

Optical fiber cable type	Wavelength (nm)	Maximum attenuation (dB/km)
50/125μm multimode	850 1300	3.5 1.5

Optical fiber cable type	Wavelength (nm)	Maximum attenuation (dB/km)
62.5/125μm multimode	850	3.5
	1300	1.5
Single-mode inside plant cable	1310	1.0
	1550	1.0
Single-mode outside plant cable	1310	0.5
	1550	0.5

Solution   The maximum attenuation for outside plant single-mode optical fiber is the same for both wavelengths. The steps to calculate the maximum attenuation are as follows:

Attenuation coefficient at 1310 or 1550nm = 0.5dB/km

Optical fiber length = 5.6km

Attenuation for the length of optical fiber = $5.6 \times 0.5 = 2.8$

Maximum attenuation for 5.6m of outside plant single-mode optical fiber is 2.8dB

Calculate the power budget and headroom for a multimode fiber-optic link.   Before we can analyze the link and do a power budget, we need to know the optical performance characteristics for transmitter and receiver.

Master It   Refer to the following graphic and two tables to calculate the power budget and headroom for the multimode fiber-optic link with 50/125μm multimode optical fiber.

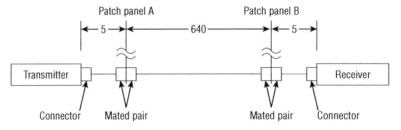

Distance in meters (not to scale)

Parameter	Symbol		Min.	Typ.	Max.	Unit
Optical output power	BOL	P_0	–19	–16.8	–14	dBm avg.
62.5/125μm, NA = 0.275 fiber	EOL		–20			
Optical output power	BOL	P_0	–22.5	–20.3	–14	dBm avg.

Parameter	Symbol	Min.	Typ.	Max.	Unit
50/125μm, NA = 0.20 fiber	EOL	−23.5			
Optical extinction ratio			0.001–50	0.03–35	%dB
Optical output power at logic "0" state	P_0 ("0")			−45	dBm avg.
Center wavelength	λ_c	1270	1300	1380	nm
Spectral width—FWHM	$\Delta\lambda$		137	170	nm
Optical rise time	t_r	0.6	1.0	3.0	ns
Optical fall time	t_f	0.6	2.1	3.0	ns
Duty cycle distortion contributed by the transmitter	DCD		0.02	0.6	ns_{p-p}
Duty cycle jitter contributed by the transmitter	DDJ		0.02	0.6	ns_{p-p}

Parameter	Symbol	Min.	Typ.	Max.	Unit
Optical input power minimum at window edge	$P_{IN\ Min.}$	(W)	−33.5	−31	dBm avg.
Optical input power maximum	$P_{IN\ Max.}$	−14	−11.8		dBm avg.
Operating wavelength	λ_c	1270		1380	nm

Solution   Using the minimum EOL optical output power from the first table and the maximum value for minimum optical input power from the second table, calculate the power budget as shown in the table that follows. Next, calculate all the link losses for optical fiber, interconnections, and splices, as shown in the following table. Next, calculate the total link loss and finally the headroom as shown in that table.

Step no.	Description	Value
1	Minimum EOL optical output power.	–23.5dBm
2	Minimum optical input power.	–31dBm
3	Subtract line 2 from line 1 to calculate the power budget.	7.5dB
4	km of optical fiber.	.65
5	Number of interconnections.	2.0
6	Number of splices.	0
7	Multiply line 4×3.5.	
8	Multiply line 4×1.5.	0.975
9	Multiply line 5×0.75.	1.5
10	Multiply line 6×0.3.	0
11	Add lines 7, 9, and 10 for total link loss at 850nm.	
12	Add lines 8, 9, and 10 for total link loss at 1300nm.	2.475dB
13	Subtract line 11 from line 3 for headroom at 850nm.	
14	Subtract line 12 from line 3 for headroom at 1300nm.	5.025dB

Calculate the power budget and headroom for a single-mode fiber-optic link. Before we can analyze the link and do a power budget, we need to know the optical performance characteristics for transmitter and receiver.

Master It Refer to the following graphic and two tables to calculate the power budget and headroom for the single-mode fiber-optic link.

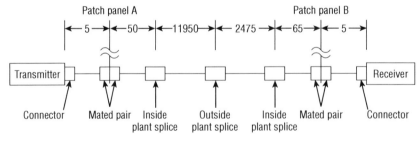

Distance in meters (not to scale)

Parameter	Symbol	Min.	Typ.	Max.	Unit
Optical output Power 9μm SMF, EOL	P_{OUT}	−5	−2	0	dBm
Center wavelength	λ_c	1260		1360	nm
Spectral width—rms	$\Delta\lambda$		1.8	4	nm rms
Optical rise time	t_r		30	70	ps
Optical fall time	t_f		150	225	ps
Extinction ratio	E_R	8.2	12		dB
Optical output eye	Compliant with eye mask Telecordia GR-253-CORE				
Back-reflection Sensitivity				−8.5	dB
Jitter generation	pk to pk			70	mUI
	RMS			7	mUI
Receiver sensitivity	$P_{IN\ MIN}$		−23	−19	dBm avg.
Receiver overload	$P_{IN\ MAX}$	0	+1		dBm avg.
Input operating wavelength	λ	1260		1550	nm
Signal detect—asserted	P_A		−27.3	−19.5	dBm avg.
Signal detect—deasserted	P_D	−35	−28.7		dBm avg.
Signal detect—hysteresis	P_H	0.5	1.4	4	dB
Reflectance			−35	−27	dB

Solution   Using the minimum EOL optical output power from the first table and the maximum value for minimum optical input power from the second table, calculate the power budget as shown in the following table. Next, calculate all the link losses for optical fiber, interconnections, and splices as shown in the table that follows. Next, calculate the total link loss and then the headroom, as shown in the table that follows.

Step No.	Description	Value
1	Minimum EOL optical output power.	–5.0dBm
2	Minimum optical input power.	–19.0dBm
3	Subtract line 2 from line 1 to calculate the power budget.	14.0dB
4	km of inside plant optical fiber.	0.125
5	km of outside plant optical fiber.	14.425
6	Number of interconnections.	2
7	Number of inside plant splices.	2
8	Number of outside plant splices.	1
9	Multiply line 4 × 1.0.	0.125
10	Multiply line 5 × 0.5.	7.2125
11	Multiply line 6 × 0.75.	1.5
12	Multiply line 7 × 0.3.	0.6
13	Multiply line 8 × 0.1.	0.1
14	Add lines 9, 10, 11, 12, and 13 for total link loss.	9.5375dB
15	Subtract line 14 from line 3 for headroom.	4.4625dB

Chapter 16: Test Equipment and Link/Cable Testing

Perform optical loss measurement testing of a cable plant.   ANSI/TIA-526-14 revisions A and B provide three methods for testing the cable plant. The difference between the three methods is how the reference power measurement is taken. Each method uses a different number of test jumpers. Method A uses two test jumpers to obtain the reference power measurement, method B uses one test jumper, and method C uses three test jumpers.

Master It   Before testing the cable plant using method B, you obtained your reference power measurement of –23.5dBm. Your test power measurement was –27.6dBm. What was the loss for the cable plant?

Solution   To obtain the optical power loss for the cable plant, subtract the test power measurement from the reference power measurement. In this case, the reference power measurement was −23.5dBm and the test power measurement was −27.6dBm, so the loss for the cable plant would be 4.1dB.

Determine the distance to a break in the optical fiber with an OTDR.   A break in an optical fiber in a fiber-optic cable is not the end of the optical fiber. However, a break in an optical fiber looks like the end of the optical fiber on the OTDR display. The same method is employed to measure the distance to a break in an optical fiber and to measure the distance to the end of an optical fiber.

Master It   You are troubleshooting fiber-optic cables with the OTDR. One optical fiber appears to be broken. Refer to the following graphic and determine the distance to the break in the optical fiber.

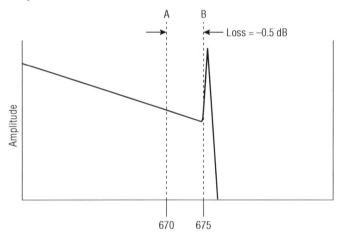

Solution   To measure the distance to the break in the optical fiber, zoom in on the back reflection, and place the A cursor on a smooth section of the trace just in front of the back reflection. Move the B cursor toward the A cursor until it is in the leading edge of the back reflection. Keeping moving the B cursor toward the A cursor until the A-B loss is −0.5dB. The distance to the break is the distance for the B cursor, which is 675m.

Measure the loss of a cable segment with an OTDR.   To find the loss for a cable segment plus the interconnections, the exact length of the segment must be known.

Master It   You are testing a new fiber-optic cable installation with the OTDR and the customer has requested the loss information for a cable segment including the interconnections. Refer to the following graphic and determine the cable segment. The A-B distance equals the length of the cable segment plus 50m.

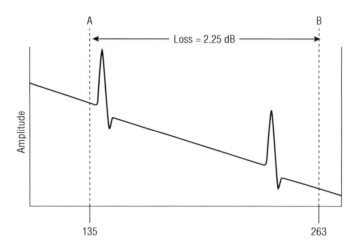

Solution The first step in finding the loss for a cable segment with interconnections is to measure the length of the cable segment. After zooming in on the cable segment, place the A cursor on a smooth section of the trace just in front of the leftmost cable segment interconnection back reflection. Place the B cursor on the smooth part of the trace after the rightmost cable segment interconnection back reflection. Looking at the A-B distance on the OTDR, move the B cursor to the right until the distance equals the length of the cable segment plus 50m. The A-B loss is 2.25dB.

To find the loss for the cable segment plus the interconnections, subtract the loss for the 50m of optical fiber from the 2.25dB loss for the cable segment and interconnections. The previously measured loss for 50m of optical fiber at 850nm is 0.14dB. The loss for the cable segment and the interconnections is 2.11dB.

Measure the loss of an interconnection with an OTDR. When measuring the loss for an interconnection with the OTDR, you must remember to subtract the loss for the optical fiber.

Master It You are testing a fiber-optic cable installation with the OTDR and the customer has requested the loss information for each interconnection. Refer to the following graphic and determine the cable segment.

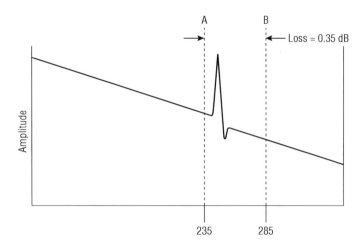

Solution   The first step in measuring interconnection loss is to horizontally zoom in on the interconnection. Position the A cursor just in front of the back reflection, then position the B cursor on a smooth area on the trace after the interconnection back reflection. The loss for the interconnection and the optical fiber between the cursors is displayed on the OTDR screen. The loss for the interconnection and 50m of optical fiber shown in the graphic is 0.35dB.

To find the loss for only the interconnection, the loss for the optical fiber between cursors A and B needs to be subtracted from the A-B loss displayed by the OTDR. This loss was previously at 0.14dB for 50m. To find the actual interconnection loss, subtract 0.14dB from the 0.35dB A-B loss. The loss for only the interconnection is approximately 0.21dB.

Measure the loss of a fusion splice or macrobend with an OTDR.   Fusion splices or macrobends do not produce a back reflection. A macrobend will always appear as a loss in the form of a dip in the smooth trace. A fusion splice may appear as a dip or a bump in the trace.

Master It   You are testing a fiber-optic cable installation with the OTDR and see a dip in the trace. Since there are no splices in this cable, you determine that this event must be a macrobend. Refer to the following graphic and determine the loss for the macrobend.

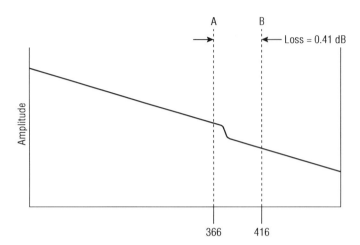

Solution To find the loss of a macrobend, horizontally zoom in on the event. The loss from a macrobend may be very small and require vertical zoom in addition to horizontal zoom. Place the A cursor on the smooth part of the trace before the dip in the trace. Place the B cursor on the smooth part of the trace after the dip, as shown in the graphic. The loss for this event is 0.41dB. The loss for the event includes the loss for the macrobend plus the 50m of optical fiber between the cursors. Subtract the loss for the 50m of optical fiber that was previously measured at 0.14dB from the event loss. The loss for this fusion splice or macrobend is 0.27dB.

Chapter 17: Troubleshooting and Restoration

Identify a high-loss interconnection with an OTDR. A bad interconnection can be quickly identified by following the sloping line as it enters and exits the interconnection back reflection. Visualize the sloping line traveling through the back reflection and exiting.

Master It You are troubleshooting a new fiber-optic cable installation with the OTDR and the trace shown in the following graphic. The back reflection is the result of an interconnection. Using only the relative decibel information on the vertical scale, approximate the loss for the interconnection and determine whether this interconnection meets the loss requirements defined by ANSI/TIA-568-C.3.

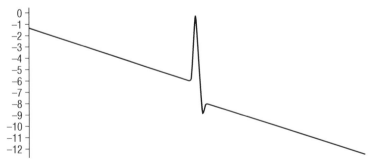

Solution   The vertical axis shows relative power in decibels. Look at the amount of energy entering the interconnection and exiting the interconnection. Using the vertical scale on the left, approximate the loss as shown in the following graphic. The approximate loss for this interconnection is 2dB, which exceeds the 0.75dB maximum interconnection loss allowed by ANSI/TIA-568-C.3.

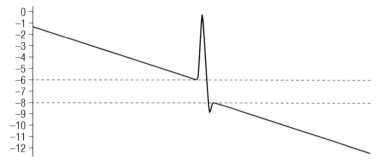

Identify a high-loss mechanical splice with an OTDR.   A high-loss mechanical splice can be quickly identified by following the sloping line as it enters and exits the splice back reflection. Visualize the sloping line traveling through the back reflection and exiting.

Master It   You are troubleshooting an older fiber-optic cable installation with the OTDR and the trace shown in the following graphic. The back reflection is the result of a mechanical splice. Using only the relative decibel information on the vertical scale, approximate the loss for the splice and determine whether this splice meets the loss requirements defined by ANSI/TIA-568-C.3.

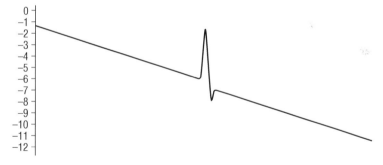

Solution   The vertical axis shows relative power in decibels. Look at the amount of energy entering the splice and exiting the splice. Using the vertical scale on the left, approximate the loss as shown in the graphic. The approximate loss for this mechanical splice is 1dB, which exceeds the 0.3dB maximum loss allowed by ANSI/TIA-568-C.3

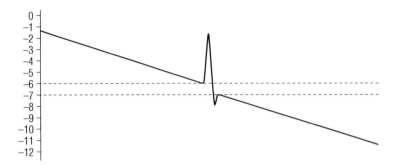

Identify a high-loss fusion splice or macrobend with an OTDR.   A high-loss fusion splice or macrobend can be quickly identified by the dip in the trace. These events do not produce a back reflection.

Master It   You are troubleshooting a new fiber-optic cable installation with the OTDR and the trace shown in the following graphic. The dip in the trace is the result of a macrobend. Using only the relative decibel information on the vertical scale, approximate the loss for the macrobend.

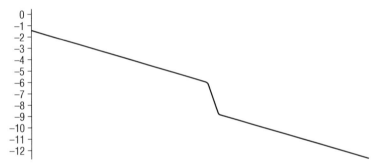

Solution   The vertical axis shows relative power in decibels. Look at the amount of energy entering the dip in the trace and exiting. Using the vertical scale on the left, approximate the loss caused by the macrobend as shown in the following graphic. The approximate loss for this macrobend is 2.9dB.

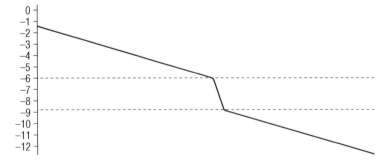

Appendix B

The Electronics Technicians Association, International (ETA) Certifications

The Electronics Technicians Association, International (ETA) is a nonprofit professional trade association established in 1978 "by technicians, for technicians" to promote the electronics service industry and to better benchmark the knowledge and skills of individual electronics technicians. The ETA currently has a membership of thousands of individuals, primarily from the electronics installation, maintenance, and repair occupations. It provides over 30 different certifications in various electronics areas. This appendix lists the competencies for ETA's top three certifications:

1. Data Cabling Installer Certification (DCI)
2. Fiber Optics Installer (FOI)
3. Fiber Optics Technician (FOT)

Data Cabling Installer (DCI) Certification 2014 Knowledge Competency Requirements

The following knowledge competency listing identifies the individual topics a person is expected to learn in preparation for the Data Cabling Installer (DCI) certification written examination:

1 Safety

1.1 Describe the various forms of protective gear that data cabling technicians have at their disposal

1.2 Explain the safety issues associated with the work area

1.3 Provide an overview of emergency response information and techniques for the workplace

2 Basic Electricity

2.1 Describe the relationships between voltage, current, resistance and power

2.2   Identify components called resistors and also non-component types of resistance in cabling technology

2.3   Use ohms law to calculate power usage and power losses in cabling circuits

2.4   Explain how noise may be generated onto communications cabling and components

2.5   Define impedance and compare impedance with resistance

2.6   Explain Signal-to-Noise Ratio

2.7   Explain the difference between inductance and inductive reactance; capacitance and capacitive reactance

2.8   Explain the importance of grounding cabling and electronics communications products

2.9   Identify wire sizes needed for grounding

2.10   Describe the types of conductor insulation used for communications wiring

2.11   Explain the difference between AC and DC circuits

3   Data Cabling Introduction

3.1   Provide a brief history of telephone and wireless communications

3.2   Describe the basic operation of the telephone system in the United States

3.3   Describe the differences between analog and digital communications signals

3.4   Identify the different categories of balanced twisted-pair cabling and components to include:

    3.4.1   Category 3

    3.4.2   Category 5e

    3.4.3   Category 6

    3.4.4   Category 6A

3.5   Understand the different types of unshielded twisted-pair (UTP) cabling

3.6   Understand the different types of shielded twisted-pair (STP) cabling

3.7   Differentiate between the uses of plenum and riser rated cabling

4   Data Communications Basics

4.1   Define audio and radio or Radio Frequency (RF) frequencies

4.2   Explain the term bandwidth

4.3   Explain the difference between frequency, bit rate and baud rate

4.4   Trace the history of the BEL and decibel and explain how and why these terms are used

4.5   Convert signals from voltage levels to their corresponding decibel equivalents decibel levels to their corresponding voltage or current levels

4.6   Convert signal gains or losses to comparative decibel readings

4.7   Define attenuation

4.8   Define crosstalk and explain how it occurs in communications cabling

4.9   Discuss how the industry has developed a comprehensive set of crosstalk measurements to ensure that permanent link cabling systems meet their intended applications in accordance with ANSI/TIA-568-C.2, section 6.3 including:

   4.9.1   Near-End Crosstalk (NEXT)

   4.9.2   Power Sum Near-End Crosstalk (PSNEXT)

   4.9.3   Attenuation to Crosstalk Ration, Far-End (ACRF)

   4.9.4   Power Sum Attenuation to Crosstalk Ration, Far-End (PSACRF)

   4.9.5   Power Sum Alien Near-End Crosstalk (PSANEXT)

   4.9.6   Power Sum Attenuation to Alien Crosstalk Ratio, Far-End (PSAACRF)

5   Cabling Specifications and Standards

5.1   Identify key industry standards necessary to specify, install, and test network cabling to include:

   5.1.1   ANSI/TIA-568-C.0, Generic Telecommunications Cabling for Customer Premises

   5.1.2   ANSI/TIA-568-C.1, Commercial Building Telecommunications Cabling Standard

   5.1.3   ANSI/TIA-568-C.2, Balanced Twisted-Pair Telecommunications Cabling and Components Standard

   5.1.4   ANSI/TIA-568-C.3, Optical Fiber Cabling Components Standard

   5.1.5   ANSI/TIA/EIA-569-B, Commercial Building Standard for Telecommunications Pathways and Spaces

   5.1.6   ANSI/TIA/EIA-570-B, Residential Telecommunications Cabling Standard

   5.1.7   ANSI/TIA/EIA-606-B, Administration Standard for the Telecommunications Infrastructure of Commercial Buildings

   5.1.8   J-STD-607-A, Commercial Building Grounding/Bonding Requirements for Telecommunications

   5.1.9   ANSI/TIA/EIA-942, Telecommunications Infrastructure Standard for Data Centers

   5.1.10   ANSI/TIA-1005, Telecommunications Infrastructure for Industrial Premises

   5.1.11   SO/IEC 11801, Generic Cabling for Customer Premises

5.1.12   ISO 11801 Class E (Augmented Category 6) Standard

5.1.13   IEEE 802.3af, Power over Ethernet (PoE) Standard

5.1.14   IEEE 802.3at, Power over Ethernet+ (Plus) Standard

5.1.15   IEEE 802.3an, Physical Layer and Management Parameters for 10 Gbps Operation, Type 10GBASE-T

5.1.16   TIA-568-B.2-ad10, Augmented Category 6 or ISO 11801 Class E Cables

5.1.17   IEEE 802.3ba Media Access Control Parameters, Physical Layers and Management Parameters for 40 Gbps and 100 Gbps Operation

5.1.18   IEEE 802.11, Wireless Standard

5.1.19   NFPA® 70, National Fire Protection Association, National Electrical Code (NEC®)

5.1.20   Canadian Electrical Code (CEC)

6   Basic Network Architectures

6.1   State that today's networking architectures fall into one of three categories:

6.1.1   Bus

6.1.2   Ring

6.1.3   Hierarchical star

6.2   Describe a network using Ethernet technologies

6.3   Describe how a Token Ring network operates

6.4   Define Fiber Distributed Data Interface (FDDI) networking specification produced by ANSI X3T9.5 committee

6.5   Explain Asynchronous Transfer Mode (ATM) designed as a high-speed communications protocol that does not depend on any specific LAN technology

6.6   Explain that in accordance with ANSI/TIA-568-C.0 generic cabling shall be installed in a hierarchal star topology

7   Cable Construction

7.1   Explain the requirements in accordance with ANSI/TIA-568-C.2 for 100 ohm category 3, 5, 5e, 6, and 6A balanced twisted-pair cabling and components

7.2   Describe broadband coaxial cabling, cords and connecting hardware to support CATV (cable television), and satellite television supported by ANSI/TIA-568-C.0 star topology in accordance with ANSI/TIA-568-C.4 Broadband Coaxial Cabling and Components Standard

7.3   Describe the differences between CAT 3, 5, 5e and 6 and 6_A twisted-pair data cables

7.4   Distinguish that horizontal cable shall consist of four balanced twisted-pairs of 22 AWG to 24 AWG thermoplastic insulated solid conductors enclosed by a thermoplastic jacket

7.5   State that the diameter of the insulated conductor shall be 1.53 mm (0.060 in) maximum

7.6   Define the ultimate breaking strength of the cable, measured in accordance with ASTM D4565, shall be 400 N (90 lbf – foot pounds) minimum

7.7   Define the pulling tension for a 4-pair balanced twisted-pair cable shall not exceed 110 N (25 lbf) during installation

7.8   Relate how twisted-pair cables shall withstand a bend radius of 4× cable diameter for UTP constructions and 8× cable diameter for screened constructions

7.9   Explain that the minimum inside bend radius for 4-pair balanced twisted-pair cord cable shall be one times the cord cable diameter

8   Cable Performance Characteristics

8.1   Describe the transmission characteristics of twisted-pair cabling in accordance with ANSI/TIA-568-C.2 to include the following:

8.1.1   Category 3: This designation applies to 100 ohm balanced twisted-pair cabling and components whose transmission characteristics are specified from 1 to 16 MHz

8.1.2   Category 5e: This designation applies to 100 ohm balanced twisted-pair cabling and components whose transmission characteristics are specified from 1 to 100 MHz

8.1.3   Category 6: This designation applies to 100 ohm balanced twisted-pair cabling and components whose transmission characteristics are specified from 1 to 250 MHz

8.1.4   Category 6A: This designation applies to 100 ohm balanced twisted-pair cabling and components whose transmission characteristics are specified from 1 to 500 MHz

8.2   Point out that category 1, 2, 4, and 5 cabling and components are not recognized as part of the new standards and, therefore, their transmission characteristics are not specified

8.3   Explain the transmission characteristics of 75 ohm coaxial cable

8.4   Explain the mechanical performance characteristics of twisted pair and coaxial cables

8.5   Describe cabling transmission performance requirements (permanent link and channel) for category 3, 5e, 6 and category 6A 100 ohm balanced twisted pair cabling as specified in ANSI/TIA-568-C.2

9   National Electrical Code – NEC & UL Requiremnst

9.1   Associate the history of the National Fire Protection Association (NFPA®) with the National Electrical Code (NEC®)

9.2   Describe that Underwriters Laboratories (UL®) is a nonprofit product safety testing and certification organization and explain the following key UL® standards:

      9.2.1   UL 444 Communications Cables

      9.2.2   NFPA 262 (formerly UL 910) Standard Method of Test for Flame Travel and Smoke of Wires and Cables for use in Air-Handling Spaces

      9.2.3   UL 1581 Reference Standard for Electrical Wires, Cables, and Flexible Cords

      9.2.4   UL 1666 Standard for Test for Flame Propagation Height of Electrical and Optical-Fiber Cables Installed Vertically in Shafts

      9.2.5   UL 1863 Standard for Communications-Circuit Accessories

  9.3   Summarize the information in the NFPA® 70's NEC® to include the following:

      9.3.1   Chapter 1 – General Requirements

      9.3.2   Chapter 2 – Wiring and Protection

      9.3.3   Chapter 3 – Wiring Methods and Materials

      9.3.4   Chapter 5 – Special Occupancies

      9.3.5   Chapter 7 – Special Conditions

      9.3.6   Chapter 8 – Communications Systems

  9.4   Relate the similarities between the Canadian Electrical Code (CEC) and the NEC

10   Telecommunications Cabling System Structure

10.1   Describe a representative model of the functional elements that comprise a generic hierarchal star topology cabling system

10.2   Point out that the hierarchal star topology specified by ANSI/TIA-568-C.0 was selected because of its acceptance and flexibility

10.3   Explain that the functional elements are equipment outlets, distributors, and cabling (subsystems) total system

10.4   Explain that in a typical commercial building where ANSI/TIA-568-C.1 applies, Distributor C represents the main cross-connect (MC), Distributor B represents the intermediate cross-connect (IC), Distributor A represents the horizontal cross-connect (HC), and the equipment outlet (EO) represents the telecommunications outlet/connector

10.5   Describe that equipment outlets (EOs) provide the outermost location to terminate the cable in a hierarchal star topology

10.6   Explain that distributors provide a location for administration, reconfigurations, connection of equipment and for testing

10.7   Discuss how distributors can be configured as interconnections or cross-connections

10.8   Explain that the function of a cabling subsystem is to provide a signal path between distributors

10.9   Identify that the recognized media in a hierarchal star topology, which shall be used individually or in combination in accordance with ANSI/TIA-568-C.0 are:

    10.9.1   100 ohm balanced twisted-pair cabling

    10.9.2   Multimode optical fiber cabling

    10.9.3   Single-mode optical fiber cabling

11   Data Cabling Installer Tools

11.1   Explain the purpose and proper usage of twisted-pair and coaxial wire strippers

11.2   Demonstrate how wire cutters and cable preparation tools are used properly to prepare cable for installation

11.3   Demonstrate the proper methods of using both twisted-pair and coaxial cable crimpers

11.4   Demonstrate a punch-down tool and show where and how it is used in cross-connect blocks (66 block), patch panels (110 block), or modular jacks that use insulation displacement connectors (IDCs)

11.5   Explain the purpose and proper use of fish tape and pull/push rod devices used in cable installation

11.6   Identify the tools for basic cable testing

11.7   Discuss the technology of pulling lubricants

11.8   Identify cable marking supplies and labels that make cable installations easy

12   Transmission Media for Networking and Telecommunications

12.1   Identify the three common bounded media types used for data transmission

    12.1.1   Twisted-pair

    12.1.2   Coaxial

    12.1.3   Fiber Optic

12.2   Discuss the common types of copper cabling and the applications that run on them

12.3   Describe the different types of twisted-pair cables and expand on their performance characteristics

    12.3.1   Category 3

    12.3.2   Category 5e

    12.3.3   Category 6

    12.3.4   Category 6A

12.4   Describe the different types coaxial cable and expand on their performance characteristics

    12.4.1   RG-59

    12.4.2   RG-6

    12.4.3   RG-62

12.5   Explain the terms hybrid and composite cable types

12.6   Describe the best practices for copper installation including following standards, not exceeding distance limits, and good installation techniques

12.7   Explain that when pulling copper cable (or wire), tension must be applied to all elements of the cable

12.8   Describe that pulling solely on the core can pull it out of the sheath and care must be taken to avoid damage to the conductors

12.9   Define the maximum allowable pulling tension is the greatest pulling force that can be applied to a cable during installation without risking damage to the conductors

12.10   Identify the minimum bend radius and maximum pulling tension in accordance with ANSI/TIA-568-C.0 section 5.3

12.11   Explain that copper data cabling and wiring systems are divided into categories or classes by the cabling standards organizations and use bandwidth needs to determine the proper customer application of each category of cabling

12.12   Discuss the importance of any telecommunications cabling system that supports data is the 110 block

12.13   Explain the usage of cross connects using punch downs in the telecommunications rooms, more common on telephone wires (66-block) than data (110-block)

12.14   Explain that the 110 block has two primary components: the 110 wiring block and the 110 connecting block

12.15   Identify the other 110 block styles including a 110-block with RJ-45 connectors and 110-blocks on the back of patch panels

12.16   Examine the different 110-block possible scenarios for use in a structured cabling system

12.17   List the common usages of the 66-block in cross-connect systems

12.18   Explain that the 66-block was used with telephone cabling for many years, but is not used in modern structured wiring installations

12.19   Describe how a 66-block is assembled using the punch-down (impact) tool and how it is terminated using a metal bridging clip

12.20   Recall that the most common type of cable connected to a 66-block is the 25-pair cable

12.21   Explain that a minimum of Category 3 cable is used for voice applications; however, Category 5e or higher is used for both data and voice

12.22   Explain that every cable run must receive a minimum level of testing and the minimum tests should determine continuity and ascertain that the wire map is correct

12.23   Demonstrate the cable testers that are used to perform the basic level of testing:

   12.23.1   Tone generator

   12.23.2   Continuity tester

   12.23.3   Wire-map tester

   12.23.4   Cable certifier

13   Work Area Telecommunications Outlet and Connectors

13.1   Explain that in accordance with ANSI/TIA-568-C.2 each four-pair horizontal cable shall be terminated in an eight-position modular jack at the work area

13.2   Explain that the telecommunications outlet/connector shall meet the requirements of clause 5.7 and the terminal marking and mounting requirements specified in ANSI/TIA-570-B Residential Telecommunications Infrastructure Standard

13.3   Describe the proper wiring scheme (pin/pair assignments) necessary to accommodate certain cabling systems to include the following:

   13.3.1   Bell Telephone Universal Service Order Code (USOC)

   13.3.2   ANSI/TIA-568-C.2 T568A and T568B

   13.3.3   ANSI X3T9.5 TP-PMD

13.4   Identify the different types of coaxial connector styles to include:

   13.4.1   F-Type RG-6 crimp-on method

   13.4.2   F-Type RG-6 twist-on method

   13.4.3   F-Type RG-6 compression method

   13.4.4   BNC RG-59

13.5   Describe the key wall plate designs

13.6   Explain the most common wall plate mounting methods in commercial applications

13.7   Describe the difference between a fixed design and modular wall plate design installation

13.8   Describe how to mount networking cables to a wall with a RJ45 surface mount biscuit jack.

13.9 Describe the use of a floor-mounted communications outlet that provides point-of-use connectivity for a broad range of applications where convenience or building requirements dictate this installation

13.10 Explain that open office design practices use multi-user telecommunications outlet assemblies (MUTOAs), consolidation points (CPs), or both to provide flexible layouts

14 Local Area Network Interconnection and Networking

14.1 Explain the basic active components of a hierarchical star network for commercial buildings and networks to include the following:

14.1.1 Network interface card (NIC)

14.1.2 Media converter

14.1.3 Repeater

14.1.4 Hub

14.1.5 Bridge

14.1.6 Switch

14.1.7 Server

14.1.8 Router

14.2 Describe the differences between a blocking and non-blocking workgroup switch and affects with the effective bandwidth performance of the switch

14.3 Identify the differences between various types of transceiver modules

15 Wireless Heterogeneous Cabling Networks

15.1 Discuss how wireless switches and routers are usually connected to the core network with some type of copper cabling media

15.2 Explain how infrared wireless systems work

15.3 Define the types of radio frequency (RF) wireless networks

15.4 Explain how microwave communication works

16 Cabling System Components

16.1 Explain that the entrance facility (EF), sometimes referred to as the demarcation point, consists of cables, connecting hardware, protection devices, and other equipment that connect to access provider (AP) cabling in accordance with ANSI/TIA-568-C.1

16.2   Explain that the equipment rooms (ERs) are considered to be distinct from telecommunications rooms (TRs) and telecommunications enclosures (TEs) because of the nature and complexity of the equipment they contain

16.3   Explain the main cross-connect (MC; Distributor C) of a commercial building are normally located in the ER

16.4   Explain that intermediate cross-connects (ICs; Distributor B), horizontal cross-connects (HCs; Distributor A), or both, of a commercial building may also be located in an ER

16.5   Describe that an ER houses telecommunications equipment, connecting hardware, splice closures, grounding and bonding facilities, and local telephone company service terminations, and premises network terminations

16.6   Explain that telecommunications rooms (TRs) and telecommunications enclosures (TEs) provide a common access point backbone and building pathways

16.7   Recognize that the TR and any TE should be located on the same floor as the work areas served

16.8   Explain that TEs may be used in addition to one TR per floor and in addition to an additional TR for each area up to 1000 m² (10,000 ft²)

16.9   Explain that the horizontal cross-connect (HC; Distributor A) of a commercial building is located in a TR or TE

16.10   Describe backbone cabling as the portion of the commercial building telecommunications cabling system that provides interconnections between entrance facilities (EFs), access provider (AP) spaces, telecommunications rooms (TRs) and telecommunications enclosures (TEs)

16.11   Explain that the horizontal cabling includes horizontal cable, telecommunications outlet/connectors in the work area, patch cords or jumpers located in a telecommunications room (TR) or telecommunications enclosure (TE)

16.12   Explain that the work area (WA) components extend from the telecommunications outlet/connector end of the horizontal cabling system to the WA equipment

17   Cabling System Design

17.1   Describe the basics of the hierarchical star, bus, ring and mesh topologies

17.2   Explain that the backbone cable length extends from the termination of the media at the main cross-connect (MC) to an intermediate cross-connect (IC) or horizontal cross-connect (HC)

17.3   Explain that the backbone cable lengths are dependent upon the application and upon the specific media chosen in accordance with ANSI/TIA-568-C.0 Annex D and specific application standard

17.4   Explain the maximum horizontal cable length shall be 90 m (295 ft), independent of media type

17.5   Describe how the telecommunications room is wired to include:

17.5.1   Local area network (LAN) wiring

17.5.2   Telephone wiring

17.5.3   Power requirements

17.5.4   HVAC considerations

17.6   Explain the concept of cabling management to include the following:

17.6.1   Physical protection

17.6.2   Electrical protection

17.6.3   Fire protection

17.7   List the lengths of the cross-connect jumpers and patch cords in the MC, IC, HC and WA in accordance with ANSI/TIA-568-C.1 to include:

17.7.1   MC or IC should not exceed 20 m (66 ft)

17.7.2   HC should not exceed 5 m (16 ft)

17.7.3   WA should not exceed 5 m (16 ft)

17.8   Explain that for each horizontal channel, the total length allowed for cords in the work area (WA), plus patch cords or jumpers, plus equipment cords in the telecommunications room (TR) or telecommunications enclosure (TE) shall not exceed 10 m (33 ft)

17.9   Describe the purpose, construction and usage of telecommunications pathways and spaces in accordance with ANSI/TIA-569-B Commercial Standard for Telecommunications Pathways and Spaces to include:

17.9.1   Entrance facility

17.9.2   Equipment room

17.9.3   Telecommunications rooms

17.9.4   Horizontal pathways

17.9.5   Backbone pathways

17.9.6   Work areas

17.10   Define the term, location and usage of both the main distribution frame (MDF) and Intermediate distribution frame (IDF)

18   Cabling Installation

18.1   Describe the steps used in installing communications cabling

18.2   Explain cable stress and the precautions for aerial construction; underground and ducts and plenum installation; define pulling tension and bend radius

18.3   Describe cabling dressing and methods of securing cabling

18.4   Explain proper labeling of cables in accordance with ANSI/TIA-606-B Administration Standard for Commercial Telecommunications Infrastructure

18.5   Identify the insulated conductor color code for 4-pair horizontal cables in accordance with ANSI/TIA-568-C.2

18.6   Demonstrate proper cable stripping for both twisted-pair and coaxial cable

18.7   Explain safety precautions for underground construction

18.8   Define Fire stopping and the different applications and types

18.9   Define the components of a grounding and bonding system for telecommunications and their purpose per J-STD-607-A Commercial Building Grounding and Bonding Requirements for Telecommunications and NEC® Article 250 Grounding and Bonding

18.10   Explain that for multipair backbone cables with more than 25 pairs, the core shall be assembled in units or sub-units of up to 25 pairs and shall be identified by a color-coded binder in accordance with ANSI/ICEA S-80-576

18.11   Explain how ducts are used for cabling installations

18.12   Outline and understand the need for firestopping

18.13   Introduce the fire-related considerations associated with installing cable runs

18.14   Describe the different components used to minimize the spread of smoke and fire throughout the structure

19   Connector Installation

19.1   Demonstrate proper installation of twisted pair connectors

19.2   Demonstrate proper installation of coaxial cable connectors

19.3   Describe the color code for telecom cabling and the pin/pair assignments

19.4   Explain the maximum pair un-twist for the balanced twisted pair cable termination shall be in accordance with ANSI/TIA-568-C.0 5.3.3.1 Table 1

20   Cabling Testing and Certification

20.1   Explain the purpose of installation testing

20.2   Describe the purpose and methods of certifying the cable plant

20.3   Define the ANSI/TIA-568-C standard performance requirements for horizontal and backbone cabling

20.4   Explain the differences between the two types of horizontal links used in copper cable certification: permanent link and channel link

20.5　Show the proper selection and use of cable testing tools and equipment

20.6　Describe cabling requirements (permanent link and channel) for Category 3, 5e, 6 and Category 6A 100 ohm balanced twisted-pair cabling as specified in ANSI/TIA-568-C.2

20.7　Explain the minimum required tests (wire map, length, insertion loss and near-end crosstalk) for all CAT 5e installations

20.8　Demonstrate how the data cabling installer thoroughly tests the newly installed cabling according to the specifications contained in the ANSI/TIA-568-C.2 standard to include:

　　20.8.1　Near-End Crosstalk (NEXT)

　　20.8.2　Power Sum Near-End Crosstalk (PSNEXT)

　　20.8.3　Attenuation to Crosstalk Ratio, Far-End (ACRF)

　　20.8.4　Power Sum Attenuation to Crosstalk Ratio, Far-End (PSACRF)

　　20.8.5　Power Sum Alien Near-End Crosstalk (PSANEXT)

　　20.8.6　Power Sum Attenuation to Alien Crosstalk Ratio, Far-End (PSAACRF)

20.9　Describe how Power over Ethernet must be checked to ensure proper wattage (up to 15.4 watts) is provided to the end device at 100 meters per the IEEE 802.3af PoE standard

21　Cabling Troubleshooting

21.1　Explain how to establish a baseline for testing or repairing a cabling system

21.2　Demonstrate a method of locating a cabling defect or problem

21.3　Describe commonly encountered cable problems and the methods used to resolve them including:

　　21.3.1　Wire-map faults

　　21.3.2　Excessive length

　　21.3.3　Opens and shorts

　　21.3.4　Excessive attenuation

　　21.3.5　Excessive crosstalk

　　21.3.6　Excessive noise

21.4　Explain that for a communications installer, the effects of the earth's magnetic field is the possibility of what is known as a ground loop and how it effects copper cabling

21.5　Define a ground fault

21.6　Explain that a troubleshooter must possess good communication skills and be able to accomplish the following:

　　21.6.1　Read technical manuals, instructions, catalogs, etc.

21.6.2   Verbal communication

21.6.3   Understand blueprints and drawings

22   Documentation

22.1   Point out that the process of troubleshooting can be greatly eased when appropriate documentation is available including:

22.1.1   Cabling diagrams

22.1.2   Description and functioning of the equipment attached to the cabling system

22.1.3   Certification test data for the network

22.2   Explain the purpose of documenting a cabling installation

22.3   Explain the required ingredients of the installation documents

22.4   Explain that the request for proposal is essential to the success of a telecommunications infrastructure project

22.5   Prepare a sample cable documentation record that meets industry standards

Fiber Optics Installer (FOI) 2014 Knowledge Competency Requirements

The following knowledge competency listing identifies the individual topics a person is expected to learn in preparation for the Fiber Optics Installer (FOI) certification written examination:

1   History of Fiber Optics

1.1   Trace the evolution of light in communication

1.2   Summarize the evolution of optical fiber manufacturing technology

1.3   Track the evolution of optical fiber integration and application

2   Principles of Fiber Optic Transmission

2.1   Describe the basic parts of a fiber-optic link

2.2   Describe the basic operation of a fiber-optic transmitter

2.3   Describe the basic operation of a fiber-optic receiver

2.4   Explain how to express gain or loss using dB

2.5   Explain how to express optical power in dBm

3   Basic Principles of Light

3.1   Describe the following:

3.1.1   Light as electromagnetic energy

3.1.2   Light as particles and waves

3.1.3   The electromagnetic spectrum; locate light frequencies within the spectrum in relation to radio and microwave communication frequencies

3.1.4   The refraction of light (Snell's Law)

3.1.5   Fresnel reflection

3.2   Explain the following:

3.2.1   How the index of refraction is used to express the speed of light through a transparent medium

3.2.2   Reflection, to include angle of incidence, critical angle, and angle of refraction

3.2.3   Fresnel reflections and their impact the performance of a fiber optic communication system

4   Optical Fiber Construction and Theory

4.1   Describe the following:

4.1.1   The basic parts of an optical fiber

4.1.2   The different materials that can be used to construct an optical fiber

4.1.3   Optical fiber manufacturing techniques

4.1.4   The tensile strength of an optical fiber

4.1.5   A mode in an optical fiber

4.1.6   The three refractive index profiles commonly found in optical fiber

4.2   Explain the propagation of light through:

4.2.1   Multimode step index optical fiber

4.2.2   Multimode graded index optical fiber

4.2.3   Single-mode optical fiber

4.3   Describe the following TIA/EIA-568-C recognized optical fibers:

4.3.1   Multimode optical fibers

4.3.2   Single-mode optical fibers

4.4   Describe the following ITU-T recognized optical fibers:

   4.4.1   ITU-T-G.652

   4.4.2   ITU-T-G.655

4.5 Describe the following ISO/IEC 11801 multimode optical fiber designations:

   4.5.1   OM1

   4.5.2   OM2

   4.5.3   OM3

   4.5.4   OM4

4.6   Describe the following types of optical fiber:

   4.6.1   Plastic cald silca (PCS)

   4.6.1   Hard clad silica (HCS)

   4.6.2   Plastic

5   Optical Fiber Characteristics

5.1   Explain dispersion in an optical fiber to include:

   5.1.1   Modal dispersion and its effects on the bandwidth of an optical fiber

   5.1.2   Material dispersion and its effects on the bandwidth of an optical fiber

   5.1.3   Chromatic dispersion and its components

5.2   Describe the causes of attenuation in an optical fiber to include:

   5.2.1   Attenuation versus wavelength in a multimode optical fiber

   5.2.2   Attenuation versus wavelength in a single-mode optical fiber

5.3   Describe the numerical aperture of an optical fiber

5.4   Explain how the number of modes in an optical fiber is defined by core diameter and wavelength

   5.4.1   Define the cutoff wavelength of a single-mode optical fiber

5.5   Describe microbends in an optical fiber

5.6   Explain macrobends in an optical fiber

6   Fiber Optic Cabling Safety

6.1   Explain how to safely handle and dispose of fiber optic cable, optical fiber chips, and debris

6.2   List the safety classifications of fiber optic light sources

6.3   Discuss the potential chemical hazards in the fiber optic environment and the purpose of the material safety data sheet (MSDS)

6.4   Cite potential electrical hazards in the fiber optic installation environment

6.5   Outline typical workplace hazards in the fiber optic environment

7   Fiber Optic Cables

7.1   Draw a cross section of a fiber optic cable and explain the purposes of each segment

7.2   Identify why and where loose tube fiber optic cable is used

7.3   Describe tight buffer fiber optic cable

7.4   Compare common strength members found in fiber optic cables

7.5   Name common jacket materials found in fiber optic cables

7.6   Describe simplex and duplex cordage and explain the difference between cordage and cable

7.7   Describe the basic characteristics of the following:

    7.7.1   Loose tube gel filled (LTGF)

    7.7.2   Loose tube gel free

        7.7.2.1   Dry water block

    7.7.3   Distribution cable

    7.7.4   Breakout cable

    7.7.5   Armored cable

    7.7.6   Messenger cable (figure 8)

    7.7.7   Ribbon cable

    7.7.8   Submarine cable

    7.7.9   Composite cable

    7.7.10   Hybrid cable

7.8   Discuss ANSI/TIA-568-C performance specifications for the optical fiber cables recognized in premises cabling standards to include:

    7.8.1   Inside plant cable

    7.8.2   Indoor-outdoor cable

    7.8.3   Outside plant cable

    7.8.4   Drop cable

7.9   Explain how and when a fan-out (furcation) kit is used

7.10   Identify how and when a breakout kit is used

7.11   List the National Electrical Code (NEC®) optical fiber cable types including:

    7.11.1   Abandoned optical fiber cable

    7.11.2   Nonconductive optical fiber cable

    7.11.3   Composite optical fiber cable

    7.11.4   Conductive optical fiber cable

7.12   Describe the NEC® listing requirements for:

    7.12.1   Optical fiber cables

    7.12.2   Optical fiber raceways

7.13   List the ANSI/TIA-598-C color code and cable markings

7.14   Describe the ISO/IEC 11801 international cabling standard

7.15   Identify the ANSI/TIA-568-C minimum bend radius specifications for inside plant, indoor/outdoor, outside plant, and drop cables

7.16   Cite the ITU-T G.652, G.657 and the Telcordia GR-20-CORE minimum bend radius specification for outside plant fiber optic cables

7.17   Describe the ANSI/TIA-568-C maximum tensile load ratings during installation for inside plant, indoor/outdoor, outside plant, and drop cables

8   Splicing

8.1   Explain the intrinsic factors that affect splice performance

8.2   Explain the extrinsic factors that affect splice performance

8.3   List the basic parts of a mechanical splicer

8.4   Describe how to perform a mechanical splice

8.5   Explain how to perform and protect a fusion splice

    8.5.1   Describe the different types of fusion splicers

8.6   List ANSI/TIA/EIA-568-C inside plant splice performance requirements

8.7   Cite ANSI/TIA-758 outside plant splice performance requirements

9   Connectors

9.1   Describe the basic parts of a fiber optic connector

9.2   Describe the physical and performance differences for the following endface finishes:

9.2.1   Flat

9.2.2   Physical contact (PC)

9.2.3   Angled physical contact (APC)

9.2.4   Ultra-Physical Contact (UPC)

9.3   Describe the basic characteristics the following connectors:

9.3.1   ANSI/TIA-568-C recognized

9.3.2   Small form factor (SFF)

9.3.3   MPO/MTP

9.3.4   Pigtail

9.4   Describe common connector ferrule materials

9.5   Describe the ANSI/TIA-568-C multimode and single-mode connector and adapter identification color code

9.6   Explain the intrinsic factors that affect connector performance

9.7   Explain the extrinsic factors that affect connector performance

9.8   Describe reflectance caused by interconnections

9.9   Identify the steps involved in an anaerobic epoxy connector termination and polish

9.10   List the steps involved in an oven cured epoxy connector termination and polish

9.11   Describe the steps involved in a pre-load epoxy connector termination and polish

9.12   Describe how to construct a no-polish, no-epoxy connector termination

9.13   Explain how to properly clean a connector

9.14   Describe how to examine the endface of a connector per IEC 61300-3-35 and ANSI/TIA-455-57B

9.15   List the ANSI/TIA-568-C connector performance requirements

9.16   List the ITU-T G.671 requirements for single-mode connectors

9.17   Compare the following fiber connectorization methods:

9.17.1   Field termination

9.17.2   Factory terminated assemblies

10   Fiber Optic Light Sources

10.1   Describe the basic operation and types of LED light sources used in fiber optic communications

10.1.1   Describe LED performance characteristics

10.1.2   Describe the performance characteristics of a LED transmitter

10.2   Describe the basic operation and types of laser light sources used with the following optical fibers:

10.2.1   Multimode

10.2.2   Single-mode

10.3   Describe laser performance characteristics

10.4   Describe the performance characteristics of a laser transmitter

10.5   Explain the benefits of using a laser light source in fiber optic communication systems

10.6   Identify which fiber type is best used for communications systems with VCSEL light sources

11   Fiber Optic Detectors and Receivers

11.1   Summarize the basic operation of a photodiode

11.2   Explain why an optical attenuator is occasionally used in a communication system

12   Cable Installation and Testing

12.1   Describe the basic cable installation parameters that may be found in manufacturer's datasheet

12.2   Explain the static and dynamic loading on a fiber optic cable during installation

12.3   List commonly used installation hardware

12.4   Describe following types of installation:

12.4.1   Tray and duct

12.4.2   Conduit

12.4.3   Direct burial

12.4.3.1   Drop cable minimum buried depth

12.4.4   Aerial installation

12.4.5   Blown fiber installation

12.4.6   Wall outlet installation

12.5   Explain cable grounding and bonding per NEC Articles 770 and 250

12.6   Summarize these types of preparation:

12.6.1   Patch panel

12.6.2   Rack mountable hardware to include:

12.6.2.1   Housings

12.6.2.2   Cable routing guides

12.6.3   Splice enclosure

12.7   Recognize that the ANSI/TIA-606 standard describes the administrative record keeping elements of a modern telecommunications infrastructure.

12.7.1   Explain that proper administration includes the timely updating of drawings, labels and records.

12.8   Explain the role of the NEC and the Canadian Electrical Code (CEC)

12.9   Explain the role of the National Electrical Safety Code (NESC)

12.10   Identify the role of the ANSI/TIA 590 standard

13   Fiber Optic System Design Considerations

13.1   Compare the following advantages of optical fiber over twisted pair and coaxial cables:

13.1.1   Bandwidth

13.1.2   Attenuation

13.1.3   Electromagnetic immunity

13.1.4   Weight

13.1.5   Size

13.1.6   Security

13.1.7   Safety

14   Test Equipment and Link/Cable Testing

14.1   Recognize that field test instruments for multimode fiber cabling should meet the requirements of ANSI/TIA-526-14 in accordance with ANSI/TIA-568-C

14.2   Recognize that field test instruments for single-mode fiber cabling should meet the requirements of TIA-526-7 in accordance with ANSI/TIA-568-C

14.3   Describe the different types of continuity testers

14.4   Explain how to use a visual fault locator (VFL) when troubleshooting the cable plant

14.5   Describe the basic operation of a single-mode and multimode light sources and optical power meters

14.5.1   Explain why a light source should not be disconnected during testing for zero calibration changes

14.6   Describe the difference between a measurement quality jumper (MQJ) and a patch cord

14.7   Define the purpose of a mode filter by having five non-overlapping wraps of multi-mode fiber on a mandrel in accordance with ANSI/TIA-568.C. (Annex E)

14.7.1   Explain that this procedure is also applicable to single-mode cabling with a single 30 mm (1.2 in) diameter loop of single-mode fiber in accordance with ANSI/TIA-526-7

14.8   Describe how to measure the optical loss in a patch cord with a light source and optical power meter using the Method A, 2-jumper reference in accordance with ANSI/TIA-526-14

14.9   Summarize the basic operation of an optical time domain reflectometer (OTDR)

Fiber Optic Technician (FOT) 2014 Knowledge Competency Requirements

The following knowledge competency listing identifies the individual topics a person is expected to learn in preparation for the Fiber Optics Technician (FOT) certification written examination:

1   Principles of Fiber Optic Transmission

1.1   Describe the basic parts of a fiber optic link

1.2   Describe the basic operation of a transmitter

1.2.1   Graphically explain how analog to digital conversion (A/D) is accomplished

1.3   Describe the basic operation of a receiver

1.3.1   Graphically explain how digital to analog conversion (D/A) is accomplished

1.4   Explain amplitude modulation (AM)

1.5   Compare digital data transmission with analog

1.6   Explain the difference between Pulse Coded Modulation (PCM) and AM

1.7   List the benefits of Multiplexing signals

1.8   Explain the how the decibel (dB) is used to express relative voltage and power levels

1.8.1   Convert voltage and power levels to and from decibel equivalents

1.8.2   Explain how to express gain or loss using dB

1.9   Explain how dBm is used to express absolute voltage and power measurements

2   Basic Principles of Light

2.1   Describe the electromagnetic spectrum and locate light frequencies within the spectrum in relation to communications frequencies

2.2   Convert various wavelengths to corresponding frequencies

2.3   Describe how the index of refraction is calculated

2.4   Explain total internal reflection

2.5   Define Fresnel reflection loss

2.6   Explain the effects of refraction

  2.6.1   Explain Snell's Law

3   Optical Fiber Construction and Theory

3.1   Describe the different materials used to manufacture optical fiber

3.2   Explain why the core and the cladding have different refractive indexes

3.3   List the different types of fiber optic coatings

3.4   Define the performance of optical fibers used in the telecommunications industry in accordance with Telecommunications Industry Association (TIA), Telcordia, and the International Telecommunications Union (ITU)

3.5   Summarize the fiber types that correspond to the designations OM1, OM2, OM3, OM4, OS1, and OS2 in accordance with ISO/IEC (the International Organization for Standardization/International Electrotechnical Commission) requirements

3.6   Describe the performance differences between single-mode and multimode optical fiber

  3.6.1   Explain why multimode optical fiber may be selected over single-mode optical fiber

3.7   Describe the basics of optical fiber manufacturing

3.8   Point out how the number modes is a characteristic used to distinguish optical fiber types

3.10   Explain the purposes for different refractive index profiles

4   Optical Fiber Characteristics

4.1   Define dispersion in an optical fiber

4.2   Explain how modal dispersion causes pulses to spread out as they travel along the fiber

  4.2.1   List the methods for overcoming modal dispersion

4.3   Explain how variations in wavelength contribute to material dispersion in an optical fiber

4.4   Explain the differences between waveguide and material dispersion in an optical fiber

4.5   Explain chromatic dispersion in an optical fiber

4.6   Explain how polarization mode dispersion (PMD) affects the two distinct polarization mode states, referred to as differential group delay (DGD)

4.7   Describe how microbends can change the performance characteristics of an optical fiber

4.8   Describe how a macrobend changes the angle at which light hits the core-cladding boundary

4.9   Explain how light that does not enter the core within the cone of acceptance may not propagate the entire length of the optical fiber

4.10   List the ANSI/TIA-568-C.3, ISO/IEC 11801, and ITU Series G minimum overfilled modal bandwidth-length product (MHz·km) limitations for multimode optical fibers

5   Safety

5.1   Explain how to safely handle and dispose of fiber optic cable

   5.1.1   Explain potential electrical hazards in a fiber optic environment

   5.1.2   Describe typical work place hazards in the fiber optic environment

5.2   Explain the three lines of defense to help you get through the day safely including:

   5.2.1   Engineering controls

   5.2.2   Personal protective equipment

   5.2.3   Good work habits

5.3   List the safety classifications of fiber optic light sources as described by the FDA, ANSI, OSHA, and IEC to prevent injuries from laser radiation

5.4   Explain the potential chemical hazards in the fiber optic environment and the purpose of the material safety data sheet (MSDS)

6   Fiber Optic Cables

6.1   Draw a cross-section of a fiber optic cable and explain the purposes of each segment

6.2   Distinguish between the two buffer type cables:

   6.2.1   Loose buffer (stranded vs. central tube)

   6.2.2   Tight buffer (distribution vs. breakout)

6.3   Identify the different types of strength members used to withstand tensile forces in an optical fiber cable

6.4   Compare jacket materials and explain how they play a crucial role in determining the environmental characteristics of a cable

6.5   Describe the following cable types:

   6.5.1   Simplex cordage

   6.5.2   Duplex cordage

6.5.3   Distribution cable

6.5.4   Breakout cable

6.5.5   Armored cable

6.5.6   Messenger cable

6.5.7   Ribbon cable

6.5.8   Submarine cable

6.5.9   Aerospace cable

6.5.10   Stranded Loose Tube cable

6.5.11   Central Loose Tube cable

6.6   Explain what hybrid cables are and where they are ordinarily used in accordance with ANSI/TIA-568-C.1

6.7   Describe a composite cable, as defined by National Electrical Code (NEC) Article 770.2

6.8   Distinguish the difference between a fanout kit (sometimes called a furcation kit) and a breakout kit

6.9   Explain how fibers can be blown through microducts instead of installing cables underground or in structures.

6.10   List the NEC optical fiber cable categories including:

6.10.1   Abandoned optical fiber cable

6.10.2   Nonconductive optical fiber cable

6.10.3   Composite optical fiber cable

6.10.4   Conductive optical fiber cable

6.11   Describe the NEC listing requirements for:

6.11.1   Optical fiber cables

6.11.2   Optical fiber raceways

6.12   Explain where the TIA-598-C color code is used and how the colors are used to identify individual cables

6.13   Describe TIA-598-C premises cable jacket colors

6.14   Explain how cable markings are used to determine the length of a cable

7   Types of Splicing

7.1   Mechanical Splicing

7.1.1   Explain the extrinsic factors that affect splice performance

7.1.2   Differentiate between the attributes and tolerances for different single-mode optical fibers as defined in ITU G series G.652, G.655, G.657, ANSI/TIA-568, ANSI/TIA-758 and Telcordia standards

7.1.3   Summarize the correct fiber preparation scoring method using a cleaver

7.1.4   Discuss the mechanical splice assembly process

7.1.5   Explain performance characteristics of index matching gel used inside the mechanical splice

7.1.6   Explain ANSI/TIA-568-C.0 (Annex E.8.3) OTDR insertion loss procedures for a reflective event mechanical splice

7.2   Fusion Splicing

7.2.1   Describe the advantages of fusion splicing over mechanical splicing

7.2.2   Summarize the correct fiber preparation scoring method using a cleaver

7.2.3   Discuss the fusion splice assembly process and splice protection

7.2.4   Explain the use of the splice closure

7.2.5   Explain ANSI/TIA-568-C.0 (Annex E.8.3) OTDR insertion loss procedures for a non-reflective event fusion splice

8   Connectors

8.1   Identify the wide variety of fiber optic connector types

8.2   Describe the three most common approaches to align the fibers including:

8.2.1   Ferrule based connector

8.2.2   V-groove assemblies for multiple fibers

8.2.3   Expanded-Beam connector

8.3   Describe the ANSI/TIA-568-C.3 section 5.2.2.4 two types of array adapters

8.3.1   Type-A MPO and MTP adapters shall mate two array connectors with connector keys key-up to key-down

8.3.2   Type-B MPO and MTP adapters shall mate two array connectors with connector keys key-up to key-up

8.4   Describe ferrule materials used with fiber optics connectors

8.5   Explain both the and extrinsic factors that affect connector performance

8.6   Define both physical contact (PC) and angled physical contact (APC) finish

8.6.1   Explain how PC and APC finishes affect both insertion loss and back reflectance

8.7   Explain how physical contact depends on connector endface geometry to include the Telcordia GR-326 three key parameters for optimal fiber contact:

  8.7.1   Radius of curvature

  8.7.2   Apex offset

  8.7.3   Fiber undercut and protrusion

8.8   Describe how and where pigtails are used in fiber cabling

8.9   Summarize connector termination methods and tools

8.10   Compare the differences between field polishing, factory polishing, and no-epoxy/no-polish connector styles

8.11   Describe how to properly perform a connector endface cleaning and visual inspection in accordance with ANSI/TIA-455-57B Preparation and Examination of Optical Fiber Endface for Testing Purposes

8.12   Explain how to guarantee insertion loss and return loss performance in accordance with the IEC 61300-3-35 the global common set of requirements for fiber optic connector endface quality

8.13   Identify both multimode and single-mode connector strain relief, connector plug body, and adapter housing following ANSI/TIA-568-C.3 section 5.2.3

9   Sources

9.1   Describe the two primary types of light sources including the light emitting diode (LED) and semiconductor laser (also called a laser diode)

9.2   Explain the basic concept, operation and address launch conditions of a LED light source

9.3   Discuss the spontaneous emission process used by LEDs to generate light

9.4   Outline the differences between the surface-emitting and the edge-emitting LEDs, which are commonly used in fiber optic communication systems

9.5   Explain the basic concept and operation of a laser diode light source

9.6   Discuss the stimulated emission process used by lasers to generate light

9.7   List the differences between the Fabry-Pérot (FP), distributed feedback (DFB), and vertical-cavity surface-emitting laser (VCSEL), which are commonly used in fiber optic communication systems

9.8   Recall the typical operational wavelengths for communication systems

9.9   Compare the performance characteristics of the LED and laser light sources to include:

  9.9.1   Output pattern (sometimes referred to as spot size)

  9.9.2   Source spectral width

9.9.3   Source output power

9.9.4   Source modulation speed

9.10   Compare the transmitter performance characteristics of the LED and laser light sources on a typical specification sheet to include:

9.10.1   Operating conditions

9.10.2   Electrical characteristics

9.10.3   Optical characteristics

9.10.4   Institute of Electrical and Electronics Engineers (IEEE) 802.3 Ethernet applications

9.11   Identify standards and federal regulations that classify the light sources used in fiber optic transmitters

9.12   Explain the differences between an overfilled launch condition and a restricted mode launch

10   Detectors and Receivers

10.1   Explain the basic concept and operation of a PN photodiode when used in an electrical circuit

10.2   Explain the use for PIN photodiodes and theory of operation

10.3   Describe the action of an avalanche photodiode (APD)

10.4   Compare the factors in photodiode performance characteristics including:

10.4.1   Responsivity

10.4.2   Quantum efficiency

10.4.3   Switching speed

10.5   Discuss how fiber optic receivers are typically packaged with the transmitter and how together, the receiver and transmitter form a transceiver

10.6   Examine a block diagram of a typical receiver and explain the function of the following:

10.6.1   Electrical subassembly

10.6.2   Optical subassembly

10.6.3   Receptacle

10.7   Describe the two key characteristics of a fiber optic receiver:

10.7.1   Dynamic Range

10.7.2   Wavelength

10.8   Describe the performance characteristics of a fiber optic receiver to include:

10.8.1   Recommended operating conditions

10.8.2   Electrical characteristics

10.8.3   Optical characteristics

10.8.4   IEEE 802.3 Ethernet applications

11   Passive Components and Multiplxers

11.1.   Discuss the different passive devices and the common parameters of each device:

11.1.1   Optical fiber and connector types

11.1.2   Center wavelength and bandwidth

11.1.3   Insertion loss

11.1.4   Excess loss

11.1.5   Polarization-dependent loss (PDL)

11.1.6   Return loss

11.1.7   Crosstalk in an optical device

11.1.8   Uniformity

11.1.9   Power handling and operating temperature

11.2   Explain how optical splitters work and describe the technologies used to include:

11.2.1   Tee splitter

11.2.2   Reflective and transmissive star splitters

11.3   Compare the different types of optical switches that open or close an optical circuit

11.4   Explain that an optical attenuator is a passive device used to reduce an optical signal's power level

11.5   Explain that an optical isolator comprises elements that only permit the forward transmission of light

11.6   Explain how wavelength division multiplexing (WDM) combines different optical wavelengths from two or more optical fibers into just one optical fiber

11.7   Explain the difference between coarse wavelength division multiplexing (CWDM) and dense wavelength division multiplexing (DWDM)

11.8   Compare the three different techniques with which to passively amplify an optical signal:

11.8.1   Erbium doped fiber amplifiers

11.8.2   Semiconductor optical amplifiers

11.8.3   Raman fiber amplifiers

11.9   Point out that an optical filter is a device that selectively permits transmission or blocks a range of wavelengths

12   Passive Optical Networks (PON)

12.1   Define the passive and active individual optical network categories

12.2   Explain that the fiber to the X (FTTX) is used to describe any optical fiber network that replaces all or part of a copper network

12.3   Discuss the major outside plant components for a fiber to the X (FTTX) passive optical network (PON) following IEEE 802.3, ITU-T G.983, ITU-T G.984, and ITU-T G.987

12.4   Compare the fundamentals of a passive optical network (PON) including fiber-to-the-home (FTTH), fiber-to-the-building (FTTB), fiber-to-the-curb (FTTC), and fiber-to-the-node (FTTN)

13   Cable Installation and Hardware

13.1   Define the physical and tensile strength requirements for optical fiber cables recognized in ANSI/TIA-568-C.3, section 4.3 to include:

13.1.1   Inside plant cables

13.1.2   Indoor-outdoor cables

13.1.3   Outside plant cable

13.1.4   Drop cables

13.2   Compare the bend radius and pull strength tensile ratings of the four common optical fiber cables recognized in ANSI/TIA-568-C.3, section 4.3

13.3   Identify some the hardware commonly used in fiber optic installation to include:

13.3.1   Pulling grips, pulling tape and pulling eyes

13.3.2   Pull boxes

13.3.3   Splice enclosures

13.3.4   Patch panels

13.4   Compare the variety of installation methods used to install a fiber optic cable such as:

13.4.1   Tray and duct

13.4.2   Conduit

13.4.3   Direct burial

13.4.4   Aerial

13.4.5   Blown optical fiber (BOF)

13.5   Describe the National Electrical Code (NEC) Article 770 and Article 250 requirements on fiber optic cables and their installation within buildings

13.5.1   Fire resistance

13.5.2   Grounding

13.6   Discuss the documentation and labeling requirements in order to follow a consistent and easily readable format as described in ANSI/TIA-606-B

13.7   Describe hardware management

14   Fiber Optic System Consuderations

14.1   List the considerations for a basic fiber optic system design

14.2   Identify the seven different performance areas within a system and evaluate performance of optical fiber to copper in the areas of:

14.2.1   Bandwidth

14.2.2   Attenuation

14.2.3   Electromagnetic immunity

14.2.4   Size

14.2.5   Weight

14.2.6   Security

14.2.7   Safety

14.3   Describe the performance of a multimode fiber optic link using the following sections of the ANSI/TIA-568-C.3

14.3.1   Section 4.2 cable transmission performance

14.3.2   Section 5.3 optical fiber splice

14.3.3   Annex A (Normative) optical fiber connector performance specifications

14.4   Explain how to prepare a multimode optical link power budget as defined in IEEE 802.3 definition 1.4.217

14.5   Calculate a multimode optical link power budget

14.6   Analyze the performance of a single-mode fiber optic link using ANSI/TIA-568-C.3, ANSI/TIA-758, and Telcordia GR-326

14.7   Explain how to prepare a single-mode optical link power budget as defined in IEEE 802.3 definition 1.4.217

14.8   Calculate a single-mode optical link power budget

15   Test Equipment and Link//Cable Testing

15.1   Compare and contrast the functional use of the following pieces of test equipment:

15.1.1   Continuity tester

15.1.2   Visual fault locator (VFL)

15.1.3   Fiber identifier

15.1.4   Optical return loss test set

15.1.5   Optical loss test set (OLTS)

15.2   Explain the proper use of the following pieces of test equipment:

15.2.1   Continuity tester

15.2.2   VFL

15.2.3   Fiber identifier

15.2.4   Optical return loss test set

15.2.5   OLTS

15.3   Compare the difference between an optical fiber patch cord and a measurement quality test jumper (MQJ)

15.4   Describe the use of a mandrel wrap or mode filter on both a multimode and single-mode source MQJ

15.6   Explain the ANSI/TIA-526-14-B optical power loss measurements of installed multi-mode fiber cable plant procedures to include:

15.6.1   Method A: Two-jumper reference

15.6.2   Method B: One-jumper reference

15.6.3   Method C: Three-jumper reference

15.7   Describe the proper setup and cable preparation for Optical Time Domain Reflectometer (OTDR) testing to include:

15.7.1   Measuring fiber length

15.7.2   Evaluating connectors and splices

15.7.3   Locating faults

15.7.4   Rayleigh scattering

15.7.5   Fresnel reflections

15.7.6   Pulse suppressor (launch fiber)

16   Troubleshooting and Restoration

16.1   Describe how to test the fiber optic cable plant using an OLTS

16.2   Describe how to test a patch cord using an OLTS

16.3   Describe how to perform OTDR unit length testing

16.4   Explain OTDR connector and splice evaluation techniques

16.5   Explain OTDR fault location techniques

16.6   Explain the purpose of acceptance testing documentation

---

ETA CONTACT INFORMATION

For more information about the ETA and its certification program, see the following information:

Address and Contact Numbers

ETA, International5 Depot StreetGreencastle, Indiana 46135Phone: (800) 288-3824 or (765) 653-8262 Fax: (765) 653-4287

Mail and Web

eta@eta-i.org

www.eta-i.org

Glossary

Numbers

1G (first generation)

Refers to the first generation of wireless-telephone technology based on analog (in contrast to digital) transmission.

2B+D

Shortcut for describing basic ISDN service (2B+D = two bearer channels and one data channel, which indicates two 64Kbps channels and one 16Kbps channel used for link management).

2G (second generation)

Refers to second-generation wireless telephone technology based on digital encryption of voice. 2G technology introduced data services starting with SMS texting.

3G (third generation)

Refers to the third generation of standards to support higher speeds for mobile telephony and broadband services.

4B/5B

Signal encoding method used in 100Base-TX/FX Fast Ethernet and FDDI standards. Four-bit binary values are encoded into 5-bit symbols.

4G (fourth generation)

Refers to the fourth generation of standards to support higher mobile telephony and broadband services.

5-4-3 Rule

A rule that mandates that between any two nodes on the network, there can only be a maximum of five segments, connected through four repeaters, or concentrators, and only three of the five segments may contain user connections.

6-around-1

A configuration used to test alien crosstalk in the lab whereby six disturber cables completely surround and are in direct contact with a central disturbed cable. Required testing for all CAT-6A cables by ANSI/TIA-568-C.2.

8B/10B

Signal encoding method used by the 1000Base-X Gigabit Ethernet standards.

8B/6T

Signal encoding method used in 100Base-T4 Fast Ethernet standard.

50-pin connector

Commonly referred to as a telco, CHAMP, or blue ribbon connector. Commonly found on telephone switches, 66-blocks, 110-blocks, and 10Base-T Ethernet hubs and used as an alternate twisted-pair segment connection method. The 50-pin connector connects to 25-pair cables, which are frequently used in telephone wiring systems and typically meet Category 3. Some manufacturers also make Category 5e–rated cables and connectors.

66-type connecting block

Connecting block used by voice-grade telephone installations to terminate twisted pairs. Not recommended for LAN use.

110-block

A connecting block that is designed to accommodate higher densities of connectors and to support higher-frequency applications. The 110-blocks are found on patch panels and cross-connect blocks for data applications.

A

abrasion mark

A flaw on an optical surface usually caused by an improperly polished termination end.

absorption

The loss of power (signal) in an optical fiber resulting from conversion of optical power (specific wavelengths of light energy into heat. Caused principally by impurities, such as water, ions of copper or chromium (transition metals), and hydroxyl ions, and by exposure to nuclear radiation. Expressed in dB/km (decibels per kilometer). Absorption and scattering are the main causes of attenuation (loss of signal) of an optical waveguide during transmission through optical fiber.

abstract syntax notation (ASN) 1

Used to describe the language interface standards for interconnection of operating systems, network elements, workstations, and alarm functions.

acceptance angle

With respect to optical-fiber cable, the angle within which light can enter an optical fiber core of a given numerical aperture and still reflect off the boundary layer between the core and the cladding. Light entering at the acceptance angle will be guided along the core rather than reflected off the surface or lost through the cladding. Often expressed as the half angle of the cone and measured from the axis. Generally measured as numerical aperture (NA); it is equal to the arcsine. The acceptance angle is also known as the cone of acceptance, or acceptance cone. *See also* numerical aperture.

acceptance cone

The cross section of an optical fiber is circular; the light waves accepted by the core are expressed as a cone. The larger the acceptance cone, the larger the numerical aperture; this means that the fiber is able to accept and propagate more light.

acceptance pattern

The amount of light transmitted from a fiber represented as a curve over a range of launch angles. *See also* acceptance angle.

access coupler

An optical device to insert or withdraw a signal from a fiber from between two ends. Many couplers require connectors on either end, and for many applications they must be APC (angled physical contact) connectors. The most popular access coupler is made by the fused biconic taper process, wherein two fibers are heated to the softening point and stretched so that the mode fields are brought into intimate contact, thus allowing a controlled portion of light to move from one core to the other.

access method

Rules by which a network peripheral accesses the network medium to transmit data on the network. All network technologies use some type of access method; common approaches include Carrier Sense Multiple Access/Collision Detect (CSMA/CD), token passing, and demand priority.

acknowledgment (ACK)

A message confirming that a data packet was received.

acrylate

A common type of optical fiber coating material that is easy to remove with hand tools.

active branching device

Converts an optical input into two or more optical outputs without gain or regeneration.

active coupler

A device similar to a repeater that includes a receiver and one or more transmitters. The idea is to regenerate the input signal and then send them on. These are used in optical fiber networks.

active laser medium

Lasers are defined by their medium; laser mediums such as gas, (CO2, helium, neon) crystal (ruby) semiconductors, and liquids are used. Almost all lasers create coherent light on the basis of a medium being activated electronically. The stimulation can be electronic or even more vigorous, such as exciting molecular transitions from higher to lower energy states, which results in the emissions of coherent light.

active monitor

In Token Ring networks, the active monitor is a designated machine and procedure that prevents data frames from roaming the ring unchecked. If a Token Ring frame passes the active monitor too many times, it is removed from the ring. The active monitor also ensures that a token is always circulating the ring.

active network

Refers to a network link that is transmitting and receiving data using electronic devices over a cabling infrastructure.

active splicing

A process performed with an alignment device, using the light in the core of one fiber to measure the transmittance to the other. Ensures optimal alignment before splicing is completed. The active splicing device allows fusion splicing to perform much better with respect to insertion losses when compared to most connectors and splicing methods. A splicing technician skilled at the use of an active splicing device can reliably splice with an upper limit of 0.03dB of loss.

ad hoc RF network

An RF network that exists only for the duration of the communication. Ad hoc networks are continually set up and torn down as needed.

adapter

With respect to optical fiber, a passive device used to join two connectors and fiber cores together. The adapter is defined by connector type, such as SC, FC, ST, LC, MT-RJ, and FDDI. Hybrid adapters can be used to join dissimilar connectors together, such as SC to FC. The adapter's key element is a "split sleeve," preferably made from zirconia and having a specific resistance force to insertion and withdrawal of a ferrule. This resistance, typically between 4 and 7 grams, ensures axial alignment of the cores.

address

An identifier that uniquely identifies nodes or network segments on a network.

address resolution protocol

A telecommunications protocol used for resolution of network layer addresses into link layer addresses, a critical function in multiple-access networks. For example, this is used to convert an IP address to an Ethernet address.

adjustable attenuator

An attenuator in which the level of attenuation is varied with an internal adjustment. Also known as a variable attenuator.

administration

With respect to structured cabling, the procedures and standards used to accurately keep track of the various circuits and connections, as well as records pertaining to them. The ANSI/TIA/EIA-606 Administration Standard for Telecommunications Infrastructure of Commercial Buildings defines specifications for this purpose.

Advanced Intelligent Network (AIN)

Developed by Bell Communications Research, a telephone network architecture that separates the service logic from the switching equipment. This allows the rapid deployment of new services with major equipment upgrades.

aerial cable

Telecommunications cable installed on aerial supporting structures such as poles, towers, sides of buildings, and other structures.

air polishing

Polishing a connector tip in a figure-eight motion without using a backing for the polishing film.

Alco Wipes

A popular brand of medicated wipes premoistened with alcohol.

alien crosstalk (AXT)

The crosstalk noise experienced by two or more adjacent twisted pair cables; measured in the lab by performing a 6-around-one test where six disturber cables impose crosstalk on the cable in the center (disturbed cable).

alignment sleeve

See mating sleeve.

all-dielectric self-supporting cable (ADSS)

An optical fiber cable with dielectric strength elements that can span long distances between supports without the aid of other strength elements.

alternating current (AC)

An electric current that cyclically reverses polarity.

American Standard Code for Information Interchange (ASCII)

A means of encoding information. ASCII is the method used by Microsoft to encode characters in text files in their operating systems.

American Wire Gauge (AWG)

Standard measuring gauge for nonferrous conductors (i.e., noniron and nonsteel). Gauge is a measure of the diameter of the conductor. The higher the AWG number, the smaller the diameter of the wire. See Chapter 1 for more information.

ampere

A unit of measure of electrical current.

amplifier

Any device that intensifies a signal without distorting the shape of the wave. In fiber optics, a device used to increase the power of an optical signal.

amplitude

The difference between high and low points of a wavelength cycle. The greater the amplitude, the stronger the signal.

amplitude modulation (AM)

A method of signal transmission in which the amplitude of the carrier is varied in accordance with the signal. With respect to optical fiber cabling, the modulation is done by varying the amplitude of a light wave, common in analog/RF applications.

analog

A signal that varies continuously through time in response to an input. A mercury thermometer, which gives a variable range of temperature readings, is an example of an analog instrument. Analog electrical signals are measured in hertz (Hz). Analog is the opposite of digital.

analog signal

An electrical signal that varies continuously (is infinitely variable within a specified range) without having discrete values (as opposed to a digital signal).

analog-to-digital (A/D) converter

A device used to convert analog signals to digital signals for storage or transmission.

ANDing

To determine whether a destination host is local or remote, a computer will perform a simple mathematical computation referred to as an AND operation.

angle of incidence

With respect to fiber optics, the angle of a ray of light striking a surface or boundary as measured from a line drawn perpendicular to the surface. Also called incident angle.

angle of reflection

With respect to fiber optics, the angle formed between the normal and a reflected beam. The angle of reflection equals the angle of incidence.

angle of refraction

With respect to fiber optics, the angle formed between the normal (a line drawn perpendicular to the interface) and a refracted beam.

angled end

An optical fiber whose end is polished to an angle to reduce reflectance.

angled physical contact (APC) connector

A single-mode optical-fiber connector whose angled end face helps to ensure low mated reflectance and low unmated reflectance. The ferrule is polished at an angle to ensure physical contact with the ferrule of another APC connector.

angular misalignment loss

The loss of optical power caused by deviation from optimum alignment of fiber-to-fiber. Loss at a connector due to fiber angles being misaligned because the fiber ends meet at an angle.

antireflection (AR)

Coating used on optical fiber cable to reduce light reflection.

apex

The highest feature on the end face of the optical fiber end.

apex offset

The displacement between the apex of the end face and the center of the optical fiber core.

AppleTalk

Apple Computer's networking protocol and networking scheme, integrated into most Apple system software, that allows Apple computing systems to participate in peer-to-peer computer networks and to access the services of AppleTalk servers. AppleTalk operates over Ethernet (EtherTalk), Token Ring (TokenTalk), and FDDI (FDDITalk) networks. *See also* LocalTalk.

application

(1) A program running on a computer. (2) A system, the transmission method of which is supported by telecommunications cabling, such as 100Base-TX Ethernet, or digital voice.

aramid strength member

The generic name for Kevlar® or Twaron® found in fiber-optic cables. A yarn used in fiber-optic cable that provides additional tensile strength, resistance to bending, and support to the fiber bundle. It is not used for data transmission.

ARCnet (Attached Resource Computer network)

Developed by Datapoint, a relatively low-speed form of LAN data link technology (2.5Mbps, in which all systems are attached to a common coaxial cable or an active or passive hub). ARCnet uses a token-bus form of medium access control; only the system that has the token can transmit.

armoring

Provides additional protection for cables against severe, usually outdoor, environments. Usually consists of plastic-coated steel corrugated for flexibility; *see also* interlock armor.

ARP (Address Resolution Protocol)

The method for finding a host's Link layer (hardware) address when only its Internet layer or some other Network layer address is known.

ARP table

A table used by the ARP protocol on TCP/IP-based network nodes that contains known TCP/IP addresses and their associated MAC (media access control) addresses. The table is cached in memory so that ARP lookups do not have to be performed for frequently accessed TCP/IP and MAC addresses.

array connector

A single ferrule connector such as the MPO that contains multiple optical fibers arranged in a row or in rows and columns.

Asymmetric Digital Subscriber Line (ADSL)

Sometimes called Universal ADSL, G.Lite, or simply DSL, ADSL is a digital communications method that allows high-speed connections between a central office (CO) and telephone subscriber over a regular pair of phone wires. It uses different speeds for uploading and downloading (hence the Asymmetric in ADSL) and is most often used for Internet connections to homes or businesses.

asynchronous

Transmission where sending and receiving devices are not synchronized (without a clock signal). Data must carry markers to indicate data division.

asynchronous transfer mode (ATM)

In networking terms, asynchronous transfer mode is a connection-oriented networking technology that uses a form of very fast packet switching in which data is carried in fixed-length units. These fixed-length units are called cells; each cell is 53 bytes in length, with 5 bytes used as a header in each cell. Because the cell size does not change, the cells can be switched very quickly. ATM networks can transfer data at extremely high speeds. ATM employs mechanisms that can be used to set up virtual circuits between users, in which a pair of users appear to have a dedicated circuit between them. ATM is defined in specifications from the ITU and Broadband Forum. For more information, see the Broadband Forum's website at www.broadband-forum.org.

attachment unit interface (AUI) port

A 15-pin connector found on older network interface cards (NICs). This port allowed you to connect the NIC to different media types by using an external transceiver. The cable that connected to this port when used with older 10Base5 media was known as a transceiver cable or a drop cable.

attenuation

A general term indicating a decrease in power (loss of signal) from one point to another. This loss can be a loss of electrical signal or light strength. In optical fibers, it is measured in decibels per kilometer (dB/km) at a specified wavelength. The loss is measured as a ratio of input power to output power. Attenuation is caused by poor-quality connections, defects in the cable, and loss due to heat. The lower the attenuation value, the better. Attenuation is the opposite of gain. See Chapter 1 for additional information on attenuation and the use of decibels.

attenuation-limited operation

In a fiber-optic link, the condition when operation is limited by the power of the received signal.

attenuation-to-crosstalk ratio (ACR)

A copper cabling measurement, the ratio between attenuation and near-end crosstalk, measured in decibels, at a given frequency. Although it is not a requirement of ANSI/TIA-568-C.2, it is used by every manufacturer as a figure of merit in denoting the signal-to-noise ratio.

attenuation-to-crosstalk ratio, far end (ACRF)

The ratio between attenuation and far-end crosstalk, measured in decibels, at a given frequency; a requirement of ANSI/TIA-568-C.2 for twisted-pair cables; formerly referred to as equal-level far-end crosstalk (ELFEXT) in ANSI/TIA/EIA-568-B.2.

attenuator

A passive device that intentionally reduces the strength of a signal by inducing loss.

audio

Used to describe the range of frequencies within range of human hearing; approximately 20 to 20,000Hz.

auxiliary AC power

A standard 110V AC power outlet found in an equipment area for operation of test equipment or computer equipment.

avalanche photodiode (APD)

With respect to optical fiber equipment, a specialized diode designed to use the avalanche multiplication of photocurrent. The photodiode multiplies the effect of the photons it absorbs, acting as an amplifier.

average picture level (APL)

A video parameter that indicates the average level of video signal, usually relative to blank and a white level.

average power

The energy per pulse, measured in joules, times the pulse repetition rate, measured in hertz (Hz). This product is expressed as watts.

average wavelength (l)

The average of the two wavelengths for which the peak optical power has dropped to half.

axial ray

In fiber-optic transmissions, a light ray that travels along the axis of a fiber-optic filament.

B

backbone

A cable connection between telecommunications or wiring closets, floor distribution terminals, entrance facilities, and equipment rooms either within or between buildings. This cable can service voice communications or data communications. In star-topology data networks, the backbone cable interconnects hubs and similar devices, as opposed to cables running between hub and station. In a bus topology data network, it is the bus cable. Backbone is also called riser cable, vertical cable, or trunk cable.

backbone wiring (or backbone cabling)

The physical/electrical interconnections between telecommunications rooms and equipment rooms. *See also* backbone.

backscatter

Usually a very small portion of an overall optical signal, backscatter occurs when a portion of scattered light returns to the input end of the fiber; the scattering of light in the direction opposite to its original propagation. Light that propagates back toward the transmitter. Also known as back reflection or backscattering.

backscatter coefficient

Also called Rayleigh backscatter coefficient, a value in dB used by an OTDR to calculate reflectance. The optical fiber manufacturer specifies backscatter coefficient.

bait

The mold or form on which silicon dioxide soot is deposited to create the optical fiber preform.

balance

An indication of signal voltage equality and phase polarity on a conductor pair. Perfect balance occurs when the signals across a twisted-pair cable are equal in magnitude and opposite in phase with respect to ground.

balanced cable

A cable that has pairs made up of two identical conductors that carry voltages of opposite polarities and equal magnitude with respect to ground. The conductors are twisted to maintain balance over a distance.

balanced coupler

A coupler whose output has balanced splits; for example, one by two is 50/50, or one by four is 25/25/25/25.

balanced signal transmission

Two voltages, commonly referred to as tip and ring, equal and opposite in phase with respect to each other across the conductors of a twisted-pair cable.

balun

A device that is generally used to connect balanced twisted-pair cabling with unbalanced coaxial cabling. The balun is an impedance-matching transformer that converts the impedance of one transmission media to the impedance of another transmission media. For example, a balun would be required to connect 100 ohm UTP to 120 ohm STP. Balun is short for balanced/unbalanced.

bandpass

A range of wavelengths over which a component will meet specifications.

bandwidth

Indicates the transmission capacity of media. For copper cables, bandwidth is defined using signal frequency and specified in hertz (Hz). For optical fiber, wavelength in nanometers (nm) defines bandwidth. Also refers to the amount of data that can be sent through a given channel and is measured in bits per second.

bandwidth-limited operation

Systems can be limited by power output or bandwidth; bandwidth-limited operation is when the total system bandwidth is the limiting factor (as opposed to signal amplitude).

barrier layer

A layer of glass deposited on the optical core to prevent diffusion of impurities into the core.

baseband

A method of communication in which the entire bandwidth of the transmission medium is used to transmit a single digital signal. The signal is driven directly onto the transmission medium without modulation of any kind. Baseband uses the entire bandwidth of the carrier, whereas broadband uses only part of the bandwidth. Baseband is simpler, cheaper, and less sophisticated than broadband.

basic rate interface (BRI)

As defined by ISDN, BRI consists of two 64Kbps B-channels used for data and one 16Kbps D-channel (used primarily for signaling). Thus, a basic rate user can have up to 128Kbps service.

battery distribution fuse bay (BDFB)

A type of DC "power patch panel" for telecommunication equipment where power feeder cables are connected to a box of fused connections. Distribution cables run from this device to the equipment.

baud

The number of signal level transitions per second. Commonly confused with bits per second, the baud rate does not necessarily transmit an equal number of bits per second. In some encoding schemes, baud will equal bits per second, but in others it will not. For example, a signal with four voltage levels may be used to transfer two bits of information for every baud.

beacon

A special frame in Token Ring systems indicating a serious problem with the ring, such as a break. Any station on the ring can detect a problem and begin beaconing.

beamsplitter

An optical device, such as a partially reflecting mirror, that splits a beam of light into two or more beams. Used in fiber optics for directional couplers.

beamwidth

For a round light beam, the diameter of a beam, measured across the width of the beam. Often specified in nanometers (nm) or millimeters (mm).

bearer channel (B-channel)

On an ISDN network, carries the data. Each bearer channel typically has a bandwidth of 64Kbps.

bel

Named for Alexander Graham Bell, this unit represents the logarithm of the ratio of two levels. See Chapter 1 for an explanation of bel and decibels.

bend insensitive

An optical fiber designed to reduce the amount of attenuation when it is bent.

bend-insensitive multimode optical fiber (BIMMF)

Multimode optical fiber designed to reduce the amount of attenuation when it is bent.

bend loss

A form of increased attenuation in a fiber where light is escaping from bent fiber. Bend loss is caused by bending a fiber around a restrictive curvature (a macrobend) or from minute distortions in the fiber (microbend). The attenuation may be permanent if fractures caused by the bend continue to affect transmission of the light signal.

bend radius (minimum)

The smallest bend a cable can withstand before the transmission is affected. UTP copper cabling usually has a bend radius that is four times the diameter of the cable; optical fiber is usually 10 times the diameter of the cable. Bending a cable any more than this can cause transmission problems or cable damage. Also referred to as cable bend radius.

biconic connector

A fiber-optic termination connector that is cone-shaped and designed for multiple connects and disconnects. The biconic connector was developed by AT&T but is not commonly used.

bidirectional attenuation testing

Refers to measuring the attenuation of a link using an OTDR in each direction.

bidirectional couplers

Couplers that operate in both directions and function in the same way in both directions.

bifurcated contact prongs

Contacts in a 66- or 110-block that are split in two so that the wire can be held better.

binary

A value that can be expressed only as one of two states. Binary values may be "1" or "0"; "high" or "low" voltage or energy; or "on" or "off" actions of a switch or light signal. Binary signals are used in digital data transmission.

binder

A tape or thread used to hold assembled cable components in place.

binder group

A group of 25 pairs of wires within a twisted-pair cable with more than 25 total pairs. The binder group has a strip of colored plastic around it to differentiate it from other binder groups in the cable.

biscuit jacks

A surface mount jack used for telephone wiring.

bistable optics

Optical devices with two stable transmission states.

bit

A binary digit, the smallest element of information in the binary number system, and thus the smallest piece of data possible in a digital communication system. A 1 or 0 of binary data.

bit error

An error in data transmission that occurs when a bit is a different value when it is received than the value at which it was transmitted.

bit error rate (BER)

In digital applications, a measure of data integrity. It is the ratio of bits received in error to bits originally sent or the ratio of incorrectly transmitted bits to total transmitted bits. BERs of one error per billion bits sent are typical.

bit error rate tester (BERT)

A device that tests the bit error rate across a circuit. One common device that does is this is called a T-BERD because it is designed to test T-1 and leased line error rates.

bit rate

The actual number of light pulses per second being transmitted through a fiber-optic link. Bit rates are usually measured in megabits per second (Mbps) or gigabits per second (Gbps).

bit stream

A continuous transfer of bits over some medium.

bit stuffing

A method of breaking up continuous strings of 0 bits by inserting a 1 bit. The 1 bit is removed at the receiver.

bit time

The number of transmission clock cycles that are used to represent one bit.

bits per second (bps)

The number of binary digits (bits) passing a given point in a transmission medium in one second.

BL

Blue. Refers to blue cable pair color in UTP twisted-pair cabling.

black body

A body or material that, in equilibrium, will absorb 100 percent of the energy incident upon it, meaning it will not reflect the energy in the same form. It will radiate nearly 100 percent of this energy, usually as heat and/or IR (infrared).

blade

A common term used to refer to a transmission card that can be inserted and plugged into an active telecommunications device.

blocking

In contrast to non-blocking, a blocking network refers to a situation where the uplink speed of a workgroup switch is less than the amount of data that is being uploaded onto the switch at a given time.

blown fiber

A method for installing optical fiber in which fibers are blown through preinstalled conduit or tube using air pressure to pull the fiber through the conduit.

BNC connector

Bayonet Neill-Concelman connector (Neill and Concelman were the inventors). A coaxial connector that uses a "bayonet"-style turn-and-lock mating method. Historically used with RG-58 or smaller coaxial cable or with 10Base-2 Ethernet thin coaxial cable.

bonding

(1) The method of permanently joining metallic parts to form an electrical contact that will ensure electrical continuity and the capacity to safely conduct any current likely to be imposed on it. (2) Grounding bars and straps used to bond equipment to the building ground. (3) Combining more than one ISDN B-channel using ISDN hardware.

bounded medium

A network medium that is used at the physical layer where the signal travels over a cable of some kind, as opposed to an unbounded medium such as wireless networking.

BR

Brown. Refers to brown cable pair color in UTP twisted-pair cabling.

braid

A group of textile or metallic filaments interwoven to form a tubular flexible structure that may be applied over one or more wires or flattened to form a strap. Designed to give a cable more flexibility or to provide grounding or shielding from EMI.

breakout cable

Multifiber cables composed of simplex interconnect cables where each fiber has additional protection by using additional jackets and strength elements, such as aramid yarn.

breakout kit

A collection of components used to add tight buffers, strength members, and jackets to individual fibers from a loose-tube buffer cable. Breakout kits are used to build up the outer diameter of fiber cable when connectors are being installed and designed to allow individual fibers to be terminated with standard connectors.

bridge

A network device, operating at the Data Link layer of the OSI model, that logically separates a single network into segments but lets the multiple segments appear to be one network to higher-layer protocols.

bridged tap

Multiple appearances of the same cable pair at several distribution points, usually made by splicing into a cable. Also known as parallel connections. Bridge taps were commonly used in coaxial cable networks and still appear in residential phone wiring installations. Their use is not allowed in any structured cabling environment.

broadband

A transmission facility that has the ability to handle a variety of signals using a wide range of channels simultaneously. Broadband transmission medium has a bandwidth sufficient to carry multiple voice, video, or data channels simultaneously. Each channel occupies (is modulated to) a different frequency bandwidth on the transmission medium and is demodulated to its original frequency at the receiving end. Channels are separated by "guard bands" (empty spaces) to ensure that each channel will not interfere with its neighboring channels. For example, this technique is used to provide many CATV channels on one coaxial cable.

broadband ISDN

An expansion of ISDN digital technology that allows it to compete with analog broadband systems using ATM or SDH.

broadcast

Communicating to more than one receiving device simultaneously.

brouter

A device that combines the functionality of a bridge and a router but can't be distinctly classified as either. Most routers on the market incorporate features of bridges into their feature set. Also called a hybrid router.

buffer (buffer coating)

A protective layer or tube applied to a fiber-optic strand. This layer of material, usually thermoplastic or acrylic polymer, is applied in addition to the optical-fiber coating, which provides protection from stress and handling. Fabrication techniques include tight or loose-tube buffering as well as multiple buffer layers. In tight-buffer constructions, the thermoplastic is extruded directly over the coated fiber. In loose-tube buffer constructions, the coated fiber "floats" within a buffer tube that is usually filled with a nonhygroscopic gel. See Chapter 10 for more information.

buffer tube

Used to provide protection and isolation for optical fiber cable. Usually hard plastic tubes, with an inside diameter several times that of a fiber, which holds one or more fibers.

building cable

Cable in a traditional cabling system that is inside a building and that will not withstand exposure to the elements. Also referred to as premises cable.

building distributor (BD)

An ISO/IEC 11801 term that describes a location where the building backbone cable terminates and where connections to the campus backbone cable may be made.

building entrance

The location in a building where a trunk cable between buildings is typically terminated and service is distributed through the building. Also the location where services enter the building from the phone company and antennas.

Building Industry Consulting Service International (BICSI)

A nonprofit association concerned with promoting correct methods for all aspects of the installation of communications wiring. More information can be found on their website at www.bicsi.org.

bundle (fiber)

A group of individual fibers packaged or manufactured together within a single jacket or tube. Also a group of buffered fibers distinguished from another group in the same cable core.

bundled cable

An assembly of two or more cables continuously bound together to form a single unit prior to installation.

bus topology

In general, a physical or logical layout of network devices in which all devices must share a common medium to transfer data.

busy token

A data frame header in transit.

bypass

The ability of a station to isolate itself optically from a network while maintaining the continuity of the cable plant.

byte

A binary "word" or group of eight bits.

C

c

The symbol for the speed of light in a vacuum.

C

The symbol for both capacitance and Celsius. In the case of Celsius, it is preceded by the degree symbol (°).

cable

(1) Copper: A group of insulated conductors enclosed within a common jacket. (2) Fiber: One or more optical fibers enclosed within a protective covering and material to provide strength.

cable assembly

A cable that has connectors installed on one or both ends. If connectors are attached to only one end of the cable, it is known as a pigtail. If its connectors are attached to both ends, it is known as a jumper. General use of these cable assemblies includes the interconnection of multimode and single-mode fiber optical cable systems and optical electronic equipment.

cable duct

A single pathway (typically a conduit, pipe, or tube) that contains cabling.

cable entrance conduits

Holes or pipes through the building foundation, walls, floors, or ceilings through which cables enter into the cable entrance facility (CEF).

cable entrance facility (CEF)

The location where the various telecommunications cables enter a building. Typically this location has some kind of framing structure (19-inch racks, or plywood sheet) with which the cables and their associated equipment can be organized.

cable management system

A system of tools, hold-downs, and apparatus used to precisely place cables and bundles of cables so that the entire cabling system is neat and orderly and that growth can be easily managed.

cable modem

A modem that transmits and receives signals usually through copper coaxial cable. Connects to your CATV connection and provides you with a 10Base-T connection for your computer. All of the cable modems attached to a cable TV company line communicate with a cable modem termination system (CMTS) at the local CATV office. Cable modems can receive and send signals to and from the CMTS only and not to other cable modems on the line. Some services have the upstream signals returned by telephone rather than cable, in which case the cable modem is known as a telco-return cable modem; these require the additional use of a phone line. The theoretical data rate of a CATV line is up to 27Mbps on the download path and about 2.5Mbps of bandwidth for upload. The overall speed of the Internet and the speed of the cable provider's access pipe to the Internet restrict the actual amount of throughput available to cable modem users. However, even at the lower end of the possible data rates, the throughput is many times faster than traditional modem connections to the Internet.

cable plant

Consists of all the copper and optical elements, including patch panels, patch cables, fiber connectors, and splices, between a transmitter and a receiver.

cable pulling lubricant

A lubricant used to reduce the friction against cables as they are pulled through conduits.

cable rearrangement facility (CRF)

A special splice cabinet used to vertically organize cables so that they can be spliced easily.

cable sheath

A covering over the core assembly that may include one or more metallic members, strength members, or jackets.

cable spool rack

A device used to hold a spool of cable and allow easy pulling of the cable off the spool.

cable tray

A shallow tray used to support and route cables through building spaces and above network racks.

cabling map

A map of how cabling is run through a building network.

campus

The buildings and grounds of a complex, such as a university, college, industrial park, or military establishment.

campus backbone

Cabling between buildings that share data and telecommunications facilities.

campus distributor (CD)

The ISO/IEC 11801 term for the main cross-connect; this is the distributor from which the campus backbone cable emanates.

capacitance

The ability of a dielectric material between conductors to store electricity when a difference of potential exists between the conductors. The unit of measurement is the farad, which is the capacitance value that will store a charge of one coulomb when a one-volt potential difference exists between the conductors. In AC, one farad is the capacitance value, which will permit one ampere of current when the voltage across the capacitor changes at a rate of one volt per second.

carrier

An electrical signal of a set frequency that can be modulated in order to carry voice, video, and/or data.

carrier detect (CD)

Equipment or a circuit that detects the presence of a carrier signal on a digital or analog network.

carrier sense

With Ethernet, a method of detecting the presence of signal activity on a common channel.

Carrier Sense Multiple Access/
Collision Avoidance (CSMA/CA)

A network media access method that sends a request to send (RTS) packet and waits to receive a clear to send (CTS) packet before sending. Once the CTS is received, the sender sends the packet of information. This method is in contrast to CSMA/CD, which merely checks to see if any other stations are currently using the media.

Carrier Sense Multiple Access/Collision Detect (CSMA/CD)

A network media access method employed by Ethernet. CSMA/CD network stations listen for traffic before transmitting. If two stations transmit simultaneously, a collision is detected and each station waits a brief (and random) amount of time before attempting to transmit again.

Category 1

Also called CAT-1. Unshielded twisted pair used for transmission of audio frequencies up to 1MHz. Used as speaker wire, doorbell wire, alarm cable, etc. Category 1 cable is not suitable for networking applications or digital voice applications. See Chapter 1 for more information. Not a recognized media by ANSI/TIA-568-C.

Category 2

Also called CAT-2. Unshielded twisted pair used for transmission at frequencies up to 4MHz. Used in analog and digital telephone applications. Category 2 cable is not suitable for networking applications. See Chapter 1 for more information. Not a recognized medium by ANSI/TIA-568.C.

Category 3

Also called CAT-3. Unshielded twisted pair with 100 ohm impedance and electrical characteristics supporting transmission at frequencies up to 16MHz. Used for 10Base-T Ethernet and digital voice applications. Recognized by the ANSI/TIA-568-C standard. See Chapters 1 and 7 for more information.

Category 4

Also called CAT-4. Unshielded twisted pair with 100 ohm impedance and electrical characteristics supporting transmission at frequencies up to 20MHz. Not used and not supported by the standards. See Chapter 1 for more information.

Category 5

Also called CAT-5. Unshielded twisted pair with 100 ohm impedance and electrical characteristics supporting transmission at frequencies up to 100MHz. Category 5 is not a recognized cable type for new installations by the ANSI/TIA-568-C standard, but its requirements are included in the standard to support legacy installations of Category 5 cable. See Chapters 1 and 7 for more information.

Category 5e

Also called CAT-5e or Enhanced CAT-5. Recognized in ANSI/TIA-568-C. Category 5e has improved specifications for NEXT, PSNEXT, Return Loss, ACRF (ELFEXT), PSACRF (PS-ELFEXT), and attenuation as compared to Category 5. Like Category 5, it consists of unshielded twisted pair with 100 ohm impedance and electrical characteristics supporting transmission at frequencies up to 100MHz. See Chapters 1 and 7 for more information.

Category 6

Also called CAT-6. Recognized in ANSI/TIA-568-C, Category 6 supports transmission at frequencies up to 250MHz over 100 ohm twisted pair. See Chapters 1 and 7 for more information.

Category 6A

Also called CAT-6A. Recognized in ANSI/TIA-568-C, Category 6A supports transmission at frequencies up to 500MHz over 100 ohm twisted pair. This is the only category cable required to meet alien crosstalk specifications. See Chapters 1 and 7 for more information.

CCIR

Consultative Committee for International Radio. Replaced by the International Telecommunications Union – Radio (ITU-R).

center wavelength (laser)

The nominal value central operating wavelength defined by a peak mode measurement where the effective optical power resides.

center wavelength (LED)

The average of two wavelengths measured at the half amplitude points of the power spectrum.

central member

The center component of a cable, which serves as an antibuckling element to resist temperature-induced stresses. Sometimes serves as a strength element. The central member is composed of steel, fiberglass, or glass-reinforced plastic.

central office (CO)

The telephone company building where subscribers' lines are joined to telephone company switching equipment that serves to interconnect those lines. Also known as an exchange center or head end. Some places call this a public exchange.

central office ground bus

Essentially a large ground bar used in a central office to provide a centralized grounding for all the equipment in that CO (or even just a floor in that CO).

centralized cabling

A cabling topology defined by ANSI/TIA-568-C whereby the cabling is connected from the main cross-connect in an equipment room directly to telecommunications outlets. This topology avoids the need for workgroup switches on individual building floors.

channel

The end-to-end transmission or communications path over which application-specific equipment is connected. Through multiplexing several channels, voice channels can be transmitted over an optical channel.

channel bank

Equipment that combines a number of voice and sometimes digital channels into a digital signal; in the case of a T-1 channel bank, it converts 24 separate voice channels into a single digital signal.

channel insertion loss

The static signal loss of a link between a transmitter and receiver for both copper and fiber systems. It includes the signal loss of the cable, connectors/splices and patch cords.

channel link

The section of a network link in between active equipment and network devices. This includes the main cable, connectors, and cross-connects.

channel service unit/digital service unit (CSU/DSU)

A hardware device that is similar to a modem that connects routers' or bridges' WAN interfaces (V.35, RS-232, etc.) to wide area network connections (Fractional-T1, T-1, Frame Relay, etc.). The device converts the data from the router or bridge to frames that can be used by the WAN. Some routers will have the CSU/DSU built directly into the router hardware, while other arrangements require a separate unit.

characteristic impedance

The impedance that an infinitely long transmission line would have at its input terminal. If a transmission line is terminated in its characteristic impedance, it will appear (electrically) to be infinitely long, thus minimizing signal reflections from the end of the line.

cheapernet

A nickname for thin Ethernet (thinnet) or 10Base-2 Ethernet systems.

cheater bracket

A bracket used for making cable installation easier.

chip

Short for microchip. A wafer of semiconducting material used in an integrated circuit (IC).

chromatic dispersion

The spreading of a particular light pulse because of the varying refraction rates of the different-colored wavelengths. Different wavelengths travel along an optical medium at different speeds. Wavelengths reach the end of the medium at different times, causing the light pulse to spread. This chromatic dispersion is expressed in picoseconds per kilometer per nanometer (of bandwidth). It is the sum of material and waveguide dispersions. *See also* waveguide dispersion, material dispersion.

churn

Cabling slang for the continual rearrangement of the various connections in a data connection frame. Office environments where network equipment and phones are frequently moved will experience a high churn rate.

circuit

A communications path between two pieces of associated equipment.

cladding

Name for the material (usually glass, sometimes plastic) that is put around the core of an optical fiber during manufacture. The cladding is not designed to carry light, but it has a slightly lower index of refraction than the core, which causes the transmitted light to travel down the core. The interface between the core and the cladding creates the mode field diameter, wherein the light is actually held reflectively captive within the core.

cladding mode

A mode of light that propagates through and is confined to the cladding.

Class A

(1) ISO/IEC 11801 channel designation using twisted-pair cabling rated to 100KHz. Used in voice and low-frequency applications. Comparable to TIA/EIA Category 1 cabling; not suitable for networking applications. (2) IP addresses that have a range of numbers from 1 through 127 in the first octet.

Class B

(1) ISO/IEC 11801 channel designation using twisted-pair cabling rated to 1MHz. Used in medium bit-rate applications. Comparable to TIA/EIA Category 2 cabling; not suitable for networking applications. (2) IP addresses that have a range of numbers from 128 through 191 in the first octet.

Class C

(1) ISO/IEC 11801 channel designation using twisted-pair cabling rated to 16MHz. Used in high bit-rate applications. Corresponds to TIA/EIA Category 3 cabling. (2) IP addresses that have a range of numbers from 192 through 223 in the first octet.

Class D

(1) ISO/IEC 11801 channel designation using twisted-pair cabling rated to 100MHz. Used in very high bit-rate applications. Corresponds to TIA/EIA Category 5 cabling. (2) IP addresses used for multicast applications that have a range of numbers from 224 through 239 in the first octet.

Class E

(1) ISO/IEC 11801 channel using twisted-pair cabling rated to 250MHz. Corresponds to TIA/EIA Category 6 cabling. (2) IP addresses used for experimental purposes that have a range of numbers from 240 through 255 in the first octet.

Class E$_A$

ISO/IEC 11801 proposed channel using twisted-pair cabling rated to 500MHz. Corresponds closely to the TIA/EIA Category 6A cabling standard, but there are differences in some internal performance requirements. (2) IP addresses used for experimental purposes that have a range of numbers from 240 through 255 in the first octet.

Class F

ISO/IEC 11801 channel using twisted-pair cabling rated to 600MHz.

Class F$_A$

ISO/IEC 11801 proposed channel using twisted-pair cabling rated to 1000MHz.

cleaving

The process of separating an unbuffered optical fiber by scoring the outside and pulling off the end. Cleaving creates a controlled fracture of the glass for the purpose of obtaining a fiber end that is flat, smooth, and perpendicular to the fiber axis. This is done prior to splicing or terminating the fiber.

clock

The timing signal used to control digital data transmission.

closet

An enclosed space for housing telecommunications and networking equipment, cable terminations, and cross-connect cabling. It contains the horizontal cross-connect where the backbone cable cross-connects with the horizontal cable. Called a telecommunications room by the TIA/EIA standards; sometimes referred to as a wiring closet.

CMG

A cable rated for general purpose by the NEC.

CMR

A cable rated for riser applications by the NEC.

coating

The first true protective layer of an optical fiber. The plastic-like coating surrounds the glass cladding of a fiber and put on a fiber immediately after the fiber is drawn to protect it from the environment. Do not confuse the coating with the buffer.

coaxial cable

Also called coax. Coaxial cable was invented in 1929 and was in common use by the phone company by the 1940s. Today it is commonly used for cable TV and by older Ethernet; twisted-pair cabling has become the desirable way to install Ethernet networks. It is called coaxial because it has a single conductor surrounded by insulation and then a layer of shielding (which is also a conductor) so the two conductors share a single axis; hence "co"-axial. The outer shielding serves as a second conductor and ground, and reduces the effects of EMI. Can be used at high bandwidths over long distances.

code division multiple access (CDMA)

In time division multiplexing (TDM), one pulse at a time is taken from several signals and combined into a single bit stream.

coefficient of thermal expansion

The measure of a material's change in size in response to temperature variation.

coherence

In light forms, characterized as a consistent, fixed relationship between points on the wave. In each case, there is an area perpendicular to the direction of the light's propagation in which the light may be highly coherent.

coherence length or time

The distance time over which a light form may be considered coherent. Influenced by a number of factors, including medium, interfaces, and launch condition. Time, all things being equal, is calculated by the coherence length divided by the phase velocity of light in the medium.

coherent communications

Where the light from a laser oscillator is mixed with the received signal, and the difference frequency is detected and amplified.

coherent light

Light in which photons have a fixed or predictable relationship; that is, all parameters are predictable and correlated at any point in time or space, particularly over an area in a lane perpendicular to the direction of propagation or over time at a particular point in space. Coherent light is typically emitted from lasers.

collimated

The expanding of light so that the rays are parallel.

collision

The network error condition that occurs when electrical signals from two or more devices sharing a common data transfer medium crash into one another. This commonly happens on Ethernet-type systems.

committed information rate (CIR)

A commitment from your service provider stating the minimum bandwidth you will get on a frame relay network averaged over time.

common mode transmission

A transmission scheme where voltages appear equal in magnitude and phase across a conductor pair with respect to ground. May also be referred to as longitudinal mode.

community antenna television (CATV)

More commonly known as cable TV, a broadband transmission facility that generally uses a 75 ohm coaxial cable to carry numerous frequency-divided TV channels simultaneously. CATV carries high-speed Internet service in many parts of the world.

compliance

A wiring device that meets all characteristics of a standard is said to be in compliance with that standard. For example, a data jack that meets all of the physical, electrical, and transmission standards for ANSI/TIA/EIA-568-B Category 5e is compliant with that standard.

composite cable

A cable that typically has at least two different types of transmitting media inside the same jacket; for example, UTP and fiber.

concatenation

The process of joining several fibers together end to end.

concatenation gamma

The coefficient used to scale bandwidth when several fibers are joined together.

concentrator

A multiport repeater or hub.

concentric

Sharing a common geometric center.

conductivity

The ability of a material to allow the flow of electrical current; the reciprocal of resistivity. Measured in "mhos" (the word ohm spelled backward).

conductor

A material or substance (usually copper wire) that offers low resistance (opposition) to the flow of electrical current.

conduit

A rigid or flexible metallic or nonmetallic raceway of circular cross section in which cables are housed for protection and to prevent burning cable from spreading flames or smoke in the event of a fire.

cone of acceptance

A cone-shaped region extending outward from an optical fiber core and defined by the core's *acceptance angle*.

Conference of European Postal and Telecommunications Administrations (CEPT)

A set of standards adopted by European and other countries, particularly defining interface standards for digital signals.

connecting block

A basic part of any telecommunications distribution system. Also called a terminal block, a punch-down block, a quick-connect block, and a cross-connect block, a connecting block is a plastic block containing metal wiring terminals to establish connections from one group of wires to another. Usually each wire can be connected to several other wires in a bus or common arrangement. There are several types of connecting blocks, such as 66 clip, BIX, Krone, or 110. A connecting block has insulation displacement connections (IDCs), which means you don't have to remove insulation from around the wire conductor before you punch it down or terminate it.

connectionless protocol

A communications protocol that does not create a virtual connection between sending and receiving stations.

connection-oriented protocol

A communications protocol that uses acknowledgments and responses to establish a virtual connection between sending and receiving stations.

connector

With respect to cabling, a device attached to the end of a cable, receiver, or light source that joins it with another cable, device, or fiber. A connector is a mechanical device used to align and join two conductors or fibers together to provide a means for attaching and decoupling it to a transmitter, receiver, or another fiber. Commonly used connectors include the RJ-11, RJ-45, BNC, FC, ST, LC, MT-RJ, FDDI, biconic, and SMA connectors.

connector-induced optical fiber loss

With respect to fiber optics, the part of connector insertion loss due to impurities or structural changes to the optical fiber caused by the termination within the connector.

connector plug

With respect to fiber optics, a device used to terminate an optical fiber.

connector receptacle

With respect to fiber optics, the fixed or stationary half of a connection that is mounted on a patch panel or bulkhead.

connector variation

With respect to fiber optics, the maximum value in decibels of the difference in insertion loss between mating optical connectors (e.g., with remating, temperature cycling, etc.). Also known as optical connector variation.

consolidation point (CP)

A location defined by the ANSI/TIA/EIA-568-B standard for interconnection between horizontal cables that extends from building pathways and horizontal cables that extend into work area pathways. Often an entry point into modular furniture for voice and data cables. ISO/IEC 11801 defines this as a transition point (TP).

consumables kit

Resupply material for splicing or terminating fiber optics.

continuity

An uninterrupted pathway for electrical or optical signals.

continuity tester

A tool that projects light from the LED or incandescent lamp into the core of the optical fiber.

controlled environmental vault (CEV)

A cable termination point in a below-ground vault whose humidity and temperature are controlled.

Copper Distributed Data Interface (CDDI)

A version of FDDI that uses copper wire media instead of fiber-optic cable and operates at 100Mbps. *See also* twisted-pair physical media dependent (TP-PMD).

cordage

Fiber-optic cable designed for use as a patch cord or jumper. Cordage may be either single-fiber (simplex) or double fiber (duplex).

core

The central part of a single optical fiber in which the light signal is transmitted. Common core sizes are 8.3 microns, 50 microns, and 62.5 microns. The core is surrounded by a cladding that has a higher refractive index that keeps the light inside the core. The core is typically made of glass or plastic.

core eccentricity

A measurement that indicates how far off the center of the core of an optical fiber is from the center of that fiber's cladding.

counter-rotating

An arrangement whereby two signal paths, one in each direction, exist in a ring topology.

coupler

With respect to optical fiber, a passive, multiport device that connects three or more fiber ends, dividing the input between several outputs or combining several inputs into one output.

coupling

The transfer of energy between two or more cables or components of a circuit. *See also* crosstalk.

coupling efficiency

How effective a coupling method is at delivering the required signal without loss.

coupling loss

The amount of signal loss that occurs at a connection because of the connection's design.

coupling ratio/loss

The ratio/loss of optical power from one output port to the total output power, expressed as a percent.

cover plate mounting bracket

A bracket within a rack that allows a plate to cover some type of hardware, such as a patch panel.

crimper, crimping, crimp-on

A device that is used to install a crimp-on connector. Crimping involves the act of using the crimping tool to install the connector.

critical angle

The smallest angle of incidence at which total internal reflection occurs; that is, at which light passing through a material of a higher refractive index will be reflected off the boundary with a material of a lower refractive index. At lower angles, the light is refracted through the cladding and lost. Due to the fact that the angle of reflection equals the angle of incidence, total internal reflection assures that the wave will be propagated down the length of the fiber. The angle is measured from a line perpendicular to the boundary between the two materials, known as normal.

cross-connect

A facility enabling the termination of cables as well as their interconnection or cross-connection with other cabling or equipment. Also known as a punch-down or distributor. In a copper-based system, jumper wires or patch cables are used to make connections. In a fiber-optic system, fiber-optic jumper cables are used.

cross-connection

A connection scheme between cabling runs, subsystems, and equipment using patch cords or jumpers that attach to connecting hardware at each end.

crossover

A conductor that connects to a different pin number at each end. *See also* crossover cable.

crossover cable

A twisted-pair patch cable wired in such a way as to route the transmit signals from one piece of equipment to the receive signals of another piece of equipment, and vice versa. Crossover cables are often used with 10Base-T or 100Base-TX Ethernet cards to connect two Ethernet devices together "back-to-back" or to connect two hubs together if the hubs do not have crossover or uplink ports. See Chapter 9 for more information on Ethernet crossover cables.

crosstalk

The coupling or transfer of unwanted signals from one pair within a cable to another pair. Crosstalk can be measured at the same (near) end or far end with respect to the signal source. Crosstalk is considered noise or interference and is expressed in decibels. Chapter 1 has an in-depth discussion of crosstalk.

crush impact

A test typically conducted using a press that is fitted with compression plates of a specified cross sectional area. The test sample is placed between the press plates and a specified force is applied to the test specimen. Cable performance is evaluated while the cable is under compression and/or after removal of load depending on the test standard specifications.

current

The flow of electrons in a conductor. *See also* alternating current (AC) and direct current (DC).

curvature loss

The macro-bending loss of signal in an optical fiber.

customer premises

The buildings, offices, and other structures under the control of an end user or customer.

cutback

A technique or method for measuring the optical attenuation or bandwidth in a fiber by measuring first from the end and then from a shorter length and comparing the difference. Usually one is at the full length of the fiber-optic cable and the other is within a few meters of the input.

cutoff wavelength

For a single-mode fiber, the wavelength above which the operation switches from multimode to single-mode propagation. This is the longest wavelength at which a single-mode fiber can transmit two modes. At shorter wavelengths the fiber fails to function as a single-mode fiber.

cut-through resistance

A material's ability to withstand mechanical pressure (such as a cutting blade or physical pressure) without damage.

cycles per second

The count of oscillations in a wave. One cycle per second equals a hertz.

cyclic redundancy check (CRC)

An error-checking technique used to ensure the accuracy of transmitting digital code over a communications channel after the data is transmitted. Transmitted messages are divided into predetermined lengths that are divided by a fixed divisor. The result of this calculation is appended onto the message and sent with it. At the receiving end, the computer performs this calculation again. If the value that arrived with the data does not match the value the receiver calculated, an error has occurred during transmission.

D

daisy chain

In telecommunications, a wiring method where each telephone jack in a building is wired in series from the previous jack. Daisy chaining is not the preferred wiring method, since a break in the wiring would disable all jacks "downstream" from the break. Attenuation and crosstalk are also higher in a daisy-chained cable. *See also* home-run cable.

dark current

The external current that, under specified biasing conditions, flows in a photo detector when there is no incident radiation.

dark fiber

An unused fiber; a fiber carrying no light. Common when extra fiber capacity is installed.

data communication equipment (DCE)

With respect to data transmission, the interface that is used by a modem to communicate with a computer.

data-grade

A term used for twisted-pair cable that is used in networks to carry data signals. Data-grade media has a higher frequency rating than voice-grade media used in telephone wiring does. Data-grade cable is considered Category 3 or higher cable.

data packet

The smallest unit of data sent over a network. A packet includes a header with addressing information, and the data itself.

data rate

The maximum rate (in bits per second or some multiple thereof) at which data is transmitted in a data transmission link. The data rate may or may not be equal to the baud rate.

data-terminal equipment (DTE)

(1) The interface that electronic equipment uses to communicate with a modem or other serial device. This port is often called the computer's RS-232 port or the serial port. (2) Any piece of equipment at which a communications path begins or ends.

datagram

A unit of data larger than or equal to a packet. Generally it is self-contained and its delivery is not guaranteed.

DB-9

Standard 9-pin connector used with Token Ring and serial connections.

DB-15

Standard 15-pin connector used with Ethernet transceiver cables.

DB-25

Standard 25-pin connector used with serial and parallel ports.

DC loop resistance

The total resistance of a conductor from the near end to the far end and back. For a single conductor, it is just the one-way measurement doubled. For a pair of conductors, it is the resistance from the near end to the far end on one conductor and from the far end to the near end on the other.

D-channel

Delta channel. On ISDN networks, the channel that carries control and signaling information at 16Kbps for BRI ISDN services or 64Kbps for PRI ISDN services.

decibel (dB)

A standard unit used to express a relative measurement of signal strength or to express gain or loss in optical or electrical power. A unit of measure of signal strength, usually the relation between a received signal and a standard signal source. The decibel scale is a logarithmic scale, expressed as the logarithmic ratio of the strength of a received signal to the strength of the originally transmitted signal. For example, every 3dB equals 50 percent of signal strength, so therefore a 6dB loss equals a loss of 75 percent of total signal strength. See Chapter 1 for more information.

degenerate waveguides

A set of waveguides having the same propagation constant for all specified frequencies.

delay skew

The difference in propagation delay between the fastest and slowest pair in a cable or cabling system. See Chapter 1 for more information on delay skew.

delta

In fiber optics, equal to the difference between the indexes of refraction of the core and the cladding divided by the index of the core.

demand priority

A network access method used by Hewlett-Packard's 100VG-AnyLAN. The hub arbitrates requests for network access received from stations and assigns access based on priority and traffic loads.

demarcation point ("dee-marc")

A point where the operational control or ownership changes, such as the point of interconnection between telephone company facilities and a user's building or residence.

demodulate

To retrieve a signal from a carrier and convert it into a usable form.

demultiplex

The process of separating channels that were previously joined using a multiplexer.

demultiplexer

A device that separates signals of different wavelengths for distribution to their proper receivers.

dense wavelength division multiplexing (DWDM)

A form of multiplexing that separates channels by transmitting them on different wavelengths at intervals of about 0.8nm (100GHz).

detector

(1) A device that detects the presence or absence of an optical signal and produces a coordinating electrical response signal. (2) An optoelectric transducer used in fiber optics to convert optical power to electrical current. In fiber optics, the detector is usually a photodiode.

detector noise limited operations

Occur when the detector is unable to make an intelligent decision on the presence or absence of a pulse due to the losses that render the amplitude of the pulse too small to be detected or due to excessive noise caused by the detector itself.

diameter mismatch loss

The loss of power that occurs when one fiber transmits to another and the transmitting fiber has a diameter greater than the receiving fiber. It can occur at any type of coupling where the fiber/coupling sizes are mismatched: fiber-to-fiber, fiber-to-device, fiber-to-detector, or source-to-fiber. Fiber-optic cables and connectors should closely match the size of fiber required by the equipment.

dichroic filter

Selectively transmits or reflects light according to selected wavelengths. Also referred to as dichromatic mirror.

dielectric

Material that does not conduct electricity, such as nonmetallic materials that are used for cable insulation and jackets. Optical fiber cables are made of dielectric material.

dielectric constant

The property of a dielectric material that determines the amount of electrostatic energy that can be stored by the material when a given voltage is applied to it. The ratio of the capacitance of a capacitor using the dielectric to the capacitance of an identical capacitor using a vacuum as a dielectric. Also called permittivity.

dielectric loss

The power dissipated in a dielectric material as the result of the friction produced by molecular motion when an alternating electric field is applied.

dielectric nonmetallic

Refers to materials used within a fiber-optic cable.

differential group delay (DGD)

The total difference in travel time between the two polarization states of light traveling through an optical fiber. This time is usually measured in picoseconds and can differ depending on specific conditions within an optical fiber and the polarization state of the light passing through it.

differential modal dispersion (DMD)

The test method used to determine the variation in arrival time of modes in a pulse of light in an optical fiber (typically for multimode fiber).

differential mode attenuation

A variation in attenuation in and among modes carried in an optical fiber.

differential mode transmission

A transmission scheme where voltages appear equal in magnitude and opposite in phase across a twisted-pair cable with respect to ground. Differential mode transmission may also be referred to as balanced mode. Twisted-pair cable used for Category 1 and above is considered differential mode transmission media or balanced mode cable.

diffraction

The deviation of a wave front from the path predicted by geometric optics when a wave front is restricted by an edge or an opening of an object. Diffraction is most significant when the aperture is equal to the order of the wavelength.

diffraction grating

An array of fine, parallel, equally spaced reflecting or transmitting lines that mutually enhance the effects of diffraction to concentrate the diffracted light in a few directions determined by the spacing of the lines and by the wavelength of the light.

diffuse infrared

Enables the use of infrared optical emissions without the need for line-of-sight between the transmitting and receiving communication entities.

digital

Refers to transmission, processing, and storage of data by representing the data in binary values (two states: on or off). On is represented by the number 1 and off by the number 0. Data transmitted or stored with digital technology is expressed as a string of 0s and 1s. Digital signals are used to communicate information between computers or computer-controlled hardware.

digital loop carrier

A carrier system used for pair gain and providing multiple next-generation digital services over traditional copper loop in local loop applications.

digital signal (DS)

A representation of a digital signal carrier in the TDM hierarchy. Each DS level is made up of multiple 64Kbps channels (generally thought of as the equivalent to a voice channel) known as DS-0 circuits. A DS-1 circuit (1.544Mbps) is made up of 24 individual DS-0 circuits. DS rates are specified by ANSI, CEPT, and the ITU.

digital signal cross-connect (DSX)

A piece of equipment that serves as a connection point for a particular digital signal rate. Each DSX equipment piece is rated for the DS circuit it services—for example, a DSX-1 is used for DS-1 signals, DSX-3 for DS-3 signals, and so on.

digital signal processor (DSP)

A device that manipulates or processes data that has been converted from analog to digital form.

digital subscriber line (DSL)

A technology for delivering high bandwidth to homes and businesses using standard telephone lines. Though many experts believed that standard copper phone cabling would never be able to support high data rates, local phone companies are deploying equipment that is capable of supporting up to theoretical rates of 8.4Mbps. Typical throughput downstream (from the provider to the customer) are rates from 256Kbps to 1.544Mbps.

DSL lines are capable of supporting voice and data simultaneously. There are many types, including HDSL (high bit-rate DSL) and VDSL (very high bit-rate DSL).

diode

A device that allows a current to move in only one direction. Some examples of diodes are light-emitting diodes (LEDs), laser diodes, and photodiodes.

direct burial

A method of placing a suitable cable directly in the ground.

direct current (DC)

An electric current that flows in one direction and does not reverse direction, unlike alternating current (AC). Direct current also means a current whose polarity never changes.

direct frequency modulation

Directly feeding a message into a voltage-controlled oscillator.

direct inside wire (DIW)

Twisted-pair wire used inside a building, usually two- or four-pair AWG 26.

directional coupler

A device that samples or tests data traveling in one direction only.

directivity

In a power tap, it is the sensitivity of forward-directed light relative to backward-directed light.

dispersion

A broadening or spreading of light along the propagation path due to one or more factors within the medium (such as optical fiber) through which the light is traveling. There are three major types of dispersion: modal, material, and waveguide. Modal dispersion is caused by differential optical path lengths in a multimode fiber. Material dispersion is caused by a delay of various wavelengths of light in a waveguide material. Waveguide dispersion is caused by light traveling in both the core and cladding materials in single-mode fibers and interfering with the transmission of the signal in the core. If dispersion becomes too great, individual signal components can overlap one another or degrade the quality of the optical signal. Dispersion is one of the most common factors limiting the amount of data that can be carried in optical fiber and the distance the signal can travel while still being usable; therefore, dispersion is one of the limits on bandwidth on fiber-optic cables. It is also called pulse spreading because dispersion causes a broadening of the input pulses along the length of the fiber.

dispersion flattened fiber

A single-mode optical fiber that has a low chromatic dispersion throughout the range between 1300nm and 1600nm.

dispersion limited operation

Describes cases where the dispersion of a pulse rather than loss of amplitude limits the distance an optical signal can be carried in the fiber. If this is the case, the receiving system may not be able to receive the signal.

dispersion shifted fiber

A single-mode fiber that has its zero dispersion wavelength at 1550nm, which corresponds to one of the fiber's points of low attenuation. Dispersion shifted fibers are made so that optimum attenuation and bandwidth are at 1550nm.

dispersion unshifted fiber

A single-mode optical-fiber cable that has its zero dispersion wavelength at 1300nm. Often called conventional or unshifted fiber.

distortion

Any undesired change in a waveform or signal.

distortion-limited operation

In fiber-optic cable, the limiting of performance because of the distortion of a signal.

distributed feedback (DFB) laser

A semiconductor laser specially designed to produce a narrow spectral output for long-distance fiber-optic communications.

distribution subsystem

A basic element of a structured distribution system. The distribution subsystem is responsible for terminating equipment and running the cables between equipment and cross-connects. Also called distribution frame subsystem.

distribution cable

An optical fiber cable used "behind-the-shelf" of optical fiber patch panels; typically composed of 900 micron tight-buffered optical fibers supported by aramid and/or glass-reinforced plastic (GRP).

disturbed pair

A pair in a UTP cable that has been disturbed by some source of noise during cable testing.

disturber

A UTP cable, or series of cables, that is disturbing a cable under test by some source of noise during cable testing.

DIX connector

Digital, Intel, Xerox (DIX) connector used to connect thinnet cables and networks.

DoD Networking Model

The Department of Defense's four-layer conceptual model describing how communications should take place between computer systems. From the bottom up, the four layers are network access, Internet, transport (host-to-host), and application.

dopants

Impurities that are deliberately introduced into the materials used to make optical fibers. Dopants are used to control the refractive index of the material for use in the core or the cladding.

download speed

See downstream.

downstream

A term used to describe the number of bits that travel from the Internet service provider to the person accessing broadband.

D-ring

An item of hardware, usually a metal ring shaped like the letter D. It may be used at the end of a leather or fabric strap, or it may be secured to a surface with a metal or fabric strap.

drain wire

An uninsulated wire in contact with a shielded cable's shield throughout its length. It is used for terminating the shield. If a drain wire is present, it should be terminated.

draw string

A small nylon cord inserted into a conduit when the conduit is installed; it assists with pulling cable through the duct.

drop

A single length of cable installed between two points.

dry nonpolish connector (DNP)

Optical fiber connector used for POF (plastic optical fiber).

DS-1

Digital Service level 1. Digital service that provides 24 separate 64Kbps digital channels.

DSX bay

A combination of the various pieces of DSX apparatus and its supporting mechanism, including its frame, rack, or other mounting devices.

DSX complexes

Any number of DSX lineups that are connected together to provide DSX functionality.

DSX lineup

Multiple contiguous DSX bays connected together to provide DSX functions.

D-type connector

A type of connector that connects computer peripherals. It contains rows of pins or sockets shaped like a sideways D. Common connectors are the DB-9 and DB-25.

DU connector

A fiber-optic connector developed by the Nippon Electric Group in Japan.

dual-attachment concentrator (DAC)

An FDDI concentrator that offers two attachments to the FDDI network that are capable of accommodating a dual (counter-rotating) ring.

dual-attachment station (DAS)

A term used with FDDI networks to denote a station that attaches to both the primary and secondary rings; this makes it capable of serving the dual (counter-rotating) ring. A dual-attachment station has two separate FDDI connectors.

dual ring

A pair of counter-rotating logical rings.

dual-tone multifrequency (DTMF)

The signal that a touch-tone phone generates when you press a key on it. Each key generates two separate tones, one in a high-frequency group of tones and one from a low-frequency group of tones.

dual-window fiber

An optical fiber cable manufactured to be used at two different wavelengths. Single-mode fiber cable that is usable at 1300nm and 1550nm is dual-window fiber. Multimode fiber cable is optimized for 850nm and 1300nm operations and is also dual-window fiber. Also known as double-window fiber.

duct

(1) A single enclosed raceway for wires or cable. (2) An enclosure in which air is moved.

duplex

(1) A link or circuit that can carry a signal in two directions for transmitting and receiving data. (2) An optical fiber cable or cord carrying two fibers.

duplex cable

A two-fiber cable suitable for duplex transmission. Usually two fiber strands surrounded by a common jacket.

duplex transmission

Data transmission in both directions, either one direction at a time (half duplex) or both directions simultaneously (full duplex).

duty cycle

With respect to a digital transmission, the product of a signal's repetition frequency and its duration.

dynamic loads

Loads, such as tension or pressure, that change over time, usually within a short period.

dynamic range

The difference between the maximum and minimum optical input power that an optical receiver can accept.

E

E-1

The European version of T-1 data circuits. Runs at 2.048Mbps.

E-3

The European version of T-3 data circuits. Runs at 34.368Mbps.

earth

A term for zero reference ground (not to mention the planet most of us live on).

edge-emitting LED

An LED that produces light through an etched opening in the edge of the LED.

effective modal bandwidth (EMB)

The bandwidth of a particular optical fiber cable, connections and splices, and transmitter combination. This can be measured in a complete channel.

effective modal bandwidth, calculated (EMBc)

The calculation of the EMB from the optical fiber's DMD and specific weighting functions of VCSELs. This is also known as CMB (calculated modal bandwidth).

effective modal bandwidth, minimum (min EMB)

The value of bandwidth used in the IEEE 802.3ae model that corresponds to the fiber specification.

EIA rack

A rack with a standard dimension per EIA. *See also* rack.

electromagnetic compatibility (EMC)

The ability of a system to minimize radiated emissions and maximize immunity from external noise sources.

electromagnetic field

The combined electric and magnetic field caused by electron motion in conductors.

electromagnetic immunity

Protection from the interfering or damaging effects of electromagnetic radiation such as radio waves or microwaves.

electromagnetic interference (EMI)

Electrical noise generated in copper conductors when electromagnetic fields induce currents. Copper cables, motors, machinery, and other equipment that uses electricity may generate EMI. Copper-based network cabling and equipment are susceptible to EMI and also emit EMI, which results in degradation of the signal. Fiber-optic cables are desirable in environments that have high EMI because they are not susceptible to the effects of EMI.

Electronic Industries Alliance (EIA)

An alliance of manufacturers and users that establishes standards and publishes test methodologies. The EIA (with the TIA and ANSI) helped to publish the ANSI/TIA/EIA-568-B cabling standard.

electro-optic

A term that refers to a variety of phenomena that occur when an electromagnetic wave in the optical spectrum travels through a material under the stress of an electric field.

electrostatic coupling

The transfer of energy by means of a varying electrostatic field. Also referred to as capacitive coupling.

electrostatic discharge (ESD)

A problem that exists when two items with dissimilar static electrical charges are brought together. The static electricity jumps to the item with lower electrical charge, which causes ESD; ESD can damage electrical and computer components.

emitter

A source of optical power.

encircled flux

A term used to describe the ratio between the transmitted power at a given distance from the center of the core of an optical fiber and the total power injected into the optical fiber.

encoding

Converting analog data into digital data by combining timing and data information into a synchronized stream of signals. Encoding is accomplished by representing digital 1s and 0s through combining high- and low-signal voltage or light states.

end finish, end-face finish

The quality of a fiber's end surface—specifically, the condition of the end of a connector ferrule. End-face finish is one of the factors affecting connector performance.

end separation

(1) The distance between the ends of two joined fibers. The end separation is important because the degree of separation causes an extrinsic loss, depending on the configuration of the connection. (2) The separation of optical fiber ends, usually taking place in a mechanical splice or between two connectors.

end-to-end loss

The optical signal loss experienced between the transmitter and the detector due to fiber quality, splices, connectors, and bends.

energy density

For radiation. Expressed in joules per square meter. Sometimes called irradiance.

entrance facility (EF)

A room in a building where antenna, public, and private network service cables can enter the building and be consolidated. Should be located as close as possible to the entrance point. Entrance facilities are often used to house electrical protection equipment and connecting hardware for the transition between outdoor and indoor cable. Also called an entrance room.

entrance point

The location where telecommunications enter a building through an exterior wall, a concrete floor slab, a rigid metal conduit, or an intermediate metal conduit.

epoxy

A glue that cures when mixed with a catalyzing agent.

epoxy-less connector

A connector that does not require the use of epoxy to hold the fiber in position.

equilibrium mode distribution

A term used to describe when light traveling through the core of the optical fiber populates the available modes in an orderly way.

equal level far-end crosstalk (ELFEXT)

ELFEXT is the name for the crosstalk signal that is measured at the receiving end and equalized by the attenuation of the cable.

equipment cable

Cable or cable assembly used to connect telecommunications equipment to horizontal or backbone cabling systems in the telecommunications room and equipment room. Equipment cables are considered to be outside the scope of cabling standards.

equipment cabling subsystem

Part of the cabling structure, typically between the distribution frame and the equipment.

equipment outlet

See telecommunications outlet.

equipment room (ER)

A centralized space for telecommunications equipment that serves the occupants of the building or multiple buildings in a campus environment. Usually considered distinct from a telecommunications closet because it is considered to serve a building or campus; the telecommunications closet serves only a single floor. The equipment room is also considered distinct because of the nature of complexity of the equipment that is contained in it.

erbium doped fiber amplifier (EDFA)

An optical amplifier that uses a length of optical fiber doped with erbium and energized with a pump laser to inject energy into a signal.

error detection

The process of detecting errors during data transmission or reception. Some of the error checking methods including CRC, parity, and bipolar variations.

error rate

The frequency of errors detected in a data service line, usually expressed as a decimal.

Ethernet

A local area network (LAN) architecture developed by Xerox that is defined in the IEEE 802.3 standard. Ethernet nodes access the network using the Carrier Sense Multiple Access/Collision Detect (CSMA/CD) access method.

European Computer Manufacturers Association (ECMA)

A European trade organization that issues its own specifications and is a member of the ISO.

excess loss

(1) In a fiber-optic coupler, the optical loss from that portion of light that does not emerge from the nominally operational pods of the device. (2) The ratio of the total output power of a passive component with respect to the input power.

exchange center

Any telephone building where switch systems are located. Also called exchange office or central office.

expanded beam

See lensed ferrule.

extrinsic attenuation

See extrinsic loss.

extrinsic factors

See extrinsic loss.

extrinsic loss

When describing a fiber-optic connection, the portion of loss that is not intrinsic to the fiber but is related to imperfect joining, which may be caused by the connector or splice, as opposed to conditions in the optical fiber itself. The conditions causing this type of loss are referred to as extrinsic factors. These factors are defects and imperfections that cause the loss to exceed the theoretical minimum loss due to the fiber itself (which is called intrinsic loss).

F

f

See frequency (f).

F-series connector

A type of coaxial RF connector used with coaxial cable.

Fabry-Pérot

A type of laser used in fiber-optic transmission. The Fabry-Pérot laser typically has a spectral width of about 5nm.

faceplate

A plate used in front of a telecommunications outlet. *See also* wall plate (or wall jack).

fanout cable

A multifiber cable that is designed for easy connectorization. These cables are sometimes sold with installed connectors or as part of a splice pigtail, with one end carrying many connectors and the other installed in a splice cabinet or panel ready for splicing or patching.

fanout kit

A collection of components used to add tight buffers to optical fibers from a loose-tube buffer cable. A fanout kit typically consists of a furcation unit and measured lengths of tight buffer material. It is used to build up the outer diameter of fiber cable for connectorization.

farad

A unit of capacitance that stores one coulomb of electrical charge when one volt of electrical pressure is applied.

faraday effect

A phenomenon that causes some materials to rotate the polarization of light in the presence of a magnetic field.

far-end

The end of a twisted pair cable that is at the far end with respect to the transmitter, i.e., at the receiver end.

far-end crosstalk (FEXT)

Crosstalk that is measured on the nontransmitting wires and measured at the opposite end from the source. See Chapter 1 for more information on various types of crosstalk.

Fast Ethernet

Ethernet standard supporting 100Mbps operation. Also known as 100Base-TX or 100Base-FX (depending on media).

FC connector

A threaded optical fiber connector that was developed by Nippon Telephone and Telegraph in Japan. The FC connector is good for single-mode or multimode fiber and applications requiring low back reflection. The FC is a screw type and is prone to vibration loosening.

Federal Communications Commission (FCC)

The federal agency responsible for regulating broadcast and electronic communications in the United States.

feeder cable

A voice backbone cable that runs from the equipment room cross-connect to the telecommunications cross-connect. A feeder cable may also be the cable running from a central office to a remote terminal, hub, head end, or node.

ferrule

A small alignment tube attached to the end of the fiber and used in connectors. These are made of stainless steel, aluminum, zirconia, or plastic. The ferrule is used to confine and align the stripped end of a fiber so that it can be positioned accurately.

fiber

A single, separate optical transmission element characterized by core and cladding. The fiber is the material that guides light or waveguides.

fiber channel

A gigabit interconnect technology that, through the 8B/10B encoding method, allows concurrent communications among workstations, mainframes, servers, data storage systems, and other peripherals using SCSI and IP protocols.

fiber curl

Occurs when there is misalignment in a mass or ribbon splicing joint. The fiber or fibers curl away from the joint to take up the slack or stress caused by misalignment of fiber lengths at the joint.

Fiber Distributed Data Interface (FDDI)

ANSI Standard X3T9.5, Fiber Distributed Data Interface (FDDI), details the requirements for all attachment devices concerning the 100Mbps fiber-optic network interface. It uses a counter-rotating, token-passing ring topology. FDDI is typically known as a backbone LAN because it is used for joining file servers together and for joining other LANs together.

fiber distribution frame (FDF)

Any fiber-optic connection system (cross-connect or interconnect) that uses fiber-optic jumpers and cables.

fiber identifier

A testing device that displays the direction of travel of light within an optical fiber by introducing a macrobend and analyzing the light that escapes the fiber.

fiber illumination kit

Used to visually inspect continuity in fiber systems and to inspect fiber connector end faces for cleanliness and light quality.

fiber protrusion

A term used to describe the distance the optical fiber end face is above the rounded connector end face.

fiber-in-the-loop (FITL)

Indicates deployment of fiber-optic feeder and distribution facilities.

fiber loss

The attenuation of light in an optical fiber transmission.

fiber-optic attenuator

An active component installed in a fiber-optic transmission system that is designed to reduce the power in the optical signal. It is used to limit the optical power received by the photodetector to within the limits of the optical receiver.

fiber-optic cable

Cable containing one or more optical fibers.

fiber-optic communication system

Involves the transfer of modulated or unmodulated optical energy (light) through optical-fiber media.

fiber-optic connector panel

A patch panel where fiber-optic connectors are mounted.

fiber-optic inter-repeater link (FOIRL)

An Ethernet fiber-optic connection method intended for connection of repeaters.

fiber-optic pigtail

Used to splice outside plant cable to the backside of a fiber-optic patch panel.

fiber-optic test procedures (FOTP)

Test procedures outlined in the EIA-RS-455 standards.

fiber-optic transmission

A communications scheme whereby electrical data is converted to light energy and transmitted through optical fibers.

fiber-optic waveguide

A long, thin strand of transparent material (glass or plastic), which can convey electromagnetic energy in the optical waveform longitudinally by means of internal refraction and reflection.

fiber optics

The optical technology in which communication signals in the form of modulated light beams are transmitted over a glass or plastic fiber transmission medium. Fiber optics offers high bandwidth and protection from electromagnetic interference and radioactivity; it also has small space needs.

fiber protection system (FPS)

A rack or enclosure system designed to protect fiber-optic cables from excessive bending or impact.

fiber test equipment

Diagnostic equipment used for the testing, maintenance, restoration, and inspection of fiber systems. This equipment includes optical attenuation meters and optical time-domain reflectometers (OTDRs).

Fiber to the X

A term used to describe any optical fiber network that replaces all or part of a copper network.

fiber undercut

A term used to describe the distance the optical fiber end face is below the rounded connector end face.

figure-8

A fiber-optic cable with a strong supporting member incorporated for use in aerial installations. *See also* messenger cable.

fill ratio

A term used to describe the percentage of conduit filled by cabling.

fillers

Nonconducting components cabled with insulated conductors or optical fibers to impart flexibility, tensile strength, roundness, or a combination of all three.

filter

A device that blocks certain wavelengths to permit selective transmission of optical signals.

firestop

Material, device, or collection of parts installed in a cable pathway (such as a conduit or riser) at a fire-rated wall or floor to prevent passage of flame, smoke, or gases through the rated barrier.

fish tape (or fish cord)

A tool used by electricians to route new wiring through walls and electrical conduit.

flex life

The average number of times a particular cable or type of cable can bend before breaking.

floating

A floating circuit is one that has no ground connection.

floor distributor (FD)

The ISO/IEC 11801 term for horizontal cross-connect. The floor distributor is used to connect between the horizontal cable and other cabling subsystems or equipment.

fluorinated ethylene-propylene (FEP)

A thermoplastic with excellent dielectric properties that is often used as insulation in plenum-rated cables. FEP has good electrical-insulating properties and chemical and heat resistance and is an excellent alternative to PTFE (Teflon®). FEP is the most common material used for wire insulation in Category 5 and better cables that are rated for use in plenums.

forward bias

A term used to describe when a positive voltage is applied to the *p* region and a negative voltage to the *n* region of a semiconductor.

four-wave mixing

The creation of new light wavelengths from the interaction of two or more wavelengths being transmitted at the same time within a few nanometers of each other. Four-wave mixing is named for the fact that two wavelengths interacting with each other will produce two new wavelengths, causing distortion in the signals being transmitted. As more wavelengths interact, the number of new wavelengths increases exponentially.

frame check sequence (FCS)

A special field used to hold error correction data in Ethernet (IEEE 802.3) frames.

frame relay

A packet-switched, wide area networking (WAN) technology based on the public telephone infrastructure. Frame relay is based on the older, analog, X.25 networking technologies.

free space optics transmission

The transmission of optical information over air using specialized optical transmitters and receivers.

free token

In Token Ring networks, a free token is a token bit set to 0.

frequency (f)

The number of cycles per second at which a waveform alternates—that is, at which corresponding parts of successive waves pass the same point. Frequency is expressed in hertz (Hz); one hertz equals one cycle per second.

frequency division multiplexing (FDM)

A technique for combining many signals onto a single circuit by dividing the available transmission bandwidth by frequency into narrower bands; each band is used for a separate communication channel. FDM can be used with any and all of the sources created by wavelength division multiplexing (WDM).

frequency hopping

Frequency hopping is the technique of improving the signal to noise ratio in a link by adding frequency diversity.

frequency modulation (FM)

A method of adding information to a sine wave signal in which its frequency is varied to impose information on it. Information is sent by varying the frequency of an optical or electrical carrier. Other methods include amplitude modulation (AM) and phase modulation (PM).

frequency response

The range of frequencies over which a device operates as expected.

Fresnel diffraction pattern

The near-field diffraction pattern.

Fresnel loss

The loss at a joint due to a portion of the light being reflected.

Fresnel reflection

Reflection of a small amount of light passing from a medium of one refractive index into a medium of another refractive index. In optical fibers, reflection occurs at the air/glass interfaces at entrance and exit ends.

Fresnel reflection method

A method for measuring the index profile of an optical fiber by measuring reflectance as a function of position on the end face.

full-duplex

A system in which signals may be transmitted in two directions at the same time.

full-duplex transmission

Data transmission over a circuit capable of transmitting in both directions simultaneously.

fundamental mode

The lowest number mode of a particular waveguide.

furcation unit

A component used in breakout kits and fanout kits for separating individual optical fibers from a cable and securing tight buffers and/or jackets around the fibers.

fusion splicing

A splicing method accomplished by the application of localized heat sufficient to fuse or melt the ends of the optical fiber, forming continuous single strand of fiber. As the glass is heated it becomes softer, and it is possible to use the glass's "liquid" properties to bond glass surfaces permanently.

G

G

Green. Used when referring to color-coding of cables.

gain

An increase in power.

gainer

A fusion splice displayed in an OTDR trace that appears to have gain instead of loss.

gamma

The coefficient used to scale bandwidth with fiber length.

gap loss

The loss that results when two axially aligned fibers are separated by an air gap. This loss is often most significant in reflectance. The light must launch from one medium to another (glass to air to glass) through the waveguide capabilities of the fiber.

gigahertz (GHz)

A billion hertz or cycles per second.

glass-reinforced plastic (GRP)

A relatively stiff dielectric, nonconducting strength element used in optical fiber cables. It is composed of continuous lengths of glass yarn formed together with thermally or UV cured high-strength epoxy. It exhibits not only excellent tensile strength, but compressive strength as well.

graded-index fiber

An optical fiber cable design in which the index of refraction of the core is lower toward the outside of the core and progressively increases toward the center of the core, thereby reducing modal dispersion of the signal. Light rays are refracted within the core rather than reflected as they are in step-index fibers. Graded-index fibers were developed to lessen the modal dispersion effects found in multimode fibers with the intent of increasing bandwidth.

ground

A common point of zero potential such as a metal chassis or ground rod that grounds a building to the earth. The ANSI/TIA/EIA-607 Commercial Building Grounding and Bonding Requirements for Telecommunications Standard is the standard that should be followed for grounding requirements for telecommunications. Grounding should never be undertaken without consulting with a professional licensed electrician.

ground loop

A condition where an unintended connection to ground is made through an interfering electrical conductor that causes electromagnetic interference. *See also* ground loop noise.

ground loop noise

Electromagnetic interference that is created when equipment is grounded at ground points having different potentials, thereby creating an unintended current path. Equipment should always be grounded to a single ground point.

guided ray

A ray that is completely confined to the fiber core.

H

half-duplex

A system in which signals may be sent in two directions, but not at the same time. In a half-duplex system, one end of the link must finish transmitting before the other end may begin.

half-duplex transmission

Data transmission over a circuit capable of transmitting in either direction. Transmission can be bidirectional but not simultaneously.

halogen

One of the following elements: chlorine, fluorine, bromine, astatine, or iodine.

hard-clad silica fiber

An optical fiber with a hard plastic cladding surrounding a step-index silica core.

hardware address

The address is represented by six sections of two hexadecimal addresses; for example, 00-03-fe-e7-18-54. This number is hard-coded into networking hardware by manufacturers. Also called the physical address; *see also* media access control (MAC) address.

hardware loopback

Connects the transmission pins directly to the receiving pins, allowing diagnostic software to test whether a device can successfully transmit and receive.

head end

(1) The central facility where signals are combined and distributed in a cable television system or a public telephone system. *See* central office (CO).

header

The section of a packet (usually the first part of the packet) where the layer-specific information resides.

headroom

The number of decibels by which a system exceeds the minimum defined requirements. The benefit of more headroom is that it reduces the bit-error rate (BER) and provides a performance safety net to help ensure that current and future high-speed applications will run at peak accuracy, efficiency, and throughput. Also called overhead or margin.

hertz (Hz)

A measurement of frequency defined as cycles per second.

heterojunction structure

An LED design in which the pn (P-type and N-type semiconductor) junction is formed from similar materials that have different refractive indices. This design is used to guide the light for directional output.

hicap service

A high-capacity communications circuit service such as a private line T-1 or T-3.

hierarchical star topology

A topology standardized in ANSI/TIA-568-C whereby all telecommunications outlets are connected to a centrally located point typically in the main equipment room.

high-order

Light rays traveling through the core of an optical fiber with a high number of reflections.

home-run cable

A cable run that connects a user outlet directly with the telecommunications or wiring closet. This cable has no intermediate splices, bridges, taps, or other connections. Every cable radiates out from the central equipment or wiring closet. This configuration is also known as star topology and is the opposite of a daisy-chained cable that may have taps or splices along its length. Home-run cable is the recommended installation method for horizontal cabling in a structured cabling system.

homojunction structure

An LED design in which the pn (P-type and N-type semiconductor) junction is formed from a single semiconductor material.

hop

A connection. In routing terminology, each router a packet passes through is counted as a hop.

horizontal cabling

The cabling between and including the telecommunications outlet and the horizontal cross-connect. Horizontal cabling is considered the permanent portion of a link; may also be called horizontal wiring.

horizontal cross-connect (HC)

A cross-connect that connects the cabling of the work area outlets to any other cabling system (like that for LAN equipment, voice equipment).

hot-swappable

A term used to describe an electrical device that may be removed or inserted without powering down the host equipment. This may be referred to as hot swapping, hot pluggable, or hot plugging.

hub

A device that contains multiple independent but connected modules of network and internetworking equipment that form the center of a hub-and-spoke topology. Hubs that repeat the signals that are sent to them are called active hubs. Hubs that do not repeat the signal and merely split the signal sent to them are called passive hubs. In some cases, hub may also refer to a repeater, bridge, switch, or any combination of these.

hybrid adapter

A mating or alignment sleeve that mates two different connector types such as an SC and ST.

hybrid cable

A cable that contains fiber, coaxial, and/or twisted-pair conductors bundled in a common jacket. May also refer to a fiber-optic cable that has strands of both single-mode and multimode optical fiber.

hybrid connector

A connector containing both fiber and electrical connectivity.

hybrid mesh network

Hybrid networks use a combination of any two or more topologies in such a way that the resulting network does not exhibit one of the standard topologies (e.g., bus, star, or ring). A hybrid topology is always produced when two different basic network topologies are connected.

hydrogen loss

Optical signal loss (attenuation) resulting from hydrogen found in the optical fiber. Hydrogen in glass absorbs light and turns it into heat and thus attenuates the light. For this reason, glass manufacturers serving the fiber-optic industry must keep water and hydrogen out of the glass and deliver it to guaranteed specifications in this regard. In addition, they must protect the glass with a cladding that will preclude the absorption of water and hydrogen into the glass.

Hypalon

A DuPont trade name for a synthetic rubber (chlorosulfonated polyethylene) that is used as insulating and jacketing material for cabling.

I

I

Symbol used to designate current.

IBM data connector

Used to connect IBM Token Ring stations using Type 1 shielded twisted-pair 150 ohm cable. This connector has both male and female components, so every IBM data connector can connect to any other IBM data connector.

IEEE 802.1 LAN/MAN Management

The IEEE standard that specifies network management, internetworking, and other issues that are common across networking technologies.

IEEE 802.2 Logical Link Control

The IEEE standard that provides specifications for the operation of the Logical Link Control (LLC) sublayer of the OSI data link layer. The LLC sublayer provides an interface between the MAC sublayer and the Network layer.

IEEE 802.3 CSMA/CD Networking

The IEEE standard that specifies a network that uses a logical bus topology, baseband signaling, and a CSMA/CD network access method. This is the standard that defines Ethernet networks. *See also* Carrier Sense Multiple Access/Collision Detect (CSMA/CD).

IEEE 802.4 Token Bus

The IEEE standard that specifies a physical and logical bus topology that uses coaxial or fiber-optic cable and the token-passing media access method.

IEEE 802.5 Token Ring

The IEEE standard that specifies a logical ring, physical star, and token-passing media access method based on IBM's Token Ring.

IEEE 802.6 Distributed Queue Dual Bus (DQDB) Metropolitan Area Network

The IEEE standard that provides a definition and criteria for a metropolitan area network (MAN), also known as a Distributed Queue Dual Bus (DQDB).

IEEE 802.7 Broadband Local Area Networks

The IEEE standard for developing local area networks (LANs) using broadband cabling technology.

IEEE 802.8 Fiber-Optic LANs and MANs

The IEEE standard that contains guidelines for the use of fiber optics on networks, including FDDI and Ethernet over fiber-optic cable.

IEEE 802.9 Integrated Services (IS) LAN Interface

The IEEE standard that contains guidelines for the integration of voice and data over the same cable.

IEEE 802.10 LAN/MAN Security

The IEEE standard that provides a series of guidelines dealing with various aspects of network security.

IEEE 802.11 Wireless LAN

The IEEE standard that provides guidelines for implementing wireless technologies such as infrared and spread-spectrum radio.

IEEE 802.12 Demand Priority Access Method

The IEEE standard that defines the concepts of a demand priority network such as HP's 100VG-AnyLAN network architecture.

impact test

Used to determine a cable's susceptibility to damage when subjected to short-duration crushing forces. Rate of impact, the shape of the striking device, and the force of the impact all are used to define the impact test procedures.

impedance

The total opposition (resistance and reactance) a circuit offers to the flow of alternating current. It is measured in ohms and designated by the symbol Z.

impedance match

A condition where the impedance of a particular cable or component is the same as the impedance of the circuit, cable, or device to which it is connected.

impedance matching transformer

A transformer designed to match the impedance of one circuit to another.

impedance mismatch

A condition where the impedance of a particular cable or component is different than the impedance of the device to which it is connected.

incident angle

The angle between the subject light wave and a plane perpendicular to the subject optical surface.

incoherent light

Light in which the electric and magnetic fields of photons are completely random in orientation. Incoherent light is typically emitted from lightbulbs and LEDs.

index matching gel

A clear gel or fluid used between optical fibers that are likely to have their ends separated by a small amount of air space. The gel matches the refractive index of the optical fiber, reducing light loss due to Fresnel reflection.

index matching material

A material in liquid, paste, gel, or film form whose refractive index is nearly equal to the core index; it is used to reduce Fresnel reflections from a fiber end face. Liquid forms of this are also called index matching gel.

index of refraction

(1) The ratio of the speed of light in a vacuum to the speed of light in a given transmission medium. This is usually abbreviated n. (2) The value given to a medium to indicate the velocity of light passing through it relative to the speed of light in a vacuum.

index profile

The curve of the refractive index over the cross section of an optical waveguide.

indoor-outdoor

Cable designed for installation in the building interior or exterior to the building.

Infiniband

A standard for a switched fabric communications link primarily used in high-performance computing.

infrared

The infrared spectrum consists of wavelengths that are longer than 700nm but shorter than 1mm. Humans cannot see infrared radiation, but we feel it as heat. The commonly used wavelengths for transmission through optical fibers are in the infrared at wavelengths between 1100nm and 1600nm.

infrared laser

A laser that transmits infrared signals.

infrared transceiver

A transmitting and receiving device that emits and receives infrared energy.

infrared transmission

Transmission using infrared transceivers or lasers.

injection laser diode (ILD)

A laser diode in which the lasing takes place within the actual semiconductor junction and the light is emitted from the edge of the diode.

innerduct

A separate duct running within a larger duct to carry fiber-optic cables.

insertion loss

Light or signal energy that is lost as the signal passes through the optical fiber end in the connector and is inserted into another connector or piece of hardware. A good connector minimizes insertion loss to allow the greatest amount of light energy through. A critical measurement for optical fiber connections. Insertion loss is measured by determining the output of a system before and after the device is inserted into the system. Loss in an optical fiber can be due to absorption, dispersion, scattering, microbending, diffusion, and the methods of coupling the fiber to the power. Usually measured in dB per item—for example, a coupler, connector, splice, or fiber. Most commonly used to describe the power lost at the entrance to a waveguide (an optical fiber is a waveguide) due to axial misalignment, lateral displacement, or reflection that is most applicable to connectors.

inside plant (IP)

(1) Cables that are the portion of the cable network that is inside buildings, where cable lengths are usually shorter than 100 meters. This is the opposite of outside-the-plant (OP or OSP) cables. (2) A telecommunications infrastructure designed for installation within a structure.

inspection microscope

A microscope designed to evaluate the end face of a fiber-optic connector.

installation load

The short-term tensile load that may be placed on a fiber-optic cable.

Institute of Electrical and Electronics Engineers (IEEE)

A publishing and standards-making body responsible for many standards used in LANs, including the 802 series of standards.

insulation

A material with good dielectric properties that is used to separate electrical components close to one another, such as cable conductors and circuit components. Having good dielectric properties means that the material is nonconductive to the flow of electrical current. In the case of copper communication cables, good dielectric properties also refer to enhanced signal-transfer properties.

insulation displacement connection (IDC)

A type of wire termination in which the wire is punched down into a metal holder that cuts into the insulation wire and makes contact with the conductor, thus causing the electrical connection to be made. These connectors are found on 66-blocks, 110-blocks, and telecommunications outlets.

integrated optical circuit

An optical circuit that is used for coupling between optoelectronic devices and providing signal processing functions. It is composed of both active and passive components.

integrated optics

Optical devices that perform two or more functions and are integrated on a single substrate; analogous to integrated electronic circuits.

integrated optoelectronics

Similar in concept to integrated optics except that one of the integrated devices on the semiconductor chip is optical and the other electronic.

Integrated Services Digital Network (ISDN)

A telecommunications standard that is used to digitally send voice, data, and video signals over the same lines. This is a network in which a single digital bit stream can carry a great variety of services. For the Internet it serves much better than analog systems on POTS (plain old telephone service), which is limited to 53Kbps. *See also* basic rate interface (BRI) and primary rate interface (PRI).

intelligence

A signal that is transmitted by imposing it on a carrier such as a beam of light to change the amplitude of the carrier.

intelligent hub

A hub that performs bridging, routing, or switching functions. Intelligent hubs are found in collapsed backbone environments.

intelligent network

A network that is capable of carrying overhead signaling information and services.

intensity

The square of the electric field amplitude of a light wave. Intensity is proportional to irradiance and may be used in place of that term if relative values are considered.

interbuilding backbone

A telecommunications cable that is part of the campus subsystem that connects one building to another.

interconnect

A circuit administration point, other than a cross-connect or an information outlet, that provides capability for routing and rerouting circuits. It does not use patch cords or jumper wires and is typically a jack-and-plug device that is used in smaller distribution arrangements or connects circuits in large cables to those in smaller cables.

interconnect cabinet

Cabinets containing connector panels, patch panels, connectors, and patch cords to interface from inside the plant to outside the plant. The interconnect cabinet is used as an access point for testing and rearranging routes and connections.

interconnection

A connection scheme that provides direct access to the cabling infrastructure and the capability to make cabling system changes using patch cords.

interface

The boundary layer between two media of different refractive indexes.

interference

(1) Fiber optic: the interaction of two or more beams of coherent or partially coherent light. (2) Electromagnetic: interaction that produces undesirable signals that interfere with the normal operation of electronic equipment or electronic transmission.

interleaving

A process where the multiplexer sends the first part of each channel, then the second part of each channel, continuing the process until all of the transmissions are completed.

interlock armor

A helical armor applied around a cable where the metal tape sides interlock giving a flexible, yet mechanically robust cable; specifically adds to the compression resistance in optical fiber cables and acts as an EMI/RFI shield in twisted-pair copper cables. The interlock armor can be made of either aluminum (more common) or steel.

intermediate cross-connect (ICC)

A cross-connect between first-level and second-level backbone cabling. This secondary cross-connect in the backbone cabling is used to mechanically terminate and administer backbone cabling between the main cross-connect and horizontal cross-connect (station cables).

intermediate distribution frame (IDF)

A metal rack (or frame) designed to hold the cables that connect interbuilding and intrabuilding cabling. The IDF is typically located in an equipment room or telecommunications room. Typically, a permanent connection exists between the IDF and the MDF.

International Organization for Standardization (ISO)

The standards organization that developed the OSI model. This model provides a guideline for how communications occur between computers. See www.iso.org for more information.

International Telecommunication Union (ITU)

The branch of the United Nations that develops communications specifications.

International Telephone and Telegraph Consultative Committee (CCiTT)

International standards committee that develops standards for interface and signal formats. Currently exists as the ITU-T.

Internet Architecture Board (IAB)

The committee that oversees management of the Internet, which is made up of several subcommittees including the Internet Engineering Task Force (IETF), the Internet Assigned Numbers Authority (IANA, which is now known as ICANN), and the Internet Research Task Force (IRTF). See www.iab.org for more information.

Internet Engineering Task Force (IETF)

An international organization that works under the Internet Architecture Board to establish specifications and protocols relating to the Internet. See www.ietf.org for more information.

Internet Research Task Force (IRTF)

An international organization that works under the Internet Architecture Board to research new Internet technologies. See www.irtf.org for more information.

Internet service provider (ISP)

A service provider or organization that offers Internet access to a user.

interoffice facility(IOF)

A communication channel of copper, fiber, or wireless media between two central offices. Often, IOF refers to telephone channels that can transport voice and/or data.

internetworking, internetwork

Involves connecting two or more computer networks via gateways using a common routing technology. The result is called an internetwork (often shortened to internet).

inter-repeater link (fiber-optic)

Defined in IEEE 802.3 and implemented over two fiber links, transmit and receive, this medium may be up to 500m and 1km long depending on the number of repeaters in the network.

inter-repeater link (in a thinnet network)

The two segments of a five-segment thinnet network that will only connect to repeaters.

intersymbol interference (ISI)

Form of distortion in digital communications systems. ISI is caused by multipath propagation and nonlinear frequency response of a channel resulting from not having enough bandwidth. ISI leads to BER and causes the loss of the channel to increase.

intersymbol loss (ISL)

Channel loss resulting from intersymbol interference.

intrabuilding backbone

Telecommunications cables that are part of the building subsystem that connect one telecommunications closet to another or a telecommunications room to the equipment room.

intrinsic absorption

A term that describes a process in which photons absorbed by the photodiode excite electrons within the photodiode.

intrinsic attenuation

Attenuation caused by the scattering or absorption of light in an optical fiber.

intrinsic factors

When describing an optical fiber connection, factors contributing to attenuation that are determined by the condition of the optical fiber itself. *See also* extrinsic loss.

intrinsic joint loss

The theoretical minimum loss that a given joint or device will have as a function of its nature. Intrinsic joint loss may also be used to describe the given theoretical minimum loss that a splice joint, coupler, or splitter may achieve.

intrinsic performance factor (IPF)

Performance specification whereby total optical channel performance is specified, rather than performance of individual components.

intrinsic splice loss

The optical signal loss arising from differences in the fibers being spliced.

intumescent

A substance that swells as a result of heat exposure, thus increasing in volume and decreasing in density.

inverter

A device that inverts the phase of a signal.

ion exchange techniques

A method for making and doping glass by ion exchange.

irradiance

The measure of power density at a surface on which radiation is directed. The normal unit is watts per square centimeter.

ISDN terminal adapter

The device used on ISDN networks to connect a local network or single machine to an ISDN network (or any non-ISDN compliant device). The ISDN terminal adapter provides line power and translates data from the LAN or individual computer for transmission on the ISDN line.

isochronous

Signals that are dependent on some uniform timing or carry their own timing information embedded as part of the signal.

isolated ground

A separate ground conductor that is insulated from the equipment or building ground.

isolation

The ability of a circuit or component to reject interference.

isolator

In fiber optics, a device that permits only forward transmission of light and blocks any reflected light.

J

jabber

A term used with Ethernet to describe the act of continuously sending data or sending Ethernet frames with a frame size greater than 1,518 bytes. When a station is jabbering, its network adapter circuitry or logic has failed, and it has locked up a network channel with its erroneous transmissions.

jack

A receptacle used in conjunction with a plug to make electrical contact between communication circuits. A variety of jacks and their associated plugs are used to connect hardware applications, including cross-connects, interconnects, information outlets, and equipment connections. Jacks are also used to connect cords or lines to telephone systems. A jack is the female component of a plug/jack connector system and may be standard, modified, or keyed.

jacket

The outer protective covering of a cable, usually made of some type of plastic or polymer.

jacketed ribbon cable

A cable carrying optical fiber in a ribbon arrangement with an elongated jacket that fits over the ribbon.

jitter

A slight movement of a transmission signal in time or phase that can introduce errors and loss of synchronization. More jitter is introduced when cable runs are longer than the network-topology specification recommends. Other causes of jitter include cables with high attenuation and signals at high frequencies. Also called phase jitter, timing distortion, or intersymbol interference.

joint

Any joining or mating of a fiber by splicing (by fusion splicing or physical contact of fibers) or connecting.

jumper

(1) A small, manually placed connector that connects two conductors to create a circuit (usually temporary). (2) In fiber optics, a simplex cable assembly that is typically made from simplex cordage.

jumper wire

A cable of twisted wires without connectors used for jumpering.

junction laser

A semiconductor diode laser.

K

Kevlar

A strong, synthetic material developed and trademarked by DuPont; the preferred strength element in cable. Also used as a material in body armor and parts for military equipment. Also known by the generic name aramid; *see also* aramid strength member.

keying

A mechanical feature of a connector system that guarantees correct orientation of a connection. The key prevents connection to a jack or an optical fiber adapter that was intended for another purpose.

kHz

Kilohertz; 1,000 hertz.

KPSI

KiloPSI. A unit of tensile strength expressed in thousands of pounds per square inch.

L

L

Symbol used to designate inductance.

ladder rack

A rack used to hold cabling that looks like a ladder.

large core fiber

Usually a fiber with a core of 200 microns or more. This type of fiber is not common in structured cabling systems.

laser

Acronym for light amplification by stimulated emission of radiation. The laser produces a coherent source of light with a narrow beam and a narrow spectral bandwidth (about 2cm). Lasers in fiber optics are usually solid-state semiconductor types. Lasers are used to provide the high-powered, tightly controlled light wavelengths necessary for high-speed, long-distance optical fiber transmissions.

laser chirp

When the laser output wavelength changes as the electron density in the semiconductor material changes.

laser diode (LD)

A special semiconductor that emits laser light when a specific amount of current is applied. Laser diodes are typically used in higher speed applications (622Mbps to 10Gbps) such as ATM, 1000Base-LX, and SONET. The mode is usually ellipse shaped and therefore requires a lens to make the light symmetrical with the mode of the fiber, which is usually round.

lasing threshold

The energy level that, when reached, allows the laser to produce mostly stimulated emissions rather than spontaneous emissions.

lateral displacement loss

The loss of signal power that results from lateral displacement from optimum alignment between two fibers or between a fiber and an active device.

lateral misalignment

When the core of the transmitting optical fiber is laterally offset from the core of the receiving optical fiber.

launch angle

In fiber-optic transmissions, the launch angle is defined as the difference between the incoming direction of the transmitting light and the alignment of the optical fiber.

launch cable

Used to connect fiber-optic test equipment to the fiber system.

launch fiber

An optical fiber used to introduce light from an optical source into another optical fiber. Also referred to as launching fiber.

lay

The axial distance required for one cabled conductor or conductor strand to complete one revolution around the axis around which it is cabled.

lay direction

The direction of the progressing spiral twist of twisted-pair wires while looking along the axis of the cable away from the observer. The lay direction can be either left or right.

Layer 2 switch

A network device that operates at the Data Link layer. The switch builds a table of MAC addresses of all connected stations and uses the table to intelligently forward data to the intended recipients.

Layer 3 switch

A network device that can route LAN traffic (Layer 3) at a speed that is nearly as quick as a Layer 2 switch device. Layer 3 switches typically perform multiport, virtual LAN, and data pipelining functions of a standard Layer 2 switch and can also perform routing functions between virtual LANs.

lbf

Abbreviation for pounds force.

LC

A small-form-factor optical fiber connector originally designed and manufactured by AT&T Bell Labs; closely resembles the RJ-11 connector.

leakage

An undesirable passage of current over the surface of or through a connector.

leased line

A private telephone line (usually a digital line) rented for the exclusive use of a leasing customer without interchange switching arrangements.

least-squares averaging

A method of measuring attenuation in a fiber-optic signal that reduces the effect of high-frequency noise on the measurement.

lensed ferrule

A ferrule with a lens on the end face, commonly known as an *expanded beam*.

light

The electromagnetic radiation visible to the human eye between 400nm and 700nm. The term is also applied to electromagnetic radiation with properties similar to visible light; this includes the invisible near-infrared radiation in most fiber-optic communication systems.

light-emitting diode (LED)

A semiconductor device used in a transmitter to convert information from electric to optical form. The LED typically has a large spectral width (that is, it produces incoherent light); LED devices are usually used on low-speed (100–256Mbps) fiber-optic communication systems, such as 100Base-FX and FDDI, that do not require long distances or high data rates.

limiting amplifier

An amplifier that amplifies the transimpedance amplifier output voltage pulses and provides a binary decision. It determines whether the electrical pulses received represent a binary 1 or a binary 0.

line build-out (LBO)

A device that amplifies a received power level to ensure that it is within proper specs.

line conditioner

A device used to protect against power surges and spikes. Line conditioners use several electronic methods to clean all power coming into the line conditioner so that clean, steady power is put out by the line conditioner.

line-of-sight transmission

The transmission of free-space-optics devices where the transmitter and receiver are in the same plane.

linear bus

See bus topology.

link

An end-to-end transmission path provided by the cabling infrastructure. Cabling links include all cables and connecting hardware that compose the horizontal or backbone subsystems. Equipment and work-area cables are not included as part of a link.

link light

A small light-emitting diode (LED) that is found on both the NIC and the hub and is usually green and labeled "Link." A link light indicates that the NIC and the hub are making a Data Link layer connection.

listed

Equipment included on a list published by an organization, acceptable to the authority having jurisdiction, that maintains periodic inspection of production of listed equipment, and whose listing states either that the equipment or material meets appropriate standards, or that it has been tested and found suitable for use in a specified manner. In the United States, electrical and data communications equipment is typically listed with Underwriters Laboratories (UL).

lobe

An arm of a Token Ring that extends from a multistation access unit (MSAU) to a workstation adapter.

local area network (LAN)

A network connecting multiple nodes within a defined area, usually within a building. The linking can be done by cable that carries optical fiber or copper. These are usually high bandwidth

(4Mbps or greater) and connect many nodes within a few thousand meters. LANs can, however, operate at lower data rates (less than 1Mbps) and connect nodes over only a few meters.

local convergence point (LCP)

An access point in a passive optical network where the feeder cables are broken out into multiple distribution cables.

local exchange carrier (LEC)

The local regulated provider of public switched telecommunications services. The LEC is regulated by the local Public Utilities Commission.

local loop

The loop or circuit between receivers (and, in two-way systems, receivers and senders), who are normally the customers or subscribers to the service's products, and the terminating equipment at the central office.

LocalTalk

A low-speed form of LAN data link technology developed by Apple Computer. It was designed to transport Apple's AppleTalk networking scheme; it uses a Carrier Sense Multiple Access/Collision Avoidance (CSMA/CA) form of media access control. Supports transmission at 230Kbps.

logical network addressing

The addressing scheme used by protocols at the OSI Network layer.

logical ring topology

A network topology in which all network signals travel from one station to another, being read and forwarded by each station. A Token Ring network is an example of a logical ring topology.

logical topology

Describes the way the information flows. The types of logical topologies are the same as the physical topologies, except that the information flow specifies the type of topology.

long wavelength

Light whose wavelength is greater than 1000nm (longer than one micron).

longitudinal conversion loss (LCL)

A measurement (in decibels) of the differential voltage induced on a conductor pair as a result of subjecting that pair to longitudinal voltage. This is considered to be a measurement of circuit balance.

longitudinal conversion transfer loss (LCTL)

Measures cable balance by the comparison of the signal appearing across the pair to the signal between ground and the pair, where the applied signal is at the opposite end of the cable from the location at which the across-pair signal is measured. LCTL is also called far-end unbalance attenuation.

longitudinal modes

Oscillation modes of a laser along the length of its cavity. Each longitudinal mode contains only a very narrow range of wavelengths; a laser emitting a single longitudinal mode has a very narrow bandwidth. The oscillation of light along the length of the laser's cavity is normally such that two times the length of the cavity will equal an integral number of wavelengths. Longitudinal modes are distinct from transverse modes.

loop

(1) A complete electrical circuit. (2) The pair of wires that winds its way from the central office to the telephone set or system at the customer's office, home, or factory.

loopback

A type of diagnostic test in which a transmitted signal is returned to the sending device after passing through a data communications link or network. This test allows the comparison of a returned signal with the transmitted signal to determine if the signal is making its way through the communications link and how much signal it is losing upon its total trip.

loose buffer

See loose-tube buffer.

loose tube

A protective tube loosely surrounding an optical fiber, often filled with gel used as a protective coating. Loose-tube cable designs are usually found in outdoor cables, not inside buildings.

loose-tube buffer

Optical fiber that is carried loosely in a buffer many times the diameter of the fiber. Loose-buffered fiber is typically terminated with a break-out kit or a fanout kit and connected to a patch panel. Also known as *loose buffer*.

loss

The attenuation of optical or electrical signal, normally measured in decibels (dB). With respect to fiber-optic cables, there are two key measurements of loss: insertion loss and return loss. Both are measured in decibels. The higher the decibel number, the more loss there is. Some copper-based and optical fiber–based materials are lossy and absorb electromagnetic radiation in one form and emit it in another—for example, heat. Some optical fiber materials are reflective and return electromagnetic radiation in the same form as it is received, usually with little or no power loss. Still others are transparent or translucent, meaning they are "window" materials; loss is the portion of energy applied to a system that is dissipated and performs no useful work. *See also* attenuation.

loss budget

A calculation and allowance for total attenuation in a system that is required in order to ensure that the detectors and receivers can make intelligent decisions about the pulses they receive.

lossy

Describes a connection having poor efficiency with respect to loss of signal.

M

MAC

(1) *See* media access control (MAC). (2) Abbreviation for moves, adds, and changes.

macrobend

A major bend in an optical fiber with a radius small enough to change the angle of incidence and allow light to pass through the interface between the core and the cladding rather than reflect off it. Macrobends can cause signal attenuation or loss by allowing light to leave the optical fiber core.

macrobending loss

Optical power loss due to large bends in the fiber.

magnetic optical isolator

See optical isolator.

main cross-connect

A cross-connect for first-level backbone cables, entrance cables, and equipment cables. The main cross-connect is at the top level of the premises cabling tree.

main distribution frame (MDF)

A wiring arrangement that connects the telephone lines coming from outside on one side and the internal lines on the other. The MDF may be a central connection point for data communications equipment in addition to voice communications. An MDF may also carry protective devices or function as a central testing point.

Manchester coding

A method of encoding a LAN in which each bit time that represents a data bit has a transition in the middle of the bit time. Manchester coding is used with 10Mbps Ethernet (10Base-2, 10Base-5, 10Base-F, and 10Base-T) LANs.

mandrel wrap

A device used to remove high-order modes caused by overfilling in a length of optical fiber for insertion loss measurements.

margin

The allowance for attenuation in addition to that explicitly accounted for in system design.

mass splicing

The concurrent and simultaneous splicing of multiple fibers at one time. Currently mass splicing is done on ribbon cable, and the standard seems to be ribbon cable with 12 fibers. Special splice protectors are made for this purpose, as well as special equipment for splicing.

material dispersion

A pulse dispersion that results from each wavelength traveling at a speed different from other wavelengths through an optical fiber. *See also* chromatic dispersion.

mating sleeve

A sleeve used to align two optical fiber ferrules.

maximum tensile rating

A manufacturer's specified limit on the amount of tension, or pulling force, that may be applied to a fiber-optic cable.

mean time between failures (MTBF)

A measurement of how reliable a hardware component is. Usually measured in thousands of hours.

measurement quality jumper (MQJ)

A term used by the U.S. Navy to describe a test jumper.

mechanical splice

With respect to fiber-optic cables, a splice in which fibers are joined mechanically (e.g., glued, crimped, or otherwise held in place) but not fused together using heat. Mechanical splice is the opposite of a fusion splice in which the two fiber ends are butted and then joined by permanently bonding the glass end faces through the softening of the glass, which is fused together.

media

Wire, cable, or conductors used for transmission of signals.

media access

The process of vying for transmission time on the network media.

media access control (MAC)

A sublayer of the OSI Data Link layer (Layer 2) that controls the way multiple devices use the same media channel. It controls which devices can transmit and when they can transmit. For most network architectures, each device has a unique address that is sometimes referred to as the MAC address. *See also* media access control (MAC) address.

media access control (MAC) address

Network adapter cards such as Ethernet, Token Ring, and FDDI cards are assigned addresses when they are built. A MAC address is 48 bits represented by six sections of two hexadecimal digits. No two cards have the same MAC address. The IEEE helps to achieve the unique addresses by assigning the first half to manufacturers so that no two manufacturers have the same first three bytes in their MAC address. The MAC address is also called the hardware address.

media filter

An impedance-matching device used to change the impedance of the cable to the expected impedance of the connected device. For example, media filters can be used in Token Ring networks to transform the 100 ohm impedance of UTP cabling to the 150 ohm impedance of media interface connections.

media interface connector (MIC)

A pair of fiber-optic connectors that links the fiber media to the FDDI network card or concentrator. The MIC consists of both the MIC plug termination of an optical cable and the MIC receptacle that is joined with the FDDI node.

medium

Any material or space through which electromagnetic radiation can travel.

medium attachment unit (MAU)

When referring to Ethernet LANs, the transceiver in Ethernet networks.

medium dependent interface (MDI)

Used with Ethernet systems; it is the connector used to make the mechanical and electrical interface between a transceiver and a media segment. An 8-pin RJ-45 connector is the MDI for Ethernet implemented using UTP.

medium independent interface (MII)

Used with 100Mbps Ethernet systems to attach MAC-level hardware to a variety of physical media systems. Similar to the AUI interface used with 10Mbps Ethernet systems. The MII is a 40-pin connection to outboard transceivers or PHY devices.

medium interface connector (MIC)

A connector that is used to link electronics and fiber-optic transmission systems.

meridian plane

Any plane that includes or contains the optical axis.

meridional ray

A light ray that passes through the axis of an optical fiber.

messenger cable

A cable with a strong supporting member attached to it for use in aerial installations. *See also* figure-8.

metropolitan area network (MAN)

An interconnected group of local area networks (LANs) that encompasses an entire city or metropolitan area.

microbend

A small radius bend in the optical fiber that changes the angle of incidence, allowing light to pass through the interface rather than reflect off it.

microbending loss

The optical power loss due to microscopic bends in the fiber.

microfarad

One millionth of a farad. Abbreviated μF and, less commonly, μfd, mf, or mfd.

micrometer

Also referred to as a micron; one millionth of a meter, often abbreviated with the symbol μ.

microwave

Wireless communication using the lower gigahertz frequencies (4–23 GHz).

midsplit broadband

A broadcast network configuration in which the cable is divided into two channels, each using a different range of frequencies. One channel is used to transmit signals and the other is used to receive.

minimum bend radius

A fiber-optic cable manufacturer's specified limit on the amount of bending that the cable can withstand without damage.

misalignment loss

The loss of optical power resulting from angular misalignment, lateral displacement, or end separation.

modal bandwidth

The bandwidth-limiting characteristic of multimode fiber systems caused by the variable arrival times of various modes.

modal dispersion

A type of dispersion or spreading that arises from differences in the amount of time that different modes take to travel through multimode fibers. Modal dispersion potentially can cause parts of a signal to arrive in a different order from the one in which they were transmitted. *See also* differential modal dispersion (DMD).

modal noise

A disturbance often measured in multimode fiber-optic transmissions that are fed by diode lasers. The higher quality the laser light feeding the fiber, the less modal noise will be measured.

mode

A single wave traveling in an optical fiber or in a light path through a fiber. Light has modes in optical fiber cable. A high-order mode is a path that results in numerous reflections off the core/cladding interface. A low-order mode results in fewer reflections. A zero-order mode is a path that goes through the fiber without reflecting off the interface at all. In a single-mode fiber, only one mode (the fundamental mode) can propagate through the fiber. Multimode fiber has several hundred modes that differ in field pattern and propagation velocity. The number of modes in an optical fiber is determined by the diameter of the core, the wavelength of the light passing through it, and the refractive indexes of the core and cladding. The number of modes increases as the core diameter increases, the wavelength decreases, or the difference between refractive indexes increases.

mode field diameter (MFD)

The actual diameter of the light beam traveling through the core and part of the cladding and across the end face in a single-mode fiber-optic cable. Since the mode field diameter is usually greater than the core diameter, the mode field diameter replaces the core diameter as a practical parameter.

mode filter

A device that can select, attenuate, or reject a specific mode. Mode filters are used to remove high-order modes from a fiber and thereby simulate EMD. *See also* mandrel wrap.

mode mixing

Coupling multiple single modes into a single multimode strand by mixing the different signals and varying their modal conditions.

mode stripper

A device that removes high-order modes in a multimode fiber to give standard measurement conditions.

modem

A device that implements modulator-demodulator functions to convert between digital data and analog signals.

modified modular jack (MMJ)

A six-wire modular jack used by the DEC wiring system. The MMJ has a locking tab that is shifted to the right-hand side.

modified modular plug (MMP)

A six-wire modular plug used by the DEC wiring system. The MMP has a locking tab that is shifted to the right side.

modular

Equipment is said to be modular when it is made of plug-in units that can be added together to make the system larger, improve the capabilities, or expand its size. Faceplates made for use with structured cabling systems are often modular and permit the use of multiple types of telecommunications outlets or modular jacks such as RJ-45, coaxial, audio, or fiber.

modular jack

A female telecommunications interface connector. Modular jacks are typically mounted in a fixed location and may have four, six, or eight contact positions, though most typical standards-based cabling systems will have an eight-position jack. Not all positions need be equipped with contacts. The modular jack may be keyed or unkeyed so as to permit only certain types of plugs to be inserted into the jack.

modulate

(1) To convert data into a signal that can be transmitted by a carrier. (2) To control.

modulation

(1) Coding of information onto the carrier frequency. Types of modulation include amplitude modulation (AM), frequency modulation (FM), and phase modulation (PM). (2) When light is emitted by a medium, it is coherent, meaning that it is in a fixed-phase relationship within fixed points of the light wave. The light is used because it is a continuous, or sinusoidal, wave (a white or blank form) on which a signal can be superimposed by modulation of that form. The modulation is a variation imposed on this white form, a variation of amplitude, frequency, or phase of the light. There are two basic forms of this modulation: one by an analog form, another by a digital signal. This signal is created in the form of the "intelligence" and superimposed on the light wave. It is then demodulation by a photodetector and converted into electrical energy.

monochromatic

Light having only one color, or more accurately only one wavelength.

Motion Pictures Experts Group (MPEG)

A standards group operating under the ISO that develops standards for digital video and audio compression.

MT-RJ connector

A duplex fiber-optic connector that looks similar to the RJ-45-type connector.

multifiber jumpers

Used to interconnect fiber-optic patch panels from point to point.

multifunction cable scanners

Also referred to as certification tools. These devices perform a series of tests for copper and fiber-optic cables used for certifying a cable system.

multimedia

An application that communicates to more than one of the human sensory receptors such as audio and video components.

multimode

Transmission of multiple modes of light. *See also* mode.

multimode depressed clad

See bend insensitive.

multimode distortion

The signal distortion in an optical waveguide resulting from the superposition of modes with differing delays.

multimode fiber

Optical fiber cable whose core has a refractive index that is graded or stepped; multimode fiber has a core diameter large enough to allow light to take more than one possible path through it. It allows the use of inexpensive LED light sources, and connector alignment and coupling is less critical than with single-mode fiber. Distances of transmission and transmission bandwidth are less than with single-mode fiber due to dispersion. The ANSI/TIA-568-C standard recognizes the use of 62.5/125-micron and 50/125-micron multimode fiber for horizontal cabling.

multimode laser

A laser that produces emissions in two or more longitudinal modes.

multiple reflection noise (MRN)

The noise at the receiver caused by the interface of delayed signals from two or more reflection points in an optical fiber span.

multiplex

The combination of two or more signals to be transmitted along a single communications channel.

multiplexer

A device that combines two or more discrete signals into a single output. Many types of multiplexing exist, including time-division multiplexing and wavelength-division multiplexing.

multiplexing

Transmitting multiple data channels in the same signal.

multipoint RF network

An RF network where the RF transmitters can access more than one receiver.

multistation access unit (MAU or MSAU)

Used in Token Ring LANs, a wiring concentrator that allows terminals, PCs, printers, and other devices to be connected in a star-based configuration to Token Ring LANs. MAU hardware can be either active or passive and is not considered to be part of the cabling infrastructure.

multiuser telecommunications outlet assembly (MuTOA)

A connector that has several telecommunications/outlet connectors in it. These are often used in a single area that will have several computers and telephones.

mutual capacitance

The capacitance (the ability to store a charge) between two conductors when they are brought adjacent to each other.

Mylar

The DuPont trademark for biaxially oriented polyethylene terephthalate (polyester) film.

MZI

Mach-Zehnder interferometer. A device used to measure the optical phase shift of various materials.

N

National Electrical Code (NEC)

The U.S. electrical wiring code that specifies safety standards for copper and fiber-optic cable used inside buildings. See Chapter 4 for more information.

National Security Agency (NSA)

The U.S. government agency responsible for protecting U.S. communications and producing foreign intelligence information. It was established by presidential directive in 1952 as a separately organized agency within the Department of Defense.

N-connector

A coaxial cable connector used for Ethernet 10Base-5 thick coaxial segments.

near-end

Defined in copper twisted-pair cables as the end of the cable where the transmitter is located.

near-end crosstalk (NEXT)

Crosstalk noise between two twisted pairs measured at the near end of the cable. Near is defined as the end of the cable where the transmission originated. See Chapter 1 for more information.

near-field radiation pattern

The distribution of the irradiance over an emitting surface (over the cross section of an optical waveguide).

near infrared

The part of the infrared spectrum near the visible spectrum, typically 700nm to 1500nm or 2000nm; it is not rigidly defined.

necking

A term used to describe a fusion splice where the diameter of the fused optical fiber is smaller near the electrodes than it was prior to being fused.

network

Ties things together. Computer networks connect all types of computers and computer-related peripherals—terminals, printers, modems, door entry sensors, temperature monitors, and so forth. The networks we're most familiar with are long-distance ones, such as phone or train networks. Local area networks (LANs) connect computer equipment within a building or campus.

network media

The physical cables that link computers in a network; also known as physical media.

NFPA 262

The fire test method that measures flame spread, peak smoke optical density, and average smoke optical density. Formerly referred to as UL 910. Cables are required to pass this test and be listed by a nationally recognized test laboratory (e.g., UL or ETL) for the cables to be allowed to be placed in plenum spaces.

NIC card (network interface card)

Also referred to as a network card. A circuit board installed in a computing device that is used to attach the device to a network. A NIC performs the hardware functions that are required to provide a computing device physical communications capabilities with a network.

NIC diagnostics

Software utilities that verify that the NIC is functioning correctly and that test every aspect of NIC operation, including connectivity to other nodes on the network.

node

Endpoint of a network connection. Nodes include any device connected to a network such as file servers, printers, and workstations.

noise

In a cable or circuit, any extraneous signal (electromagnetic energy) that interferes with the desired signal normally present in or passing through the system.

noise equivalent power (NEP)

The optical input power to a detector needed to generate an electrical signal equal to the inherent electrical noise.

Nomex

A DuPont trademark for a temperature-resistant, flame-retardant nylon.

nominal velocity of propagation (NVP)

The speed that a signal propagates through a cable expressed as a decimal fraction of the speed of light in a vacuum. Typical copper cables have an NVP value of between 0.6c and 0.9c.

non-blocking network

In contrast to blocking, a non-blocking network refers to a situation where the uplink speed of a workgroup switch is greater than the amount of data that is being uploaded onto the switch at a given time.

non-current-carrying conductive members

Conductive members of a cable such as metallic strength members, metallic vapor barriers, metallic armor, or a metallic sheath whose primary function is not to carry electrical current.

non-return to zero (NRZ)

A digital code in which the signal level is low for a 0 bit and high for a 1 bit and which does not return to zero volts between successive 1 bits or between successive 0 bits.

non-zero-dispersion-shifted fiber

A type of single-mode optical fiber designed to reduce the effects of chromatic dispersion while minimizing four-wave mixing.

normal

A path drawn perpendicular to the *interface*, or boundary layer between two media, that is used to determine the *angle of incidence* of light reaching the interface.

normal angle

The angle that is perpendicular to a surface.

NT-1

Used to terminate ISDN at the customer premises. It converts a two-wire ISDN U interface to a four-wire S/T interface.

numerical aperture (NA)

The light-gathering ability of a fiber, defining the maximum angle to the fiber axis at which light will be accepted and propagated through the fiber. The numerical aperture is determined by the refractive indexes of the core and cladding. The numerical aperture is also used to determine the fiber's acceptance angle.

Nyquist Minimum

The calculated minimum effective sampling rate for a given analog signal based on its highest expected frequency. The Nyquist Minimum requires sampling to take place at a minimum of twice the expected highest frequency of an analog signal. For example, if an analog signal's highest frequency is expected to be 10kHz, it must be sampled at a rate of at least 20kHz.

O

O

Orange, when used in conjunction with color-coding for twisted-pair cabling.

OC-1

Optical carrier level one, equal to 51.84Mbps. This is a SONET channel, whose format measures 90 bytes and is composed of the transport overhead and the synchronous payload envelope.

OC-3

SONET Channel of 155.52Mbps.

OC-12

SONET channel of 622.08Mbps.

OC-48

SONET channel of 2.5Gbps.

OC-192

SONET channel of 10Gbps, currently the highest level now available.

octet

Eight bits (also called a byte).

OFCP

An optical fiber cable that has conducting (metal) elements in its construction and that meets the plenum test requirements of NFPA 262; examples of conducting elements in the cable include the copper wire or interlock aluminum armoring.

OFCR

An optical fiber cable that has conducting (metal) elements in its construction and that meets the riser test requirements of UL 1666; examples of conducting elements in the cable include the copper wire or interlock aluminum armoring.

OFNP

An optical fiber cable that has no conducting (metal) elements and meets the plenum test requirements of NFPA 262 (UL910).

OFNR

An optical fiber cable that has no conducting (metal) elements and meets the riser test requirements of UL 1666.

off-hook

The handset's state of being lifted from its cradle. The term originated from when the early handsets were actually suspended from a metal hook on the phone. With modern telephones, when the handset is removed from its hook or cradle, it completes the electrical loop, thus signaling the central office to provide a dial tone. Opposite of on-hook.

office principle ground point (OPGP)

The main grounding point in a central office. Usually connects directly to an earth ground like a water pipe.

ohm

A unit of electrical resistance. The value of resistance through which a potential of one volt will maintain a current of one ampere.

on-hook

The telephone handset's state of resting in its cradle. The phone is not connected to any particular line. Only the bell is active—that is, it will ring if a call comes in. Opposite of off-hook.

open circuit

An incomplete circuit. It can be either a break in a cable or a switch that's turned off.

open fault

A break in the continuity of a circuit. This means that the circuit is not complete or the cable/fiber is broken. This condition is also called unmated, open, or unterminated.

Open Systems Interconnection (OSI)

A model defined by the ISO to categorize the process of communication between computers in terms of seven layers. *See also* International Organization for Standardization (ISO).

operational load

See static load.

operating wavelength

The wavelength at which a fiber-optic receiver is designed to operate. Typically, an operating wavelength includes a range of wavelengths above and below the stated wavelength.

operations, administration, maintenance, and provisioning (OAM&P)

A telecommunications term for the support functions of a network.

optical amplifier

Increases the power of an optical signal without converting any of the signals from optical to electrical energy and then back to optical so that the amplification processes the optical signal wholly within optical amplification equipment. The two most common optical amplifiers are semiconductor laser amplifiers and those made from doped fiber, such as the EDFA (erbium doped fiber

amplifier), which amplifies with a laser pump diode and a section of erbium doped fiber.

optical attenuator

Reduces the intensity of light waves, usually so that the power is within the capacity of the detector. There are three basic forms of attenuators: fixed optical attenuators, stepwise variable optical attenuators, and continuous variable optical attenuators. Attenuation is normally achieved either by a doped fiber or an offset or core misalignment. *See also* attenuator.

optical bandpass

The range of optical wavelengths that can be transmitted through a component.

optical carrier n

Optical signal standards. The *n* indicates the level where the respective data rate is exactly *n* times the first level OC-1. OC-1 has a data rate of 51.84Mbps. OC-3 is three times that rate, or 155.52Mbps, and so on. Associated with SONET. OC levels are medium-dependent on fiber.

optical combiner

A device used to combine fiber-optic signals.

optical continuous wave reflectometer (OCWR)

A device that measures optical return loss, or the loss of optical signals due to reflection back toward the transmitter.

optical detector

A transducer that generates an electronic signal when excited by an optical power source.

optical directional coupler (ODC)

A directional coupler used to combine or separate optical power.

optical fiber cable

An assembly consisting of one or more optical fibers. These optical fibers are thin glass or plastic filaments used for the transmission of information via light signals. The individual optical fibers are the signal carrying part of a fiber-optic cable. *See also* single-mode fiber (SMF) and multimode fiber.

optical fiber duplex adapter

A mechanical media termination device designed to align and join two duplex connectors.

optical fiber duplex connector

A mechanical media termination device designed to transfer optical power between two pairs of optical fibers.

optical isolator

A component used to block out reflected and other unwanted light.

optical link budget

Also referred to as the optical loss budget. This is the maximum amount of signal loss permitted for your network for a given application protocol. Typically verified at multiple wavelengths.

optical loss test set

An optical power meter and a light source calibrated for use together to detect and measure loss of signal on an optical cable.

optical polarization

A term used to describe the orientation in space of a time varying field vector of an optical signal.

optical power meter

A meter designed to measure absolute optical power.

optical receiver

An optoelectronic circuit that converts an incoming signal to an electronic signal. The optical receiver will include a transducer in the form of a detector, which might be a photodiode or other device. When irradiated by an optical power device, it will be able to translate the optical signal into an electronic signal.

optical reference plane

Defines the optical boundary between the MIC (media interface connector) plug and the MIC receptacle.

optical repeater

An optoelectronic device, which could include an amplifier, that receives a signal and amplifies it, especially in the case of analog signals. In the case of a digital signal, the optical repeater reshapes or retimes the signal and then retransmits it.

optical return loss (ORL)

ORL is a ratio expressed in decibels. The reflection is caused by a component or an assembly.

optical spectrum

Starts with red, then orange, yellow, green, blue, indigo, and finally violet. Each color represents a wavelength or frequency of electromagnetic energy; the spectrum is between 400nm and 700nm. 400nm is the ultraviolet portion of the spectrum and 700nm is the infrared portion of the spectrum.

optical splitter

A device used to split fiber-optic signals.

optical subassembly

The portion of a fiber-optic receiver that guides light from the optical fiber to the photodiode.

optical time-domain reflectometer (OTDR)

A device used to test a fiber-optic link, including the optical fiber and connectors, by launching an optical signal (light pulse) through the link and measuring the amount of energy that is reflected back. The OTDR is a troubleshooting device that can pinpoint faults throughout a fiber-optic link.

optical transmitter

An optoelectronic circuit that converts an electronic signal into an optical signal.

optical trench

A term used to describe a layer of glass that surrounds the core of the optical fiber. It has a lower refractive index than the core or the cladding. The low refractive index acts as a barrier and prevents light from escaping the core of the optical fiber when it is bent, reducing the amount of attenuation.

optical waveguide

Any structure that can guide light; the optical waveguide is a nonconductive material with a central core of optically transparent material (usually silica glass) surrounded by a transparent cladding material that has a lower refractive index than the core.

optoelectronic

Any device that uses or responds to optical power in its internal operation.

optomechanical

A term used to describe a device that moves optical fiber or bulk optic elements by means of mechanical devices.

outlet box

A metallic or nonmetallic box mounted within a wall, floor, or ceiling used to hold outlet, connector, or transition devices.

output

The useful signal or power delivered by a circuit or device.

outside plant

A telecommunications infrastructure designed for installation outside of any structure.

outside plant (OSP) cables

Typically used outside of the wire center but also may be routed into the CEF. Since OSP cables are more flammable than premises (indoor) cables, the distance of penetration into the building must be limited.

overfilled launch

The process of an LED transmitter overfilling a multimode optical fiber.

oversampling

A method of synchronous bit synchronization. The receiver samples the signal at a much faster rate than the data rate. This permits the use of an encoding method that does not add clocking transitions.

over-voltage threshold

The level of over-voltage that will trip the circuit breaker in a surge protector.

P

P region

The area in a semiconductor that is doped to have an abundance of electron acceptors in which vacancies in the valence electron level are the dominant current carriers.

packet

Bits grouped serially in a defined format containing a command or data message sent over a network. The packet is the major structure of data sent over a network.

packet switching

The process of breaking messages into packets. Each packet is then routed optimally across the network. Packet sequence numbers are used at the destination node to reassemble packets.

packing fraction

At a cut end, the fraction of the face surface area of a fiber-optic bundle that is fiber core.

PAM5x5

The signal-encoding technique used in the Ethernet 100Base-T2 and 1000Base-T media systems.

Part 68 requirements

Specifications established by the FCC as the minimum acceptable protection that communications equipment must provide to the telephone network.

passive branching device

A device that divides an optical input into two or more optical outputs.

passive coupler

Divides light without generating new light.

passive network

A network that does not use electrically powered equipment or components excluding the transmitter to get the signal from one place to another.

passive optical network

An optical network that does not use electrically powered equipment or components excluding the transmitter to get the signal from one place to another.

patch cable

Any flexible piece of cable that connects one network device to the main cable run or to a patch panel that in turn connects to the main cable run; also called *patch cord*. Used for interconnecting circuits on a patch panel or cross-connect. Patch cables are short distance, usually have connectors preinstalled on both ends, are used to connect equipment, and are generally between 3 and 6 meters long.

patch panel

A connecting hardware that typically provides a means to connect horizontal or backbone cables to an arrangement of fixed connectors that may be accessed using patch cords or equipment cords to form cross-connections or interconnections. Patch panels may connect either copper or optical fiber cables.

patching

A means of connecting circuits via cords and connectors that can be easily disconnected and reconnected at another point. May be accomplished by using modular patch cords connected between jack fields or by patch cord assemblies that plug onto connecting blocks.

pathway

A facility (e.g., conduit, cable tray, raceway, ducting, or plenum) for the placement and protection of telecommunications cables.

peak

The maximum instantaneous value of a varying current or voltage.

peak wavelength

The optical wavelength at which the power output of a source is at its maximum level.

pedestal

A device, usually mounted on the floor or ground, which is used to house voice/data jacks or power outlets at the point of use. Also commonly referred to as a monument, tombstone, above-floor fitting, or doghouse.

periodicity

Uniformly spaced variations in the insulation diameter of a transmission cable that result in reflections of a signal.

permanent link

The portion of a link that contains the horizontal cabling, one transition point, telecommunications outlet, and one cross-connect.

permanent virtual circuit (PVC)

Technology used by frame relay (as well as other technologies like X.25 and ATM) that allows virtual data circuits to be set up between the sender and receiver over a packet-switched network.

phase

An angular relationship between waves or the position of a wave in its oscillation cycle.

phase modulation (PM)

One of three basic methods of adding information to a sine wave signal in which its phase is varied to impose information on it. *See also* amplitude modulation (AM) and frequency modulation (FM).

phase shift

A change in the phase relationship between two alternating quantities.

photo-bleaching

A reduction in added loss that occurs when a fiber is exposed to light. Ionizing radiation causes added loss. This loss can be reduced by transmitting light through the fiber during normal operation or by exposing the fiber to sunlight.

photodetector

An optoelectronic transducer, such as a pin photodiode or avalanche photodiode, that acts as a light detector.

photodiode

A component that converts light energy into electrical energy. The photodiode is used as the receiving end of a fiber-optic link.

photon

A basic unit of light; the smallest quantum particulate component of light.

photonic

A term coined to describe devices using photons, analogous to electronic, describing devices working with electrons.

physical bus topology

A network that uses one network cable that runs from one end of the network to the other. Workstations connect at various points along this cable. These networks are easy to run cable for, but they are typically not as reliable as a star topology. 10Base-2 Ethernet is a good example of a network architecture that uses a physical bus topology.

physical contact (PC)

Description for a connector that places an optical fiber end in direct physical contact with the optical fiber end of another connector.

physical mesh topology

A network configuration that specifies a link between each device in the network. A physical mesh topology requires a lot of cabling and is difficult to reconfigure.

physical ring topology

A network topology that is set up in a circular fashion. Data travels around the ring in one direction, and each device on the ring acts as a repeater to keep the signal strong as it travels. Each device incorporates a receiver for the incoming signal and a transmitter to send the data on to the next device in the ring. The network is dependent on the ability of the signal to travel around the ring. Cabling a physical ring topology is difficult because of the amount of cable that must be run. FDDI is an example of a network that can be wired to use a physical ring topology.

physical star topology

A network in which a cable runs from each network device to a central device called a hub. The hub allows all devices to communicate as if they were directly connected. The network may logically follow another type of topology such as bus or ring topology, but the wiring is still a star topology.

physical topology

The physical layout of a network, such as bus, star, ring, or mesh.

picofarad

One millionth of one millionth of a farad. Abbreviated pf.

picosecond (PS)

One trillionth of a second.

piercing tap

A specially designed connector used to connect a thicknet cable to a node. Also called a vampire tap.

pigtail

(1) A short length of fiber with a permanently attached device, usually a connector, on one end. (2) A fiber-optic cable assembly consisting of a connector or hardware device (such as a light source package installed by a manufacturer) on one end and an unterminated fiber at the other end. Normally found in applications wherein a splice is convenient for terminating a device with a connector. Also used when the loss characteristics of the connector must be known precisely. For instance, a splice of .03dB might be reliably predicted and controlled, but the variability of most commercially available terminations is unacceptable, so a pre-characterized cable assembly is cut into a pigtail and attached to the device through splicing.

PIN photodiode

A photodiode that works like a PN photodiode; however, it is manufactured to offer better performance.

pinout scheme

The pinout scheme is the pattern that identifies which wires in a UTP cable are connected to each pin of a connector.

plain old telephone service (POTS)

The basic service that supplies standard single-line telephones, telephone lines, and access to the public switched network; it only receives and places calls and has no added features like call waiting or call forwarding.

planar waveguide

A waveguide fabricated in a flat material such as a thin film.

plastic-clad silica (PCS) fiber

A step-index multimode fiber that has a silica core and is surrounded by a lower index plastic cladding.

plastic fiber

Optical fiber having a plastic core and plastic cladding rather than using glass.

plasticizer

A chemical added to plastics to make them softer and more flexible.

plenum

The air-handling space between the walls, under structural floors, and above drop ceilings when used to circulate and otherwise handle air in a building. Plenum-grade cable can be run through these spaces if local building codes permit it.

plenum cable

Cable whose flammability and smoke characteristics allow it to be routed in plenum spaces without being enclosed in a conduit; all cables with this rating must pass NFPA 262 (formerly UL 910).

plug

The male component of a plug/jack connector system. In premises cabling, a plug provides the means for a user to connect communications equipment to the communications outlet.

PN diode

A basic photodiode.

point-to-point transmission

Carrying a signal between two endpoints without branching to other points.

polarity

In an electrical circuit, it identifies which side is positive and which is negative. In a fiber-optic network, it is the positioning of connectors and adapters to ensure that there is an end-to-end transmission path between the transmitter and the receiver of a channel.

polarization

(1) Alignment of the electric and magnetic fields that make up an electromagnetic wave. Normally refers to the electric field. If all light waves have

the same alignment, the light is polarized. (2) The direction of vibration of the photons in the light wave.

polarization maintaining fiber

Optical fiber that maintains the polarization of light that enters it.

polarization mode dispersion

Spreading of the light wave caused by imperfections in a single-mode optical fiber that slow down a polarization mode of the signal. When one polarization mode lags behind another, the signal spreads out and can become distorted.

polarization stability

The degree of variation in insertion loss as the polarization state of the input light is varied.

polling

A media access control method that uses a central device called a controller, which polls each device in turn and asks if it has data to transmit. 100VG-AnyLAN hubs poll nodes to see if they have data to transmit.

polybutadiene

A type of synthetic rubber often blended with other synthetic rubbers to improve their dielectric properties.

polyethylene (PE)

A thermoplastic material with excellent electrical properties. PE is used as an insulating material and as jacket material where flame-resistance requirements allow.

polyimide

A polymer that is used to coat optical fiber for harsh environment applications.

polymer

A substance made of repeating chemical units or molecules. The term is often used as a synonym for plastic, rubber, or elastomer.

polypropylene

A thermoplastic material that is similar to polyethylene, which is somewhat stiffer and has a higher softening point (temperature), with comparable electrical properties.

polyurethane (PUR)

A broad class of polymers that are noted for good abrasion and solvent resistance. Not as common as PVC (polyvinyl chloride).

polyvinyl chloride (PVC)

A general-purpose thermoplastic used for wire and cable insulation and jackets.

potting

The process of sealing by filling with a substance to exclude moisture.

pot life

The amount of time after mixing where epoxy can be used before it begins to harden and must be discarded.

power brownout

Occurs when power drops below normal levels for several seconds or longer.

power level

The ratio between the total power delivered to a circuit, cable, or device and the power delivered by that device to a load.

power overage

Occurs when too much power is coming into a piece of equipment. *See also* power spike and power surge.

power ratio

The ratio of power appearing at the load to the input power. Expressed in decibels.

power sag

Occurs when the power level drops below normal and rises to normal in less than one second.

power spike

Occurs when the power level rises above normal and drops back to normal for less than a second. *See also* power overage and power surge.

power sum

A test method for cables with multiple pairs of wire whereby the mathematical sum of pair-to-pair crosstalk from a reference wire pair is measured while all other wire pairs are carrying signals. Power-sum tests are necessary on cables that will be carrying bidirectional signals on more than two pairs. Most commonly measured on four-pair cables.

power-sum alien near-end crosstalk (PSANEXT)

The power-sum measurement of alien crosstalk in the near end (at the transmitter); a newly established requirement in ANSI/TIA-568-C.2 for Category 6A cables.

power-sum alien far-end crosstalk (PSAFEXT)

The power-sum measurement of alien crosstalk in the far end (away from the transmitter); the result from this test is used with the attenuation to calculate power-sum alien attenuation-to-crosstalk ratio (far end) (PSAACRF), a requirement of ANSI/TIA-568-C.2 for Category 6A cables.

power-sum alien attenuation-to-crosstalk ratio, far end (PSAACRF)

The ratio of the attenuation (signal) to the power-sum measurement of alien crosstalk in the far end (noise); a requirement of ANSI/TIA-568-C.2 for Category 6A cables.

power-sum attenuation-to-crosstalk ratio (PSACR)

The ratio of attenuation to power-sum near-end crosstalk; while not a requirement of ANSI/TIA-568-C.2, it is a figure of merit used by all manufacturers to denote the signal-to-power-sum noise ratio in the near end.

power-sum attenuation-to-crosstalk ratio, far end (PSACRF)

The ratio of attenuation to power-sum far-end crosstalk; a requirement of ANSI/TIA-568-C.2; formerly known as power-sum equal-level far-end crosstalk (PSELFEXT)

power surge

Occurs when the power level rises above normal and stays there for longer than a second. *See also* power overage and power spike.

power underage

Occurs when the power level drops below the standard level. Opposite of power overage.

preform

A short, thick glass rod that forms the basis for an optical fiber during the manufacturing process. The preform is created first, and then melted and drawn under constant tension to form the long, thin optical fiber.

prefusing

Fusing the end of a fiber-optic cable with a low current to clean the end; it precedes fusion splicing.

preload

A connector with built-in adhesive that must be preheated before the fiber can be installed.

premises

A telecommunications term for the space occupied by a customer or an authorized/joint user in a building on continuous or contiguous property that is not separated by a public road or highway.

premises wiring system

The entire wiring system on a user's premises, especially the supporting wiring that connects the communications outlets to the network interface jack.

prewiring

Wiring that is installed before walls and ceilings are enclosed. Prewiring is usually easier than waiting until the walls are built to install wire.

primary coating

The specialized coating applied to the surface of the fiber cladding during manufacture.

primary rate interface (PRI)

As defined by the ISDN standard, consists of 23 B-channels (64Kbps each) and one 64Kpbs D-channel (delta channel) in the United States, or 30 B-channels and one D-channel in Europe.

private branch exchange (PBX)

A telephone switching system servicing a single customer, usually located on that customer's premises. It switches calls both inside a building and outside to the telephone network, and it can sometimes provide access to a computer from a data terminal. Now used interchangeably with PABX (private automatic branch exchange).

profile alignment system (PAS)

A fiber splicing technique for using non-electrooptical linked access technology for aligning fibers for splicing.

propagation delay

The difference in time between when a signal is transmitted and when it is received.

protector

A device that limits damaging voltages on metallic conductors by protecting them against surges and transients.

protocol

A set of predefined, agreed-upon rules and message formats for exchanging information among devices on a network.

protocol analyzer

A software and hardware troubleshooting tool used to decode protocol information to try to determine the source of a network problem and to establish baselines.

public data network

A network established and operated for the specific purpose of providing data transmission services to the public. *See also* public switched network.

public switched network

A network provided by a common carrier that provides circuit switching between public users, such as the public telephone network.

public switched telephone network (PSTN)

The basic phone service provided by the phone company. *See also* plain old telephone service (POTS).

puck

In fiber optics, a metal disc that holds a connector in the proper position against an abrasive for polishing the connector end face.

pull load

See tensile load.

pull strength

The pulling force that can be applied to a cable without damaging a cable or affecting the specified characteristics of the cable. Also called pull tension.

pull string

A string that is tied to a cable and is used to pull cables through conduits or over racks. Similar to fish tape.

pulse

A current or voltage that changes abruptly from one value to another and back to the original value in a finite length of time.

pulse code modulation (PCM)

The most common method of converting an analog signal, such as speech, to a digital signal by sampling at a regular rate and converting each sample to an equivalent digital code. The digital data is transmitted sequentially and returned to analog format after it is received.

pulse dispersion

The dispersion of pulses as they travel along an optical fiber.

pulse spreading

The dispersion of an optical signal with time as it propagates through an optical fiber.

pulse width

The amount of energy a time domain reflectometer transmits as its test pulse. This may be a selectable feature allowing for a time domain reflectometer to assess faults at various extents of distance along a cable.

punch-down block

A generic name for any cross-connect block where the individual wires in UTP are placed into a terminal groove and "punched down" with a special tool. The groove pierces the insulation and makes contact with the inner conductor. The punch-down operation may also trim the wire as it terminates. Punch-downs are performed on telecommunications outlets, 66-blocks, and 110-blocks. Also called cut down.

Q

quadrature amplitude modulation (QAM)

The modulation of two separate signals onto carriers at a single frequency and kept separate by having the two signals 90 degrees out of phase.

quality of service (QoS)

Data prioritization at the Network layer of the OSI model. QoS results in guaranteed throughput rates.

quantizer IC

An integrated circuit (IC) that measures received optical energy and interprets each voltage pulse as a binary 1 or 0.

quantizing error

Inaccuracies in analog to digital conversion caused by the inability of a digital value to match an analog value precisely. Quantizing error is reduced as the number of bits used in a digital sample increases, since more bits allow greater detail in expressing a value.

quantum

A basic unit, usually used in reference to energy. A quantum of light is called a photon.

quartet signaling

The encoding method used by 100VG-AnyLAN, in which the 100Mbps signal is divided into four 25Mbps channels and then transmitted over different pairs of a cable. Category 3 cable transmits one channel on each of four pairs.

R

R

Symbol for resistance.

raceway

Any channel or structure used in a building to support and guide electrical and optical fiber wires or cables. Raceways may be metallic or nonmetallic and may totally or partially enclose the wiring (e.g., conduit, cable trough, cellular floor, electrical metallic tubing, sleeves, slots, under-floor raceways, surface raceways, lighting fixture raceways, wireways, busways, auxiliary gutters, and ventilated flexible cableways). *See also* pathway.

rack

A frame-like structure where patch panels, switches, and other network equipment are installed. The typical dimension is 19 inches.

radial refractive index profile

The refractive index measured in a fiber as a function of the distance from the axial core or center.

radiant flux

Radiant flux is the measured amount of energy on a surface per unit time.

radiation (rad) hardened

Used to describe material that is not sensitive to the effects of nuclear radiation; such material is typically used for military applications.

radio frequency (RF)

The frequencies in the electromagnetic spectrum that are used for radio communications.

radio frequency interference (RFI)

The interference on copper cabling systems caused by radio frequencies.

radius of curvature

A term used to describe the roundness of the connector end face, measured from the center axis of the connector ferrule.

Raman amplification

An amplification method using a pump laser to donate energy to a signal to amplify it without using a doped length of fiber.

ray

A beam of light in a single direction. Usually a representation of light traveling in a particular direction through a particular medium.

Rayleigh scattering

The redirection of light caused by atomic structures and particles along the light's path. Rayleigh scattering is responsible for some attenuation in optical fiber, because the scattered light is typically absorbed when it passes into the cladding.

reactance

A measure of the combined effects of capacitance and inductance on an alternating current. The amount of such opposition varies with the frequency of the current. The reactance of a capacitor decreases with an increase in frequency. The opposite occurs with an inductance.

receive cable

A cable used to measure insertion loss with an OTDR at the far end of the cable plant.

receive jumper

The test jumper connected to the optical power meter.

receiver

A device whose purpose is to capture transmitted signal energy and convert that energy for useful functions. In fiber-optic systems, an electronic component that converts light energy to electrical energy.

receiver sensitivity

In fiber optics, the amount of optical power required by a particular receiver in order to transmit a signal with few errors. Can be considered a measure of the overall quality of receiver. The more sensitive the receiver, the better its quality.

receptacle

The part of a fiber-optic receiver that accepts a connector and aligns the ferrule for proper optical transmission.

reference jumper

A test jumper.

reflectance

A percentage that represents the amount of light that is reflected back along the path of transmission from the coupling region, the connector, or a terminated fiber.

reflection

(1) A return of electromagnetic energy that occurs at an impedance mismatch in a transmission line, such as a LAN cable. *See also* return loss. (2) The immediate and opposite change in direction that happens to a light beam when it strikes a reflective surface. Reflection causes several spectral problems, including high optical distortion and enhanced intensity noise.

refraction

The bending of a beam of light as it enters a medium of different density. Refraction occurs as the velocity of the light changes between materials of two different refractive indexes.

refractive index gradient

The change in refractive index with respect to the distance from the axis of an optical fiber.

refractive index profile

A graphical description of the relationship between the refractive indexes of the core and the cladding in an optical fiber.

regenerator

A receiver-transmitter pair that detects a weak signal, cleans it up, then sends the regenerated signal through another length of fiber.

Registered Communications Distribution Designer (RCDD)

A professional accreditation granted by BICSI (the Building Industry Consulting Service International). RCDDs have demonstrated a superior level of knowledge of the telecommunications wiring industry and associated disciplines.

registered jack (RJ)

Telephone and data jacks/applications that are registered with the FCC. Numbers such as RJ-11 and RJ-45 are widely misused in the telecommunications industry—the RJ abbreviation was originally used to identify a type of service and wiring pattern to be installed, not a specific jack type. A much more precise way to identify a jack is to specify the number of positions (width of opening) and number of conductors. Examples include the eight-position, eight-conductor jack and the six-position, four-conductor jack.

remodel box

An electrical box that is designed to clamp to the wallboard, as opposed to a "new construction" box, which is nailed to a wall stud.

repeater

(1) A device that receives, amplifies (and reshapes), and retransmits a signal. It is used to overcome attenuation by boosting signal levels, thus extending the distance over which a signal can be transmitted. Repeaters can physically extend the distance of a LAN or connect two LAN segments.

resistance

In DC (direct current) circuits, the opposition a material offers to current flow, measured in ohms. In AC (alternating current) circuits, resistance is the real component of impedance and may be higher than the value measured at DC.

resistance unbalance

A measure of the inequality of the resistance of the two conductors of a transmission line.

responsivity

(1) The ratio of a detector's output to input, usually measured in units of amperes per watt (or microamperes per microwatt). (2) The measure of how well a photodiode converts a wavelength or range of wavelengths of optical energy into electrical current.

restricted mode launch (RML)

A launch that limits a laser's launch condition so that fewer modes are used.

retermination

The process of disconnecting, then reconnecting a cable to a termination point (possibly moving the cable in the process).

retractile cord

A cord with a specially treated insulation or jacket that causes it to retract like a spring. Retractile cords are commonly used between a telephone and a telephone handset.

return loss

The ratio of reflected power to inserted power. Return loss is a measure of the signal reflections occurring along a channel or basic link and is related to various electrical mismatches along the cabling. This ratio, expressed in decibels, describes the ratio of optical power reflected by a component, for instance a connector, to the optical power introduced to that component.

return reflection

Light energy that is reflected from the end of a fiber through Fresnel reflection.

return to zero (RZ)

A digital coding scheme where the signal level is low for a 0 bit and high for a 1 bit during the first half of a bit interval; in either case, the bit returns to zero volts for the second half of the interval.

reversed biased

A term used to describe when the n region of the semiconductor is connected to a positive electrical potential and the p region is connected to a negative electrical potential.

reversed pair

A wiring error in twisted-pair cabling where the conductors of a pair are reversed between connector pins at each end of a cable. A cabling tester can detect a reversed pair.

RF access point

A wireless receiver and transmitter used in a wireless network. This point accesses incoming RF signals and transmits RF signals.

RF noise

Signal noise experienced on copper cables that is caused by radio frequency interference.

RFP

A request for proposal. A documented set of requirements used by a designer to estimate the cost of a project.

RG-58

The type designation for the coaxial cable used in thin Ethernet (10Base-2). It has a 50 ohm impedance rating and uses BNC connectors.

RG-62

The type designation for the coaxial cable used in ARCnet networks. It has a 93 ohm impedance and uses BNC connectors.

RG/U

Radio grade/universal. RG is the common military designation for coaxial cable.

ribbon

Multiple conductors or optical fibers clad in a single, flat, ribbon-like cable.

ring

(1) A polarity designation of one wire of a pair indicating that the wire is that of the secondary color of a five-pair cable (which is not commonly used anymore) group (e.g., the blue wire of the blue/white pair). (2) A wiring contact to which the ring wire is attached. (3) The negative wiring polarity (*see also* tip). (4) Two or more stations in which data is passed sequentially between active stations, each in turn examining or copying the information before finally returning it to the source. *See also* ring topology.

ring conductor

A telephony term used to describe one of the two conductors that is in a cable pair used to provide telephone service. This term was originally coined from its position as the second (ring) conductor of a tip-ring-sleeve switchboard plug. *See also* ring.

ring topology

A network topology in which terminals are connected in a point-to-point serial fashion in an unbroken circular configuration. Many logical ring topologies such as Token Ring are wired as a star for greater reliability.

ripcord

A length of string built into optical fiber cables that is pulled to split the outer jacket of the cable without using a blade.

riser

(1) A designation for a type of cable run between floors Fire-code rating for indoor cable that is certified to pass through the vertical shaft from floor to floor. (2) A space for indoor cables that allow cables to pass between floors, normally a vertical shaft or space.

riser cable

A type of cable used in vertical building shafts, such as telecommunications and utility shafts. Riser cable typically has more mechanical strength than general use cable and has an intermediate fire protection rating.

RJ-45

A USOC code identifying an eight-pin modular plug or jack used with unshielded twisted-pair cable. Officially, an RJ-45 connector is a telephone connector designed for voice-grade circuits. Only RJ-45-type connectors with better signal handling characteristics are called eight-pin connectors in most standards documents, though most people continue to use the RJ-45 name for all eight-pin connectors.

RJ-connector

A modular connection mechanism that allows for as many as eight copper wires (four pairs). Commonly found in phone (RJ-11) or 10Base-T (RJ-45) connections.

rope strand

A conductor composed of groups of twisted strands.

router

A device that connects two networks and allows packets to be transmitted and received between them. A router may also determine the best path for data packets from source to destination. Routers primarily operate on Layer 3 (the Network layer) of the OSI model.

routing

A function of the network layer that involves moving data throughout a network. Data passes through several network segments using routers that can select the path the data takes.

RS-232C

The EIA's registered standard that defines an interface that computers use to talk to modems and other serial devices such as printers or plotters.

S

sample-and-hold circuit

A circuit that samples an analog signal such as a voltage level and then holds the voltage level long enough for the analog-to-digital (A/D) converter to change the level to a numerical value.

SC connector

An optical-fiber connector made from molded plastic using push-pull mechanics for joining to a fiber adapter. The SC connector has a 2.5mm ferrule push-pull latching mechanism and can be snapped together to form duplex and multifiber connectors. SC connectors are the preferred fiber-optic cable for premises cabling and are recognized by the ANSI/TIA/EIA-568-B standard for structured cabling.

scanner

A cable-testing device that uses TDR methods to detect cable transmission anomalies and error conditions.

scattering

(1) A property of glass that causes light to deflect from the fiber and contributes to losses. (2) The redirection of light caused by atomic structures and particles along the light's path. *See also* Rayleigh scattering.

score

To lightly nick the optical fiber with a blade.

screened twisted-pair (ScTP) cable

A balanced four-pair UTP with a single foil or braided screen surrounding all four pairs in order to minimize EMI radiation or susceptibility. Screened twisted-pair is also sometimes called foil twisted-pair (FTP). ScTP is a shielded version of Category 3, 5, 5e, and 6 UTP cables; they are less susceptible to EMI than UTP cables but are more susceptible than STP cables.

scribe

A tool used to mark an object prior to some type of drilling or cutting.

secondary coating

The acrylate coating layer applied over the primary coating that provides a hard outer surface.

segment

A portion of a network that uses the same length of cable (electrically contiguous). Also the portion of a network that shares a common hub or set of interconnected hubs.

SELFOC Lens

A trade name used by the Nippon Sheet Glass Company for a graded-index fiber lens. A segment of graded-index fiber made to serve as a lens.

semiconductor

In wire industry terminology, a material possessing electrical conductivity that falls somewhere between that of conductors and insulators. Usually made by adding carbon particles to an insulator.

This is not necessarily the same as semiconductor materials such as silicon or germanium.

semiconductor laser

A laser in which the injection of current into a semiconductor diode produces light by recombination of holes and electrons at the junction between p- and n-doped regions. Also called a semiconductor diode laser.

semiconductor optical amplifier (SOA)

A laser diode with optical fibers at each end instead of mirrors. The light from the optical fiber at either end is amplified by the diode and transmitted from the opposite end.

sensitivity

For a fiber-optic receiver, the minimum optical power required to achieve a specified level of performance, such as BER.

separator

Pertaining to wire and cable, a layer of insulating material such as textile, paper, or Mylar, which is placed between a conductor and its dielectric, between a cable jacket and the components it covers, or between various components of a multiple conductor cable. It can be used to improve stripping qualities or flexibility, or it can offer additional mechanical or electrical protection to the components it separates.

sequential markings

Markings on the outside of a fiber-optic cable to aid in measuring the length of the cable.

serial

When applied to digital data transmission, it means that the binary bits are sent one after another in the order they were generated.

series wiring

See daisy chain.

service loop

A loop or slack left in a cable when the cable is installed and terminated. This loop allows future trimming of the cable or movement of equipment if necessary.

service profile identifier (SPID)

The ISDN identification number issued by the phone company that identifies the ISDN terminal equipment attached to an ISDN line.

sheath

An outer protective layer of a fiber-optic or copper cable that includes the cable jacket, strength members, and shielding.

shield

A metallic foil or multiwire screen mesh that is used to reduce electromagnetic fields from penetrating or exiting a transmission cable. Also referred to as a screen.

shield coverage

The physical area of a cable that is actually covered by shielding material, often expressed as a percentage.

shield effectiveness

The relative ability of a shield to screen out undesirable interference. Frequently confused with the term shield coverage.

shielded twisted pair (STP)

A type of twisted-pair cable in which the pairs are enclosed in an outer braided shield, although individual pairs may also be shielded. STP most often refers to the 150 ohm IBM Type 1, 2, 6, 8, and 9 cables used with Token Ring networks. Unlike UTP cabling, the pairs in STP cable have an individual shield, and the individual shielded cables are wrapped in an overall shield. The primary advantages of STP cable are that it has less attenuation at higher frequencies and is less susceptible to EMI. Since the advent of standards-based structured wiring, STP cable is rarely used in the United States.

short-link phenomenon

Cross-talk signal errors occurring in excessively short lengths of cables. This depends on application speed and is corrected by using longer cable runs.

short wavelength

In reference to light, a wavelength shorter than 1000nm.

SI units

The standard international system of metric units.

signal

The information conveyed through a communication system.

signal encoding

The process whereby a protocol at the Physical layer receives information from the upper layers and translates all the data into signals that can be transmitted on a transmission medium.

signaling

The process of transmitting data across the medium. Types of signaling include digital and analog, baseband and broadband.

signal-to-noise ratio (SNR or S/N)

The ratio of received signal level to received noise level, expressed in decibels and abbreviated SNR or S/N. A higher SNR ratio indicates better channel performance. The relationship between the usable intended signal and the extraneous noise present. If the SNR limit is exceeded, the signal transmitted will be unusable.

silica glass

Glass made mostly of silicon dioxide used in conventional optical glass that is used commonly in optical fiber cables.

Silicone

A General Electric trademark for a material made from silicon and oxygen. Can be in thermosetting elastomer or liquid form. The thermosetting elastomer form is noted for high heat resistance.

silver satin cable

The silver-gray voice-grade patch cable used to connect a telephone to a wall jack such as that used by home telephones. Silver satin cables are unsuitable for use in LAN applications because they do not have twisted pairs, and this results in high levels of crosstalk and capacitance.

simplex

(1) A link that can only carry a signal in one direction. (2) A fiber-optic cable or cord carrying a single fiber. Simplex cordage is mainly used for patch cords and temporary installations.

simplex cable

A term sometimes used to describe a single-fiber cable.

simplex transmission

Data transmission over a circuit capable of transmitting in only one direction.

single attachment station (SAS)

With FDDI networks, denotes a station that attaches to only one of two rings in a dual-ring environment.

single-board computer

A circuit board containing the components needed for a computer that performs a prescribed task. A single-board computer is often the basis of a larger piece of equipment that contains it.

single-ended line

An unbalanced circuit or transmission line, such as a coaxial cable (see also balanced line and unbalanced line).

single-frequency laser

A laser that emits a range of wavelengths small enough to be considered a single frequency.

single-mode depressed clad

See bend insensitive.

single-mode fiber (SMF)

Optical-fiber cable with a small core, usually between two and nine microns, which can support only one wavelength. It requires a laser source for the input because the acceptance cone is so small. The small core radius approaches the wavelength of the source. Single-mode optical-fiber cable is typically used for backbones and to transmit data over long distances.

single polarization fibers

Optical fibers capable of carrying light in only one polarization.

sintering

A process in optical fiber manufacturing in which the soot created by heating silicon dioxide is compressed into glass to make the fiber preform.

sinusoidal

A signal that varies over time in proportion to the sine of an angle. Alternating current (AC) is sinusoidal.

skew ray

A light ray that does not intersect the fiber axis and generally enters the fiber at a very high angle.

skin effect

The tendency of alternating current to travel on the surface of a conductor as the frequency increases.

slash sheet

Within SAE International, a family of documents where the main document refers to the overall/general aspect and the slash sheet refers to specific items.

slitting cord (or slitting string)

See ripcord.

small form factor (SFF) connector

A type of optical fiber connector that is designed to take up less physical space than a standard-sized connector. An SFF connector provides support for two strands of optical fiber in a connector enclosure that is similar to an RJ-45. There is currently no standard for SFF connectors; types include the LC and the MT-RJ connectors.

sneak current

A low-level current that is of insufficient strength to trigger electrical surge protectors and thus may be able to pass between these protectors undetected. The sneak current may result from contact between communications lines and AC power circuits or from power induction. This current can cause equipment damage unless secondary protection is used.

solar cell

A device used to convert light energy into electrical current.

solid-state laser

A laser whose active medium is a glass or crystal.

soliton

A special type of light pulse used in fiber-optic communications in combination with optical amplifiers to help carry a signal longer distances.

source

In fiber optics, the device that converts the information carried by an electrical signal to an optical signal for transmission over an optical fiber. A fiber-optic source may be a light-emitting diode or laser diode.

source address

The address of the station that sent a packet, usually found in the source area of a packet header. In the case of LAN technologies such as Ethernet and Token Ring, the source address is the MAC (media access control) address of the sending host.

spectral bandwidth

(1) The difference between wavelengths at which the radiant intensity of illumination is half its peak intensity. (2) Radiance per unit wavelength interval.

spectral width

A measure of the extent of a spectrum. The range of wavelengths within a light source. For a source, the width of wavelengths contained in the output at one half of the wavelength of peak power. Typical spectral widths are between 20nm and 170nm for an LED and between 1nm and 5nm for a laser diode.

spectrum

Frequencies that exist in a continuous range and have a common characteristic. A spectrum may be inclusive of many spectrums; the electromagnetic radiation spectrum includes the light spectrum, the radio spectrum, and the infrared spectrum.

speed of light (c)

In a vacuum, light travels 299,800,000 meters per second. This is used as a reference for calculating the index of refraction.

splice

(1) A permanent joint between two optical waveguides. (2) Fusing or mechanical means for joining two fiber ends.

splice enclosure

A cabinet used to organize and protect splice trays.

splice tray

A container used to organize and protect spliced fibers.

split pair

A wiring error in twisted-pair cabling where one of a pair's wires is interchanged with one of another pair's wires. Split pair conditions may be determined with simple cable testing tools (simple continuity tests will not reveal the error because the correct pin-to-pin continuity exists between ends). The error may result in impedance mismatch, excessive crosstalk, susceptibility to interference, and signal radiation.

splitter

A device with a single input and two or more outputs.

splitting ratio

The ratio of power emerging from two output ports of a coupler.

spontaneous emission

The emission of random photons (incoherent light) at the junction of the p and n regions in a light-emitting diode when current flows through it. Occurs when a semiconductor accumulates spurious electrons. Spontaneous emission interferes with coherent transmission.

S/T interface

The four-wire interface of an ISDN terminal adapter. The S/T interface is a reference point in ISDN.

ST connector

A fiber-optic connector with a bayonet housing; it was developed by AT&T but is not in favor as much as SC or FC connectors. It is used with older Ethernet 10Base-FL and fiber-optic inter-repeater links (FIORLs).

stabilized light source

An LED or laser diode that emits light with a controlled and constant spectral width, central wavelength, and peak power with respect to time and temperature.

standards

Mutually agreed-upon principles of protocol or procedure. Standards are set by committees working under various trade and international organizations.

star coupler

A fiber-optic coupler in which power at any input port is distributed to all output ports. Star couplers may have up to 64 input and output ports.

star network

See hierarchical star topology.

star topology

(1) A method of cabling each telecommunications outlet/connector directly to a cross-connect in a horizontal cabling subsystem. (2) A method of cabling each cross-connect to the main cross-connect in a backbone cabling subsystem. (3) A topology in which each outlet/connector is wired directly to the hub or distribution device.

static charge

An electrical charge that is bound to an object.

static load

Load such as tension or pressure that remains constant over time, such as the weight of a fiber-optic cable in a vertical run.

station

A unique, addressable device on a network.

stay cord

A component of a cable, usually of high tensile strength, used to anchor the cable ends at their points of termination and keep any pull on the cable from being transferred to the electrical conductors.

step index fiber

An optical fiber cable, usually multimode, that has a uniform refractive index in the core with a sharp decrease in index at the core/cladding interface. The light is reflected down the path of the fiber rather than refracted as in graded-index fibers. Step-index multimode fibers generally have lower bandwidths than graded-index multimode fibers.

step-index single-mode fiber

A fiber with a small core that is capable of carrying light in only one mode. Sometimes referred to as single-mode optical fiber cable.

step insulated

A process of applying insulation to a cable in two layers. Typically used in shielded networking cables so that the outer layer of insulation can be removed and the remaining conductor and insulation can be terminated in a connector.

stimulated emission

The process in which a photon interacting with an electron triggers the emission of a second photon with the same phase and direction as the first. Stimulated emission is the basis of a laser.

stitching

The process of terminating multiconductor cables on a punch-down block such as a 66-block or 110-block.

STP-A

Refers to the enhanced IBM Cabling System specifications with the Type A suffix. The enhanced Type 1A, 2A, 6A, and 9A cable specifications were designed to support operation of 100Mbps FDDI signals over copper. See Type 1A, Type 2A, Type 6A, or Type 9A.

strength member

The part of a fiber-optic cable composed of Kevlar aramid yarn, steel strands, or fiberglass filaments that increases the tensile strength of the cable.

Strike termination device

A lightning rod system engineered to prevent harm to a network caused by lightning strikes.

structural return loss (SRL)

A measurement of the impedance uniformity of a cable. It measures energy reflected due to structural variations in the cable. The higher the SRL number, the better the performance; this means more uniformity and lower reflections.

structured cabling system

Telecommunications cabling that is organized into a hierarchy of wiring termination and interconnection structures. The concept of structured wiring is used in the common standards from the TIA and EIA. See Chapter 1 for more information on structured cabling and Chapter 2 for more information on structured cabling standards.

stud cavity

A type of frame used in building construction.

submarine cable

A cable designed to be laid underwater.

subminiature D-connector

The subminiature D-connector is a family of multipin data connectors available in 9-, 15-, 25-, and 37-pin configurations. Sometimes referred to as DB9, DB15, DB25, and DB37 connectors, respectively.

subnetwork

A network that is part of another network. The connection is made through a router.

subnet mask

Subnet masks consist of 32 bits, usually a block of 1s followed by a block of 0s. The last block of 0s designate that part as being the host identifier. This allows a classful network to be broken down into subnets.

supertrunk

A cable that carries several video channels between the facilities of a cable television company.

surface-emitting LED

An LED in which incoherent light is emitted at all points along the PN junction.

surface light-emitting diode (SLED)

A light-emitting diode (LED) that emits light from its flat surface rather than its side. These are simple and inexpensive and provide emission spread over a wide range.

surface-mount assembly (SMA) connector

An optical fiber cable connector that is a threaded type connector. The SMA 905 version is a straight ferrule design, whereas the SMA 906 is a stepped ferrule design.

surface-mount box

A telecommunications outlet that is placed in front of the wall or pole.

surface-mount panel

A panel for electrical or data communications connections placed on a wall or pole.

surge

A rapid rise in current or voltage, usually followed by a fall back to a normal level. Also referred to as a transient.

surge protector

A device that contains a special electronic circuit that monitors the incoming voltage level and then trips a circuit breaker when an over-voltage reaches a certain level, called the over-voltage threshold.

surge suppression

The process by which transient voltage surges are prevented from reaching sensitive electronic equipment. *See also* surge protector.

switch

A mechanical, optical, or optomechanical device that completes or breaks an optical path or routes an optical signal.

switched network

A network that routes signals to their destinations by switching circuits or packets. Two different packets of information may not take the same route to get to the same destination in a packet-switched network.

synchronous

Transmission in which the data is transmitted at a fixed rate with the transmitter and receiver being synchronized.

Synchronous Optical Network (SONET)

The underlying architecture in most systems, which uses cells of fixed length.

T

T-1

A standard for digital transmission in North America. A digital transmission link with a capacity of 1.544Mbps (1,619,001 bits per second), T-1 lines are used for connecting networks across remote distances. Bridges and routers are used to connect LANs over T-1 networks.

T-3

A 44.736Mbps multichannel digital transmission system for voice or data provided by long-distance carriers. T-3C operates at 90Mbps.

tap

(1) A device for extracting a portion of the optical fiber. (2) On Ethernet 10Base-5 thick coaxial cable, a method of connecting a transceiver to the cable by drilling a hole in the cable, inserting a contact to the center conductor, and clamping the transceiver onto the cable at the tap. These taps are referred to as vampire taps.

tap loss

In a fiber-optic coupler, the ratio of power at the tap port to the power at the input port. This represents the loss of signal as a result of tapping.

tapered fiber

An optical fiber whose transverse dimensions vary monotonically with length.

T-carrier

A carrier that is operating at one of the standard levels in the North American Digital Hierarchy, such as T-1 (1.544Mbps) or T-3 (44.736Mbps).

T-coupler

A coupler having three ports.

TDMM

Abbreviation for telecommunications distribution methods manual. A manual used by BICSI for accreditation.

tee coupler

A device used for splitting optical power from one input port to two output ports.

Teflon

DuPont Company trademark for fluorocarbon resins. *See also* fluorinated ethylene propylene (FEP) and tetrafluoroethylene (TFE).

telco

An abbreviation for telephone company.

telecommunications

Any transmission, emission, or reception of signs, signals, writings, images, sounds, or information of any nature by cable, radio, visual, optical, or other electromagnetic systems.

telecommunications bus bar

Refers to thick strips of copper or aluminum that conduct electricity within a switchboard, distribution board, substation, or other electrical apparatus in a telecommunications network.

Telecommunications Industry Association (TIA)

The standards body that helped to author the ANSI/TIA/EIA-568-B Commercial Building Telecommunications Cabling Standard in conjunction with EIA and that continues to update it, along with standards for pathways, spaces, grounding, bonding, administration, field testing, and other aspects of the telecommunications industry. See Chapter 2 for more information.

telecommunications infrastructure

A collection of telecommunications components that together provide the basic support for the distribution of all information within a building or campus. This excludes equipment such as PCs, hubs, switches, routers, phones, PBXs, and other devices attached to the telecommunications infrastructure.

telecommunications outlet

A fixed connecting device where the horizontal cable terminates that provides the interface to the work area cabling. Typically found on the floor or in the wall. Sometimes referred to as a telecommunications outlet/connector or a wall plate.

telecommunications closet

See telecommunications room.

telecommunications room

An enclosed space for housing telecommunications equipment, cable terminations, and cross-connect cabling used to serve work areas located on the same floor. The telecommunications room is the typical location of the horizontal cross-connect and is distinct from an equipment room because it is considered to be a floor-serving (as opposed to building- or campus-serving) facility.

Telecommunications Systems Bulletin (TSB)

A document released by the TIA to provide guidance or recommendations for a specific TIA standard.

tensile load

The amount of pulling force placed on a cable.

tensile strength

Resistance to pulling or stretching forces.

terminal

(1) A point at which information may enter or leave a communications network. (2) A device by means of which wires may be connected to each other.

terminal adapters

ISDN customer-premise equipment that is used to connect non-ISDN equipment (computers and phones) to an ISDN interface.

terminate

(1) To connect a wire conductor to something, typically a piece of equipment like patch panel, cross-connect, or telecommunications outlet. (2) To add a component such as a connector or a hardware connection to a bare optical fiber end.

terminator

A device used on coaxial cable networks that prevents a signal from bouncing off the end of the network cable, which would cause interference with other signals. Its function is to absorb signals on the line, thereby keeping them from bouncing back and being received again by the network or colliding with other signals.

termini

A term used to describe multiple fiber-optic contacts.

terminus

A single fiber-optic contact.

test fiber box

A box that contains optical fiber and is used as a launch or receive cable when testing the cable plant with an OTDR.

test jumper

A single- or multi-fiber cable used for connections between an optical fiber and test equipment.

tetrafluoroethylene (TFE)

A thermoplastic material with good electrical insulating properties and chemical and heat resistance.

theoretical cutoff wavelength

The shortest wavelength at which a single light mode can be propagated in a single-mode fiber. Below the cutoff several modes will propagate; in this case, the fiber is no longer single-mode but multimode.

thermal rating

The temperature range in which a material will perform its function without undue degradation such as signal loss.

thermo-optic

Using temperature to alter the refractive index.

thermoplastic

A material that will soften, flow, or distort appreciably when subjected to sufficient heat and pressure. Examples are polyvinyl chloride and polyethylene, which are commonly used in telecommunications cable jackets.

thicknet

Denotes a coaxial cable type (similar to RG-8) that is commonly used with Ethernet (10Base-5) backbones. Originally, thicknet cabling was the only cabling type used with Ethernet, but it was replaced as backbone cabling by optical fiber cabling. Thicknet cable has an impedance of 50 ohms and is commonly about 0.4 inches in diameter.

thinnet

Denotes a coaxial cable type (RG-58) that was commonly used with Ethernet (10Base-2) local area networks. This coaxial cable has an impedance of 50 ohms and is 0.2 inches in diameter. It is also called cheapernet due to the fact that it was cheaper to purchase and install than the bulkier (and larger) thicknet Ethernet cabling. The maximum distance for a thinnet segment is 180 meters.

tight buffer

A type of optical fiber cable construction where each glass fiber is buffered tightly by a protective thermoplastic coating to a diameter of 900 microns. High tensile strength rating is achieved, providing durability and ease of handling and connectorization.

tight-buffered

An optical fiber with a buffer that matches the outside diameter of the optical fiber, forming a tight outer protective layer; typically 900 microns in diameter.

time division multiple access (TDMA)

A method used to divide individual channels in broadband communications into separate time slots, allowing more data to be carried at the same time.

time division multiplexing (TDM)

Digital multiplexing that takes one pulse at a time from separate signals and combines them in a single stream of bits.

time domain reflectometry (TDR)

A technique for measuring cable lengths by timing the period between a test pulse and the reflection of the pulse from an impedance discontinuity on the cable. The returned waveform reveals undesired cable conditions, including shorts, opens, and transmission anomalies due to excessive bends or crushing. The length to any anomaly, including the unterminated cable end or cable break, may be computed from the relative time of the wave return and nominal velocity of propagation of the pulse through the cable. For optical fiber cables, *see also* optical time-domain reflectometry.

timing circuit

In an optical time-domain reflectometer, the circuit used to coordinate and regulate the testing process.

tinsel

A type of electrical conductor composed of a number of tiny threads, each having a fine, flat ribbon of copper or other metal closely spiraled about it. Used for small-sized cables requiring limpness and extra-long flex life.

tip

(1) A polarity designation of one wire of a pair indicating that the wire is that of the primary (common) color of a five-pair cable (which is not commonly used anymore) group (e.g., the white/blue wire of the blue pair). (2) A wiring contact to which the tip wire is connected. (3) The positive wiring polarity. *See also* ring.

tip conductor

A telephony term used to describe the conductor of a pair that is grounded at the central office when the line is idle. This term was originally coined from its position as the first (tip) conductor of a tip-ring-sleeve switchboard plug. *See also* tip.

TNC

A threaded connector used to terminate coaxial cables. TNC is an acronym for Threaded Neill-Concelman (Neill and Concelman invented the connector).

token passing

A media access method in which a token (data packet) is passed around the ring in an orderly fashion from one device to the next. A station can transmit only when it has the token. The token continues around the network until the original sender receives the token again. If the host has more data to send, the process repeats. If not, the original sender modifies the token to indicate that the token is free for anyone else to use.

Token Ring

A ring topology for a local area network (LAN) in which a supervisory frame, or token, must be received by an attached terminal or workstation before that terminal or workstation can start transmitting. The workstation with the token then transmits and uses the entire bandwidth of whatever communications media the token ring network is using. The most common wiring scheme is called a star-wired ring. Only one data packet can be passed along the ring at a time. If the data packet goes around the ring without being claimed, it eventually makes its way back to the sender. The IEEE standard for Token Ring is 802.5.

tone dial

A push-button telephone dial that makes a different sound (in fact, a combination of two tones) for each number pushed. The technically correct name for tone dial is dual-tone multifrequency (DTMF) since there are two tones generated for each button pressed.

tone generator

A small electronic device used to test network cables for breaks and other problems that sends an electronic signal down one set of UTP wires. Used with a tone locator or probe.

tone locator

A testing device or probe used to test network cables for breaks and other problems; designed to sense the signal sent by the tone generator and emit a tone when the signal is detected in a particular set of wires.

topology

The geometric physical or electrical configuration describing a local communication network, as in network topology; the shape or arrangement of a system. The most common topologies are bus, ring, and star.

total attenuation

The loss of light energy due to the combined effects of scattering and absorption.

total internal reflection

The reflection of light in a medium of a given refractive index off of the interface with a material of a lower refractive index at an angle at or above the critical angle. Total internal reflection occurs at the core/cladding interface within an optical fiber cable.

tracer

The contrasting color-coding stripe along an insulated conductor of a wire pair.

tractor

A machine used to pull a preform into an optical fiber using constant tension.

transceiver

The set of electronics that sends and receives signals on the Ethernet media system. Transceivers may be small outboard devices or they may be built into an Ethernet port. Transceiver is a combination of the words *transmitter* and *receiver*.

transducer

A device for converting energy from one form to another, such as optical energy to electrical energy.

transfer impedance

For a specified cable length, relates to a current on one surface of a shield to the voltage drop generated by this current on the opposite surface of the shield. The transfer impedance is used to determine shield effectiveness against both ingress and egress of interfering signals. Shields with lower transfer impedance are more effective than shields with higher transfer impedance.

transient

A high-voltage burst of electrical current. If the transient is powerful enough, it can damage data transmission equipment and devices that are connected to the transmission equipment.

transimpedance amplifier

In a fiber-optic receiver, a device that receives electrical current from the photodiode and amplifies it before sending it to the *quantizer IC*.

transition point (TP)

An ISO/IEC 11801 term that defines a location in the horizontal cabling subsystem where flat under-carpet cabling connects to round cabling or where cable is distributed to modular furniture. The ANSI/TIA/EIA-568-B equivalent of this term is consolidation point.

transmission line

An arrangement of two or more conductors or an optical waveguide used to transfer a signal from one location to another.

transmission loss

The total amount of signal loss that happens during data transmission.

transmission media

Anything such as wire, coaxial cable, fiber optics, air, or vacuum that is used to carry a signal.

transmit jumper

The test jumper connected to the light source.

transmitter

With respect to optical fiber cabling, a device that changes electrical signals to optical signals using a laser and associated electronic equipment such as modulators. Among various types of light transmitters are light-emitting diodes (LEDs), which are used in lower speed (100 to 256Mbps) applications such as FDDI, and laser diodes, which are used in higher-speed applications (622Mbps to 10Gbps) such as ATM and SONET.

transverse modes

In the case of optical fiber cable, light modes across the width of the waveguide.

tree coupler

A passive fiber-optic coupler with one input port and three or more output ports, or with three or more input ports and one output port.

tree topology

A LAN topology similar to linear bus topology, except that tree networks can contain branches with multiple nodes.

triaxial cable

Coaxial cable with an additional outer copper braid insulated from signal carrying conductors. It has a core conductor and two concentric conductive shields. Also called triax.

triboelectric noise

Electromagnetic noise generated in a shielded cable due to variations in capacitance between the shield and conductor as the cable is flexed.

trunk

(1) A phone carrier facility such as phone lines between two switches. (2) A telephone communication path or channel between two points, one of them usually a telephone company facility.

trunk cable

The main cable used in thicknet Ethernet (10Base-5) implementations.

trunk line

A transmission line running between telephone switching offices.

T-series connections

A type of digital connection leased from the telephone company or other communications provider. Each T-series connection is rated with a number based on speed. T-1 and T-3 are the most popular.

TSI

Time slot interchanger. A device used in networking switches to provide non-port blocking.

turn-key agreement

A contractual arrangement in which one party designs and installs a system and "turns over the keys" to another party who will operate the system. A system may also be called a turn-key system if the system is self-contained or simple enough that all the customer has to do is "turn the key."

twinaxial cable

A type of communications cable consisting of two center conductors surrounded by an insulating spacer, which is in turn surrounded by a tubular outer conductor (usually a braid, foil, or both). The entire assembly is then covered with an insulating and protective outer layer. Twinaxial is often thought of as dual-coaxial cable. Twinaxial cable was commonly used with Wang VS terminals, IBM 5250 terminals on System/3x, and AS/400 minicomputers. Also called twinax.

twin lead

A transmission line used for television receiving antennas having two parallel conductors separated by insulating material. Line impedance is determined by the diameter and spacing of the conductors and the insulating material. The conductors are usually 300 ohms.

twisted pair

Two insulated copper wires twisted around each other to reduce induction (thus interference) from one wire to the other. The twists, or lays, are varied in length from pair to pair to reduce the potential for signal interference between pairs. Several sets of twisted-pair wires may be enclosed in a single cable. In cables greater than 25 pairs, the twisted pairs are grouped and bound together in groups of 25 pairs.

twisted-pair–physical media dependent (TP-PMD)

Technology developed by the ANSI X3T9.5 working group that allows 100Mbps transmission over twisted-pair cable.

Type 1

150-ohm shielded twisted-pair (STP) cabling conforming to the IBM Cabling System specifications. Two twisted pairs of 22 AWG solid conductors for data communications are enclosed in a braided shield covered with a sheath. Type 1 cable has been tested for operation up to 16MHz. Available in plenum, nonplenum, riser, and outdoor versions.

Type 1A

An enhanced version of IBM Type 1 cable rated for transmission frequencies up to 300MHz.

Type 2

150 ohm shielded twisted-pair (STP) cabling conforming to the IBM Cabling System specifications. Type 2 cable is popular with those who insist on following the IBM Cabling System because there are two twisted pairs of 22 AWG solid conductors for data communications that are enclosed in a braided shield. In addition to the shielded pairs, there are four pairs of 22 AWG solid conductors for telephones that are included in the cable jacket but outside the braided shield. Tested for transmission frequencies up to 16MHz. Available in plenum and nonplenum versions.

Type 2A

An enhanced version of IBM Type 2 cable rated for transmission speeds up to 300MHz.

Type 3

100 ohm unshielded twisted-pair (UTP) cabling similar to ANSI/TIA/EIA 568-B Category 3 cabling. 22 AWG or 24 AWG conductors with a minimum of two twists per linear foot. Typically four twisted pairs enclosed within cable jacket.

Type 5

100/140-micron optical fiber cable conforming to the IBM Cabling System specifications. Type 5 cable has two optical fibers that are surrounded by strength members and a polyurethane jacket. There is also an IBM Type 5J that is a 50/125-micron version defined for use in Japan.

Type 6

150 ohm shielded twisted-pair (STP) cabling that conforms to the IBM Cabling System specifications. Two twisted pairs of 26 AWG stranded conductors for data communications. Flexible for use in making patch cables. Tested for operation up to 16MHz. Available in nonplenum version only.

Type 6A

An enhanced version of IBM Type 6 cable rated for transmission speeds up to 300MHz.

Type 8

150 ohm under-carpet cable conforming to the IBM Cabling System specifications. Two individually shielded parallel pairs of 26 AWG solid conductors for data communications. The cable includes "ramped wings" to make it less visible when installed under carpeting. Tested for transmission speeds up to 16MHz. Type 8 cable is not very commonly used.

Type 9

150 ohm shielded twisted-pair (STP) cabling that conforms to the IBM Cabling System specifications. A plenum-rated cable with two twisted pairs of 26 AWG solid or stranded conductors for data communications enclosed in a braided shield covered with a sheath. Tested for transmission speeds up to 16MHz.

Type 9A

An enhanced version of IBM Type 9 cable rated for transmission speeds up to 300MHz.

U

UL Listed

If a product carries this mark, it means UL found that representative samples of this product met UL's safety requirements.

UL Recognized

If a product carries this mark, it means UL found it acceptable for use in a complete UL Listed product.

ultraviolet

The electromagnetic waves invisible to the human eye with wavelengths between 10nm and 400nm.

upload speed

The speed used to save or send data upstream of the point of transmission.

unbalanced line

A transmission line in which voltages on the two conductors are unequal with respect to ground; one of the conductors is generally connected to a ground point. An example of an unbalanced line is a coaxial cable. This is the opposite of a balanced line or balanced cable.

underground cable

Cable that is designed to be placed beneath the surface of the ground in ducts or conduit. Underground cable is not necessarily intended for direct burial in the ground.

Underwriters Laboratories, Inc. (UL)

A privately owned company that tests to make sure that products meet safety standards. UL also administers a program for the certification of category-rated cable with respect to flame ratings. See Chapter 4 for more information on Underwriters Laboratories.

unified messaging

The integration of different streams of communications (email, SMS, fax, voice, video, etc.) into a single interface, accessible from a variety of different devices.

uniformity

The degree of insertion loss difference between ports of a coupler.

uninterruptible power supply (UPS)

A natural line conditioner that uses a battery and power inverter to run the computer equipment that plugs into it. The battery charger continuously charges the battery. The battery charger is the only thing that runs off line voltage. During a power problem, the battery charger stops operating, and the equipment continues to run off the battery.

Universal Service Order Code (USOC)

Developed by AT&T/the Bell System, the Universal Service Order Code (pronounced "U-sock") identifies a particular service, device, or connector wiring pattern. Often used to refer to an old cable color-coding scheme that was current when USOC codes were in use. USOC is not used in wiring LAN connections and is not supported by current standards due to the fact that high crosstalk is exhibited at higher frequencies. See Chapter 9 for more information.

unmated

Optical fiber connectors in a system whose end faces are not in contact with another connector, resulting in a fiber that is launching light from the surface of the glass into air. Also called unterminated or open.

unshielded twisted pair (UTP)

(1) A pair of copper wires twisted together with no electromagnetic shielding around them. (2) A cable containing multiple pairs of UTP wire. Each wire pair is twisted many times per foot (higher-grade UTP cable can have more than 20 twists per foot). The twists serve to cancel out electromagnetic

interference that the transmission of electrical signal through the pairs generates. An unshielded jacket made of some type of plastic then surrounds the individual twisted pairs. Twisted-pair cabling includes no shielding. UTP most often refers to the 100 ohm Categories 3, 5e, and 6 cables specified in the ANSI/TIA/EIA-568-B standard.

upstream

A term used to describes the number of bits being sent to the Internet service provider; often referred to as *upload speed*.

uptime

The portion of time a network is running.

V

vampire tap

See piercing tap.

velocity of propagation

The transmission speed of electrical energy in a length of cable compared to speed in free space. Usually expressed as a percentage. Test devices use velocity of propagation to measure a signal's transit time and thereby calculate the cable's length. *See also* nominal velocity of propagation (NVP).

vertical-cavity surface-emitting laser (VCSEL)

A type of laser that emits coherent energy along an axis perpendicular to the PN junction.

very high frequency (VHF)

Frequency band extending from 30MHz to 300MHz.

very low frequency (VLF)

Frequency band extending from 10KHz to 30KHz.

video

A signal that contains visual information, such as a picture in a television system.

videoconferencing

The act of conducting conferences via a video telecommunications system, local area network, or wide area network.

videophone

A telephone-like device that provides a picture as well as sound.

visible light

Electromagnetic radiation visible to the human eye at wavelengths between 400nm and 700nm.

visual fault locator (VFL)

A testing device consisting of a red laser that fills the fiber core with light, allowing a technician to find problems such as breaks and macrobends by observing the light through the buffer, and sometimes the jacket, of the cable.

voice circuit

A telephone company circuit capable of carrying one telephone conversation. The voice circuit is the standard unit in which telecommunications capacity is counted. The U.S. analog equivalent is 4KHz. The digital equivalent is 56Kbps in the U.S. and 64Kbps in Europe. In the U.S., the Federal Communications Commission restricts the maximum data rate on a voice circuit to 53Kbps.

voice-grade

A term used for twisted-pair cable used in telephone systems to carry voice signals. Usually Category 3 or lower cable, though voice signals can be carried on cables that are higher than Category 3.

volt

A unit of expression for electrical potential or potential difference. Abbreviated as V.

voltage drop

The voltage developed across a component by the current flow through the resistance of the component.

volt ampere (VA)

A designation of power in terms of voltage and current.

W

W

(1) Symbol for watt or wattage. (2) Abbreviation for white when used in conjunction with twisted-pair cable color codes; may also be Wh.

wall plate (or wall jack)

A wall plate is a flat plastic or metal that usually mounts in or on a wall. Wall plates include one or more jacks.

watt

A unit of electrical power.

waveform

A graphical representation of the amplitude of a signal over time.

waveguide

A structure that guides electromagnetic waves along their length. The core fiber in an optical fiber cable is an optical waveguide.

waveguide couplers

A connection in which light is transferred between waveguides.

waveguide dispersion

That part of the chromatic dispersion (spreading) that occurs in a single-mode fiber as some of the light passes through the cladding and travels at a higher velocity than the signal in the core, due to the cladding's lower refractive index. For the most part, this deals with the fiber as a waveguide structure. Waveguide dispersion is one component of chromatic dispersion.

waveguide scattering

The variations caused by subtle differences in the geometry and fiber index profile of an optical fiber.

wavelength

(1) The distance between two corresponding points in a series of waves. (2) With respect to optical fiber communications, the distance an electromagnetic wave travels in the time it takes to oscillate through a complete cycle. Wavelengths of light are measured in nanometers or micrometers. Wavelength is preferred over the term frequency when describing light.

wavelength division multiplexing (WDM)

A method of carrying multiple channels through a fiber at the same time (multiplexing) whereby signals within a small spectral range are transmitted at different wavelengths through the same optical-fiber cable. *See also* frequency division multiplexing (FDM).

wavelength isolation

A wave division multiplexer's isolation of a light signal from the unwanted optical channels in the desired optical channel.

wavelength variance

The variation in an optical parameter caused by a change in the operating wavelength.

wide area network (WAN)

A network that crosses local, regional, and international boundaries. Some types of Internetwork technology such as a leased-line, ATM, or frame relay connect local area networks (LANs) together to form WANs.

Wi-Fi

The technology that allows an electronic device to exchange data or connect to the Internet wirelessly using radio waves. Also known as wireless LAN.

window

In optical transmission, a wavelength at which attenuation is low, allowing light to travel greater distances through the fiber before requiring a repeater.

wire center

(1) Another name for a wiring or telecommunications closet. (2) A telephone company building where all local telephone cables converge for service by telephone switching systems. Also called central office or exchange center.

wire cross-connect

A piece of equipment or location at which twisted-pair cabling is terminated to permit reconnection, testing, and rearrangement. Cross-connects are usually located in equipment rooms and telecommunications closets and are used to connect horizontal cable to backbone cable. Wire cross-connects typically use a 66- or 110-block. These blocks use jumpers to connect the horizontal portion of the block to the backbone portion of the block.

wire fault

A break in a segment or cable that causes an error. A wire fault might also be caused by a break in the cable's shield.

wire map

See cabling map.

wireless bridge

A wireless bridge is a hardware component used to connect two or more network segments (LANs or parts of a LAN) that are physically and logically (by protocol) separated. It does not necessarily always need to be a hardware device, as some operating systems (such as Windows, Linux, Mac OS X, and FreeBSD) provide software to bridge different protocols.

wiring closet

See telecommunications room.

work area

The area where horizontal cabling is connected to the work area equipment by means of a telecommunications outlet. A telecommunications outlet serves a station or desk. *See also* work area telecommunications outlet.

work area cable

A cable used to connect equipment to the telecommunications outlet in the user work area. Sometimes called a patch cable or patch cord.

work area telecommunications outlet

Sometimes called a wall plate or workstation outlet, a connecting device located in a work or user area where the horizontal cabling is terminated. A work-area telecommunications outlet provides connectivity for work area patch cables, which in turn connect to end-user equipment such as computers or telephones. The telecommunications outlet can be recessed in the wall, mounted on the wall or floorboard, or recessed in the floor or a floor monument.

workgroup

A collection of workstations and servers on a LAN that are designated to communicate and exchange data with one another.

workstation

A computer connected to a network at which users interact with software stored on the network. Also called a PC (personal computer), network node, or host.

X

X

(1) Symbol for reactance. (2) Symbol often used on wiring diagrams to represent a cross-connect.

xDSL

A generic description for the different DSL technologies such as ADSL, HDSL, RADSL. *See also* digital subscriber line (DSL).

XTC

An optical fiber connector developed by OFTI; not in general use.

Y

Y-coupler

In fiber optics, a variation on the T-coupler, where input light is split between two channels that branch out like a "Y" from the input.

Z

z

Symbol for impedance.

zero-dispersion slope

In single-mode fiber, the chromatic dispersion slope at the fiber's zero-dispersion wavelength.

zero-dispersion point

In an optical fiber of a given refractive index, the narrow range of wavelengths within which all wavelengths travel at approximately the same speed. The zero-dispersion point is useful in reducing chromatic dispersion in single-mode fiber.

zero-dispersion-shifted-fiber

See dispersion shifted fiber.

zero-dispersion wavelength

In single-mode fiber, the wavelength where waveguide dispersion cancels out material dispersion and total chromatic dispersion is zero.

zero loss reference

Some power meters allow the user to set the meter to read 0 dB while measuring the reference test cable. When you then test the installed cable run, the meter will display only the loss of the cable run minus the loss of the reference cable. This function is similar to the tare function of a scale.

zero-order

When the light rays entering the core of an optical fiber travel in a straight line through the fiber.

zipcord

Duplex fiber-optic interconnect cable consisting of two tight-buffered fibers with outer jackets bonded together, resembling electrical wiring used in lamps and small appliances.

Index

Note to the Reader: Throughout this index **boldfaced** page numbers indicate primary discussions of a topic. *Italicized* page numbers indicate illustrations.

losses
    bending, **79–81**, *79*
    cable segment and interconnection, **484**, *484*
    fusion splice and macrobends, **483–484**, *483–484*
    insertion. *See* insertion loss
    optical, **458–462**, *459–461*
    passive components, **314–315**
    power, **24–25**
low-order mode, 60, *60*
LSA (least-squares averaging), 478
LTGF (loose tube, gel-filled) cable, 119
Lucite, 6

M

machine polishing, **234**
macrobending loss (MBL), 81
macrobends, **80–81**
    attenuation, 76
    loss measurements, **483–484**, *483–484*
    OTDR for, 526, *526*
    testing, 520, *520*
magnetic fields, 33–34, *34*
magnetic optical isolators, **331**, *331*
mandrel wrap, **455–457**, *456*
MANs (metropolitan area networks)
    IEEE 802.3 for, 410
    SOA in, 339
markings on fiber-optic cables, **138–143**, *141–143*
    color, 139–140, 254, 491–492
    external, **138–139**, *139*
    jacket, **141**, **490–491**
    numbers, **141**, *141*
    sequential markings, **142**, *142*
material dispersion, **70**, *70*
Material Safety Data Sheet (MSDS), 108, 155–156
materials
    optical fiber, **52–54**, *54*
    refractive index, **40**
mating sleeves
    cleaning, 238, 503–505, *503–507*
    concentricity, 195
    connectors, 191–192, *192*
    diameter mismatch, 195
    inspecting, 502
    noncircularity, 196
    patch panels, 375, *375*
    terminus, 216
MBL (macrobending loss), 81
MCVD (Modified Chemical Vapor Deposition), **56–57**, *57*
measurement quality jumpers (MQJ), 454, *455*

mechanical splices, **165**, *165*
    procedure, **170–174**, *171–173*
    specialty, **166**, *166*
mechanical transfer (MT) ferrules
    description, **211**, *211*
    endface cleaning, 503, *503*
medium, light, 39–41
medium interface connectors, 209
messenger cable, **127**, *127*
metal ferrules, 188
meters
    optical fiber, 54
    optical power, **449–453**, *450–451*
Method A loss measurement, **460**, *460*
Method B loss measurement, **460–461**
Method C loss measurement, **461–462**, *461*
metropolitan area networks (MANs)
    IEEE 802.3 for, 410
    SOA in, 339
microbends
    attenuation, 76
    description, **79**, *79*
    polarization-mode dispersion, 74
microscopes
    connector termination, 229
    endfaces, **247–249**, *247–250*
microwave range, 37
MIL-DTL-38999 four-optical-fiber connectors, 216, *216*
Mini-BNC Connectors, **208**, *208*
minimum bandwidth-length product, 75
minimum bend radius, 55
minimum power in multimode link analysis, 421–422
mirroring performance specifications, **86–87**
misalignment
    connectors, 196–197
    splices, **153–154**, *153–154*
mismatch
    connectors, 195, **491–492**
    core, **148–149**, *149*
    splices, **150**, *150*
modal dispersion, **61–63**, **69–70**
modal noise in design, 409
mode-conditioning patch cords, 410, *410*
mode field diameter
    connectors, 195
    mismatch, **149–150**, *150*
    single-mode fibers, 62
mode filters, **455–457**, *456–458*
mode partition noise, 409
modes, optical fiber, **59**
Modified Chemical Vapor Deposition (MCVD), **56–57**, *57*
modulation, amplitude, 16